THIRD EDITION

REMOTE SENSING

PRINCIPLES AND INTERPRETATION

FLOYD F. SABINS

Remote Sensing Enterprises, Incorporated

and

*University of California,
Los Angeles*

W. H. FREEMAN AND COMPANY

New York

COVER IMAGE: Landsat thematic mapper (TM) satellite image of southeast Utah. TM band 2
(green) is shown in blue, band 4 (reflected IR) in green, and band 7 (reflected IR) in red. The
irregular body of water in the southeast (lower right) portion of the image is Lake Powell, which is
impounded by the Glen Canyon Dam on the Colorado River. The dark oval feature southeast
of Lake Powell is Navajo Mountain, which is one of the four sacred mountains of the Navajo
Indians who inhabit this region. The large northwest-trending oval with a blue signature in the
center is the Waterpocket fold. To the northeast, the dark mountains with patches of green
vegetation are the Henry Mountains where G. K. Gilbert of the U. S. Geological Survey first
recognized the flat-floored igneous intrusions that he named *laccoliths*. At the north-central
margin, the truncated oval feature is the anticline called the San Rafael Swell. In the northwest,
the dark areas are volcanic rocks of the Fish Lake Mountains and the Sevier and Awapa Plateaus.
The bright-blue patches are snow on the highest elevations. Most of the region is underlain by
strata of sandstone and shale. Late Cretaceous rocks have magenta and brown hues.
Jurassic and late Triassic are yellow and tan. Early Triassic and Permian rocks are bluish gray.
This image was digitally processed at the Chevron remote sensing facility. Prints of this and
other Landsat images are available from Chevron Petroleum Technology Corporation,
Attention: Linda Fry, P. O. Box 446, La Habra, CA 90633, Fax 310-694-7122.

Acquisitions Editor: Holly Hodder
Project Editor: Christine Hastings
Text Designer: Marsha Cohen
Cover Designer: Blake Logan
Illustration Coordinator: Bill Page
Production Coordinator: Sheila Anderson
Composition: W. H. Freeman Electronic Publishing Center/Andrew Kudlacik
Manufacturing: Quebecor Semline

Library of Congress Cataloging-in-Publication Data   (to come)

Sabins, Floyd F.
    Remote sensing : principles and interpretation / Floyd F. Sabins
    —3d ed.
        p.   cm.
    ISBN 0-7167-2442-1 (hardcover)
    1. Remote sensing   I. Title
    G70.4.S15   1996
    621.36'78—dc20   96-31940
                                                          CIP

Printed in the United States of America

First printing, 1996

To

Janice and our grandsons, Connor, Eric, Robert, and Spencer

CONTENTS

5 THERMAL INFRARED IMAGES

6 RADAR TECHNOLOGY AND TERRAIN INTERACTIONS

7 SATELLITE RADAR SYSTEMS AND IMAGES

8 DIGITAL IMAGE PROCESSING

PREFACE

Much has happened since the previous edition of this book appeared in 1986. Major advances in the past decade indicate the dynamic nature of the science of remote sensing and its expanding applications. I have attempted to express this dynamism in the third edition.

This book is designed for an introductory university course in remote sensing. No prior training in remote sensing is required. Courses in introductory physics, physical geography, and physical geology are useful background, but are not essential, for users of this book. This text follows the format of the remote sensing course I teach in the Earth and Space Sciences Department at UCLA, which emphasizes the interpretation of images and their application to a range of disciplines.

ORGANIZATION OF THE BOOK

The first chapter introduces the major remote sensing systems and the interactions between electromagnetic energy and materials that are the basis for remote sensing. The six following chapters describe the major imaging system: photographs, Landsat, earth resource and environmental satellites, thermal infrared, and radar. For each system the following topics are covered:

1. physical properties of materials and their interactions with electromagnetic energy that determine signatures on images
2. design and operation of the imaging system
3. characteristics of images
4. guidelines and examples for interpreting images

A chapter on digital image processing describes computer methods for restoring and enhancing images and extracting information. The remaining chapters describe practical applications of remote sensing to environmental monitoring, oil and mineral exploration, land-use and geographic information systems, and natural hazards. The final chapter compares different images for a test site and describes the advantages and limitations of different images for various applications. Each chapter includes a series of questions. A glossary provides definitions of remote sensing terms and acronyms. The Appendix, "Basic Geology for Remote Sensing," describes earth science concepts that are employed in some image interpretations. The answers to the end-of-chapter questions are available via e-mail (http://www.whfreeman.com).

RECENT ADVANCES IN REMOTE SENSING

Recent technological advances have remarkably improved and expanded the techniques of remote sensing. Over 50 percent of the illustrations and text are new to this third edition. The index maps, inside the front and back covers, show that the images represent much of the world.

The following new satellite imaging systems, which have been deployed since the second edition, are described:

JERS-1 by Japan
ERS-1 by the European Space Agency
Radarsat by Canada
SIR-C by NASA
SPOT-1, 2, 3
Magellan by NASA, which has acquired
  complete radar coverage of Venus
IRS by India
Geosat by the U.S. Navy

Both spatial and spectral resolution of images have greatly improved. Examples of these new images are illustrated and evaluated.

Important nonrenewable resources were discovered in the past decade through remote sensing. The first commercial oil fields in Papua New Guinea were discovered using aircraft radar images. In Saudi Arabia a new oil production trend was discovered following Landsat interpretations. Digitally processed Landsat images of Collahuasi, Chile, defined a world-class copper deposit that is being developed. These exploration success stories are documented.

Images from the advanced very high resolution radiometer (AVHRR) are now routinely used for global mapping of vegetation. Geographic information systems (GIS) merge remote sensing images with other data sets to provide new insights for many applications. Satellite radar altimetry data provide bathymetric maps with details and accuracy that greatly surpass the maps of a decade ago.

This is a very exciting time for the dynamic field of remote sensing. We are the stewards of our planet and remote sensing provides valuable tools and resources to manage our world responsibly.

ACKNOWLEDGMENTS

Colleagues in industry, government, and universities around the world provided many of the illustrations and are acknowledged in the figure captions and text. The major source of images is the Chevron Corporation, my employer for 38 years before my retirement in 1992. The experience, research opportunities, and recognition that Chevron provided me are gratefully acknowledged. I have taught remote sensing in the Earth and Space Sciences Department at UCLA since 1975. The interaction with and feedback from my students have improved the instructional aspects of the book.

John P. Ford (Jet Propulsion Laboratory, Ret.) reviewed the entire manuscript. Sections were reviewed by Dann Halverson (University of Southwestern Louisiana), Ronald W. Marrs (University of Wyoming), Stuart E. Marsh (University of Arizona), Scott M. Robeson (Indiana University), and Stephen R. Yool (University of Arizona). Joyce Quinn drafted most of the new figures. Susan Middleton did the final editing. Holly Hodder, Christine Hastings, Sheila Anderson, Blake Logan, Bill Page, and Andrew Kudlacik guided the book through the production process at W. H. Freeman and Company. I welcome comments and suggestions on this edition, which may be addressed to me at 1724 Celeste Lane, Fullerton, CA 92633, Fax 310-694-7122.

LABORATORY MANUAL

I have prepared a "Remote Sensing Laboratory Manual" that provides interpretation projects for each chapter of this third edition. The manual is available from

Kendall/Hunt Publishing Company
P. O. Box 1840
Dubuque, Iowa 52004-1840
Telephone 800-228-0180
Fax 800-772-9165

Fullerton, California
July 1996

REMOTE SENSING

INTRODUCTION TO CONCEPTS AND SYSTEMS

Prior to the 1960s our view of the earth and the universe was restricted to observations and photographs using visible light. Distant views were obtained only from aircraft and telescopes. Today the science of remote sensing provides instruments that view our universe at wavelengths far greater than those of visible light. The instruments are deployed in satellites and aircraft to record images of the earth and solar system that can be digitally analyzed with our personal computers to provide information on a wide range of topics. Some of these include the global environment, land use, renewable and nonrenewable resources, natural hazards, and the geology of Venus.

In this book *remote sensing* is defined as the science of

- acquiring,
- processing, and
- interpreting

images, and related data, obtained from aircraft and satellites that record the interaction between matter and electromagnetic radiation. *Acquiring* images refers to the technology employed, such as an electro-optical scanning system. *Processing* refers to the procedures that convert the raw data into images. *Interpreting* the images is, in my opinion, the most important step because it converts an image into information that is meaningful and valuable for a wide range of users. The *interaction between matter and electromagnetic energy* is determined by

- the physical properties of the matter, and
- the wavelength of electromagnetic energy that is remotely sensed.

A later section in this chapter describes the various wavelength regions and the types of interactions.

The term *remote sensing* refers to methods that employ electromagnetic energy, such as light, heat, and radio waves, as the means of detecting and measuring target characteristics. Underwater surveys that use pulses of sonic energy for imaging (*sonar*) are considered a remote sensing method. The science of remote sensing excludes geophysical methods such as electrical, magnetic, and gravity surveys that measure force fields rather than electromagnetic radiation.

Aerial photography is the original form of remote sensing and remains a widely used method. Interpretations of aerial photographs have led to the discovery of many oil and mineral deposits. These successes, using only the visible region of the electromagnetic spectrum, suggested that additional discoveries could be made by using other wavelength regions. In the 1960s, technology was developed to acquire images in the infrared (IR) and microwave regions, which greatly expanded the scope and applications of remote sensing. The development and deployment of manned and unmanned earth satellites began in the 1960s and provided an orbital vantage point for acquiring images of the earth. For a review of the history of remote sensing, see Fischer and others (1975). Most remote sensing data, except for aerial photographs, are now acquired in digital format and processed by computers to produce images for interpretation.

This chapter introduces the basic physical concepts and the imaging systems that are employed in remote sensing. Subsequent chapters describe major types of remote sensing and are followed by chapters that describe applications of the technology.

Table 1-1 Metric nomenclature for distance

Unit	Symbol	Equivalent
Kilometer	km	$1000 \text{ m} = 10^{-3} \text{ m}$
Meter[a]	m	$1.0 \text{ m} = 10^0 \text{ m}$
Centimeter	cm	$0.01 \text{ m} = 10^{-2} \text{ m}$
Millimeter	mm	$0.001 \text{ m} = 10^{-3} \text{ m}$
Micrometer[b]	μm	$0.000001 \text{ m} = 10^{-6} \text{ m}$
Nanometer	nm	10^{-9} m

[a]Basic unit

[b]Formerly called micron (μ).

UNITS OF MEASURE

This book employs the metric system with the following standard units and abbreviations:

meter	m
second	sec
kilogram	kg
gram	g
radian	rad
hertz	Hz
watt	W

Distance is expressed in the multiples and fractions of meters shown in Table 1-1. Where appropriate for clarity, English units for distance will be used, with metric equivalents shown in parentheses.

Frequency (ν) is the number of wave crests passing a given point in a specified period of time. Frequency was formerly expressed as "cycles per second," but today we use *hertz* (Hz) as the unit for a frequency of one cycle per second. The terms for designating frequencies are shown in Table 1-2.

Temperature is given in degrees Celsius (°C) or in degrees Kelvin (°K). (The Kelvin scale is also known as the *absolute temperature scale*.) A temperature of 273°K is equivalent to 0°C. (The metric system omits the degree symbol for Kelvin

Table 1-2 Terms used to designate frequencies

Unit	Symbol	Frequency, cycles · sec^{-1}
Hertz	Hz	1
Kilohertz	kHz	10^3
Megahertz	MHz	10^6
Gigahertz	GHz	10^9

temperatures; however, the letter K is also used to designate other constants; so °K is used in this text.) A few temperatures commonly given in degrees Fahrenheit (°F) will remain in that scale where conversion to degrees Celsius is inconvenient.

In fractional statements, units in the denominator are identified by a fractional superscript. For example, the property called *thermal inertia (P)* for a particular rock type is expressed as

$$P = 0.53 \text{ cal} \cdot \text{cm}^{-2} \cdot \text{sec}^{-1/2} \cdot {}^\circ\text{C}^{-1}$$

This expression means that for the particular rock type, thermal inertia *P* equals 0.53 calories per square centimeter per second to the square root per degree Celsius.

ELECTROMAGNETIC ENERGY

Electromagnetic energy refers to all energy that moves with the velocity of light in a harmonic wave pattern. A harmonic pattern consists of waves that occur at equal intervals in time. The wave concept explains how electromagnetic energy propagates (moves), but this energy can only be detected as it interacts with matter. In this interaction, electromagnetic energy behaves as though it consists of many individual bodies called *photons* that have such particle-like properties as energy and momentum. When light bends (refracts) as it propagates through media of different optical densities, it is behaving like waves. When a light meter measures the intensity of light, however, the interaction of photons with the light-sensitive photodetector produces an electrical signal that varies in strength proportional to the number of photons. Suits (1983) describes the characteristics of electromagnetic energy that are significant for remote sensing.

Properties of Electromagnetic Waves

Electromagnetic waves can be described in terms of their velocity, wavelength, and frequency. All electromagnetic waves travel at the same velocity (*c*). This velocity is commonly referred to as the *speed of light,* since light is one form of electromagnetic energy. For electromagnetic waves moving through a vacuum, $c = 299{,}793 \text{ km} \cdot \text{sec}^{-1}$ or, for practical purposes, $c = 3 \times 10^8 \text{ m} \cdot \text{sec}^{-1}$.

The *wavelength* (λ) of electromagnetic waves is the distance from any point on one cycle or wave to the same position on the next cycle or wave. The *micrometer* (μm) is a convenient unit for designating wavelength of both visible and IR radiation. In order to avoid decimal numbers, optical scientists commonly employ nanometers (nm) for measurements of very short wavelengths, such as visible light.

Unlike velocity and wavelength, which change as electromagnetic energy is propagated through media of different den-

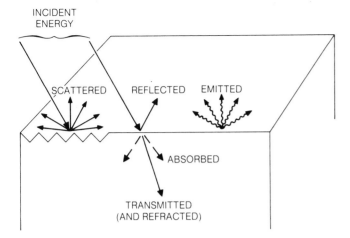

Figure 1-1 Interaction processes between electromagnetic energy and matter.

sities, frequency remains constant and is therefore a more fundamental property. Electronic engineers use frequency nomenclature for designating radio and radar energy regions. This book uses wavelength rather than frequency to simplify comparisons among all portions of the electromagnetic spectrum. Velocity (c), wavelength (λ), and frequency (ν) are related by

$$c = \lambda\nu \tag{1-1}$$

Interaction Processes

Electromagnetic energy that encounters matter, whether solid, liquid, or gas, is called *incident* radiation. Interactions with matter can change the following properties of the incident radiation: intensity, direction, wavelength, polarization, and phase. The science of remote sensing detects and records these changes. We then interpret the resulting images and data to determine the characteristics of the matter that interacted with the incident electromagnetic energy.

During interactions between electromagnetic radiation and matter, mass and energy are conserved according to basic physical principles. Figure 1-1 illustrates the five common results of these interactions. The incident radiation may be

1. *Transmitted*, that is, passed through the substance. Transmission of energy through media of different densities, such as from air into water, causes a change in the velocity of electromagnetic radiation. The ratio of the two velocities is called the *index of refraction (n)* and is expressed as

$$n = \frac{c_a}{c_s} \tag{1-2}$$

where c_a is the velocity in a vacuum and c_s is the velocity in the substance.

2. *Absorbed*, giving up its energy largely to heating the matter.
3. *Emitted* by the substance, usually at longer wavelengths, as a function of its structure and temperature.
4. *Scattered*, that is, deflected in all directions. Surfaces with dimensions of *relief*, or roughness, comparable to the wavelength of the incident energy produce scattering. Light waves are scattered by molecules and particles in the atmosphere whose sizes are similar to the wavelengths of light.
5. *Reflected*, that is, returned from the surface of a material with the angle of reflection equal and opposite to the angle of incidence. Reflection is caused by surfaces that are smooth relative to the wavelength of incident energy. *Polarization*, or direction of vibration, of the reflected waves may differ from that of the incident wave.

Emission, scattering, and reflection are called *surface phenomena* because these interactions are determined primarily by properties of the surface, such as color and roughness. Transmission and absorption are called *volume phenomena* because they are determined by the internal characteristics of matter, such as density and conductivity. The particular combination of surface and volume interactions with any particular material depend on both the wavelength of the electromagnetic radiation and the specific properties of that material.

These interactions between matter and energy are recorded on remote sensing images, from which one may interpret the characteristics of matter. Individual interaction mechanisms are described more completely in later chapters. For example, scattering of light by the atmosphere is described in Chapter 2. Absorbed and emitted thermal IR energy are described in Chapter 5. Scattering of radar energy by surfaces is described in Chapter 6.

ELECTROMAGNETIC SPECTRUM

The *electromagnetic spectrum* is the continuum of energy that ranges from meters to nanometers in wavelength, travels at the speed of light, and propagates through a vacuum such as outer space. All matter radiates a range of electromagnetic energy such that the peak intensity shifts toward progressively shorter wavelengths with increasing temperature of the matter.

Wavelength Regions and Bands

Figure 1-2 shows the electromagnetic spectrum, which is divided on the basis of wavelength into *regions* described in Table 1-3. The electromagnetic spectrum ranges from the very short wavelengths of the gamma-ray region (measured in fractions of nanometers) to the long wavelengths of the radio region (measured in meters). The horizontal scale in Figure 1-2 is logarithmic in order to portray adequately the shorter wavelengths. Notice that the visible region (0.4 to 0.7 μm) occupies only a small portion of the spectrum. Energy reflected from the

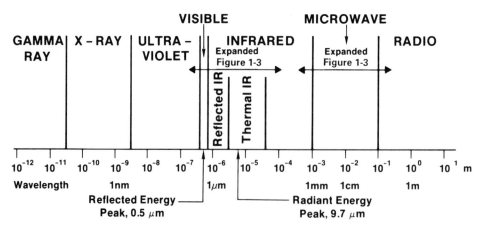

Figure 1-2 Electromagnetic spectrum. Expanded versions of the visible, infrared, and microwave regions are shown in Figure 1-3.

Table 1-3 Electromagnetic spectral regions

Region	Wavelength	Remarks
Gamma-ray region	< 0.03 nm	Incoming radiation completely absorbed by the upper atmosphere and not available for remote sensing.
X-ray region	0.03 to 30 nm	Completely absorbed by the atmosphere. Not employed in remote sensing.
Ultraviolet region	0.03 to 0.4 μm	Incoming wavelengths less than 0.3 μm completely absorbed by ozone in the upper atmosphere.
Photographic UV band	0.3 to 0.4 μm	Transmitted through the atmosphere. Detectable with film and photodetectors, but atmospheric scattering is severe.
Visible region	0.4 to 0.7 μm	Imaged with film and photodetectors. Includes reflected energy peak of earth at 0.5 μm.
Infrared region	0.7 to 100 μm	Interaction with matter varies with wavelength. Atmospheric transmission windows are separated by absorption bands.
Reflected IR band	0.7 to 3.0 μm	Reflected solar radiation that contains no information about thermal properties of materials. The interval from 0.7 to 0.9 μm is detectable with film and is called the photographic IR band.
Thermal IR band	3 to 5 μm, 8 to 14 μm	Principal atmospheric windows in the thermal region. Images at these wavelengths are acquired by optical-mechanical scanners and special vidicon systems but not by film.
Microwave region	0.1 to 100 cm	Longer wavelengths that can penetrate clouds, fog, and rain. Images may be acquired in the active or passive mode.
Radar	0.1 to 100 cm	Active form of microwave remote sensing. Radar images are acquired at various wavelength bands.
Radio	>100 cm	Longest-wavelength portion of electromagnetic spectrum.

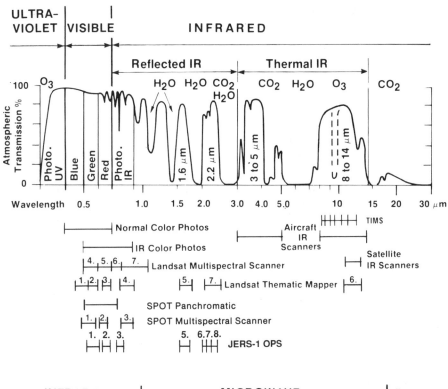

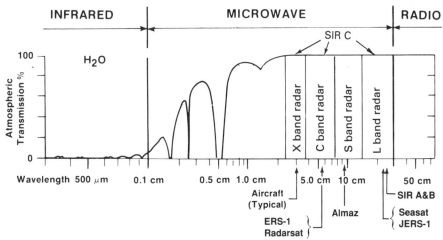

Figure 1-3 Expanded diagrams of the visible and infrared regions (upper) and microwave regions (lower) for transmission through the atmosphere. Gases responsible for atmospheric absorption bands are indicated. Wavelength bands recorded by commonly used remote sensing systems are shown (middle).

earth during daytime may be recorded as a function of wavelength. The maximum amount of energy is reflected at the 0.5-μm wavelength, which corresponds to the green wavelengths of the visible region and is called the *reflected energy peak* (Figure 1-2). The earth also radiates energy both day and night, with the maximum energy radiating at the 9.7-μm wavelength. This *radiant energy peak* occurs in the thermal portion of the IR region (Figure 1-2).

The earth's atmosphere absorbs energy in the gamma-ray, X-ray, and most of the ultraviolet (UV) regions; therefore, these regions are not used for remote sensing. Terrestrial remote sensing records energy in the microwave, infrared, and visible regions, as well as the long wavelength portion of the UV region. Figure 1-3 shows details of these regions. The horizontal axes show wavelength on a logarithmic scale; the vertical axes show the percentage of electromagnetic energy that is transmitted through the earth's atmosphere. Wavelength intervals with high transmission are called *atmospheric windows* and are used to acquire remote sensing images. The major remote sensing regions (visible, infrared, and microwave) are further subdivided into *bands*, such as the blue, green, and red bands of the visible region (Figure 1-3). Horizontal lines in the center of the diagram show wavelength bands in the UV through thermal IR regions recorded by major imaging systems such as cameras and scanners. For the Landsat systems, the numbers identify specific bands recorded by these systems. The characteristics of the remote sensing regions are summarized in Table 1-3.

Passive remote sensing systems record the energy that naturally radiates or reflects from an object. An *active* system supplies its own source of energy, directing it at the object in order to measure the returned energy. Flash photography is an example of active remote sensing, in contrast to available light photography, which is passive. Another common form of active remote sensing is radar (Table 1-3), which provides its own source of electromagnetic energy at microwave wavelengths. Sonar systems transmit pulses of sonic energy.

Atmospheric Effects

Our eyes inform us that the atmosphere is essentially transparent to light, and we tend to assume that this condition exists for all electromagnetic energy. In fact, the gases of the atmosphere absorb electromagnetic energy at specific wavelength intervals called *absorption bands*. Figure 1-3 shows these absorption bands together with the gases in the atmosphere responsible for the absorption.

Wavelengths shorter than 0.3 μm are completely absorbed by the ozone (O_3) layer in the upper atmosphere (Figure 1-3). This absorption is essential to life on earth, because prolonged exposure to the intense energy of these short wavelengths destroys living tissue. For example, sunburn occurs more readily at high mountain elevations than at sea level. Sunburn is caused by UV energy, much of which is absorbed by the atmosphere at sea level. At higher elevations, however, there is less atmosphere to absorb the UV energy.

Clouds consist of aerosol-sized particles of liquid water that absorb and scatter electromagnetic radiation at wavelengths less than about 0.1 cm. Only radiation of microwave and longer wavelengths is capable of penetrating clouds without being scattered, reflected, or absorbed.

IMAGE CHARACTERISTICS

In general usage, an *image* is any pictorial representation, irrespective of the wavelength or imaging device used to produce it. A *photograph* is a type of image that records wavelengths from 0.3 to 0.9 μm that have interacted with light-sensitive chemicals in photographic film. Images can be described in terms of certain fundamental properties regardless of the wavelength at which the image is recorded. These properties are scale, brightness, contrast, and resolution. The tone and texture of images are functions of the fundamental properties.

Scale

Scale is the ratio of the distance between two points on an image to the corresponding distance on the ground. A common scale on U.S. Geological Survey topographic maps is 1:24,000, which means that one unit on the map equals 24,000 units on the ground. Thus 1 cm on the map represents 24,000 cm (240 m) on the ground, or 1 in. represents 24,000 in. (2000 ft). The maps and images of this book show scales graphically as bars.

The deployment of imaging systems on satellites has changed the concepts of image scale. In this book, scales of images are designated as follows:

Small scale (greater than 1:500,000)	1 cm = 5 km or more (1 in. = 8 mi or more)
Intermediate scale (1:50,000 to 1:500,000)	1 cm = 0.5 to 5 km (1 in. = 0.8 to 8 mi)
Large scale (less than 1:50,000)	1 cm = 0.5 km or less (1 in. = 0.8 mi or less)

These designations differ from the traditional scale concepts of aerial photography. Forty years ago, 1:62,500 was the minimum scale of commercially available photographs and was considered small-scale. Today sensing systems on high-altitude aircraft and satellites can acquire photographs and images of excellent quality at much smaller scales. Optimum image scale is determined by how the images are to be interpreted. With the advent of satellite images, many investigators have been surprised at the amount and types of information that can be interpreted from very small scale images.

Brightness and Tone

Remote sensing systems detect the intensity of electromagnetic radiation that an object reflects, emits, or scatters at particular wavelength bands. Variations in intensity of electromagnetic radiation from the terrain are displayed as variations in brightness on images. On positive images, such as those in this book, the brightness of objects is directly proportional to the intensity of electromagnetic radiation that is detected from that object.

Brightness is the magnitude of the response produced in the eye by light; it is a subjective sensation that can be determined only approximately. *Luminance* is a quantitative measure of the intensity of light from a source and is measured with a device called a photometer, or light meter. People who interpret images rarely, if ever, make quantitative measurements of brightness variations on an image. Variations in brightness may be calibrated with a gray scale such as the one in Figure 1-4. Each distinguishable shade from black to white is a separate *tone*. In practice, most interpreters do not use an actual gray scale the way one would use a centimeter scale; they character-

Figure 1-4 Gray scale.

ize areas on an image as light, intermediate, or dark in tone using their own mental concept of a gray scale.

On aerial photographs the tone of an object is primarily determined by the ability of the object to reflect incident sunlight, although atmospheric effects and the spectral sensitivity of the film are also factors. On images acquired at other wavelength regions, tone is determined by other physical properties of objects. On thermal IR images the tone of an object is proportional to the heat radiating from the object. On radar images the tone of an object is determined by the intensity at which the transmitted beam of radar energy is scattered back to the receiving antenna.

Contrast Ratio

Contrast ratio (*CR*) is the ratio between the brightest and darkest parts of the image and is defined as

$$CR = \frac{B_{max}}{B_{min}} \tag{1-3}$$

where B_{max} is the maximum brightness of the scene and B_{min} is the minimum brightness. Figure 1-5 shows images of high, medium, and low contrast, together with profiles of brightness variation across each image. On a brightness scale of 0 to 10, these images have the following contrast ratios:

A. High contrast $CR = \frac{9}{2} = 4.5$

B. Medium contrast $CR = \frac{5}{2} = 2.5$

C. Low contrast $CR = \frac{3}{2} = 1.5$

Note that when $B_{min} = 0$, *CR* is infinity; when $B_{min} = B_{max}$, *CR* is unity. This discussion is summarized from the extensive review by Slater (1983), which describes other terms for contrast. In addition to describing an entire scene, contrast ratio is

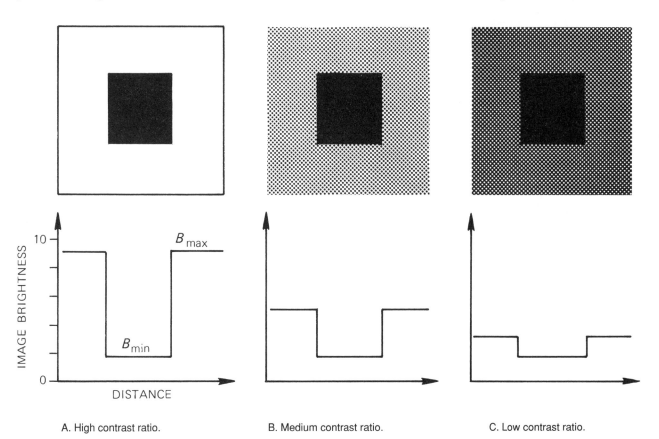

A. High contrast ratio.   B. Medium contrast ratio.   C. Low contrast ratio.

Figure 1-5 Images of different contrast ratios (upper) with corresponding brightness profiles (lower).

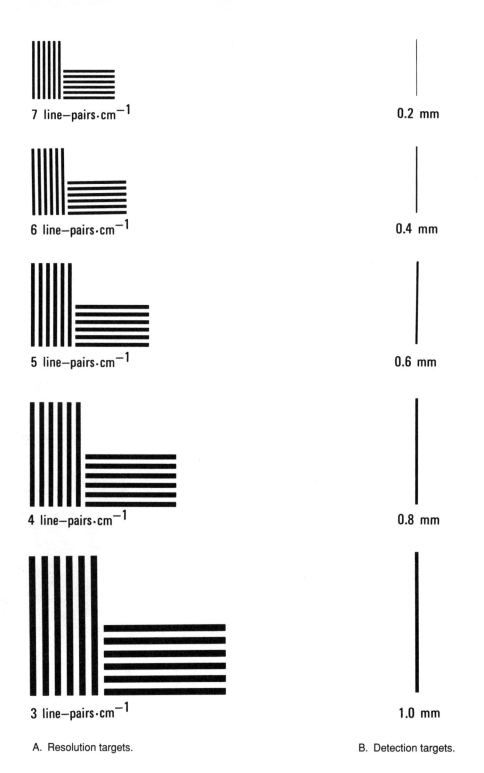

7 line—pairs·cm^{-1}

0.2 mm

6 line—pairs·cm^{-1}

0.4 mm

5 line—pairs·cm^{-1}

0.6 mm

4 line—pairs·cm^{-1}

0.8 mm

3 line—pairs·cm^{-1}

1.0 mm

A. Resolution targets.

B. Detection targets.

Figure 1-6 Resolution and detection targets with high contrast ratio. View this chart from a distance of 5 m (16.5 ft). For A, determine the most closely spaced set of bars you can resolve. For B, determine the narrowest line you can detect.

also used to describe the ratio between the brightness of an object on an image and the brightness of the adjacent background. Contrast ratio is a vital factor in determining the ability to resolve and detect objects.

Images with a low contrast ratio are commonly referred to as "washed out," with monotonous, nearly uniform tones of gray. Low contrast may result from the following causes:

1. The objects and background of the scene may have a nearly uniform electromagnetic response at the particular wavelength band that the remote sensing system recorded. In other words, the scene has an inherently low contrast ratio.
2. Scattering of electromagnetic energy by the atmosphere can reduce the contrast of a scene. This effect is most pronounced in the shorter-wavelength portions of the photographic remote sensing band, as described in Chapter 2.
3. The remote sensing system may lack sufficient sensitivity to detect and record the contrast of the terrain. Incorrect recording techniques can also result in low contrast images even though the scene has a high contrast ratio when recorded by other means.

A low contrast ratio, regardless of the cause, can be improved by digital enhancement methods, as described in Chapter 8.

Spatial Resolution and Resolving Power

This book defines *spatial resolution* as the ability to distinguish between two closely spaced objects on an image. More specifically, it is the minimum distance between two objects at which the images of the objects appear distinct and separate. Objects spaced together more closely than the resolution limit will appear as a single object on the image. Forshaw and others (1983) discuss alternative definitions of spatial resolution.

Resolving power and spatial resolution are two closely related concepts. The term *resolving power* applies to an imaging system or a component of the system, whereas *spatial resolution* applies to the image produced by the system. For example, the lens and film of a camera system each have a characteristic resolving power that, together with other factors, determines the resolution of the photographs.

Spatial resolution of a photographic system is customarily determined by photographing a standard resolution target, such as the one shown in Figure 1-6A, under specified conditions of illumination and magnification. The resolution targets, or *bar charts*, consist of alternating black and white bars of equal width called *line-pairs*. Spacing of resolution targets is expressed in line-pairs. For the target with 5 line-pairs · cm^{-1}, each black bar is 0.1 cm wide and separated by a white bar of the same width. The photograph is viewed under magnification, and the observer determines the most closely spaced set of line-pairs for which the bars and spaces are discernible. Spatial resolution of the photographic system is stated as the number of line-pairs per millimeter of the resolved target.

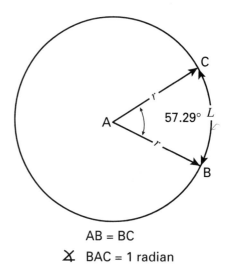

$$AB = BC$$
$$\measuredangle \; BAC = 1 \; radian$$

Figure 1-7 Radian system of angular measurement.

Human judgment and visual characteristics are critical components in this analysis, which therefore is not completely objective and reproducible. Spatial resolution is different for objects of different shape, size, arrangement, and contrast ratio. An alternative method of describing resolution is the *modulation transfer function* (MTF), which employs a bar chart with progressively closer spacing of the bars (McKinney, 1980).

An alternative method of measuring spatial resolution is *angular resolving power*, which is defined as the angle subtended by imaginary lines passing from the imaging system and two targets spaced at the minimum resolvable distance. Angular resolving power is commonly measured in radians. As shown in Figure 1-7, a *radian* (rad) is the angle subtended by an arc BC of a circle having a length equal to the radius AB of the circle. Because the circumference of a circle has a length equal to 2π times the radius, there are 2π, or 6.28, rad in a circle. A radian corresponds to 57.3° or 3438 min, and one milliradian (mrad) is 10^{-3} rad. In the radian system of angular measurement,

$$Angle = \frac{L}{r} \; rad \tag{1-4}$$

where L is the length of the subtended arc and r is the radius of the circle. A convenient relationship is that at a distance r of 1000 units, 1 mrad subtends an arc L of 1 unit. Figure 1-8 illustrates the angular resolving power of a remote sensing system (the eye) that can resolve the center bar chart of Figure 1-6 at a distance of 5 m. This chart has 5 line-pairs · cm^{-1}, and the bars are separated by 1 mm. For these targets with a high contrast ratio, the angular resolving power is 0.2 mrad.

Resolving power and spatial resolution will be discussed for each remote sensing system described in this book, but you should remember the following points:

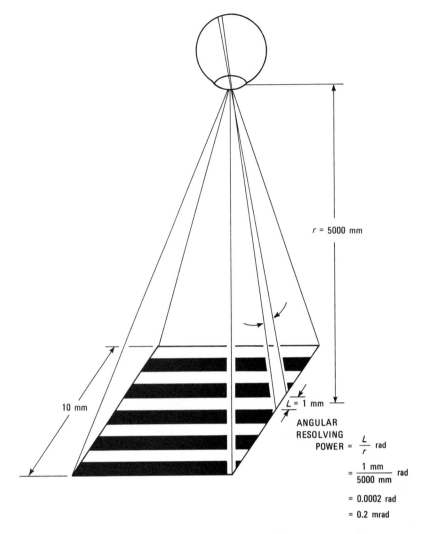

r = 5000 mm

10 mm

L = 1 mm

ANGULAR
RESOLVING
POWER $= \dfrac{L}{r}$ rad

$= \dfrac{1 \text{ mm}}{5000 \text{ mm}}$ rad

$= 0.0002$ rad

$= 0.2$ mrad

Figure 1-8 Angular resolving power (in milliradians) for a remote sensing system that can resolve 5 line-pairs · cm⁻¹ at a distance of 5 m.

1. Theoretical resolving power of a system is rarely achieved in actual operation.
2. Resolution alone does not adequately determine whether an image is suitable for a particular application.
3. Resolution is the minimum separation between two objects for which the images appear distinct and separate; it is *not* the size of the smallest object that can be seen. By knowing the resolution and scale of an image, however, one can estimate the size of the smallest detectable object.

Other Characteristics of Images

Detectability is the ability of an imaging system to record the presence or absence of an object, although the identity of the object may be unknown. An object may be detected even though it is smaller than the theoretical resolving power of the imaging system.

Recognizability is the ability to identify an object on an image. Objects may be detected and resolved and yet not be recognizable. For example, roads on an image appear as narrow lines that could also be railroads or canals. Unlike resolution, there are no quantitative measures for recognizability and detectability. It is important for the interpreter to understand the significance and correct use of these terms. Rosenberg (1971) summarizes the distinctions between them.

A *signature* is the expression of an object on an image that enables the object to be recognized. Characteristics of an object that control its interaction with electromagnetic energy determine its signature. For example, the spectral signature of an object is its brightness measured at a specific wavelength of energy.

Texture is the frequency of change and arrangement of tones on an image. *Fine, medium,* and *coarse* are qualitative terms used to describe texture.

An *interpretation key* is a characteristic or combination of characteristics that enables an object to be identified on an image. Typical keys are size, shape, tone, and color. The associations of different characteristics are valuable keys. On images of cities, one may recognize single-family residential areas by the association of a dense street network, lawns, and small buildings. The associations of certain landforms and vegetation species are keys for identifying different types of rocks.

VISION

Of our five senses, two (touch and vision) detect electromagnetic radiation. Some of the nerve endings in our skin detect thermal IR radiation as heat but do not form images. Vision is the most important sense and accounts for most of the information input to our brain. Vision is not only an important remote sensing system in its own right, but it is also the means by which we interpret the images produced by other remote sensing systems. The following section analyzes the human eye as a remote sensing system. Much of the information is summarized from Gregory (1966).

Structure of the Eye

For such a complex structure, the human eye (Figure 1-9) appears deceptively simple. Light enters through the clear *cornea*, which is separated from the lens by fluid called the *aqueous humor*. The *iris* is the pigmented part of the eye that controls the variable aperture called the *pupil*. It is commonly thought that variations in pupil size allow the eye to function over a wide range of light intensities. However, the pupil varies in area over a ratio of only 16:1 (that is, the maximum area is 16 times the minimum area), whereas the eye functions over a brightness range of about 100,000:1. The pupil contracts to limit the light rays to the central and optically best part of the lens, except when the full opening is needed in dim light. The pupil also contracts for near vision, increasing the depth of field for near objects.

A common misconception is that the lens *refracts* (bends) the incoming rays of light to form the image. The amount that light bends when passing through two adjacent media is determined by the difference in the refractive indices (n) of the two media; the greater the difference, the greater the bending. For the eye, the maximum difference is between air ($n = 1.0$) and the cornea ($n = 1.3$), and this interface is where the maximum light refraction occurs. Although the lens is relatively unimportant for forming the image, it is important in *accommodating*, or focusing, for near and far vision. In cameras, this accommodation is done by changing the position of the lens relative to the film. In the human eye, the shape rather than the position of the lens is changed by muscles that vary the tension on the lens. For near-vision tension, the muscles release, allowing the lens to become thicker in the center and assume a more convex cross section. With age, the cells of the lens harden and the lens becomes too rigid to accommodate for different distances; this is the time in life when bifocal glasses may become necessary to provide for near and far vision.

An inverted image is focused on the *retina*, a thin sheet of interconnected nerve cells that includes the light receptor cells called rods and cones, which convert light into electrical impulses. The rods and cones receive their names from their longitudinal shapes when viewed microscopically. The cones function in daylight conditions to give color, or *photopic*, vision. The rods function under low illumination and give vision only in tones of gray, called *scotopic* vision. Rods and cones are not uniformly distributed throughout the retinal surface. The maximum concentration and organization of receptor cells is in the *fovea* (Figure 1-9), a small region at the center of the retina that provides maximum visual acuity. You can demonstrate the existence and importance of the fovea by concentrating on a single letter on this page. The rest of the page and even the nearby words and letters will appear indistinct because they are outside the field of view of the fovea. The eye is in continual motion to bring the fovea to bear on all parts of the page or scene. Close to the fovea is the blind spot, where the optic nerve joins the eye and there are no receptor cells. The electrical impulses from the receptor cells are transmitted to the brain, which interprets them as visual perception.

Resolving Power of the Eye

The diameter of the largest receptor cells in the fovea (3 μm) determines the resolving power of the eye. Multiplying this diameter maximum by the refractive index of the vitreous humor ($n = 1.3$) determines an effective diameter (4 μm) for the receptor cells. The *image distance,* or distance from the retina to the lens, is about 20 mm, or 20,000 μm. The effective width of the receptors is 4/20,000 (1/5000) of the image distance. Image distance is proportional to *object distance*, which is the distance from the eye to the object. An object forms an image that

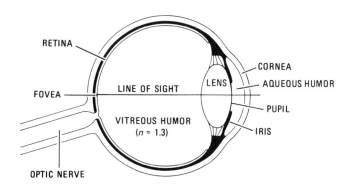

Figure 1-9 Structure of the human eye.

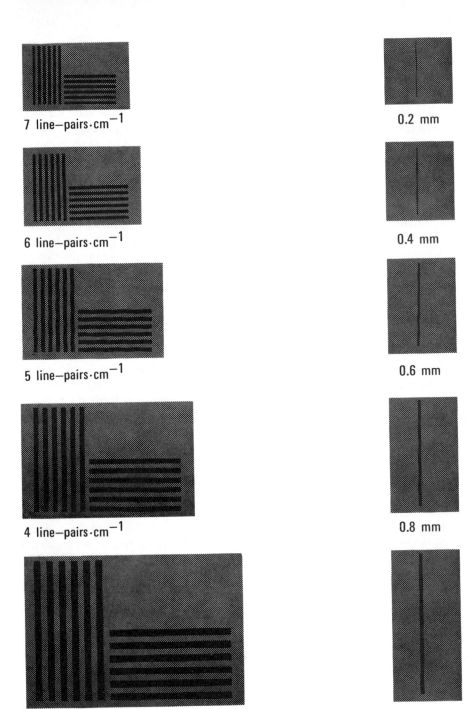

7 line–pairs·cm^{-1}

6 line–pairs·cm^{-1}

5 line–pairs·cm^{-1}

4 line–pairs·cm^{-1}

3 line–pairs·cm^{-1}

A. Resolution targets.

0.2 mm

0.4 mm

0.6 mm

0.8 mm

1.0 mm

B. Detection targets.

Figure 1-10
Resolution and detection targets with medium contrast ratio. View this chart from a distance of 5 m (16.5 ft). For A, determine the most closely spaced set of bars you can resolve. For B, determine the narrowest line you can detect. Compare these values with those determined for Figure 1-6.

fills the width of a receptor if the object width is 1/5000 the object distance. Therefore, adjacent objects must be separated by 1/5000 the object distance for their images to fall on alternate receptors and be resolved by the eye.

You can estimate the resolving power of your eyes in the following manner. View the resolution targets of Figure 1-6A at a distance of 5 m (16.4 ft), and determine the most closely spaced set of line-pairs that you can resolve. Also determine the narrowest of the bars in Figure 1-6B that you can detect. Make these determinations now, before reading further, because the following text may influence your perception of the targets.

For the high-contrast resolution targets of Figure 1-6A at a distance of 5 m, the normal eye should be able to resolve the middle set that has 5 line-pairs $\cdot$ cm^{-1}. The black and white bars are 1 mm wide. The *instantaneous field of view* (IFOV) of any detector is the solid angle through which a detector is sensitive to radiation. Equation 1-4 is used to calculate the IFOV of the eye, where the radius (r) is 5000 mm and the length of the subtended arc (L) is 1 mm:

$$IFOV = \frac{L}{r} \text{ rad (1-4)}$$

$$= \frac{1 \text{ mm}}{5000 \text{ mm}} \text{ rad}$$

$$= 0.2 \times 10^{-3} \text{ rad}$$

$$= 0.2 \text{ mrad}$$

Figure 1-8 shows the relationships of the resolution targets to the IFOV of the eye. The 0.2-mrad IFOV of the eye means that at a distance of 1000 units, the eye can resolve high-contrast targets that are spaced no closer than 0.2 units.

Detection Capability of the Eye

When the detection targets of Figure 1-6B are viewed from a distance of 5 m, most readers can detect the narrowest bar, which is 0.2 mm wide. Recall, however, that at this distance the minimum separation at which bar targets can be resolved is 1.0 mm. This test illustrates the difference between resolution and detection.

Detection is influenced not only by the size of objects but also by their shape and orientation. For example, if dots are used in place of lines in Figure 1-6B, the diameter of the smallest detectable dot would be considerably larger than 0.2 mm.

Effect of Contrast Ratio on Resolution and Detection

The resolution and detection targets in Figure 1-10 have the same spacing as those in Figure 1-6, but the contrast ratio has been reduced by the addition of a gray background. To evalu-

ate the effect of the lower contrast ratio, view Figure 1-10 from a distance of 5 m, and determine which targets can be resolved and detected. Using this figure, most readers can resolve only 3 line-pairs $\cdot$ cm^{-1}, and the smallest detectable target is the 0.6-mm-wide line. These dimensions are larger than the 5 line-pairs $\cdot$ cm^{-1} and the 0.2-mm line of the high-contrast target and demonstrate the effect of a lower contrast ratio on resolution and detection.

REMOTE SENSING SYSTEMS

The eye is a familiar example of a remote sensing system. The inorganic remote sensing systems described in this book belong to the two major categories: framing systems and scanning systems.

Framing Systems

Framing systems instantaneously acquire an image of an area, or *frame,* on the terrain. Cameras and vidicons are common examples of such systems (Figure 1-11). A *camera* employs a

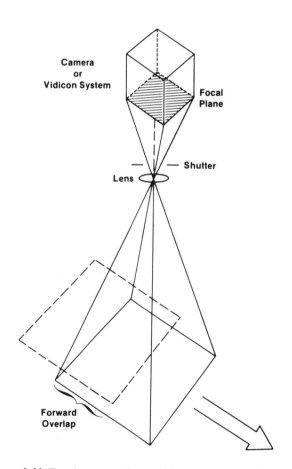

Figure 1-11 Framing system for acquiring remote sensing images. Cameras and vidicons are framing systems.

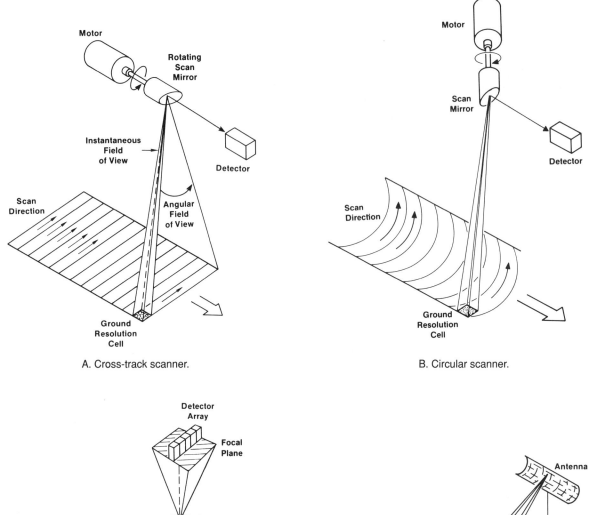

A. Cross-track scanner.

B. Circular scanner.

C. Along-track scanner.

D. Side-scanning system.

Figure 1-12 Scanning systems for acquiring remote sensing images.

lens to form an image of the scene at the *focal plane,* which is the plane at which the image is sharply defined. A shutter opens at selected intervals to allow light to enter the camera, where the image is recorded on photographic film. A *vidicon* is a type of television camera that records the image on a photosensitive electronically charged surface. An electron beam then sweeps the surface to detect the pattern of charge differences that constitute the image. The electron beam produces a signal that may be transmitted and recorded on magnetic tape for eventual display on film.

Successive frames of camera and vidicon images may be acquired with *forward overlap* (Figure 1-11). The overlapping portions of the two frames may be viewed with a stereoscope to produce a three-dimensional view, as described in Chapter 2. Film is sensitive only to portions of the UV, visible, and reflected IR regions (0.3 to 0.9 μm). Special vidicons are sensitive into the thermal band of the IR region. A framing system can instantaneously image a large area because the system has a dense array of detectors located at the focal plane: The retina of the eye has a network of rods and cones, the emulsion of camera film contains tiny grains of silver halide, and a vidicon surface is coated with sensitive phosphors.

Scanning Systems

A *scanning system* employs a single detector with a narrow field of view that is swept across the terrain to produce an image. When photons of electromagnetic energy radiated or reflected from the terrain encounter the detector, an electrical signal is produced that varies in proportion to the number of photons. The electrical signal is amplified, recorded on magnetic tape, and played back later to produce an image. All scanning systems sweep the detector's field of view across the terrain in a series of parallel scan lines. Figure 1-12 shows the four common scanning modes: cross-track scanning, circular scanning, along-track scanning, and side scanning.

Cross-Track Scanners The widely used *cross-track scanners* employ a faceted mirror that is rotated by an electric motor, with a horizontal axis of rotation aligned parallel with the flight direction (Figure 1-12A). The mirror sweeps across the terrain in a pattern of parallel scan lines oriented *normal* (perpendicular) to the flight direction. Energy radiated or reflected from the ground is focused onto the detector by secondary mirrors (not shown). Images recorded by cross-track scanners, and other scanner systems, are described by two characteristics: spectral resolution and spatial resolution.

Spectral resolution refers to the wavelength interval that is recorded by a detector. In Figure 1-13 the vertical scale shows the response, or signal strength, of a detector as a function of wavelength, shown in the horizontal scale. As the wavelength increases, the detector response increases to a maximum and then decreases. Spectral resolution, or *bandwidth,* is defined as the wavelength interval recorded at 50 percent of the peak

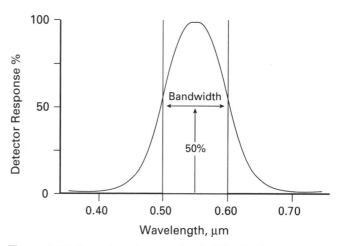

Figure 1-13 Spectral resolution, or bandwidth, of a detector. Bandwidth of this detector is 0.10 μm.

response of a detector. In Figure 1-13 the 50 percent limits occur at 0.50 and 0.60 μm, corresponding to a bandwidth of 10 μm. The section on Reflectance Spectra from Hyperspectral Data describes the effects of different bandwidths on image data.

Spatial resolution was defined earlier, using the eye as an example. The physical dimensions of a detector determine its spatial resolution which is expressed as angular resolving power, measured in milliradians (mrad). Angular resolving power determines the *IFOV* (defined earlier). As shown in Figure 1-12A, the *IFOV* subtends an area on the terrain called a *ground resolution cell.* Dimensions of a ground resolution cell are determined by the detector IFOV and the altitude of the scanning system. A detector with an *IFOV* of 1 mrad at an altitude of 10 km subtends a ground resolution cell of 10 by 10 m. Figure 1-14 shows the important relationship between size of ground resolution cells and spatial resolution of images. The different cell sizes for Figure 1-14B–D were produced by computer processing of the original 10-m data of Figure 1-14A. The images cover a portion of the town of Victorville, California, and the adjacent desert terrain (Figure 1-15). Resolving fine spatial detail of the urban area requires 10-by-10-m cells. Larger linear features, such as the fault and the Mojave River are detectable with 30-by-30-m cells. The images in Figure 1-14 are shown at a large scale (1:24,000), where the 10-m version is obviously optimum. At smaller scales (1:100,00), however, larger cells, such as 30 m (Figure 1-14C), produce readily interpretable images.

The *angular field of view* (Figure 1-12A) is that portion of the mirror sweep, measured in degrees, that is recorded as a scan line. The angular field of view and the altitude of the system determine the *ground swath,* which is the width of the

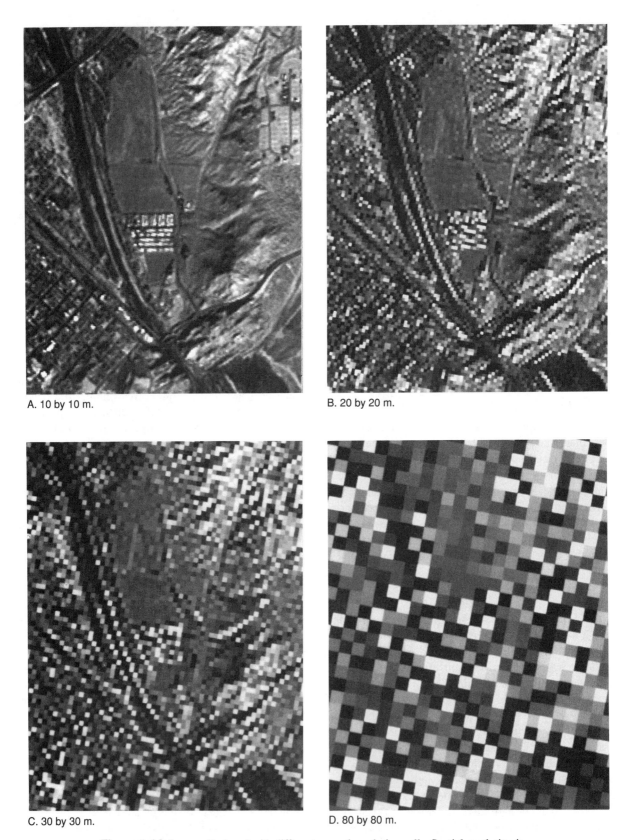

A. 10 by 10 m.

B. 20 by 20 m.

C. 30 by 30 m.

D. 80 by 80 m.

Figure 1-14 Images displayed with different ground resolution cells. Spatial resolution is determined by the size of the cell. The area is a portion of Victorville in southern California.

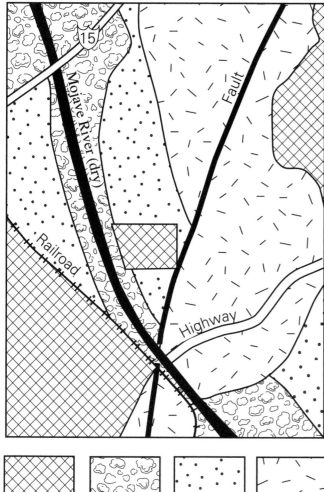

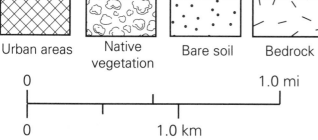

| Urban areas | Native vegetation | Bare soil | Bedrock |

0   1.0 mi

0   1.0 km

Figure 1-15 Location map of Victorville, California, showing land-cover categories.

terrain strip represented by the image. Ground swath is calculated as

$$\text{Ground swath} = \tan\left(\frac{\begin{array}{c}\text{angular field}\\\text{of view}\end{array}}{2}\right) \times \text{altitude} \times 2 \qquad (1\text{-}6)$$

The distance between the scanner and terrain is greater at the margins of the ground swath than at the center of the swath. As a result, ground resolution cells are larger toward the margins than at the center, which results in a geometric distortion that is characteristic of cross-track scanner images. This distortion is corrected by digital processing, as shown in Chapter 8.

At the high altitude of satellites, a narrow angular field of view is sufficient to cover a broad swath of terrain. For this reason the rotating mirror is replaced by a flat mirror that oscillates back and forth through an angle of approximately 15°. An example is the scanner of Landsat described in Chapter 3.

The strength of the signal generated by a detector is a function of the following factors:

Energy flux The amount of energy reflected or radiated from terrain is the *energy flux*. For visible detectors, this flux is lower on a dark day than on a sunny day.

Altitude For a given ground resolution cell, the amount of energy reaching the detector is inversely proportional to the square of the distance. At greater altitudes the signal strength is weaker.

Spectral bandwidth of the detector The signal is stronger for detectors that respond to a broader bandwidth of energy. For example, a detector that is sensitive to the entire visible range will receive more energy than a detector that is sensitive to a narrow band, such as visible red.

Instantaneous field of view Both the physical size of the sensitive element of the detector and the effective focal length of the scanner optics determine the *IFOV*. A small *IFOV* is required for high spatial resolution but also restricts the *signal strength* (amount of energy received by the detector).

Dwell time The time required for the detector *IFOV* to sweep across a ground resolution cell is the *dwell time*. A longer dwell time allows more energy to impinge on the detector, which creates a stronger signal.

For a cross-track scanner, the dwell time is determined by the detector *IFOV* and by the velocity at which the scan mirror sweeps the *IFOV* across the terrain. As shown in Figure 1-16A a typical airborne scanner with a detector *IFOV* of 1 mrad, a 90° angular field of view, and operating at 2×10^{-2} sec per scan line at an altitude of 10 km has a dwell time of 1×10^{-5} sec per ground resolution cell. It is instructive to compare the dwell time with the ground speed of the aircraft. At a typical ground speed of 720 km · h^{-1}, or 200 m · sec^{-1}, the aircraft crosses the 10 m of a ground resolution cell in 5×10^{-2} sec. The cross-track scanner time of 1×10^{-5} is 5×10^3 times faster than the ground velocity of the aircraft. The high scanner speed relative to ground speed is required to prevent gaps between adjacent scan lines.

The short dwell time of cross-track scanners imposes constraints on the other factors that determine signal strength. For example, the *IFOV* and spectral bandwidth must be large enough to produce a signal of sufficient strength to overcome the inherent electronic noise of the system. The signal-to-noise ratio must be sufficiently high for the signal to be recognizable.

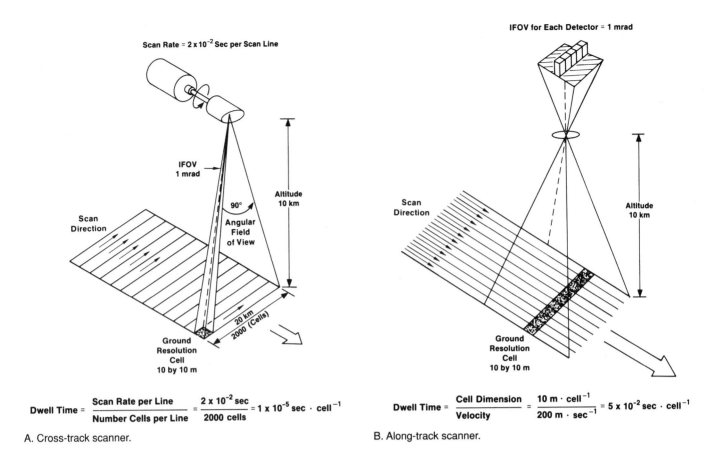

Scan Rate = 2 x 10^{-2} Sec per Scan Line

IFOV
1 mrad

90°

Angular
Field
of View

Altitude
10 km

Scan
Direction

20 km
2000 (Cells)

Ground
Resolution
Cell
10 by 10 m

$$\text{Dwell Time} = \frac{\text{Scan Rate per Line}}{\text{Number Cells per Line}} = \frac{2 \times 10^{-2} \text{ sec}}{2000 \text{ cells}} = 1 \times 10^{-5} \text{ sec} \cdot \text{cell}^{-1}$$

A. Cross-track scanner.

IFOV for Each Detector = 1 mrad

Scan
Direction

Altitude
10 km

Ground
Resolution
Cell
10 by 10 m

$$\text{Dwell Time} = \frac{\text{Cell Dimension}}{\text{Velocity}} = \frac{10 \text{ m} \cdot \text{cell}^{-1}}{200 \text{ m} \cdot \text{sec}^{-1}} = 5 \times 10^{-2} \text{ sec} \cdot \text{cell}^{-1}$$

B. Along-track scanner.

Figure 1-16 Dwell time calculated for cross-track and along-track scanners.

Circular Scanners In a *circular scanner* the scan motor and mirror are mounted with a vertical axis of rotation that sweeps a circular path on the terrain (Figure 1-12B). Only the forward portion of the sweep is recorded to produce images. An advantage of this system is that the distance between scanner and terrain is constant and all the ground resolution cells have the same dimensions. The major disadvantage is that most image processing and display systems are designed for linear scan data; therefore the circular scan data must be extensively reformatted prior to processing.

Circular scanners are used for reconnaissance purposes in aircraft. The axis of rotation is tilted forward to acquire images of the terrain well in advance of the aircraft position. The images are displayed in real time on a screen in the cockpit to guide the pilot. Airborne circular scanners with IR detectors are called FLIR (forward looking IR) systems.

Along-Track Scanners For scanner systems to achieve finer spatial and spectral resolution, the dwell time for each ground resolution cell must be increased. One method is to eliminate the scanning mirror and provide an individual detector for each ground resolution cell across the ground swath (Figure 1-12C).

The detectors are placed in a linear array in the focal plane of the image formed by a lens system.

The long axis of the linear array is oriented normal to the flight path, and the *IFOV* of each detector sweeps a ground resolution cell along the terrain parallel with the flight track direction (Figure 1-12C). *Along-track scanning* refers to this movement of the ground resolution cells. These systems are also called pushbroom scanners because the detectors are analogous to the bristles of a broom pushed along the floor.

For along-track scanners, the dwell time of a ground resolution cell is determined solely by the ground velocity, as Figure 1-16B illustrates. For a jet aircraft flying at 720 km · h^{-1}, or 200 m · sec^{-1}, the along-track dwell time for a 10-m cell is 5×10^{-2} sec, which is 5×10^{3} times greater than the dwell time for a comparable cross-track scanner. The increased dwell time allows two improvements: (1) detectors can have a smaller *IFOV,* which provides finer spatial resolution, and (2) detectors can have a narrower spectral bandwidth, which provides higher spectral resolution. Some experimental airborne along-track scanners operate with a spectral bandwidth of 0.01 μm. Typical cross-track scanners have bandwidths of 0.10 μm, which is a spectral resolution coarser by one order of magnitude.

Side-Scanning Systems The cross-track, circular, and along-track scanners just described are passive systems, since they detect and record energy naturally reflected or radiated from the terrain. Active systems, which provide their own energy sources, operate primarily in the *side-scanning mode*. The example in Figure 1-12D is a radar system that transmits pulses of microwave energy to one side of the flight path (range direction) and records the energy scattered from the terrain back to the antenna, as described in Chapter 6. Another system is side-scanning sonar, which transmits pulses of sonic energy in the ocean to map bathymetric features (Chapter 9).

Scanner Systems Compared Cross-track and along-track scanners have different characteristics that are summarized in the following chart.

	Cross-track scanner	*Along-track scanner*
Angular field of view	Wider	Narrower
Mechanical system	Complex	Simple
Optical system	Simple	Complex
Spectral range of detectors	Wider range	Narrower range, but expanding
Dwell time	Shorter	Longer

The selection of a scanner system involves a number of choices, or trade-offs. Cross-track scanners are generally preferred for reconnaissance surveys because the wider angular field of view records images with a wide swath width. Along-track scanners are preferred for recording detailed spectral and spatial information. The longer dwell time can accommodate detectors with a narrow bandwidth or a small *IFOV*.

SPECTRAL REFLECTANCE CURVES

The various framing and scanning systems record images. Another important aspect of remote sensing is the acquisition of nonimaging data. *Spectral reflectance curves,* or reflectance spectra, record the percentage of incident energy, typically sunlight, that is reflected by a material as a function of wavelength of the energy. Figure 1-17 shows reflectance spectra of vegetation and typical rocks. The horizontal axis shows the wavelength of incident energy, which ranges from the visible into reflected IR spectral regions. The vertical axis shows the percentage of incident energy reflected at the different wavelengths. The curves are offset vertically to prevent confusing overlaps. Downward excursions of a curve are called *absorption features* because they represent absorption of incident energy. Upward excursions are called *reflectance peaks*. These spectral features are valuable clues for recognizing materials, such as the rocks in Figure 1-17, on remote sensing images. For this reason reflectance spectra of many materials have

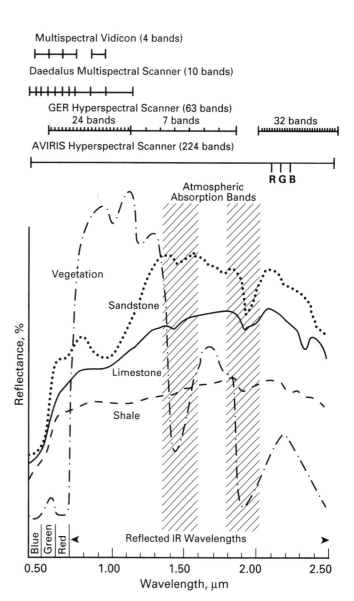

Figure 1-17 Reflectance spectra of rocks and vegetation. Spectral bands of typical multispectral and hyperspectral systems are shown.

been recorded and published. Hunt (1980) published a number of mineral spectra and explained the interactions between energy and matter that cause the spectral features at different wavelengths. Hunt (1980) also provided references to his extensive publications of spectra of rocks and minerals. Clark and others (1990) described laboratory spectroscopy and the causes of absorption features in mineral spectra. They also provide an extensive collection of mineral spectra. Grove and others (1992) published laboratory spectra of 150 minerals. Price (1995) describes 11 published collections of reflectance spectra in the visible and reflected IR regions (0.4 to 2.5 μm) that include soils, agriculture, grasses, shrubs, rocks, minerals, fabrics, metals, and building materials. Price assembled the 3417

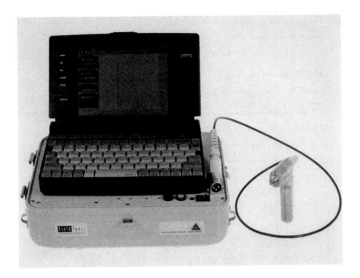

Figure 1-18 Spectrometer for recording reflectance spectra in the field or laboratory. Courtesy Analytical Spectral Devices, Inc., Boulder, CO.

spectra into a standardized digital format on a personal computer diskette. Copies of the diskette are available for research purposes from

J. C. Price
Beltsville Agricultural Research Center
USDA Agricultural Research Service
Beltsville, MD 20705

These spectra should be useful for comparing and identifying spectra recorded from hyperspectral scanners, which are described in a following section.

Reflectance spectra are recorded by instruments called *reflectance spectrometers*. Figure 1-18 is a typical spectrometer that may be used in the field. In the laboratory it uses a light source that matches the characteristics of sunlight. Modern spectrometers record spectra in a digital format that may be displayed directly in real time on the screen of a laptop computer. Some spectrometer systems provide a digital reference library of spectra of known materials, such as rocks, soils, and vegetation. Optional software can compare the spectrum of an unknown material with the library and provide a possible identification. This spectral matching technique is subject to misidentifications, however, as analyzed by Price (1994).

Some manufacturers of spectrometers suitable for remote sensing are listed below; readers may contact them for information on specifications, optional equipment, and cost.

Analytical Spectral Devices (ASD)
4760 Walnut Street, Suite 105
Boulder, CO 80301
Phone: 303-444-6522

Geophysical & Environmental Research Corp. (GER)
One Bennett Common
Millbrook, NY 12545
Phone 914-677-6100

Integrated Spectronics Pty. Ltd.
P.O. Box 437
Baulkham Hill
NSW, 2153, Australia
Phone: 02-887-8760

The ground resolution cell of a handheld spectrometer covers only a few square centimeters. The ground resolution cell of a scanner in an aircraft or satellite covers tens to hundreds of square meters. Because of this difference in coverage one must be cautious when using spectra to evaluate scanner images. Longshaw (1976) analyzed and described the problems of using laboratory spectra to interpret images of rock outcrops.

MULTISPECTRAL IMAGING SYSTEMS

A *multispectral image* is an array of simultaneously acquired images that record separate wavelength intervals, or bands. Multispectral images differ from images such as conventional photographs, which record a single image that spans a broad spectral range. Much of this book deals with multispectral images that are acquired in all the remote sensing spectral regions. Slater (1985) provides a survey of multispectral systems.

Multispectral systems differ in the following characteristics:

- Imaging technology—framing or scanning method
- Total spectral range recorded
- Number of spectral bands recorded
- Range of wavelengths recorded by each spectral band (bandwidth)

The upper portion of Figure 1-17 shows the wavelength bands recorded by representative aircraft multispectral systems, which are listed in Table 1-4 and described in the following sections. Multispectral systems are also widely used in satellites, as described in subsequent chapters. Multispectral images are acquired by two methods: framing systems or scanning systems.

Multispectral Framing Systems

The simplest multispectral framing systems consist of several cameras or vidicons mounted together and aligned to acquire simultaneous multiple images of an area. Today the preponderance of multispectral imagery is acquired by scanners, but vidicon framing systems serve a useful purpose.

Multispectral Cameras *Multispectral cameras* employ a range of film and filter combinations to acquire black-and-

Table 1-4 Representative aircraft multispectral imaging systems

System	Technology	Spectral range, μm	Bands	Bandwidth, μm
Vidicon	Framing (vidicon)	0.40 to 0.90	4	0.10
Daedalus scanner	Scanning (cross-track)	0.38 to 1.10	10	0.03 to 0.20
GER scanner	Scanning (cross-track)	0.50 to 2.50	63	0.025 to 0.175
AVIRIS scanner	Scanning (cross-track)	0.40 to 2.50	224	0.010
SFSI scanner	Scanning (along-track)	1.22 to 2.42	115	0.010

white photographs that record narrow spectral bands. The shutters are linked together and triggered simultaneously. These systems are also called *multiband cameras*. These were the original multispectral systems and are mentioned for historical purposes because they have essentially been replaced by vidicon systems. Lowman (1969) and the National Aeronautics and Space Administration (NASA, 1977) show examples of multispectral photographs.

Multispectral Vidicons *Multispectral vidicons* operate in two modes: (1) two or more individual systems record images at different wavelength bands, or (2) a single system records multiple bands. Several multispectral vidicon systems have been configured; these range from two to four black-and-white vidicons that record narrow bands in the visible and reflected

IR regions. Marsh and others (1991) used a two-vidicon system that records red and reflected IR bands to analyze a hazardous waste site in Arizona. The system was also used to assess land cover in the Mato Grosso, Brazil (Marsh, Walsh, and Sobrevila, 1994). Neale and Crowther (1994) assembled a system from three commercial vidicons that records green, red, and reflected IR bands.

Table 1-4 lists the characteristics of a commercial aircraft vidicon system that acquires four spectral bands (blue, green, red, and reflected IR) that can be composited into color images. Monday and others (1994) used such images to map the city of Irving, Texas (70 mi^2), in 8 months, rather than the 3 years estimated for conventional mapping. King (1995) describes the technology and applications of multispectral vidicon systems and provides references to related publications.

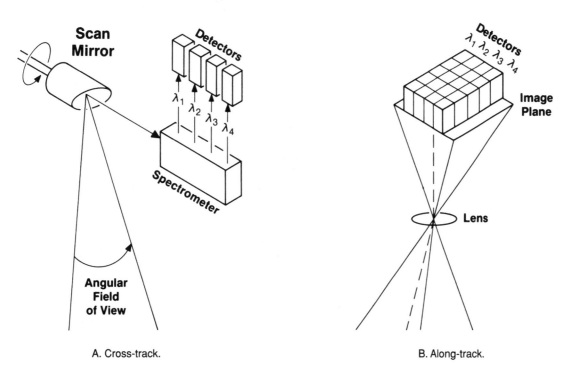

A. Cross-track.

B. Along-track.

Figure 1-19 Multispectral scanner systems.

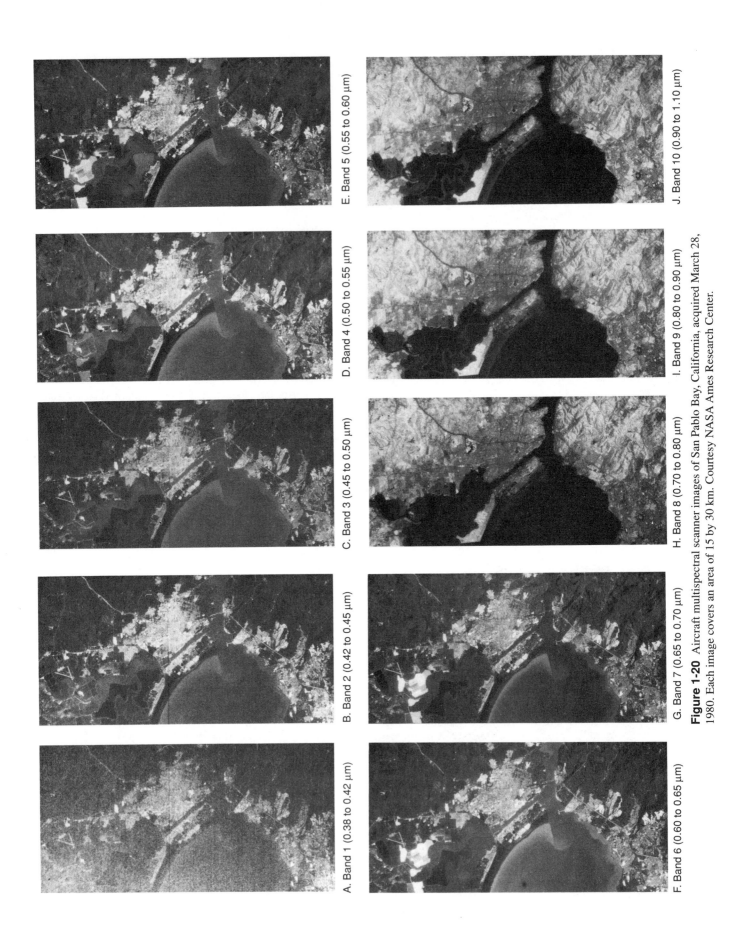

A. Band 1 (0.38 to 0.42 μm)

B. Band 2 (0.42 to 0.45 μm)

C. Band 3 (0.45 to 0.50 μm)

D. Band 4 (0.50 to 0.55 μm)

E. Band 5 (0.55 to 0.60 μm)

F. Band 6 (0.60 to 0.65 μm)

G. Band 7 (0.65 to 0.70 μm)

H. Band 8 (0.70 to 0.80 μm)

I. Band 9 (0.80 to 0.90 μm)

J. Band 10 (0.90 to 1.10 μm)

Figure 1-20 Aircraft multispectral scanner images of San Pablo Bay, California, acquired March 28, 1980. Each image covers an area of 15 by 30 km. Courtesy NASA Ames Research Center.

Multispectral Scanning Systems

Multispectral scanner systems are widely used to acquire images from aircraft and satellites. Both cross-track and along-track systems are used.

Cross-Track Multispectral Scanner Images

Cross-track scanners employ a spectrometer to disperse the incoming energy into a spectrum (Figure 1-19A). Detectors are positioned to record specific wavelength bands of energy (denoted $\lambda_1 \cdots \lambda_4$ in the figure). Figure 1-20 shows images of San Pablo Bay, California, acquired by a cross-track multispectral scanner manufactured by the Daedalus Corporation. Figure 1-21 shows the categories of land use and land cover in the San Pablo Bay area, which is the northern extension of San Francisco Bay. Table 1-5 lists the characteristics of the scanner and the 10 multispectral images. Plate 1A is a color composite image that was prepared by projecting bands 2 (blue), 4 (green), and 7 (red) in blue, green, and red, respectively. Many other color combinations can easily be created.

The black-and-white images of the individual spectral bands (Figure 1-20) demonstrate the relationships among wavelength, atmospheric scattering, contrast ratio, and spatial resolution. Band 1 in the UV and blue region records the shortest wavelengths of all the bands and has the maximum atmospheric scattering, resulting in a low contrast ratio and poor spatial resolution. The network of streets in the city of Vallejo is a useful resolution target; as the wavelength of the images increases, the ability to discern the streets improves and reaches a maximum in the reflected IR region (bands 8, 9, and 10).

TABLE 1-5 Daedalus aircraft multispectral scanner and images

Aircraft altitude	19.5 km
Scanner *IFOV*	1.25 mrad
Ground resolution cell	24 by 24 m
Scan angle	42°
Image swath width	14.7 km

Band*	Wavelength, μm	Spectral band
1	0.38 to 0.42	UV and blue
2	0.42 to 0.45	Blue
3	0.45 to 0.50	Blue
4	0.50 to 0.55	Green
5	0.55 to 0.60	Green
6	0.60 to 0.65	Red
7	0.65 to 0.70	Red
8	0.70 to 0.80	Reflected IR
9	0.80 to 0.90	Reflected IR
10	0.90 to 1.10	Reflected IR

*Combining bands 2, 4, and 7 in blue, green, and red produces a normal color image.

Water, vegetation, and urban areas are the major types of land cover and land use in the San Pablo Bay area. In Figure 1-17 the bandwidth for each multispectral image can be compared to the spectral reflectance curve for vegetation. Vegetation has a higher reflectance in the green band than in the blue and red bands where chlorophyll absorbs energy. The reflectance of vegetation increases abruptly in the reflected IR region. These spectral characteristics of vegetation are also seen in the brightness signatures of vegetated hills in the images at corresponding wavelengths. The signature of water is also different in the various spectral bands. In San Pablo Bay, patterns of suspended silt are obvious in the visible bands; in the IR bands, however, water has a uniform dark signature because these wavelengths are completely absorbed. Some of the salt evaporating ponds, identified in Figure 1-21, have red and pink signatures in Plate 1A because of red microorganisms.

Along-Track Multispectral Scanner Images

Along-track multispectral scanners employ multiple linear arrays of detectors with each array recording a separate band of energy (Figure 1-19B). Because of the extended dwell time, the detector bandwidth may be narrow and produce an adequate signal. Along-track scanner images acquired by SPOT and other satellite systems are illustrated in Chapter 4.

Side-Scanning Multispectral Images

Aircraft and satellite radar systems have been developed that record two or more wavelengths of microwave energy. Typical images are shown in Chapter 7.

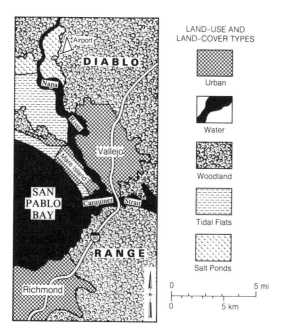

Figure 1-21 Land-use and land-cover types of the San Pablo Bay area, California, interpreted from aircraft multispectral scanner images.

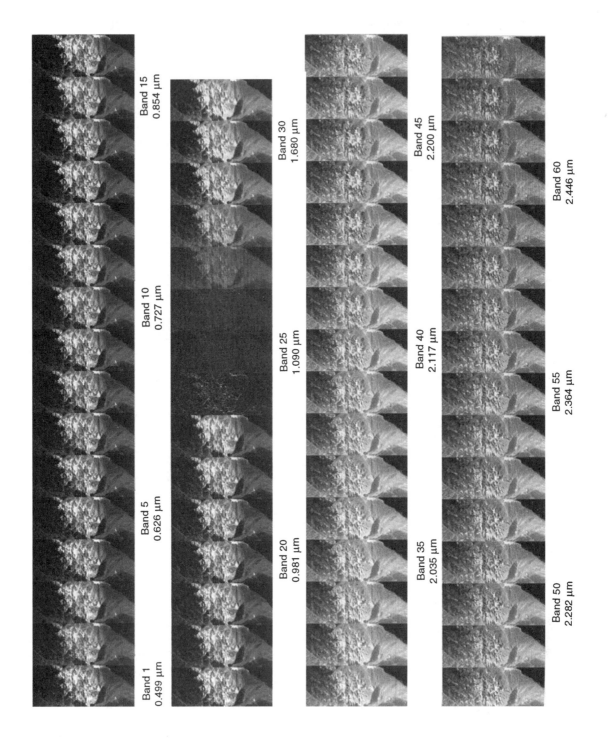

Figure 1-22 Hyperspectral scanner images of Cuprite, Nevada, acquired with a GER 63-band aircraft system.

HYPERSPECTRAL SCANNING SYSTEMS

From the beginning of remote sensing, imaging technology has advanced in two major ways:

1. Improvement in the spatial resolution of images—accomplished primarily by decreasing the *IFOV* of detectors.
2. Improvement in the spectral resolution of images—accomplished by increasing the number of spectral bands and decreasing the bandwidth of each band.

Conventional multispectral scanners record up to 10, or so, spectral bands with bandwidths on the order of 0.10 µm. *Hyperspectral scanners* are a special type of multispectral scanner that records many tens of bands with bandwidths on the order of 0.01 µm. Today these systems are used only on aircraft, but eventually they will be carried on satellites. Table 1-4 lists characteristics of some current hyperspectral scanners.

GER Hyperspectral Scanner

Figure 1-22 shows the 63 hyperspectral bands recorded by a system developed by Geophysical & Environmental Research, Incorporated (GER). The area is the Cuprite mining district in west-central Nevada (Figure 1-23). The band number and wavelength are shown for every fifth image. In Figure 1-17 the tick marks show the spectral distribution of the GER image bands. Only seven bands are located in the interval of 1.00 to 2.00 µm, which is dominated by absorption bands caused by water vapor; 24 bands are in the region of 0.50 to 1.00 µm; 32 bands are in the region of 2.00 to 2.50 µm. The poor quality of bands 24 to 28 (Figure 1-22) is due to the absorption of energy by atmospheric water vapor.

Hyperspectral images are recorded in digital format which results in a large volume of data that can be analyzed using the image-processing techniques described in Chapter 8. Daedalus Enterprises, Inc., of Ann Arbor, Michigan, also manufactures an aircraft hyperspectral scanner that records 102 bands of imagery in the visible, reflected IR, and thermal IR regions.

AVIRIS Hyperspectral Scanner

Jet Propulsion Laboratory (JPL) developed a hyperspectral scanner system called the *airborne visible/infrared imaging spectrometer* (AVIRIS), which acquires 224 images each with a spectral bandwidth of 10 nm in the region of 0.4 to 2.5 µm (Figure 1-17). AVIRIS is carried in a NASA U-2 aircraft at an altitude of 20 km. The images have a swath width of 10.5 km and a spatial resolution of 20 m (Vane and others, 1993). Plate 1B is a color image of the Cuprite mining district and was prepared from the following AVIRIS bands: The band at 2.21 µm is shown in blue, 2.138 µm in green, and 2.088 µm in red. In Figure 1-17 the spectral positions of these bands are indicated by the letters R, G, and B along the AVIRIS range. In the color image of Plate 1B the blue-gray tones are volcanic rocks that

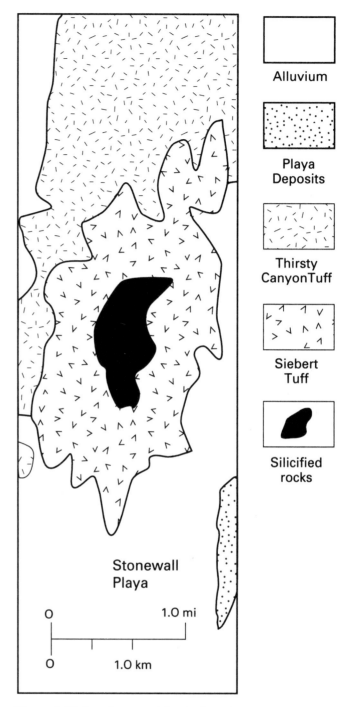

Figure 1-23 Location map of Cuprite, Nevada.

have been replaced by silica (Figure 1-23). The surrounding orange and red tones are volcanic rocks that have been replaced to various degrees by clays and other minerals. The blue tones are younger volcanic rocks that have not been replaced by other minerals. The ability to recognize replacement minerals on AVIRIS images is important for mineral exploration, as described in Chapter 11.

AVIRIS is currently an experimental system, but similar systems will become commercially available in the future. Specialized computer systems are required to process the large volumes of data in the 224 bands of imagery. JPL has convened a series of workshops where investigators have reported results of AVIRIS projects. Vane (1988) and Green (1990, 1991) have prepared proceedings volumes for these workshops. The journal *Remote Sensing of the Environment* devoted one issue (vol. 44, nos. 2/3, May/June, 1993) to a series of papers that describe a wide range of applications of AVIRIS images.

Reflectance Spectra from Hyperspectral Data

Hyperspectral scanners are also called *imaging spectrometers*. The narrow spectral bands of hyperspectral images may be converted into reflectance spectra (Van der Meer, 1994).

Figure 1-24 shows spectra that were calculated from AVIRIS data of Cuprite, Nevada. Each solid spectral curve represents an array of 5-by-5 ground resolution cells, which is an area of 100 by 100 m on the ground. The three curves represent areas where three different minerals (kaolinite, alunite, and budding-

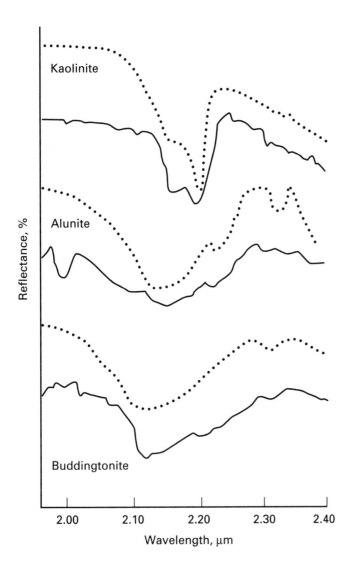

Figure 1-24 Spectra of minerals derived from AVIRIS data (solid lines) and measured by laboratory spectrometer (dotted lines). From Van de Meer (1994, Figure 3).

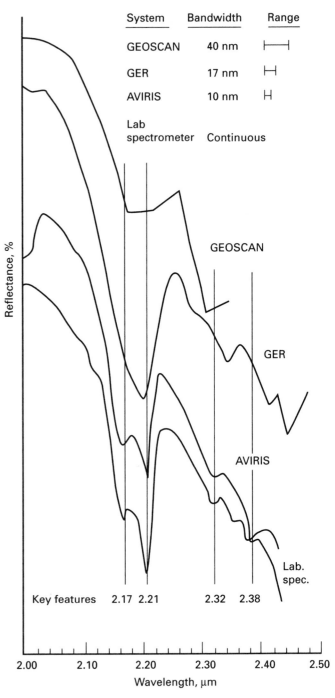

Figure 1-25 Spectra of the clay mineral kaolinite recorded by hyperspectral scanners and by a laboratory spectrometer. Bandwidths of the scanners are shown. Vertical lines indicate absorption features that are keys for recognizing the kaolinite spectrum. Compiled from data in Kruse (1996).

tonite) occur. Kaolinite is a clay mineral, alunite is aluminum sulfate, and buddingtonite is an ammonium feldspar. For each area the percentage of reflectance for each AVIRIS band is plotted as a function of wavelength. The values are connected to produce the solid curves in Figure 1-24. The dotted lines are laboratory spectra for the three minerals, which are similar to the AVIRIS spectra. The differences between the AVIRIS spectra and the spectrometer spectra are explained as follows: The ground resolution cells of AVIRIS include a variety of materials in addition to the predominant mineral, and these contaminate an AVIRIS spectrum, whereas a laboratory spectrum represents a pure sample of each mineral.

Table 1-4 lists characteristics of a Canadian along-track scanner called the SFSI, which is an acronym for SWIR (short wave IR) Full Spectrum Imager. Neville and others (1995) describe the system and show spectra and images.

Figure 1-25 shows reflectance spectra for kaolinite that were produced by Kruse (1996) from data acquired by the GER (63 bands) and AVIRIS (224 bands) hyperspectral scanners. Also shown is a spectrum from the Geoscan multispectral scanner, which records up to 24 bands that are selected from 46 available bands. Figure 1-25 lists the bandwidths of these systems

and shows the bandpass range graphically. For comparison, the bottom kaolinite spectrum was recorded by a laboratory spectrometer and provides the highest spectral resolution. The vertical lines show four key absorption features from the laboratory spectrum that are diagnostic for identifying the kaolinite spectrum. Only AVIRIS, with a 10-nm bandwidth, records all the key features. Similar results were obtained for spectra of buddingtonite and alunite. Despite their lower spectral resolution, Kruse (1996) notes that the GER and Geoscan systems are useful for projects that do not require the identification of specific minerals.

In summary, the spectral resolution (bandwidth) of a remote sensing system (Figure 1-13) determines the ability to distinguish and identify materials based on their spectral characterisitics. Spatial resolution (Figure 1-14) determines the ability to distinguish objects based on their geometric characteristics.

SOURCES OF REMOTE SENSING INFORMATION

Table 1-6 lists scientific journals devoted to remote sensing that are published by technical societies or commercial publishers.

Table 1-6 Remote sensing journals and societies

Journal	*Publisher*
Canadian Journal of Remote Sensing	Canadian Aeronautics and Space Institute Saxe Building 60-75 Sparks Street Ottawa, Canada K1P 5A5
Earth Observation Magazine	EOM, Inc. 13741 E. Rice Place, Suite 125 Aurora, CO 80015
Geo Abstracts—G. Remote Sensing, Photogrammetry, and Cartography	Geo Abstracts Regency House 34 Duke Street Norwich NR3 3AP United Kingdom
Geocarto International	Geocarto International Centre G.P.O. Box 4122 Hong Kong
IEEE Transactions on Geoscience and Remote Sensing	IEEE Remote Sensing and Geoscience Society Institute of Electrical and Electronics Engineers 445 Hoes Lane Piscataway, NJ 08854
International Journal of Remote Sensing	Remote Sensing Society c/o Taylor and Francis, Ltd. Rankine Road Basingstoke, Hants. RG24 OPR United Kingdom

(concluded on the following page)

Table 1-6 *(concluded)*

Journal	Publisher
Photogrammetric Engineering and Remote Sensing	American Society for Photogrammetry and Remote Sensing 210 Little Falls Street Falls Church, VA 22046
Reflections	Radarsat International 3851 Shell Road, Suite 200 Richmond, BC V6X 2W2 Canada
Remote Sensing in Canada	Canadian Remote Sensing Society 130 Slater Street, Suite 818 Ottawa, Ontario K1P 6E2 Canada
Remote Sensing Newsletter	Geological Remote Sensing Group c/o Dr. Stuart Marsh British Geological Survey Keyworth, Nottingham NG12 5GG United Kingdom
Remote Sensing of Environment	Elsevier Science Publishing Company 655 Avenue of the Americas New York, NY 10010-5107
Remote Sensing Reviews	Gordon and Breach Science Publishers P.O. Box 786, Cooper Station New York, NY 10276
Washington Remote Sensing Letter	M. Felscher, Publisher P.O. Box 2075 Washington, DC 20013

Table 1-7 Remote sensing organizations and conferences.

Organization	Address	Conference
Environmental Research Institute of Michigan	P.O. Box 134001 Ann Arbor, MI 48113	"Remote Sensing of Environment" "Thematic Conferences on Remote Sensing"
EROS Data Center of U.S. Geological Survey	Sioux Falls, SD 57198	"Pecora Symposium on Remote Sensing" (annual conference)
Jet Propulsion Laboratory	4800 Oak Grove Drive Pasadena, CA 91103	Publishes reports; conducts conferences and workshops
IEEE Geoscience and Remote Sensing Society	345 East 47th Street New York, NY 10017	"IEEE International Geoscience and Remote Sensing Symposium" (annual symposium)

Table 1-8 Remote sensing images and information on Internet and the Worldwide Web

Facility	Data	Address
Canada Center for Remote Sensing	Landsat 5 and SPOT	http://www.ccrs.emr.ca/cdql/.html
EOSAT	Landsat, IRS, JERS, ERS	http://www.eosat.com
Goddard Space Flight Center	AVHRR	http://xtreme.gsfc.nasa.gov/
Japan NASDA	JERS-1	http://hdsn.eoc.nasda.go.jp/
Johnson Space Center	Images from manned satellites	http://images.jsc.nasa.gov/html/home.htm
Jet Propulsion Laboratory	Educational Outreach Center	http://www.jpl.nasa.gov/education.html
Jet Propulsion Laboratory	Imaging radar	http://southport.jpl.nasa.gov/
Jet Propulsion Laboratory	Public Information Office	http://www.jpl.nasa.gov
NASA	Starting point	http://hypatia.gsfc.nasa.gov/NASA.homepage.html
NASA	Scientific and tech. info.	http://www.sti.nasa.gov
National Oceanic and Atmospheric Administration	Image catalog	http://www.esdim.noaa.gov/NOAA.Catalog/NOAA.Catalog.html
National Oceanic and Atmospheric Administration	Weather satellite data	http://www.ncdc.noaa.gov
Radarsat Canada	Radarsat	http/radarsat.sou.gc.ca
Syracuse University	SIR-C teachers guide	http://ericir.syr.edu/NASA/nasa.htm1
U. S. Geological Survey	Global Land Information System	http://edcwww.cr.usgs.gov/glis/glis.html
U. S. Geological Survey	U.S. spy satellite images	http://edcwww.cr.usgs.gov/dclass.dclass.html
United Kingdom	Weather images	http://web/nexor.co.uk/users/jpo/weather/weather.html

Membership in these societies is open to all investigators. Cracknell (1992) has published a directory of remote sensing journals and societies that are based in Europe. In addition to these journals, many articles on remote sensing are published in journals devoted to other disciplines such as geology, geography, and oceanography. Table 1-7 lists remote sensing conferences conducted by various organizations.

Under the editorship of R. N. Colwell (1983), The American Society of Photogrammetry and Remote Sensing (ASPRS) published the second edition of the *Manual of Remote Sensing*, which is a useful reference. ASPRS is currently preparing a third edition of the manual.

Table 1-8 is a partial listing of remote sensing images and information that are accessible through Internet and the Worldwide Web. Both Internet and the Worldwide Web are dynamic environments in which new sites are being opened and existing sites are being modified or discontinued. Therefore Table 1-8 is a sample of sites that were available in early 1996.

COMMENTS

Remote sensing is defined as the science of acquiring, processing, and interpreting images and related data obtained from aircraft and spacecraft that record the interaction between matter and electromagnetic energy.

The electromagnetic spectrum is divided into wavelength regions. The regions employed in remote sensing range from short-wavelength UV energy to the long-wavelength microwave and radio energy. The electromagnetic regions are further subdivided into narrow wavelength bands. Electromagnetic energy interacts with matter by being scattered, reflected, transmitted, absorbed, or emitted. Subsequent chapters of this book describe these interactions for radiation of the different wavelength bands, together with the technology employed in sensing the radiation.

The interpretation of an image depends upon its scale, tone, texture, contrast ratio, and spatial resolution. In this text, spatial resolution refers to the minimum distance between two objects at which they can be distinguished. Remote sensing systems operate in the framing mode or the scanning mode. Multispectral images consist of two or more simultaneously recorded spectral bands of imagery. Any three bands can be combined, one each in blue, green, and red, to produce color images.

QUESTIONS

1. Use Equation 1-1 to calculate the wavelength in centimeters of radar energy at a frequency of 10 GHz. What is the frequency in gigahertz of radar energy at a wavelength of 25 cm?

2. What is the temperature of boiling water at sea level in degrees Kelvin?

3. Distinguish between the earth's radiant energy peak and the reflected energy peak.

4. The atmosphere is essential for life on earth, but it causes problems for remote sensing. Describe these problems.

5. Use Equation 1-3 to calculate the contrast ratio between a target with a brightness of 17 and a background with a brightness of 8.

6. On images acquired from a satellite (at a 910-km altitude), targets on the ground separated by 80 m can be resolved. Use Equation 1-4 to calculate the angular resolving power (in milliradians) of the scanning system.

7. Assume that your eyes have the normal resolving power (0.2 mrad) and that you are an airline passenger at an altitude of 9 km. For targets on the ground with high contrast ratio, what is the minimum separation (in meters) at which you can resolve these targets?

8. An airborne cross-track scanner has the following characteristics: $IFOV$ = 1.5 mrad; angular field of view = 45°; scan mirror rotates at 4000 rpm (revolutions per minute). The aircraft altitude is 10 km. Calculate the following:

   Size of ground resolution cell = ____ by ____ m

   Width of ground swath = ____ km

   Dwell time for a ground resolution cell = ____ sec.

9. An along-track scanner has detectors with a 2-mrad $IFOV$. The scanner is carried in an aircraft at an altitude of 15 km and a ground speed of 600 km · h^{-1}. Calculate the following:

   Ground resolution cell = ____ by ____ m

   Dwell time for a ground resolution cell = ____ sec.

10. Refer to the aircraft multispectral scanner images and map of San Pablo Bay (Figures 1-20, 1-21). Select the three images that show maximum differences in brightness (reflectance) for the following terrain categories:

    Vegetation in northeast corner of the scene: bands ____, ____, ____.

    Silty water in San Pablo Bay adjacent to Mare Island: bands ____, ____, ____.

    Urban areas of Vallejo: bands ____, ____, ____.

    Salt ponds in northwest portion of the image: bands ____, ____.

REFERENCES

Clark, R. N., and others, 1990, High spectral resolution reflectance spectroscopy of minerals: Journal of Geophysical Research, v. 95, p. 12,653–12,680.

Colwell, R. N., ed., 1983, Manual of remote sensing, second edition: American Society of Photogrammetry, Falls Church, VA.

Cracknell, A. P., 1992, Learned societies, learned journals and other publications: International Journal of Remote Sensing, v. 13, p. 1217–1228.

Fischer, W. A. and others, 1975, History of remote sensing in Reeves, R. G., ed., Manual of remote sensing: ch. 2, p. 27–50, American Society of Photogrammetry, Falls Church, VA.

Forshaw, M. R., A. Haskell, P. F. Miller, D. J. Stanley, and J. R. G. Townshend, 1983, Spatial resolution of remotely sensed imagery—a review paper: International Journal Remote Sensing, v. 4, p. 497–520.

Green, R. O., 1990, Proceedings of the second Airborne Visible/Infrared Imaging Spectrometer (AVIRIS) workshop: Jet Propulsion Laboratory Publication 90-54, Pasadena, CA.

Green, R. O., 1991, Proceedings of the third Airborne Visible/Infrared Imaging Spectrometer (AVIRIS) workshop: Jet Propulsion Laboratory Publication 91-28, Pasadena, CA.

Gregory, R. L., 1966, Eye and brain, the psychology of seeing: World University Library, McGraw-Hill Book Co., New York, NY.

Grove, C. I., S. J. Hook, E. D. Paylor, 1992, Laboratory reflectance spectra of 150 minerals, 0.4 to 2.5 micrometers: Jet Propulsion Laboratory Publication 92-2, Pasadena, CA.

Hunt, G. L., 1980, Electromagnetic radiation—the communication link in remote sensing, in Siegal, B. S., and Gillespie, A. R., eds., Remote sensing in geology: John Wiley & Sons, New York, NY.

King, D. L., 1995, Airborne multispectral digital camera and video sensors—a critical review of system designs and applications: Canadian Journal of Remote Sensing, v. 21, p. 245–273.

Kruse, F. A., 1996, Cuprite Nevada—supplemental field trip information, in Shaulis, L., ed., Remote sensing field trip of Red Rock Canyon, Death Valley, Goldfield, and Cuprite: Eleventh Thematic Conference on Applied Geologic Remote Sensing, Environmental Research Institute of Michigan, Ann Arbor, MI.

Longshaw, T. G., 1976, Application of an analytical approach to field spectroscopy in geological remote sensing: Modern geology, v. 5, p. 93–107.

Lowman, P. D., 1969, Apollo 9 multispectral photography—geologic analysis: NASA Goddard Space Flight Center, Report X-644-69-423, Greenbelt, MD.

Marsh, S. E., J. L. Walsh, C. T. Lee, and L. A. Graham, 1991, Multitemporal analysis of hazardous waste sites through the use of a new bi-spectral video remote sensing system and standard color-IR photography: Photogrammetric Engineering and Remote Sensing, v. 57, p. 1221–1226.

Marsh, S. E., J. L. Walsh, and C. Sobrevila, 1994, Evaluation of airborne video data for land-cover classification accuracy assessment in an isolated Brazilian forest: Remote sensing of environment, v. 48, p. 61–69.

McKinney, R. G., 1980, Photographic materials and processing in Slama, C. C., ed., Manual of photogrammetry, fourth edition: ch. 6, p. 305–366, American Society of Photogrammetry, Falls Church, VA.

Monday, H. M., J. S. Urban, D. Mulawa, and C. A. Benkelman, 1994, City of Irving utilizes high resolution multispectral imagery for N. P. D. E. S. compliance: Photogrammetric Engineering and Remote Sensing, v. 60, p. 411–416.

NASA, 1977, Skylab explores the earth: NASA SP-250, Washington, DC.

Neale, C. M. U. and B. G. Crowther, 1994, An airborne multispectral video/radiometer remote sensing system—development and calibration: Remote Sensing of Environment, v. 49, p. 187–194.

Neville, R. A., N. Rowlands, R. Marois, and I. Powell, 1995, SFSI—Canada's first airborne SWIR imaging spectrometer: Canadian Journal of Remote Sensing, v. 21, p. 328–336.

Price, J. C., 1994, How unique are spectral signatures?: Remote Sensing of Environment, v. 49, p. 181–186.

Price, J. C., 1995, Examples of high resolution visible to near-infrared reflectance spectra and a standardized collection for remote sensing studies: International Journal of Remote Sensing, v. 16, p. 993–1000.

Rosenberg, P., 1971, Resolution, detectability, and recognizability: Photogrammetric Engineering, v. 37, p. 1244–1258.

Slater, P. N., 1983, Photographic systems for remote sensing *in* Colwell, R. N., ed., Manual of remote sensing, second edition: ch. 6, p. 231–291, American Society Photogrammetry, Falls Church, VA.

Slater, P. N., 1985, Survey of multispectral imaging systems for earth observation, in R. N. Colwell, ed., Manual of remote sensing, second edition: ch. 6, p. 231–291, American Society for Photogrammetry and Remote Sensing, Falls Church, VA.

Suits, G. H., 1983, The nature of electromagnetic radiation *in* Colwell, R. N., ed., Manual of remote sensing, second edition: ch. 2, p. 37–60, American Society of Photogrammetry, Falls Church, VA.

Van der Meer, F., 1994, Extraction of mineral absorption features from high-spectral resolution data using non-parametric geostatistical techniques: International Journal of Remote Sensing, v. 15, p. 2193–2214.

Vane, G., ed., 1988, Proceedings of the Airborne Visible/Infrared Imaging Spectrometer (AVIRIS) performance evaluation workshop: Jet Propulsion Laboratory Publication 88-38, Pasadena, CA.

Vane, G., R., O. Green, T. G. Chrien, H. T. Enmark, E. G. Hansen, and W. M. Porter, 1993, The Airborne Visible/Infrared Imaging Spectrometer (AVIRIS): Remote Sensing of Environment, v. 44, p. 127–143.

ADDITIONAL READING

Avery, T. E. and G. L. Berlin, 1992, Fundamentals of remote sensing and airphoto interpretation, fifth edition: Macmillan, New York, NY.

Beaumont, E. A. and Foster, N. H., eds., 1992, Remote sensing: American Association of Petroleum Geologists, Treatise of Petroleum Geology Reprint Series, no. 19, Tulsa, OK.

Ben-Dor, E. and F. A. Kruse, 1995, Surface mineral mapping of Makhtesh Ramon Negev, Israel using GER 63 channel scanner data: International Journal of Remote Sensing, v. 16, p. 3529–3553.

Colwell, R. N. and others, 1963, Basic matter and energy relationships involved in remote reconnaissance: Photogrammatric Engineering and Remote Sensing, v. 29, p. 761–799.

Elachi, C., 1987, Introduction to the physics and techniques of remote sensing: John Wiley & Sons, New York, NY.

Everitt, J. H., and others, 1995, A three-camera multispectral digital video imaging system: Remote Sensing of Environment, v. 54, p. 333–337.

Hook, S. J., C. D. Elvidge, M. Rast, and H. Watanabe, 1991, An evaluation of short-wavelength-infrared (SWIR) data from the AVIRIS and GEOSCAN instruments for mineralogic mapping at Cuprite, Nevada: Geophysics, v. 56, p. 1432–1440.

Hyatt, E., 1988, Keyguide to information sources in remote sensing: Mansell, London, England.

Lillesand, T. M. and R. W. Kiefer, 1994, Remote sensing and image interpretation, third edition: John Wiley & Sons, New York, NY.

Rees, W. G., 1990, Physical principles of remote sensing: Cambridge University Press, Cambridge, England.

Southworth, C. S., 1985, Characteristics and availability of data from earth imaging satellites: U.S. Geological Survey, Bulletin 1631.

Vane, G., ed., 1990, Imaging spectroscopy of the terrestrial environment: SPIE—The International Society for Optical Engineering, v. 1298.

Vane, G. and A. F. H. Goetz, 1993, Terrestrial imaging spectrometry—current status, future trends: Remote Sensing of Environment, v. 44, p. 117–126.

PHOTOGRAPHS FROM AIRCRAFT AND SATELLITES

Photographs acquired from aircraft (aerial photographs) were the first form of remote sensing imagery, and they remain the most widely used images today. Knowing the techniques for interpreting aerial photographs is essential background for understanding other remote sensing images. Indeed, aerial photographs are used throughout this text to aid in explaining thermal IR, radar, and other kinds of images. In the enthusiasm for satellite images and new forms of airborne remote sensing, one should not overlook the advantages of aerial photographs. Topographic maps are made from aerial photographs, and many engineering projects use aerial photographs. Soil conservation studies, agricultural crop inventories, and city planning all employ aerial photographs. Geologic mapping and exploration commonly begin with an analysis of photographs. In the early 1970s, interpretation of aerial photographs led to the discovery of several valuable oil fields in Irian Jaya, Indonesia. Philipson (1996) edited a manual of photographic interpretation.

Many photographs have been acquired from earth-orbiting satellites and are described in this chapter.

INTERACTIONS BETWEEN LIGHT AND MATTER

As with other forms of electromagnetic energy, light may be reflected, absorbed, or transmitted by matter. Aerial photographs record the light reflected by a surface, which is determined by the property called albedo. *Albedo* is the ratio of the energy reflected from a surface to the energy incident on the surface. Dark surfaces have a low albedo, and bright surfaces have a high albedo. Light that is not reflected is transmitted or absorbed by the material. During its transmission through the atmosphere, light interacts with the gases and particulate matter in a process called scattering, which has a strong effect on aerial photographs.

Atmospheric Scattering

Atmospheric scattering results from multiple interactions between light rays and the gases and particles of the atmosphere, as shown in Figure 2-1. The two major processes, selective scattering and nonselective scattering, are related to the size of particles in the atmosphere. In *selective scattering* the shorter wavelengths of UV energy and blue light are scattered more severely than the longer wavelengths of red light and IR energy. Selective scattering is caused by fumes and by gases such as nitrogen, oxygen, and carbon dioxide. The selective scattering of blue light causes the blue color of the sky. The red skies at sunrise and sunset are due to sunlight passing horizontally through the atmosphere, which scatters blue and green wavelengths so only red light reaches the viewer.

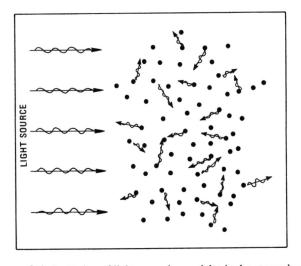

Figure 2-1 Scattering of light waves by particles in the atmosphere.

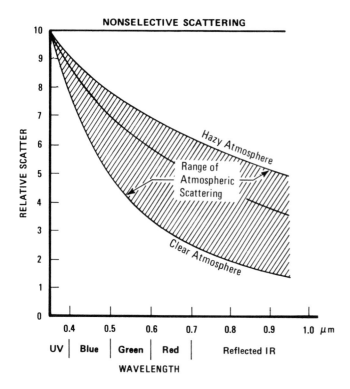

Figure 2-2 Atmospheric scattering as a function of wavelength. The shaded region shows the range of scattering caused by typical atmospheres. From Slater (1983, Figure 6-15).

In *nonselective scattering* all wavelengths of light are equally scattered. Nonselective scattering is caused by dust, clouds, and fog in which the particles are much larger than the wavelengths of light. Clouds and fog are aerosols of very fine water droplets; they are white because the droplets scatter all wavelengths equally. The curves in Figure 2-2 show relative scattering as a function of wavelength. Nonselective scattering is shown by the horizontal line.

Scattering in the atmosphere results from a combination of selective and nonselective processes. The range of atmospheric scattering is shown by the shaded area in Figure 2-2. The lower curve of the shaded area represents a clear atmosphere and the upper curve a hazy atmosphere. Typical atmospheres have scattering characteristics that are intermediate between these extremes. The important point for aerial photography is that the earth's atmosphere scatters UV and blue wavelengths at least twice as strongly as red light.

Light scattered by the atmosphere illuminates shadows, which are never completely dark but are bluish in color. This scattered illumination is referred to as *skylight* to distinguish it from direct sunlight. A striking characteristic of photographs taken by Apollo astronauts on the surface of the moon is the black appearance of the shadows. The lack of atmosphere on the moon precludes any scattering of light into the shadowed areas.

Effects of Scattering on Aerial Photographs

Scattered light that enters the camera is a source of illumination but contains no information about the terrain. This extra illumination reduces the contrast ratio (*CR*) of the scene, thereby reducing the spatial resolution and detectability of the photograph. Figure 2-3 diagrams the effect of scattered light on the contrast ratio of a scene in which a dark area (brightness = 2) is surrounded by a brighter background (brightness = 5). For the original scene with no scattered light (Figure 2-3A,B), the contrast ratio is determined from Equation 1.3 as follows:

$$CR = \frac{B_{max}}{B_{min}}$$

$$= \frac{5}{2}$$

$$= 2.5$$

Figure 2-3C shows the appearance of the scene in conditions of heavy haze, where the atmosphere contributes 5 brightness

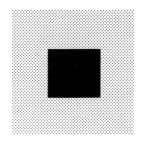

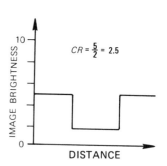

A. Original scene.

B. Brightness profile of image with no scattered light.

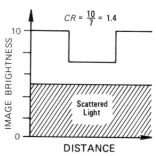

C. Profile of image with 5 brightness units added by scattered light.

D. Brightness profile and contrast ratio of image with scattered light.

Figure 2-3 Effect of scattered light on the contrast ratio of an image.

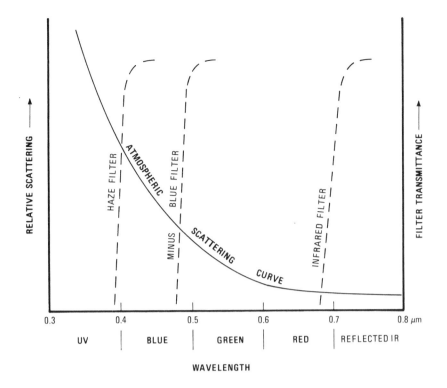

Figure 2-4 Atmospheric scattering diagram and transmission curves of filters used in aerial photography. Shorter wavelengths to the left of each filter curve are absorbed. From Slater (1983, Figure 6-57).

units of scattered light. As the brightness profile of Figure 2-3D shows, scattered light adds uniformly to all parts of the scene and results in a contrast ratio of

$$CR = \frac{10}{7}$$
$$= 1.4$$

Thus atmospheric scattering has reduced the contrast ratio of the scene from 2.5 to 1.4, which lowers the spatial resolution on a photograph of that scene. Chapter 1 demonstrated this relationship between contrast ratio and resolving power. The effect of atmospheric scattering on aerial images is illustrated later in this chapter.

Filtering out the selectively scattered shorter wavelengths before they reach the film reduces the effects of atmospheric scattering. Superposed on the scattering curve of Figure 2-4 are the spectral transmittance curves of typical filters used in aerial photography showing the wavelengths absorbed and transmitted. There is a trade-off with filters: although they reduce haze, they also remove the spectral information contained in the wavelengths that are absorbed.

FILM TECHNOLOGY

Photographic film consists of a flexible transparent base coated with a layer of light-sensitive emulsion approximately 100 μm in thickness (Figure 2-5A). The *emulsion* is initially a suspension in solidified gelatin of grains of silver halide (a salt) a few micrometers or less in diameter. The grains have been precipitated from solution rather rapidly to make them irregular, with numerous points of imperfection on the surface. After the emulsion is deposited on the film base, further processing of the grains increases their sensitivity to light. *Photographic exposure* is the photochemical reaction between photons of light incident upon silver halide grains in the emulsion. When a photon strikes one of the grains, an electron in the silver halide crystal is given enough energy to move freely about and may be trapped at an imperfection in the grain. By combining with a silver ion lacking one electron, the electron may then convert the ion into a silver atom (Jones, 1968). This atom cannot remain an atom for long by itself, but if two electrons in the grain are liberated within about a second, a stable combination of silver atoms will form at the imperfection (Figure 2-5B).

The success of the photographic method depends upon the requirement that a silver halide grain receive more than one

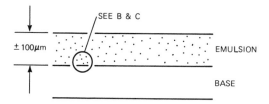

A. Cross section of film.

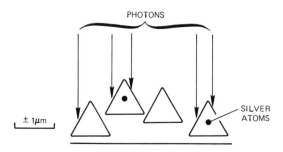

B. Exposure of silver halide grains.

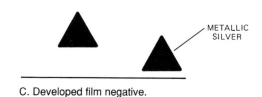

C. Developed film negative.

Figure 2-5 Film technology.

photon within a short time. If only one photon were needed, random photons caused by normal ionic vibration would soon convert all the grains to silver. The stable combinations of silver atoms are large enough to trigger the conversion of the entire silver halide grain to metallic silver when the film is chemically developed. Before chemical development, however, exposed silver halide grains have the same appearance as unexposed grains, so at this stage the film is said to contain a *latent image.*

Developing is the chemical process of changing the latent image into a real image by converting the exposed silver halide grains into opaque grains of silver. The film is immersed in the developer, which is a water solution containing a reducing agent that does not interact with unexposed silver halide grains. For exposed grains with silver atoms at a point of imperfection, however, the agent starts reducing the silver halide to silver. As Figure 2-5C shows, the entire grain converts to metallic silver once the reduction process begins.

Fixing is the next process, which removes unexposed grains, leaving clear areas in the emulsion. The resulting film is called a *negative film* because bright targets form dark images on the film. When the film is printed onto photographic paper, the dark negative images are reversed and bright targets appear bright on the print.

One advantage of photographic remote sensing is the enormous amplification that occurs in the development process. A few photons absorbed in a grain of silver halide with a volume of 1 μm^3 will produce more than 10^{10} atoms of developed silver, which is an amplification of more than 1 billion times. Other advantages of photographic remote sensing are high resolving power, low cost, versatility, and ease of operation. Another advantage is the capacity of film to store large amounts of information. On the developed film, each grain, whether exposed or unexposed, records information about the scene. There are more than 150 million (1.5×10^8) such grains on a 6.5-cm² (1-in.²) piece of film. James (1966) is a standard reference on the photographic process.

The three major disadvantages of photographic remote sensing are the following:

1. It is restricted to the spectral region of 0.3 to 0.9 μm.
2. It is restricted by weather, lighting conditions, and atmospheric effects.
3. Information is recorded in a nondigital format. In order to be computer-processed, photographs must be converted into digital format, as described in Chapter 8.

CHARACTERISTICS OF AERIAL PHOTOGRAPHS

Aerial photographs are acquired with a variety of cameras, films, and filters. Characteristics such as resolution, scale, and relief displacement are common to all aerial photographs. The sections that follow discuss these characteristics in some detail.

Spatial Resolution of Photographs

Spatial resolution, or *resolving power,* of aerial photographs is influenced by several factors:

1. Atmospheric scattering, which was discussed earlier.
2. Vibration and motion of the aircraft, which are minimized by vibration-free camera mounts and motion compensation devices.
3. Resolving power of lenses.
4. Resolving power of films.

All of these factors combine to determine the spatial resolution of a photograph.

Resolving Power of Lenses

The resolving power of a lens is determined by its optical quality and size. If a lens is used to photograph a resolution target, such as those shown in Figure 1-6A, there is an upper limit to the number of line-pairs within the space of a millimeter that can be resolved on the resulting photograph. This maximum number of resolvable line-pairs per millimeter is a measure of the resolving power of the lens.

Resolving Power of Film

The resolving power of film is determined by several factors, the most important of which is *granularity*. The two factors that largely determine granularity are the size distribution of silver halide grains in the emulsion and the nature of the development process. Films with high granularity have lower resolving power than those with low granularity. There is a trade-off between granularity and the *speed of film*: films with high granularity are faster, meaning they are more sensitive to light.

One method for expressing the resolving power of film is to photograph a resolution target and determine the maximum number of line-pairs per millimeter one can distinguish on the developed film. As illustrated earlier in Figures 1-6 and 1-10, targets with high contrast ratios produce better resolution than those with low contrast ratios; terrain features typically have low contrast ratios. Film resolving power is commonly given for targets with both high contrast ratios and low contrast ratios. A widely used black-and-white film, Kodak Panatomic X aerial film, has a resolving power of 300 line-pairs $\cdot$ mm^{-1} for high-contrast targets and 80 line-pairs $\cdot$ mm^{-1} for low-contrast targets. *System resolution* (R_s) of a camera and film combination results from the resolving powers of the lens and film and typically ranges from about 25 to 100 line-pairs $\cdot$ mm^{-1}.

Ground Resolution

Ground resolution expresses the ability to resolve ground features on aerial photographs. System resolution is converted into ground resolution by the formula

$$R_g = \frac{R_s f}{H} \tag{2-1}$$

where

R_g = ground resolution in line-pairs per millimeter

H = camera height above ground in meters. (Do not confuse this with aircraft altitude above mean sea level.)

R_s = system resolution in line-pairs per millimeter

f = camera focal length in millimeters

Figure 2-6 shows the geometric basis for this relationship. For a camera lens with a focal length of 152 mm producing photographs with a system resolution of 20 line-pairs $\cdot$ mm^{-1} acquired at a camera height of 6100 m, the ground resolution, using Equation 2-1, is

$$R_g = \frac{R_s f}{H} \tag{2-1}$$

$$= \frac{20 \text{ line-pairs} \cdot \text{mm}^{-1} \times 152 \text{ mm}}{6100 \text{ m}}$$

$$= 0.5 \text{ line-pairs} \cdot \text{m}^{-1}$$

Under the conditions specified in this example, the most closely spaced resolution target on the ground that can be resolved on the photograph consists of 2.0 line-pairs $\cdot$ m^{-1}. The width of an individual line-pair in meters is determined by the reciprocal

$$\frac{1.0 \text{ line-pairs}}{R_g}$$

and is 0.5 m in this example. *Minimum ground separation* is the minimum distance between two objects on the ground at which they can be resolved on the photograph. As Figure 2-6 shows, it is the separation between lines or bars in the resolution target and is determined by

$$\text{Minimum ground separation} = \frac{1.0 \text{ line-pairs} / R_g}{2} \tag{2-2}$$

$$= \frac{1.0 \text{ line-pairs} / 2.0 \text{ line-pairs} \cdot \text{m}^{-1}}{2} = 0.25 \text{ m}$$

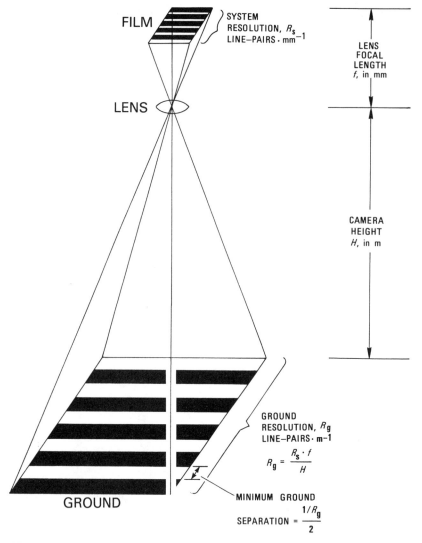

FILM

SYSTEM RESOLUTION, R_s LINE—PAIRS · mm^{-1}

LENS

LENS FOCAL LENGTH f, in mm

CAMERA HEIGHT H, in m

GROUND RESOLUTION, R_g LINE—PAIRS · m^{-1}

$$R_g = \frac{R_s \cdot f}{H}$$

MINIMUM GROUND SEPARATION = $\dfrac{1/R_g}{2}$

GROUND

Figure 2-6 Ground resolution and minimum ground separation on aerial photographs.

TABLE 2-1 Minimum ground separation on typical aerial photographs acquired at different heights (focal length of camera lens is 152 mm)

| | | Minimum ground separation for system resolution R_s of | |
Camera height (H), m	Scale of photographs	40 line-pairs · mm^{-1}, m	100 line-pairs · mm^{-1}, m
1525	1:10,000	0.12	0.05
3050	1:20,000	0.25	0.10
4575	1:30,000	0.37	0.15
6100	1:40,000	0.50	0.20

Table 2-1 lists minimum-ground-separation values for typical aerial photographs acquired with camera systems of medium and high system resolutions. The camera heights and lens focal length of Table 2-1 correspond to the medium-resolution aerial photographs in Figure 2-7. Inspection of the photographs with a magnifier indicates that these values for minimum ground separation are appropriate, although the photographs lack a ground resolution target, which is necessary for precise measurement.

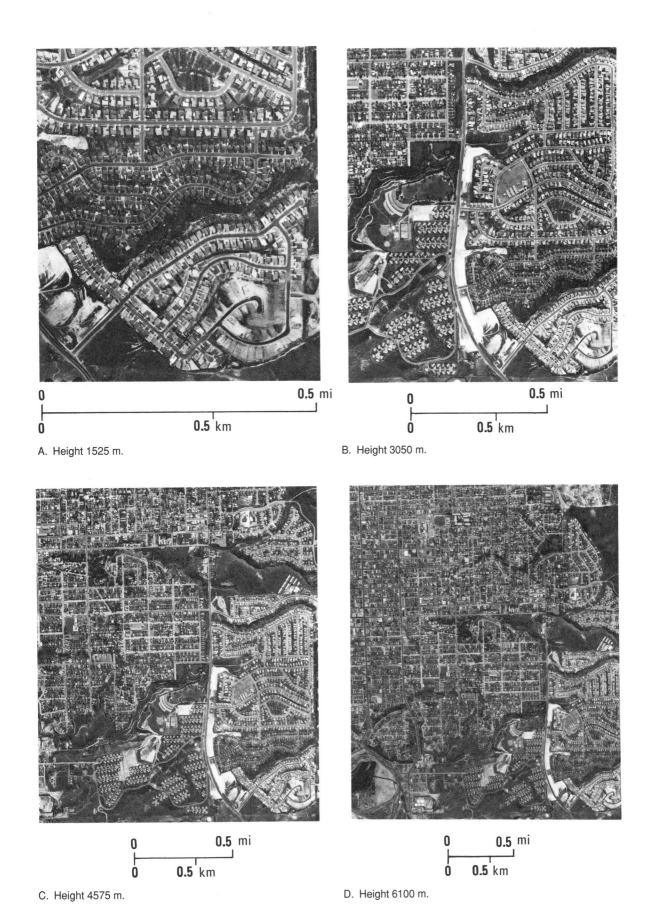

0   0.5 mi

0   0.5 km

A. Height 1525 m.

0   0.5 mi

0   0.5 km

B. Height 3050 m.

0   0.5 mi

0   0.5 km

C. Height 4575 m.

0   0.5 mi

0   0.5 km

D. Height 6100 m.

Figure 2-7 Aerial photographs of Palos Verde Peninsula, California, acquired at different camera heights with a 152-mm-focal-length lens. Table 2-1 lists minimum ground separation values for this medium-resolution system. The southeastern corner is common to all photographs.

Table 2-2 Features recognizable on aerial photographs at different minimum-ground-separation values

Minimum ground separation, m	Recognizable features
15.0	Identify geographic features such as shorelines, rivers, mountains, and water.
4.50	Differentiate settled areas from undeveloped land.
1.50	Identify roadways.
0.15	Distinguish front from the rear of automobiles.
0.05	Count people, particularly if there are shadows and if the individuals are not in crowds.

Source: After Rosenblum (1968, Table 2).

Table 2-2 lists features that may be identified on photographs with different ground separation values. These are only guidelines to illustrate the general relationship between ground resolution and recognition.

Photographic Scale

The scale of aerial photographs is determined by the relationship

$$\text{Scale} = \frac{1}{H/f} \qquad (2\text{-}3)$$

Both H and f must be given in the same units, typically meters. For example, the scale of a photograph acquired at a camera height of 3050 m with a 152-mm lens (Figure 2-7B) is

$$\frac{1}{3050 \text{ mm} / 0.152 \text{ mm}} = \frac{1}{20,000} \text{ or } 1{:}20{,}000$$

A scale of 1:20,000 means that 1 cm on the photograph represents 20,000 cm (or 200 m) on the ground (1 in. = 20,000 in. = 1667 ft). Figure 2-7 illustrates the different scales that result from photographing the same area at different altitudes with the same camera.

Relief Displacement

Figure 2-8 illustrates the geometric distortion called *relief displacement,* which is present on all vertical aerial photographs that are acquired when the camera is aimed directly down. The tops of objects such as buildings appear to "lean" away from the *principal point,* or optical center, of the photograph. The amount of displacement increases at greater radial distances from the center and reaches a maximum at the corners of the photograph. Figure 2-9A shows the geometry of image displacement, where light rays are traced from the terrain through the camera lens and onto the film. Prints made from the film appear as though they were in the position shown by the plane of photographic print in Figure 2-9A. The vertical arrows on the terrain represent objects of various heights located at various distances from the principal point. The light ray reflected from the base of object A intersects the plane of the photographic print at position A, and the ray from the top intersects the print at A'. The distance A-A' is the relief displacement (d) shown in the plan view (Figure 2-9B).

The amount of relief displacement (d) on an aerial photograph is

1. directly proportional to the height (h) of the object. For objects A and C (Figure 2-9A) at equal distances from the principal point, d is greater for A, which is the taller object.
2. directly proportional to the radial distance (r) from the principal point to the top point on the displaced image corresponding to the top of the object (Figure 2-9B). For objects A and B, which are of equal height, d is greater for A because it is located farther from the principal point.
3. inversely proportional to the height (H) of the camera above the terrain.

These relationships are expressed mathematically as

$$d = \frac{h \cdot r}{H}$$

which may be transposed to

$$h = \frac{H \cdot d}{r} \qquad (2\text{-}4)$$

This equation may be used to determine the height of an object from its relief displacement on an aerial photograph. For the building in the lower right corner of Figure 2-8, d and r are

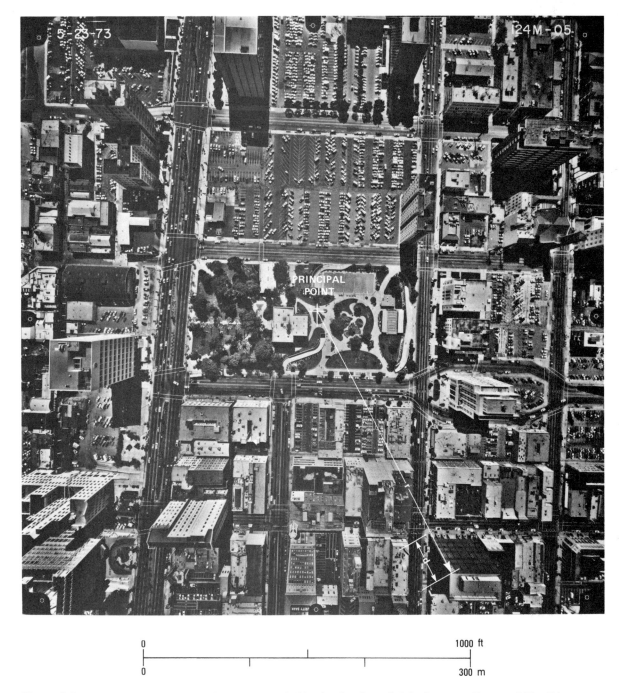

Figure 2-8 Vertical aerial photograph of Long Beach, California, showing relief displacement. Courtesy J. Van Eden.

measured using the scale of the photograph, and the height is calculated from Equation 2-4 as

$$h = \frac{212 \text{ m} \times 40 \text{ m}}{260 \text{ m}} = 32.6 \text{ m}$$

Orthophotographs are aerial photographs that have been scanned into a digital format and computer-processed to remove the radial distortion. These photographs have a consistent scale throughout the image and may be used as maps.

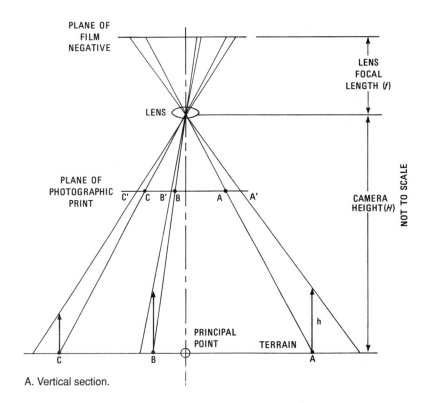

A. Vertical section.

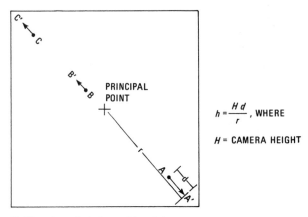

$h = \dfrac{H d}{r}$, WHERE

H = CAMERA HEIGHT

B. Plan view of photographic print.

Figure 2-9 Geometry of relief displacement on a vertical aerial photograph.

PHOTOMOSAICS

Aerial photographs are typically acquired at scales of 1:80,000 or larger and therefore cover relatively small areas. Taking photographs on a series of parallel flight lines provides broader coverage. Along a flight line, successive pho-tographs are acquired with 60 percent forward overlap (Figure 1-11). Flight lines are spaced to provide 30 percent *sidelap*, which is the overlap between adjacent strips of pho-tographs. A *photomosaic* is a composite of these individual photographs that covers an extended area. Figure 2-10A is a photomosaic of the northern Coachella Valley in southern

A. Photomosaic. From Sabins (1973A, Figure 6).

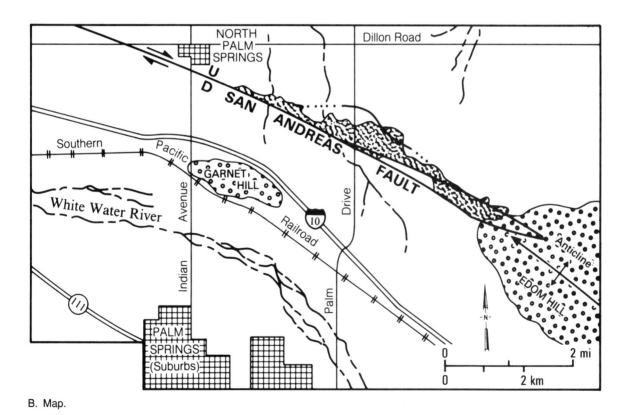

B. Map.

Figure 2-10 Photomosaic and map of Coachella Valley, California.

California. Flight lines are oriented north-south. One can recognize that this is a "homemade" mosaic because the borders of individual photographs are visible. Professionals who make mosaics process the individual photographs in the dark room to a uniform tone and contrast; after these are assembled, the borders between adjacent photographs are almost impossible to recognize.

The Coachella Valley photomosaic (Figure 2-10) includes a variety of natural and man-made features. Windblown sand, which covers much of the area, has a bright tone. Outcrops of sedimentary bedrock at Garnet Hill and Edom Hill are eroded into ridges and canyons. Vegetation has a dark signature as

seen along the White Water River and the golf course on the northern edge of Palm Springs. The San Andreas fault strikes northwest through the area and is marked by a pronounced linear feature in the vicinity of Palm Drive, where the fault forms the boundary between windblown sand on the south and vegetated terrain on the north. The fault is a barrier to the southward movement of groundwater in the subsurface. The water table is shallower on the north side of the fault and thus supports the growth of native vegetation. This expression of geologic features by vegetation patterns is called a *vegetation anomaly*. The transportation network, urban areas, and other cultural features are clearly visible.

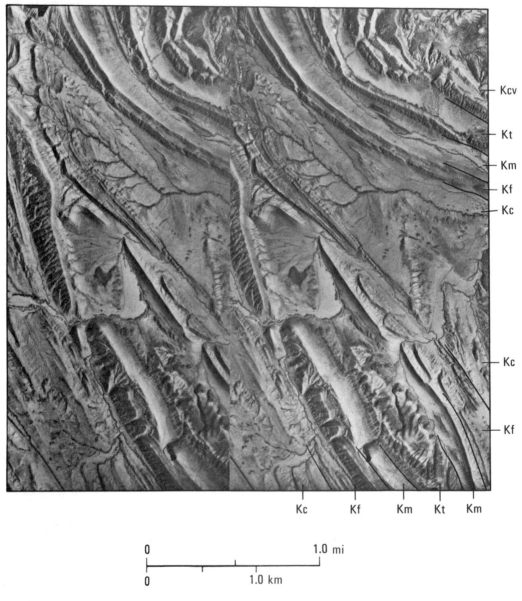

Figure 2-11 Stereo pair of aerial photographs of the Alkali anticline, Bighorn Basin, Wyoming, with formation contacts indicated. Formation symbols are explained in Figure 2-14.

Figure 2-12 Stereoscope for viewing stereo pairs of aerial photographs.

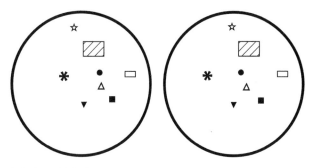

A. Left and right images.

Highest

1. ●
2. ✳
3. △
4. ■
5. ▼
6. ▭
7. ▨
8. ☆

Lowest

B. Ranking of apparent height of targets in A.

Figure 2-13 Test of stereo vision.

STEREO PAIRS OF AERIAL PHOTOGRAPHS

A pair of successive overlapping photographs along a flight line constitutes a *stereo pair,* which may be viewed with a stereoscope to produce a three-dimensional image called a *stereo model.* Figure 2-11 is a stereo pair of the Alkali anticline in the eastern part of the Bighorn Basin, Wyoming, that is suitable for viewing with a simple stereoscope (Figure 2-12). Before viewing the photographs in stereo, it is advisable to test one's ability to see in stereo.

Test of Stereo Vision

Figure 2-13 is a test of stereo vision. The two large circles (Figure 2-13A) are not exact duplicates. Within each circle the symbols are radially offset to produce different amounts of relief displacement. As a result, when viewed with a stereoscope, the symbols will appear to float in space at different levels above and below the plane of the page. Adjust the interpupillary distance on the stereoscope to suit your eyes. For most people approximately 6.4 cm is adequate. In normal vision the two eyes converge at a point in the scene that is being viewed. The stereoscope, however, causes the eyes to diverge, with the left eye viewing the left circle and the right eye viewing the right circle. The brain then merges the two images to produce the stereo model in which the individual symbols "float" at different heights. Complete the test by ranking the symbols on the basis of relative height, starting with 1 for the highest. Compare your results with the ranks shown in Figure 2-13B.

The next step is to view the photographs in Figure 2-11 with the stereoscope. When the stereoscope is properly adjusted, the image will appear as a series of alternating ridges and valleys. Figure 2-14 provides a topographic and geologic map of the area. The viewer will be impressed by the apparent extreme vertical relief, which is caused by a characteristic of stereo models called vertical exaggeration.

Vertical Exaggeration

Vertical exaggeration (*VE*) results because the perceived vertical scale is larger than the horizontal scale in a stereo model. The amount of exaggeration may be approximately calculated by

$$VE = \left(\frac{AB}{H} \right) \left(\frac{AVD}{EB} \right) \tag{2-5}$$

where

AB = air base, or ground distance, which is the distance on the ground between the centers of successive overlapping photographs.

H = height of the camera above terrain.

AVD = apparent stereoscopic viewing distance. This is the distance not from the stereoscope lenses to the photographs but from the lenses to the plane in space where the image appears to be. The stereo model always appears to be somewhere below the tabletop on which the stereoscope sits.

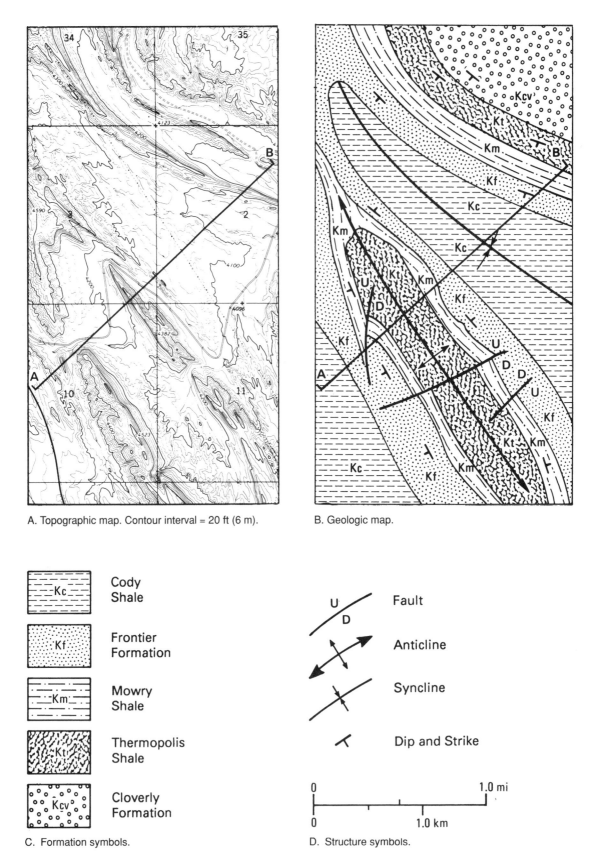

A. Topographic map. Contour interval = 20 ft (6 m).

B. Geologic map.

Kc — Cody Shale

Kf — Frontier Formation

Km — Mowry Shale

Kt — Thermopolis Shale

Kcv — Cloverly Formation

C. Formation symbols.

U / D — Fault

Anticline

Syncline

Dip and Strike

0   1.0 mi

0   1.0 km

D. Structure symbols.

Figure 2-14 Topographic and geologic maps of the Alkali anticline, Bighorn Basin, Wyoming.

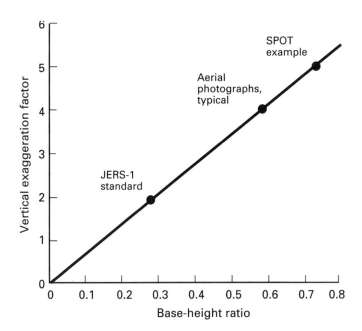

Figure 2-15 Relationship between the base–height ratio and the vertical exaggeration factor for stereo models. From Thurrell (1953, Figure 5).

scale four times that of the horizontal scale (4×), which is the same vertical exaggeration as that of the stereo model. Comparing the topographic profiles demonstrates the effect of vertical exaggeration.

Although vertical relief is exaggerated by a factor of 4 on a stereo model, the angles of topographic slope and structural dip do not increase four times. Figure 2-17 illustrates the slope exaggeration of a hill with a height (BC) of 270 m and a width (AB) of 1000 m. The tangent of the true-slope angle (CAB) is BC/AB, which is 0.27, or 15°. On the stereo model, the exaggerated height DB is 1080 m, and the tangent of the exaggerated-slope angle (DAB) is BD/AB, which is 1.08, or 47°. The effect of the 4× vertical exaggeration on topographic slopes is illustrated by the cross section in Figure 2-16B. La Prade (1972, 1973) describes the geometry of stereoscopic photographs and reviews various hypotheses for explaining vertical exaggeration.

Interpretation of Stereo Pairs

Geologic maps may be prepared by interpreting stereo pairs. For readers who are unfamiliar with geology, the Appendix is a condensed description of geologic concepts employed in interpreting remote sensing images. The first step in interpreting an area underlain by sedimentary rocks, such as the Alkali anticline, is to define the geologic units, or formations. Figure 2-14 lists the formations and their map symbols, which are mapped in Figure 2-14B. The formations have the following signatures in the photograph:

Cody Shale A thick unit of shale with medium gray tone that is readily eroded and forms broad valleys cut by numerous streams. Cody Shale is the youngest formation exposed at the Alkali anticline.

Frontier Formation Medium gray sandstone with shale interbeds. The sandstone is resistant to erosion and forms ridges. The nonresistant shale forms narrow valleys. The prominent dipslopes of the sandstone are incised by numerous closely spaced minor stream channels, producing a distinctive serrated appearance.

Mowry Shale Siliceous shale that weathers to a slope with a very light gray tone.

Thermopolis Shale Very dark gray shale with thin, light-toned interbeds that cause a banded appearance.

Cloverly Formation Alternating beds of resistant sandstone and nonresistant shale that weather to ridges and valleys. The Cloverly is the oldest formation exposed in the area.

Several interpreters using a variety of stereoscopes estimated an average value for *AVD* of 45 cm (Wolf, 1974).

EB = eye base, which is the distance between the interpreter's eyes. For the average adult, *EB* is 6.4 cm.

For the stereo pair of Figure 2-11, with an air base of 1700 m and height of 3000 m, vertical exaggeration may be estimated from Equation 2-5 as

$$VE = \left(\frac{1700 \text{ m}}{3000 \text{ m}} \right) \left(\frac{45 \text{ cm}}{6.4 \text{ cm}} \right)$$

$$= 4.0\times$$

On viewing Figure 2-11 with a stereoscope, the average interpreter should perceive vertical distances to be exaggerated four times the equivalent horizontal distances. In Equation 2-5 the terms *AVD* and *EB* are constant; therefore, the amount of vertical exaggeration is determined by the ratio *AB/H*, which is called the *base–height ratio*. Figure 2-15 shows the vertical exaggeration for stereo pairs with various base–height ratios.

The effect of vertical exaggeration is illustrated in Figure 2-16, which shows two topographic profiles constructed along line AB in the contour map of Figure 2-14A. The profile in Figure 2-16A has the same vertical and horizontal scales (1×), which is equivalent to no vertical exaggeration. The profile in Figure 2-16B was constructed with the vertical

These characteristics are used to map the contacts between formations.

At several localities, the contacts are interrupted and offset by normal faults (see the Appendix), which are mapped with

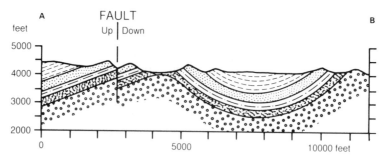

A. No vertical exaggeration.

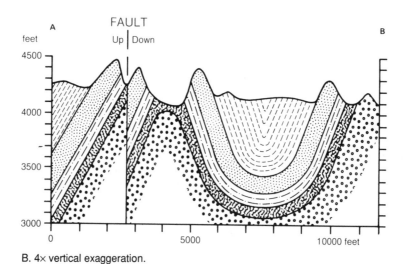

B. 4× vertical exaggeration.

Figure 2-16 Topographic profiles and geologic cross sections of the Alkali anticline, showing effects of vertical exaggeration.

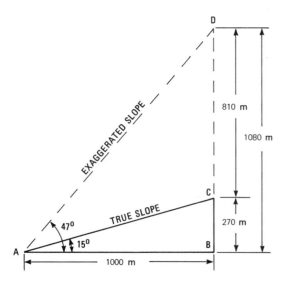

Figure 2-17 Slope exaggeration on a stereo model with 4× vertical exaggeration.

heavy lines and symbols to show the displacement along the faults (Figure 2-14B). Attitudes of the beds are readily seen in the stereo model and are recorded with symbols showing strike and dip. The amount of dip is vertically exaggerated in the same fashion as topographic slopes shown in Figure 2-17. Techniques for converting exaggerated dips to true dips are described by Miller (1961). Outcrop patterns and the attitude of beds are used to locate the axial trace of the Alkali anticline and the adjacent syncline.

As mentioned earlier, the topographic map in Figure 2-14A was used to draw the topographic profile for the cross section with no vertical exaggeration (Figure 2-16A). The geologic cross section was filled in by adding the faults, formation contracts, and strike-and-dip information from the geologic map (Figure 2-14B). Simple geometric techniques were used to produce the cross section with 4× vertical exaggeration (Figure 2-16B). This photogeologic interpretation should not be considered complete until the map has been checked in the field.

LOW-SUN-ANGLE PHOTOGRAPHS

Aerial photographs are normally acquired between 10:00 a.m. and 3:00 p.m., when the sun is at a high angle above the horizon and shadows have a minimum extent. These photographs are desirable for topographic mapping, which requires unobscured terrain, but for geologic interpretation photographs acquired with lower sun angles are often valuable. The photograph in Figure 2-18A was acquired shortly after sunrise, when the sun was only 15° above the horizon. A photograph of the same area acquired at midday (Figure 2-18B) is shown for comparison. The area is a gravel-covered slope on the east flank of Carson Range, south of Reno, Nevada. The low-sun-angle photograph includes several north-trending linear features with prominent bright or dark signatures. As shown on the map and cross section (Figure 2-19), these features are low topographic *scarps,* which are caused by active normal faults that cut the surface. The east-facing scarps are strongly illuminated by the early morning sun and have bright signatures, or highlights. The west-facing scarps are shadowed and have dark signatures. Orientation of the highlights and shadows provides information on the sense of displacement along the faults, shown in the cross section (Figure 2-19B). Note, however, that if the scarps trended east-west, essentially parallel with the sun azimuth, the shadows and highlights would not be present in the photograph.

In the midday photograph, highlights and shadows are minimal and the fault scarps are inconspicuous. There are some subtle linear gray traces along the scarps that are not due to illumination, but are local concentration of sagebrush. Runoff from rainfall is concentrated at the foot of the scarps and supports a higher density of vegetation. This pattern is one of several types of vegetation anomaly that aid in recognizing geologic features.

Acquiring good low-sun-angle photographs in the summer is complicated by the limited number of hours in the morning and evening when the desired illumination occurs. Illumination values are low and change rapidly during these times; therefore, proper camera exposures may be difficult to achieve. At middle and high latitudes, the sun is at relatively low elevations throughout the day in the winter, which may be an optimum season for acquiring these photographs. Low-sun-angle photographs are widely used to recognize subtle topographic features associated with active faults.

BLACK-AND-WHITE PHOTOGRAPHS

Several types of black-and-white films are available for acquiring aerial photographs at wavelengths ranging from UV, through visible, and into the photographic portion of the IR region (Figure 1-2).

Panchromatic Black-and-White Photographs

Black-and-white photographs exposed by visible light are called *panchromatic photographs.* Panchromatic aerial photographs are normally acquired with a minus-blue filter over the lens; as Figure 2-4 shows, this filter eliminates the UV and blue wavelengths selectively scattered by the atmosphere. Examples of these minus-blue photographs are shown in Figures 2-7, 2-8, and 2-11.

These photographs are a widely used and readily available remote sensing product. Stereo coverage of most of the United States is available at modest prices from the agencies listed later in this chapter in the section "Sources of Aerial Photographs." These photographs are used to compile topographic maps, geologic surveys, engineering studies, and crop inventories. Color photographs are superior for many applications, but panchromatic photographs are still a major source of remote sensing information.

IR Black-and-White Photographs

By using IR-sensitive film and a filter such as the Kodak Wratten 89B, which transmits only reflected IR energy (Figure 2-20), one can obtain photographs in the portion of the IR spectral region at wavelengths of 0.7 to 0.9 μm. This reflected solar radiation is called *photographic IR energy* and should not be confused with thermal IR energy, which occurs at wavelengths of 3 to 14 μm. Figure 2-21 illustrates simultaneously acquired panchromatic and IR black-and-white photographs of the Massachusetts coast. This example demonstrates the following advantages of IR photographs:

1. Haze penetration improves because the filter eliminates the severe atmospheric scattering that occurs in the visible and UV regions. Eliminating most scattered light results in a higher contrast ratio and therefore higher spatial resolution on the IR photograph, as discussed earlier.
2. Maximum reflectance from vegetation occurs in the photographic IR region, as shown by the bright tones in the IR photograph. In addition, maximum spectral differences between vegetation types, such as hardwoods and conifers, show up in the photographic IR region, which is advantageous for mapping plant communities.
3. IR energy is almost totally absorbed by water, which causes water to have a dark tone on IR photographs. For this reason, boundaries between land and water show up more clearly on IR photographs than on panchromatic photographs. Note the tidal-flat area in the upper part of each photograph in Figure 2-21. The shoreline and individual tidal channels are clearly distinguishable on the IR photograph (Figure 2-21B). Such distinctions are not possible on the panchromatic photograph (Figure 2-21A) because light penetrates the shallow water and does not differentiate submerged areas from the land.

IR color film, described in a later section, combines these properties of IR black-and-white film with the advantages of color.

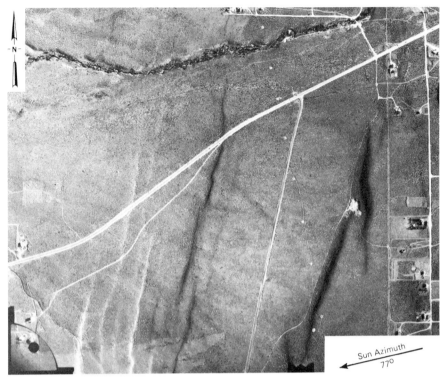

A. Low-sun-angle photograph acquired June 23, 1972, at 5:30 a.m. local sun time with sun elevation of 15°. From Walker and Trexler (1977, Figure 3).

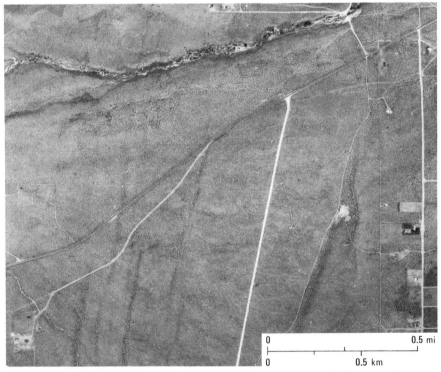

B. High-sun-angle photograph acquired May 21, 1966, at midday by U.S. Geological Survey.

Figure 2-18 Low-sun-angle photograph and high-sun-angle photograph of east flank of Carson Range, Nevada.

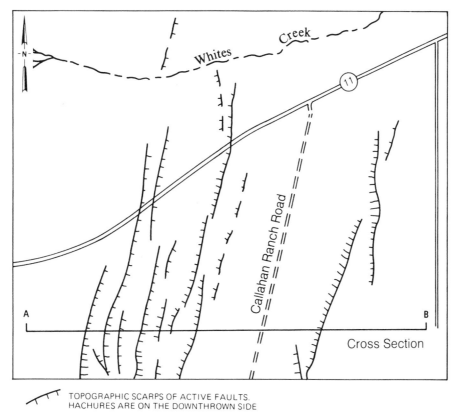

TOPOGRAPHIC SCARPS OF ACTIVE FAULTS.
HACHURES ARE ON THE DOWNTHROWN SIDE

A. Interpretation map.

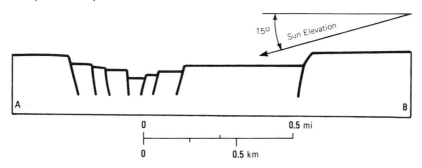

B. Vertically exaggerated cross section along line AB.

Figure 2-19 Interpretation map and cross section of the low-sun-angle photograph in Figure 2-18A.

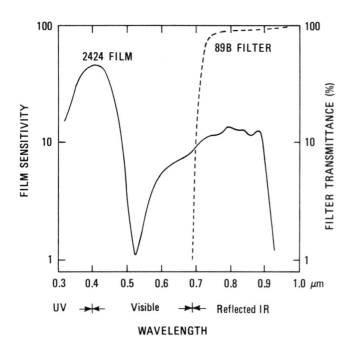

Figure 2-20 Combinations of films and filters for IR black-and-white photographs. From Vizy (1974, Figure 7).

UV Photographs

The UV spectral region extends from 3 nm to 0.4 μm; however, the atmosphere only transmits UV wavelengths from 0.3 to 0.4 μm, which is known as the *photographic UV region*. Photographs may be acquired in the photographic UV region with film and filter combinations such as Kodak Plus-X Aerographic film 2402 and the Kodak Wratten 18A filter (Figure 2-22). The Kodak Wratten 39 filter transmits both UV and blue energy and, for most applications, is almost as useful as the Wratten 18A filter. Most camera lenses absorb UV energy of wavelengths less than about 0.35 μm, but special quartz lenses transmit shorter wavelengths.

UV photographs have low contrast ratios and poor spatial resolution because of severe atmospheric scattering (Figure 2-2). As a result, the UV spectral region is rarely employed in remote sensing, except for special applications such as monitoring oil films on water (illustrated in Chapter 9). UV photographs have largely been replaced by images acquired with multispectral scanners, as described in Chapter 3.

COLOR SCIENCE

The average human eye can discriminate many more shades of color than it can tones of gray. This greatly increased information content is a major factor favoring the use of color photographs.

The visible region is divided into the blue, green, and red bands (or colors) shown in Figure 2-23A. These are called *additive primary colors* because white light is formed when equal amounts of blue, green, and red light are added. Most colors of the visible spectrum may be formed by combining various amounts of the additive primary colors.

Additive Primary Colors

Figure 2-23B shows the characteristics of three filters, each of which transmits one additive primary color and absorbs the other two. A color image can be produced by acquiring three separate black-and-white pictures of a subject using the three primary filters. Positive films of the three pictures can then be *registered* (superposed to align precisely) on a screen with three projectors, each projector using the primary color filter appropriate for that black-and-white photograph. The result is a color image.

For most purposes it is impractical to use three projectors to produce color photographs, and additive primary colors cannot be mixed by directly superposing the films. As Figure 2-23B shows, because each additive primary filter absorbs two-thirds of the spectrum, no light is transmitted where any two filters are superposed. Color television employs additive primaries not by superposing but by juxtaposing blue, green, and red specks small enough to blend together when observed by the eye.

Subtractive Primary Colors

In order to mix colors by superposition of films, the three *subtractive primary colors*—yellow, magenta, and cyan—must be used. As Figure 2-23C shows, each subtractive primary filter absorbs one-third of the visible spectrum and transmits the remaining two-thirds.

The yellow subtractive primary filter absorbs blue light. The magenta primary is a bluish red filter that absorbs green. Cyan is a bluish green filter that absorbs red. Figure 2-24 is a color triangle that relates the additive primaries at the corners to the subtractive primaries along the sides. Each subtractive primary absorbs the additive primary at the opposite corner and transmits the additive primaries at the adjacent corners. When any two different subtractive primary filters are superposed and illuminated with white light, the color transmitted will be the additive color located at their common corner on the triangle. For example, white light projected through overlapping yellow and cyan filters appears green. Superposition of all three subtractive filters absorbs all light and the result is perceived as black. Complementary colors are pairs of colors that produce white light when added together, such as magenta and green. On the color triangle (Figure 2-24), complementary colors are located opposite each other. A color film system using a single projector is possible with subtractive primary colors, as the following section describes.

A. Panchromatic photograph.

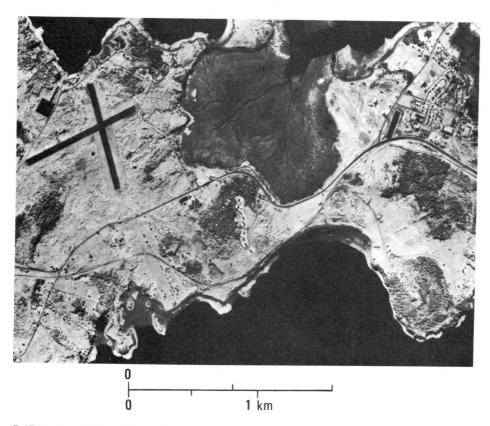

B. IR black-and-white photograph.

Figure 2-21 Conventional and IR black-and-white aerial photographs of the Massachusetts coast.

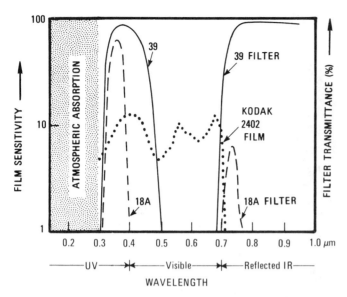

Figure 2-22 Combinations of films and filters for UV photographs. From Vizy (1974, Figure 6).

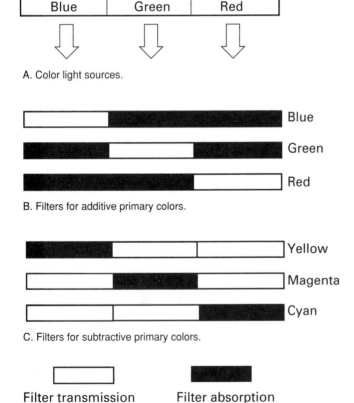

A. Color light sources.

B. Filters for additive primary colors.

C. Filters for subtractive primary colors.

Filter transmission   Filter absorption

Figure 2-23 Transmission and absorption characteristics of filters for additive and subtractive colors.

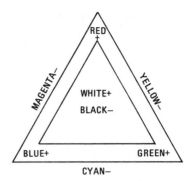

Figure 2-24 Color triangle showing relationship of additive (+) and subtractive filters (–).

NORMAL COLOR PHOTOGRAPHS

Normal color photographs record a scene in its true colors. The photographs are recorded on *color film,* which is a transparent medium that may be either positive or negative. The films are used to produce color prints on an opaque base, such as the aerial photograph in Plate 1C. Haze filters are used when acquiring normal color aerial photographs to absorb UV wavelengths that are strongly scattered by the atmosphere (Figure 2-2). A minus-blue filter is not used because it would destroy the color balance of the film. Color aerial photographs typically have a bluish appearance because of blue light scattered by the atmosphere.

Negative Color Film

Figure 2-25A shows a cross section of *color negative film,* which consists of a transparent base coated with three emulsion layers with the following characteristics:

1. Each layer is sensitive to one of the additive primary colors: blue, green, or red.
2. During developing, each emulsion layer forms a color dye complementary to the primary color that exposed the layer: the blue-sensitive emulsion layer forms a yellow negative image; the green-sensitive layer forms a magenta negative image; and the red-sensitive layer forms a cyan negative image.

The upper portion of Figure 2-25 shows a color subject. Figure 2-25A shows how an image of the color subject is formed on negative film. The bottom emulsion layer is sensitive to red light and produces a complementary cyan image of a red subject. Green and blue subjects produce magenta and yellow images, respectively. The white subject exposes all three layers, resulting in an image that transmits no light; the black subject results in a clear image because none of the layers are exposed. As Figure 2-25A shows, the film records the color of a subject

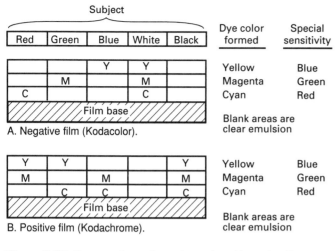

Subject					Dye color formed	Special sensitivity
Red	Green	Blue	White	Black		

A. Negative film (Kodacolor).

		Y		Y	Yellow	Blue
	M			M	Magenta	Green
C				C	Cyan	Red
Film base						

Blank areas are clear emulsion

A. Negative film (Kodacolor).

Y	Y			Y	Yellow	Blue
M			M		M	Magenta	Green
		C	C		C	Cyan	Red
Film base						

Blank areas are clear emulsion

B. Positive film (Kodachrome).

Figure 2-25 Cross sections of negative and positive color film, showing how images are formed on the three emulsion layers.

as its complementary color on the negative. The image on a negative color film is projected onto photographic paper, coated with sensitive emulsions, that is developed to produce a color print.

Positive Color Film

Figure 2-25B shows a cross section of positive color film, which records a scene in its true colors. A red subject forms a clear image on the red-sensitive emulsion, which becomes cyan where it is not exposed by red light. The red subject forms a magenta image on the green-sensitive layer and a yellow image on the blue-sensitive layer. When viewed with transmitted white light, the yellow and magenta images absorb blue and green, respectively, and allow a red image to be projected. A white subject forms clear images on all three layers.

The original positive film transparency may be viewed on a light table, which provides maximum resolution. However, the film rolls require special handling and viewing equipment and are not suitable for use in the field. Black-and-white prints, color prints, and color transparencies can be made from negative color film. Paper prints, despite their slightly lower resolution, are more versatile and easily used in the field.

IR COLOR PHOTOGRAPHS

In *IR color film* the spectral sensitivities of the emulsion layers are changed to record energy of other wavelengths, including photographic IR (0.7 to 0.9 µm). This film is sold as Kodak Aerochrome Infrared film, type 2443, which is available only as positive film. Plate 1D is an IR color photograph of the area covered by the normal color photograph in Plate 1C. IR color

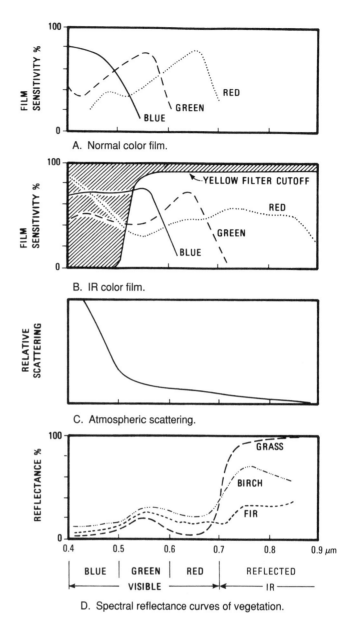

A. Normal color film.

B. IR color film.

C. Atmospheric scattering.

D. Spectral reflectance curves of vegetation.

Figure 2-26 Spectral sensitivity of normal and IR color film, together with an atmospheric scattering diagram and vegetation reflectance spectra. From Sabins (1973B, Figure 2).

film was originally designed for military reconnaissance and was called camouflage detection film. The name *false color film* is occasionally used, but *IR color film* is the preferred name.

IR color film is best described by comparing it with normal color positive film. Figure 2-26A,B shows the spectral sensitivity of the three emulsion layers that produce blue, green, and red images. In normal color film (Figure 2-26A) each layer is exposed by the corresponding wavelength band of light. The blue-imaging layer is exposed by blue light, the green-imaging layer by green light, and the red-imaging layer by red light. In

IR color film (Figure 2-26B) the photochemistry of each layer is changed, and they are sensitive to different wavelengths of light: the blue-imaging layer is exposed by green light; the green-imaging layer is exposed by red light; and the red-imaging layer is exposed by reflected IR energy. All three layers are also sensitive to blue light, which is eliminated by placing a yellow (minus-blue) over the camera lens. The shading in Figure 2-26B highlights the curve showing blue wavelengths removed by the Wratten 12 yellow filter. Removing these strongly scattered wavelengths (Figure 2-26C) improves the contrast ratio and spatial resolution of IR color film.

Because the term *infrared* suggests heat, some users mistakenly assume that the red tones on IR color film record variations in temperature. A few moments' thought will show that this is not the case. If the IR-sensitive layer were sensitive to ambient heat, it would be exposed by the warmth of the camera body itself. As pointed out in Chapter 1, thermal radiation occurs at wavelengths longer than 3 µm, which is beyond the sensitivity range of IR film (0.7 to 0.9 µm). To repeat, the red-imaging layer of IR color film is exposed by reflected IR energy, not by thermal IR energy.

In addition to large sizes for aerial cameras, IR color film is available in 35-mm size for use in ordinary cameras. The cost of the 20-exposure cassettes and processing is comparable to that of normal color films. A user can evaluate IR color film at minimal expense with this format. It is useful to acquire normal color photographs with a second camera to compare with IR color photographs. (IR color film may deteriorate with time and excessive heat; if keeping the film for more than a few weeks, store it in a freezer. Allow the frozen film to reach room temperature before opening the sealed container to prevent moisture from condensing on the cold film.)

A yellow (minus-blue) filter, such as the Kodak Wratten 12, is used with IR color film. This film-and-filter combination has an approximate speed of ASA 100. Some experimentation will be necessary to determine the optimum exposure because conventional light meters do not measure the same spectral region to which the film is sensitive. Some cameras have an IR setting on the focusing ring that is intended for IR black-and-white film. Do not use this setting for IR color film because two of the three emulsion layers are sensitive to visible wavelengths.

NORMAL COLOR AND IR COLOR PHOTOGRAPHS COMPARED

Plate 1C,D shows a normal color photograph and an IR color photograph that were simultaneously acquired of the UCLA campus in the western part of Los Angeles. Figure 2-27 is a location map. Comparing these photographs is a useful way to understand their different characteristics. Table 2-3 compares the color signatures of common subjects on the two types of photographs.

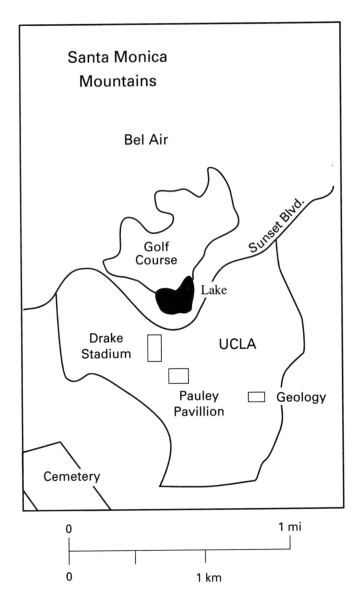

Figure 2-27 Location map of the UCLA area, Los Angeles, California.

Signatures of Vegetation

The most striking difference between the photographs in Plate 1C,D is the red color of healthy vegetation in the IR color photograph, which is explained by the spectral reflectance curves of vegetation shown in Figure 2-26D. Spectral reflectance curves show the percentage of incident energy reflected by a material as a function of wavelength. Blue and red light are absorbed by foliage. Up to 20 percent of the incident green light is reflected, causing the familiar green color of leaves on normal color photographs. The spectral reflectance of vegetation increases abruptly in the photographic IR region, which includes the wavelengths that expose the red-imaging layer in IR color film.

Table 2-3 Terrain signatures on normal color film and IR color film

Subject	Normal color film	IR color film
Healthy vegetation:		
Broadleaf type	Green	Red to magenta
Needle-leaf type	Green	Reddish brown to purple
Stressed vegetation:		
Previsual stage	Green	Pink to blue
Visual stage	Yellowish green	Cyan
Autumn leaves	Red to yellow	Yellow to white
Clear water	Blue-green	Dark blue to black
Silty water	Light green	Light blue
Damp ground	Slightly darker than dry soil	Distinctly darker than dry soil
Shadows	Blue with details visible	Black with few details visible
Water penetration	Good	Moderate to poor
Contacts between land and water	Poor to fair discrimination	Excellent discrimination
Red bed outcrops	Red	Yellow

Figure 2-28 is a diagrammatic cross section of a leaf that explains these spectral signatures of vegetation. The transparent epidermis allows incident sunlight to penetrate into the mesophyll, which consists of two layers: (1) the palisade parenchyma of closely spaced cylindrical cells, and (2) the spongy parenchyma of irregular cells with abundant interstices filled with air. Both types of mesophyll cells contain chlorophyll, which reflects part of the incident green wavelengths and absorbs all of the blue and red energy for photosynthesis. The longer wavelengths of photographic IR energy penetrate into the spongy parenchyma, where the energy is strongly scattered and reflected by the boundaries between cell walls and air spaces. The high IR reflectance of leaves is caused not by chlorophyll but by the internal cell structure. Gausman (1985) gives details of optical properties of plant leaves in the visible and reflected IR regions. Buschmann and Nagel (1993) describe the roles of chlorophyll and cell structure in spectral reflectance of leaves.

Detection of Stressed Vegetation

Vegetation may be stressed because of drought, disease, insect infestation, or other factors that deprive the leaves of water. Figure 2-29 compares the internal structure of nonstressed and stressed leaves. The nonstressed leaf (Figure 2-29A) has a cell structure and reflectance characteristics comparable to those in Figure 2-28. In the stressed leaf (Figure 2-29B), the shortage of water causes the mesophyll cells to collapse, which strongly

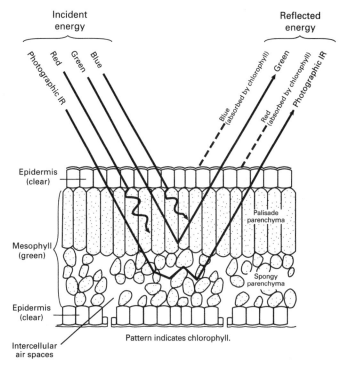

Figure 2-28 Diagrammatic cross section of a leaf, showing interaction with incident energy. Incident blue and red wavelengths are absorbed by chlorophyll in the process of photosynthesis. Incident green wavelengths are partially reflected by chlorophyll. Incident IR energy is strongly scattered and reflected by cell walls in the mesophyll. Modified from Buschmann and Nagel (1993, Figure 9).

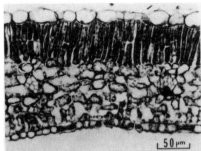

A. Nonstressed.

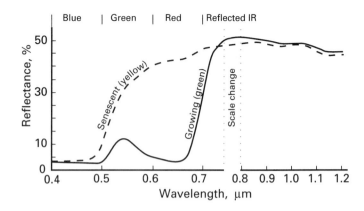

Figure 2-30 Reflectance spectra of green and senescent foliage. In the autumn, chlorophyll deteriorates, which reduces the absorption of incident red energy. The development of anthocyanin and tannin causes the yellow-red fall colors. From Schwaller and Tkach (1985, Figure 2).

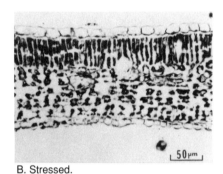

B. Stressed.

Figure 2-29 Photomicrographs of cross sections of nonstressed and stressed leaves. Collapse of cells in the mesophyll layer strongly reduces reflectance of incident IR energy. From Everitt and Nixon (1986, Figure 1). Courtesy J. H. Everitt, U.S. Department of Agriculture.

reduces IR reflectance from the spongy parenchyma. This decreased reflectance diminishes the red signature in IR color photographs. Chlorophyll is still present, and the foliage may have a green signature in normal color photographs for some time after the onset of stress. In IR color photographs, however, stressed foliage has a distinctive blue signature. The loss of IR reflectance is a *previsual symptom* of plant stress because it often occurs days or even weeks before the visible green color begins to change. The previsual effect may be used for early detection of disease and insect damage in crops and forests. Evidence of plant stress is seen in the intramural playing field east of Drake Stadium (Figure 2-27 and Plate 1C,D), which is watered by a sprinkler system. In the normal color photograph the field is entirely green, but in the IR color photograph the red signature is interrupted by blue strips that indicate inadequately watered turf.

Autumn Senescence of Vegetation

In the autumn, leaves of deciduous trees undergo senescence and turn red, yellow, and brown. Figure 2-30 compares spectra of green and senescent foliage. The green chlorophyll has de-

cayed, and red wavelengths are no longer absorbed. The organic compounds anthocyanin and tannin are formed, causing the familiar autumn colors (Boyer and others, 1988). The spectrum for senescent foliage (Figure 2-30) shows nearly equal reflectance values in the green, red, and photographic IR bands, which results in a white signature in IR color photographs. Boyer and others also describe the changes in leaf physiology and spectral reflectance during senescence.

Signatures of Other Terrain Features

The small lake north of UCLA has a dark green signature in the normal color photograph that blends with the vegetation. In the IR color photograph the lake has a dark blue signature that contrasts with the red signature of vegetation. This ability to enhance the difference between vegetation and water is especially valuable for mapping drainage patterns in heavily forested terrain. Silty water has a light blue signature in IR color photographs. One can recognize damp ground on IR color photographs by its relatively darker signature, caused by absorption of IR energy. Shadows are darker in IR color photographs than in normal color photographs because the yellow filter eliminates blue light.

The IR color photograph in Plate 1D has a better contrast ratio than the normal color photograph, for two reasons:

1. The yellow filter eliminates blue light, which is preferentially scattered by the atmosphere, as shown by the curve in Figure 2-26C. Eliminating much of the scattering improves the contrast ratio.
2. For vegetation, soils, and rocks, reflectance differences are commonly greater in the photographic IR region than in the visible region.

The higher contrast ratio of the IR color photograph results in improved spatial resolution, which is evident when one compares finer details of the two photographs. On the slopes of the Santa Monica Mountains (upper left corner), for example, closely spaced shrubs may be separated more readily in the IR color photograph. In the urban areas, individual buildings are more distinctly separate in the IR color example.

HIGH-ALTITUDE AERIAL PHOTOGRAPHS

Aerial photographs have traditionally been acquired at altitudes of approximately 6000 m or less, resulting in scales of 1:40,000 or larger (Table 2-1). In the 1970s, improvements in cameras and film enabled acquisition of photographs at higher altitudes, which provide adequate resolution for many applications. The advantage of high-altitude, smaller-scale photographs is that fewer photographs are required to cover an area.

NASA High-Altitude Photographs

For a number of years, NASA has been acquiring photographs of the United States from U-2 and RB-57 reconnaissance aircraft at altitudes of 18 km above terrain with standard aerial cameras (152-mm focal length) on film with a 23-by-23-cm format. The resulting photographs cover 839 km^2 at a scale of 1:120,000. Black-and-white, normal color, or IR color film is used; many missions employ two cameras to acquire photographs with two different film types.

Coverage of NASA photographs is concentrated over numerous large regional test sites for which repeated coverage over several years may be available. Many areas lack this coverage.

National High Altitude Photography Program

The National High Altitude Photography (NHAP) program, coordinated by the U.S. Geological Survey, began in 1978 to acquire coverage of the United States with a uniform scale and format. From aircraft at an altitude of 12 km, two cameras (23-by-23-cm format) acquire black-and-white photographs and IR color photographs. The black-and-white photographs are acquired using a camera with a 152-mm focal length to produce photographs at a scale of 1:80,000, which cover 338 km^2. Figure 2-31 is an NHAP photograph of Washington, D.C. A stereo pair of these photographs (not illustrated) covers the area of a standard U.S. Geological Survey topographic quadrangle (1:24,000 scale). The stereo pair can be used to produce new maps or update existing maps.

The IR color photographs are acquired using a camera with a 210-cm focal length to produce photographs at a scale of 1:58,000, which cover 178 km^2.

SOURCES OF AERIAL PHOTOGRAPHS

The distribution center for aerial photographs acquired by the U.S. Geological Survey and NASA is

U.S. Geological Survey
EROS Data Center
Sioux Falls, SD 57198

An inquiry to the EROS Data Center (EDC) should specify the latitude and longitude boundaries of the desired area and the type of photography required. The major categories are as follows:

1. **Aerial mapping photographs:** typically at 1:40,000 scale or larger
2. **National High Altitude Photographs:** 1:80,000-scale black-and-white and 1:58,000 color IR
3. **NASA aircraft photographs:** 1:120,000-scale black-and-white, normal color, and IR color

The EDC will provide computer listings of available photographs, price lists, and instructions for selecting and ordering the desired coverage.

The Agricultural Stabilization and Conservation Service (ASCS) has also photographed much of the United States. One can obtain a set of state index maps and ordering instructions from

Western Aerial Photography Laboratory
ASCS–USDA
P.O. Box 30010
Salt Lake City, UT 84130

Black-and-white photographs of U.S. national forests are available from regional offices of the U.S. Forestry Service. Many local aerial photography contractors have negatives and can furnish prints of areas over which they have flown. If necessary, an aerial contractor can be hired to acquire needed photographs.

NEW TECHNOLOGY

For several decades new technology for aerial photography has been incremental in nature. New cameras have been developed. Films with higher spatial resolution and faster speeds have been introduced. Two recent developments have the potential to change the science.

Aerial photographs are being digitized and distributed on CD-ROMs that are compatible with desktop computers and image-processing software (see Chapter 8). Many photographs are stored on a single CD-ROM and are readily available for viewing, processing, and reproduction. This technology may replace film and photographic prints as media for storing and distributing photographs.

0   4 mi.

0   4 km.

Figure 2-31 National High Altitude Photography program photograph of Washington, D.C., acquired from an altitude of 12 km.

The second trend has the long-term potential to eliminate film as the medium for acquiring photographs. In digital cameras film is replaced with an array of *charge-coupled devices* (CCDs), which are tiny light-sensitive detectors. The image is recorded electronically as an array of digital numbers suitable for computer processing. Normal color and IR color photographs are recorded by means of internal filters. A number of digital photographs are stored in the camera and are transferred to other storage media. Digital cameras are presently available only in small formats comparable to 35-mm film cameras. Within a few years this technology may be available for aerial photography. King (1995) reviews the characteristics and applications of digital cameras and photographs.

PHOTOGRAPHS FROM SATELLITES

Extensive collections of photographs have been acquired from manned and unmanned earth-orbiting satellites. The cameras, film, and filters are modified from those used in aircraft. Beginning in 1962 the United States conducted a series of manned satellite programs in preparation for the Apollo missions that landed humans on the moon. These programs also acquired photographs of the earth that were summarized in Sabins (1986, Chapter 3). The current Space Shuttle program has acquired many satellite photographs of the earth.

Space Shuttle

The Space Shuttle program, or *Space Transportation System* (STS), began in 1981. The vehicle is similar in size to a medium commercial jet airliner and accommodates a crew of up to seven on missions that last up to 9 days. Figure 2-32 shows the profile of a Shuttle mission. When launched from Cape Canaveral, Florida, the Shuttle is attached to two solid-propellant rockets, plus a large liquid-fuel tank that feeds the three engines. Shortly after launch the solid-propellant rockets are expended and return on parachutes to the ocean, where they are retrieved for future use. Later the external liquid-fuel tank is jettisoned and disintegrates on reentering the atmosphere. Once the Shuttle is in orbit, it maneuvers with two small rocket engines. Doors to the cargo bay are opened, and the Shuttle inverts to aim remote sensing systems at the earth. When a mission is completed, the orbital maneuvering system (OMS) engines fire in retrograde fashion to cause reentry and an unpowered landing.

Two Shuttle missions have carried the *modular optoelectronic multispectral scanner* (MOMS) of the German Aerospace Research Establishment (DFVLR). The MOMS is an along-track scanner that records two spectral bands (0.575 to 0.625 μm and 0.825 to 0.975 μm) with ground resolution cells of 20 by 20 m. The MOMS has acquired relatively few images, which are available only to investigators selected by the DFVLR.

Figure 2-32 Profile of a Space Shuttle mission.

Many photographs have been acquired from the Shuttle by handheld cameras and by the large format camera (to be described shortly).

Handheld-Camera Photographs The Shuttle has several windows for acquiring photographs with handheld cameras. Most photographs are recorded on normal color film at formats of 35, 70, and 140 mm, using a variety of cameras and lenses. After each mission NASA prepares a catalog listing the location, features, and time for each photograph. For example, the catalog for STS Mission 37 in April 1991 lists 4283 photographs of the earth. Photographs are available from

U.S. Geological Survey
EROS Data Center
Sioux Falls, SD 57198
Telephone: 605-594-6151

Technology Applications Center
University of New Mexico
Albuquerque, NM 87131
Telephone: 505-277-3622

Media Services Branch
Still Photography Library
NASA Lyndon B. Johnson Space Center
P.O. Box 58425, Mail Code AP3
Houston, TX 77258
Telephone: 713-483-4231

Large-Format-Camera Photographs The large format camera (LFC) was fabricated specifically for the Space Shuttle. "Large format" refers to the film size of 23 by 46 cm, with the longer dimension oriented in the orbit direction. The LFC was carried on the October 1984 Shuttle mission and acquired black-and-white, normal color, and IR color photographs. The focal length of the lens is 30.5 cm, and the spatial resolution is 80 line-pairs $\cdot$ mm^{-1}. Table 2-4 lists additional characteristics of the LFC. Figure 2-33 is an enlarged portion of an LFC photograph of Boston, Massachusetts, that shows the high spatial resolution of these photographs and their potential for interpreting patterns of land use and land cover.

A number of LFC photographs were acquired with forward overlap for stereo viewing. Figure 2-34 is a portion of a stereo pair with 4× vertical exaggeration of an area in the Mojave Desert of southeastern California and adjacent Nevada. The mountains are fault blocks surrounded by valleys filled with detritus eroded from the mountains. Geologic features are particularly well expressed in this stereo model. The Spring Mountains in the east portion (upper part) of the photographs consist of generally west-dipping sedimentary rocks. The light-toned rocks that form prominent cliffs along the eastern front of the Spring Mountains are sandstone of Jurassic age. The sandstone is overlain on the west by dark-toned carbonate rocks of Paleozoic age that have been thrust eastward for many kilometers over the sandstone. The Death Valley and Garlock fault systems are clearly seen at the north flank of the Avawatz Mountains. Reproductions of LFC photographs are no longer available, but the system demonstrated the utility of photographs acquired from orbital altitudes.

Declassified Photographs from Intelligence Satellites

For many years the United States has employed satellites to acquire photographs of strategic areas, mainly the Sino-Soviet bloc, for intelligence purposes. These photographs have been highly classified and unavailable. In 1995 the United States announced the declassification of intelligence photographs acquired from 1960 to 1972 by the CORONA camera, which is now obsolete. The CORONA missions acquired over 800,000 photographs. Each photograph covers approximately 16 by 195 km at spatial resolutions ranging from 2 to 8 m. The transparencies will be duplicated, indexed, and transferred to the EDC for unrestricted sale to the public by late 1996. Contact the EDC for information on prices and availability. McDonald (1995A, 1995B) has reviewed the history and specifications of the classified U.S. satellite photography programs and has published a number of photographs.

Table 2-4 Camera systems on satellites

Characteristics	*LFC* *United States*	*KVR-1000* *Russia*	*TK-350* *Russia*
Satellite altitude, km	239 to 370	220	220
Terrain coverage, km	Variable	34 by 57	175 by 175
Spatial resolution, m	Variable	2 to 3	5 to 10
Stereo overlap, %	20 to 80	Minimal	60 to 80
Spectral range, μm	Normal color IR color Panchromatic	0.51 to 0.76	0.51 to 0.76

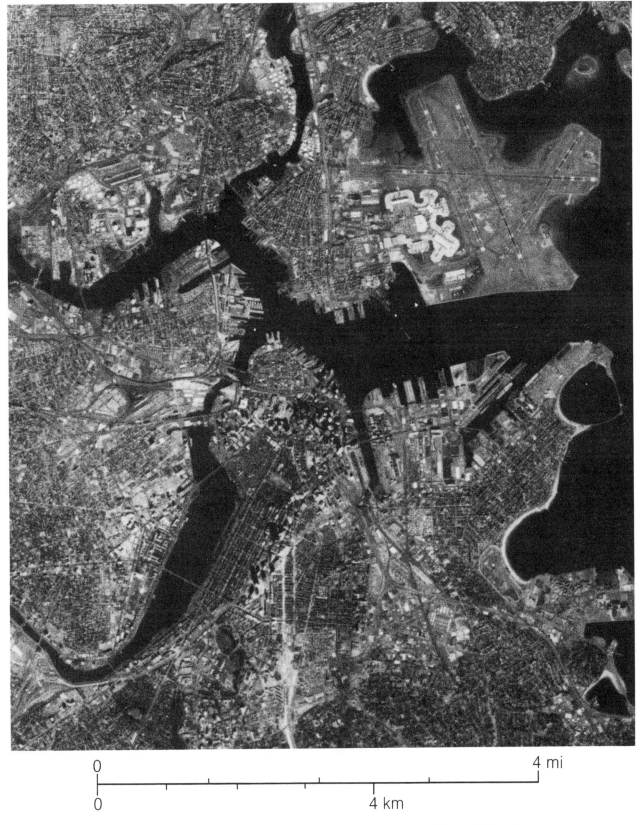

0   4 mi

0   4 km

Figure 2-33 Portion of a large-format-camera photograph of Boston, Massachusetts.

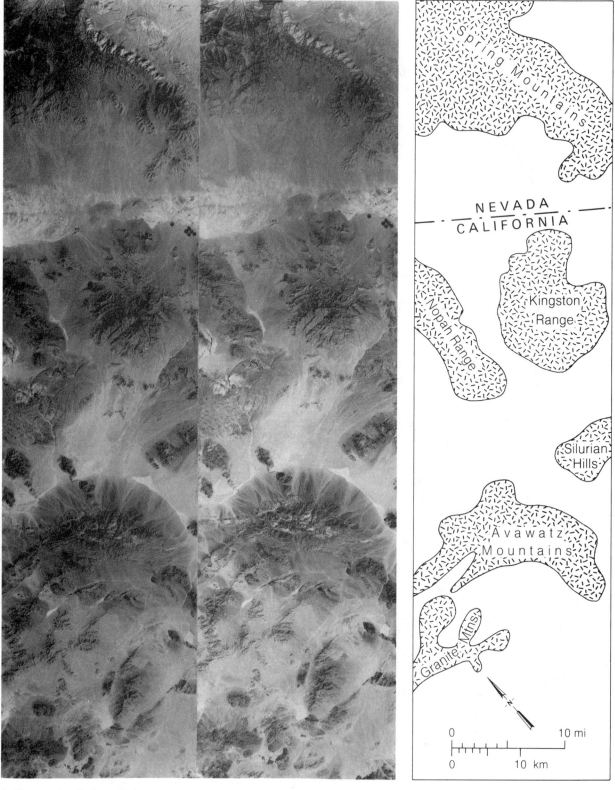

A. Stereo pair with 4× vertical exaggeration.

B. Location map.

Figure 2-34 Stereo pair of large-format-camera photographs in the Mojave Desert of southeastern California and Nevada.

0   2 mi

0   2 km

Figure 2-35 Enlarged portion of Russian KVR-1000 satellite photograph of Washington, D.C. Courtesy EOSAT Co.

Photographs from Russian Satellites

For at least two decades Russia has acquired photographs from unmanned earth-orbiting satellites. Many of these photographs are available for general distribution. Table 2-4 lists characteristics of photographs acquired by the KVR-1000 and TK-350 cameras. These cameras are carried on Kosmos satellites. Each mission lasts approximately 45 days and can photograph nearly 50 percent of the earth's land area, depending upon weather conditions. Upon completion of a mission the capsule containing the camera and exposed film is returned to earth.

Figure 2-35 is an enlarged portion of a KVR-1000 photograph of Washington, D.C. The Pentagon is located at the extreme southwest corner. The white circle in the southeast corner is the JFK stadium. It is instructive to compare this KVR photograph (acquired from a height of 220 km) with the high-altitude aircraft photograph in Figure 2-31 (12 km height). The KVR subscene covers the northeast portion of the area in the aircraft photograph. Scale of the KVR photograph is approximately twice as large as that of the aircraft photograph. This comparison demonstrates the potential of satellite photographs for detailed interpretation. The KVR subscene shows only 36 percent of the original photograph, which covers 2000 km^2.

TK-350 photographs (not shown) are usually acquired in tandem with the KVR-1000 photographs. The overlapping TK-350 photographs can be analyzed in stereo to obtain elevation data for topographic mapping. Digitized versions of KVR-1000 and TK-350 photographs are available from:

EOSAT Company
Customer Service Department
4300 Forbes Boulevard
Lanham, MD 20706
Telephone: 800-344-9933

EOSAT can provide information on prices and available coverage.

COMMENTS

Photographs are a versatile and useful form of remote sensing for the following reasons:

1. The film provides excellent spatial resolution and has a high information content.
2. Photographs cost relatively little.
3. Different films provide a sensitivity range from the UV spectral region through the visible and into the reflected IR region.
4. Low-sun-angle photographs enhance subtle topographic features that are suitably oriented with respect to the sun's azimuth.
5. Stereo photographs are valuable aids for many types of interpretation.

The principal drawbacks of aerial photographs are that

1. Daylight and good weather are necessary to acquire them.
2. In the shorter wavelength regions, atmospheric scattering reduces their contrast ratio and resolving power.
3. Information is recorded in the analog mode. Film must be digitized in order to digitally process the data (Chapter 8), although digital cameras are being developed.
4. The longest wavelength recorded is 0.9 μm, which omits the valuable spectral information at longer wavelengths.

The advantages often outweigh the disadvantages, and one should evaluate aerial photographs as a possible data source for any remote sensing investigation.

QUESTIONS

1. Normal color photographs taken of subjects in shaded areas have a bluish cast. Explain why.
2. Calculate the contrast ratio for a scene in which the brightest and darkest areas have brightness values of 6 and 2, respectively.
3. Suppose the scene in question 2 is covered by an atmosphere that contributes 4 brightness values of scattered light. What is the resulting contrast ratio? Panchromatic aerial photographs will be acquired of this scene. How can their contrast ratio be improved?
4. What is the ground resolution for aerial photographs acquired at a height of 5000 m with a camera having a system resolution of 30 line-pairs · mm^{-1} and a focal length of 304 mm?
5. What is the minimum ground separation in the photographs of question 4?
6. What is the scale of the photographs of question 4?
7. For Figure 2-8, calculate the height of the highest portion of the building in the extreme lower left corner.
8. The airbase for two overlapping photographs is 1500 m. The photographs were acquired from a height of 3000 m. What is base–height ratio of this stereo pair? What is the vertical exaggeration of the stereo model?
9. Photographs can be acquired from satellites, using the same films and filters as aerial photographs. Describe the advantages and disadvantages of satellite photographs relative to aerial photographs.

REFERENCES

Boyer, M., J. Miller, M. Berlanger, and E. Hare, 1988, Senescence and spectral reflectance in leaves of northern pin oak (*Quercus palustris* Muenchh.): Remote Sensing of Environment, v. 25, p. 71–87.

Buschmann, C. and E. Nagel, 1993, *In vivo* spectroscopy and internal optics of leaves as basis for remote sensing of vegetation: International Journal of Remote Sensing, v. 14, p. 711–722.

Everitt, J. H. and P. R. Nixon, 1986, Canopy reflectance of two drought-stressed shrubs: Photogrammetric Engineering and Remote Sensing, v. 52, p. 1189–1192.

Gausman, H. W., 1985, Plant leaf optical properties in visible and near-infrared light: Texas Tech University Graduate Studies, No. 29, Lubbock, TX.

James, T. H., 1966, The theory of the photographic process, third edition: Macmillan Co., New York, NY.

Jones, R. C., 1968, How images are detected: Scientific American, v. 219, p. 111–117.

Kienko, Y. P., 1991, Resource-F subsystem: Geodesy and Cartography, v. 7, p. 9–15, Moscow, Russia.

King, D. J., 1995, Airborne multispectral digital camera and video sensors—a critical review of system designs and applications: Canadian Journal of Remote Sensing, v. 21, p. 245–273.

La Prade, G. L., 1972, Stereoscopy—a more general theory: Photogrammetric Engineering and Remote Sensing, v. 38, p. 1177–1187.

La Prade, G. L., 1973, Stereoscopy—will data or dogma prevail? Photogrammetric Engineering and Remote Sensing, v. 39, p. 1271–1275.

McDonald, R. A., 1995A, Opening the cold war sky to the public—declassifying satellite reconnaissance imagery: Photogrammetric Engineering and Remote Sensing, v. 61, p. 385–390.

McDonald, R. A., 1995B, CORONA: Photogrammetric Engineering and Remote Sensing, v. 61, p.689–720.

Miller, C. V., 1961, Photogeology: McGraw-Hill Book Co., New York, NY.

Philipson, W. R., ed., 1996, The manual of photographic interpretation: American Society for Photogrammetry and Remote Sensing, Falls Church, VA.

Rosenblum, L., 1968, Image quality in aerial photography: Optical Spectra, v. 2, p. 71–73.

Sabins, F. F. , 1973A, Aerial camera mount for 70-mm stereo: Photogrammetric Engineering and Remote Sensing, v. 39, p. 579–582.

Sabins, F. F., 1973B, Engineering geology applications of remote sensing *in* Moran, D. E., ed., Geology, seismicity, and environmental impact: Association of Engineering Geologists, Special Publication, p. 141, 155, Los Angeles, CA.

Sabins, F. F., 1986, Remote sensing—principles and interpretation, second edition: W. H. Freeman and Co., New York, NY.

Schwaller, M. R. and S. J. Tkach, 1985, Premature leaf senescence—remote sensing detection and utility for geobotanical prospecting: Economic Geology, v. 80, p. 250–255.

Slater, P. N., 1983, Photographic systems for remote sensing *in* Colwell, R. N., ed., Manual of remote sensing, second edition: ch. 6, p. 231–291, American Society for Photogrammetry and Remote Sensing, Falls Church, VA.

Thurrell, R. F., 1953, Vertical exaggeration in stereoscopic models: Photogrammetric Engineering and Remote Sensing, v. 19, p. 579–588.

Vizy, K. N., 1974, Detecting and monitoring oil slicks with aerial photos: Photogrammetric Engineering and Remote Sensing, v. 40, p. 697–708.

Walker, P. M. and D. T. Trexler, 1977, Low sun-angle photography: Photogrammetric Engineering and Remote Sensing, v. 43, p. 493–505.

Wolf, P. R., 1974, Elements of photogrammetry: McGraw-Hill Book Co., New York, NY.

ADDITIONAL READING

Avery, T. E. and G. L. Berlin, 1992, Fundamentals of remote sensing and airphoto interpretation, fifth edition: Macmillan Publishing Co., New York, NY.

Cravat, H. R. and R. Glaser, 1971, Color aerial stereograms of selected coastal areas in the United States: U.S. Department of Commerce, National Oceanic and Atmospheric Administration, Washington, DC.

DeMarsh, L. E. and E. J. Giorgianni, 1989, Color science for imaging systems: Physics Today, n. 42, p. 44–52.

Falkner, E., 1995, Aerial mapping methods and applications: Lewis Publishers, Boca Raton, FL.

Foster, N. H. and E. H. Beaumont, eds., 1992, Photogeology and photogeomorphology: American Association of Petroleum Geologists, Treatise of Petroleum Geology Reprint Series, No. 18, Tulsa, OK.

Light, D. L., 1996, Film cameras or digital sensors—the challenge ahead for aerial imaging: Photogrammetric Engineering and Remote Sensing, v. 62, p. 285–291.

Lynch, D. K. and W. Livingston, 1995, Color and light in nature: Cambridge Press, Cambridge, MA.

Mollard, J. D. and J. R. Jones, 1984, Airphoto interpretation and the Canadian landscape: Canadian Government Publishing Center, No. M52-60/1984E, Hull, Canada.

Phillipson, W. R., ed., 1966, The manual of photographic interpretation, second edition: American Society for Photogrammetry and Remote Sensing, Falls Church, VA.

Rasher, M. E. and W. Weaver, 1990, Basic photo interpretation: Soil Conservation Service, U.S. Department of Agriculture, Washington, DC.

Ray, R. G., 1960, Aerial photographs in geologic mapping and interpretation: U.S. Geological Survey Professional Paper 373.

Smith, J. T. and A. Anson, eds., 1968, Manual of color aerial photography: American Society for Photogrammetry and Remote Sensing, Falls Church, VA.

CHAPTER

3

LANDSAT IMAGES

Landsat is an unmanned system that prior to 1974 was called ERTS (Earth Resources Technology Satellite). Initially NASA operated Landsat, but in 1985 responsibility for operating the system transferred to the EOSAT Company, a private corporation. Landsat operates in the international public domain, which means that

1. under an "open skies" policy, images are acquired of the entire earth without obtaining permission from any government;
2. users anywhere in the world may purchase all images at uniform prices and priorities.

The Landsat program has been a major contributor to the growth and acceptance of remote sensing as a scientific discipline. Landsat provided the first repetitive worldwide database with adequate spatial and spectral resolution for many applications. Landsat data are available in digital format, which has promoted the science of digital image processing. Present and future generations of remote sensing specialists are indebted to the late William T. Pecora and William Fischer of the U.S. Geological Survey, who did so much to make Landsat a reality.

Landsat satellites have been placed in orbit using Delta rockets launched from Vandenberg Air Force Base on the California coast between Los Angeles and San Francisco. The five Landsats belong to two generations of technology with different satellites, orbital characteristics, and imaging systems. Freden and Gordon (1983) give details of the Landsat program.

LANDSATS 1, 2, AND 3

Table 3-1 lists orbital characteristics of the three satellites of the first generation, which were launched in 1972, 1975, and 1978; all have ceased operation, but they produced hundreds of thousands of valuable images. The multispectral scanner (MSS) was the primary imaging system in this first generation of Landsat. A return-beam vidicon system was also carried (Sabins, 1987, Chapter 4), but those images were of limited

Table 3-1 Orbit patterns and imaging systems of first and second generations of Landsat

Generation	Landsats 1, 2, and 3	Landsats 4 and 5
Altitude	918 km	705 km
Orbits per day	14	14.5
Number of orbits (paths)	251	233
Repeat cycle	18 days	16 days
Image sidelap at equator	14.0 percent	7.6 percent
Crosses 40°N latitude at (local sun time, approx.)	9:30 a.m.	10:30 a.m.
Operational from	1972 to 1984	1982 to future
On-board data storage	Yes	No
Imaging systems:		
Multispectral scanner	Yes	Yes
Thematic mapper	No	Yes

Table 3-2 Characteristics of Landsat imaging systems

	Multispectral scanner (MSS)	*Thematic mapper (TM)*
Spectral region		
Visible and reflected IR	0.50 to 1.10 μm	0.45 to 2.35 μm
Thermal IR	—	10.5 to 12.5 μm
Spectral bands	4	7
Terrain coverage		
East-west direction	185 km	185 km
North-south direction	185 km	170 km
Instantaneous field of view		
Visible and reflected IR	0.087 mrad	0.043 mrad
Thermal IR	—	0.17 mrad
Ground resolution cell		
Visible and reflected IR	79 by 79 m	30 by 30 m
Thermal IR	—	120 by 120 m

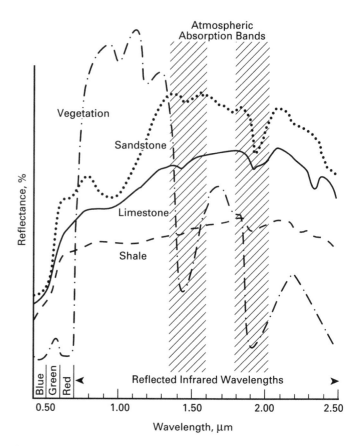

Figure 3-1 Reflectance spectra of vegetation and sedimentary rocks, showing spectral ranges of Landsat MSS and TM bands.

value. Table 3-2 lists characteristics of the MSS, a cross-track scanning system that records four spectral bands of imagery with a ground resolution cell of 79 by 79 m. Figure 3-1 shows spectral ranges of the MSS bands together with reflectance spectra of vegetation and sedimentary rocks. Table 3-3 lists the wavelengths recorded by the MSS bands. In this book, MSS bands are designated 1, 2, 3, 4 in accordance with current terminology.

Figure 3-2 shows the four spectral bands for an MSS scene of the Los Angeles region. Beginning at the north margin of the image, successive scan lines are offset to the west to compensate for the earth's rotation during the approximately 25 sec required to scan the terrain. This offset accounts for the slanted parallelogram outline of the images. An IR color image can be prepared by assigning MSS band 1 (green) to the color blue, band 2 (red) to green, and band 4 (reflected IR) to red. A major advantage of Landsat images is the 185-by-185-km (34,000 km²) coverage that facilitates regional interpretations. The Los Angeles scene covers parts of four major physiographic provinces, as shown in the location map (Figure 3-3). The Central Valley, in the northwest portion of the scene, is a major agricultural area, as shown by the rectangular field patterns. Rugged mountains of the Sierra Nevada separate the Central Valley from the Antelope Valley, which is the westernmost extension of the Mojave Desert. The Antelope Valley is bounded on the north by the Garlock fault and on the south by the San Andreas fault. Both of these active strike-slip faults are clearly expressed as linear valleys. The desert terrain of the Antelope Valley is bright on all the MSS bands. Along the south margin are alluvial fans of gravel eroded from bedrock of the Transverse Ranges. The light- to dark-gray signatures of the fans are determined by characteristics of the parent bedrock.

The Sheep Canyon Fan, identified in Figure 3-3, is dark in all MSS bands because it consists of gravel eroded from the Pelona Schist, which is very dark.

Irrigated fields in the Antelope Valley and Central Valley are dark in band 2 (red) and bright in bands 3 and 4 (reflected IR). These signatures are explained by the spectral reflectance curve for vegetation in Figure 3-1. Vegetation has low reflectance in band 2 because red wavelengths are absorbed by chlorophyll; reflectance is high in bands 3 and 4 because the internal structure of leaves strongly reflects these IR wavelengths (Figure 2-28).

Mountains of the Transverse Ranges trend westward and form the northern border of the Los Angeles and Ventura Basins, which are part of the Peninsular Range Province. The mountains are cut by numerous faults, such as the San Gabriel fault. A conspicuous dark patch north of Ventura was burned in a brushfire. Bright patches in the extreme eastern portion of the range are clouds lodged against the high mountain ridges.

Table 3-3 Landsat multispectral scanner (MSS) spectral bands

MSS band*	Wavelength, μm	Color	Projection color for IR color composite image
1 (4)	0.5 to 0.6	Green	Blue
2 (5)	0.6 to 0.7	Red	Green
3 (6)	0.7 to 0.8	Reflected IR	—
4 (7)	0.8 to 1.1	Reflected IR	Red

*Numbers in parentheses were used for images acquired by Landsats 1, 2, and 3. For Landsats 4 and 5 the MSS bands are designated 1, 2, 3, and 4.

The Los Angeles and Ventura Basins are lowlands underlain by deep depressions filled with sedimentary rocks that generated vast reserves of oil. The giant oil fields of the region are largely depleted and are now being developed into real estate ventures. The Ventura Basin is still largely agricultural, which causes the bright signatures in bands 3 and 4. The Los Angeles Basin is completely urbanized. The central city has a dark signature on bands 3 and 4 because vegetation is absent. The surrounding suburbs have gray signatures due to landscaping mingled with buildings and pavement. Scattered bright patches are parks, golf courses, and cemeteries.

Water in the Pacific Ocean is uniformly dark on all bands. The MSS data were digitally processed to emphasize land features. Other processing methods could enhance spectral variations caused by turbidity and shallow bathymetric features.

LANDSATS 4 AND 5

The second generation of Landsat consists of two satellites launched July 16, 1982, and March 1, 1984. Landsat 4 has ceased functioning. Landsat 5 was still functioning as of March 1996, but its ultimate lifetime is unpredictable. Landsat 6 was launched in September 1993 but failed to reach orbit, which was a major loss to the remote sensing community. When Landsat 5 fails there will be a gap in the formerly continuous coverage of Landsat images, dating back to 1972.

Satellites

Figure 3-4 shows the second generation of Landsat satellites, which carry an improved imaging system called the thematic mapper (TM) and the MSS. The solar array generates electrical power to operate the satellite. The microwave antenna receives instructions and transmits image data to the ground receiving stations, shown in Figure 3-5. When a satellite is within the receiving range of a station, TM images are scanned and transmitted simultaneously. Images of areas beyond receiving ranges are transmitted to *Tracking and Data Relay Satellites* (TDRS), which are placed in geostationary orbits. The TDRS system relays the image data to a receiving station at Norman, Oklahoma, which then relays the data via a communication satellite to the EOSAT facility in Maryland. At EOSAT the data are archived on high-density tapes (HDT). Data in the HDT format are converted into computer-compatible tapes (CCT) that are used for digital image processing and to generate master film transparencies. All MSS data are transmitted directly to ground receiving stations.

Thematic Mapper (TM) Imaging System

The TM is a cross-track scanner with an oscillating scan mirror and arrays of 16 detectors for each of the visible and reflected IR bands. Data are recorded on both eastbound and westbound sweeps of the mirror, which allows a slower scan rate, longer dwell time, and higher signal-to-noise ratio than with MSS images. At the satellite altitude of 705 km the 14.9° angular field of view covers a swath 185 km wide (Figure 3-5). Spectral ranges of the six visible and reflected IR bands are shown in Figure 3-1. Band 6 (10.4 to 12.5 μm) records thermal IR energy, which is beyond the range covered by Figure 3-1. The TM was originally designed to include bands 1 through 5 (visible and reflected IR) and band 6 (thermal IR). Users pointed out that information in the spectral band from 2.1 to 2.4 μm had great value for geologic mapping and mineral exploration. Band 7 was added to acquire these data, which are widely used, as shown in Chapter 11. The original system for numbering TM bands remained the same, however, which explains why band 7 is out of sequence on a spectral basis. Table 3-4 lists the characteristics of the TM bands.

Preparing TM Color Images

TM bands are generally made into color images for interpretation (Table 3-5). Band 6 is rarely used because of its coarse spatial resolution (120 m), but it is employed in thermal

A. Band 1 (0.5 to 0.6 μm).

B. Band 2 (0.6 to 0.7 μm).

C. Band 3 (0.7 to 0.8 μm).

D. Band 4 (0.8 to 1.1 μm).

Figure 3-2 Spectral bands of MSS images of the Los Angeles region. Images cover 185 by 185 km.

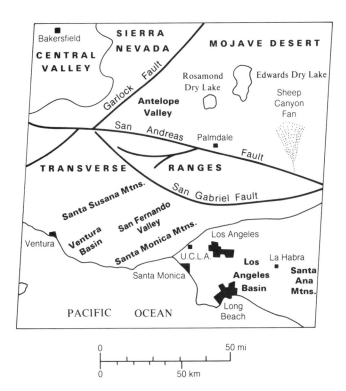

Figure 3-3 Location map of the Los Angeles region.

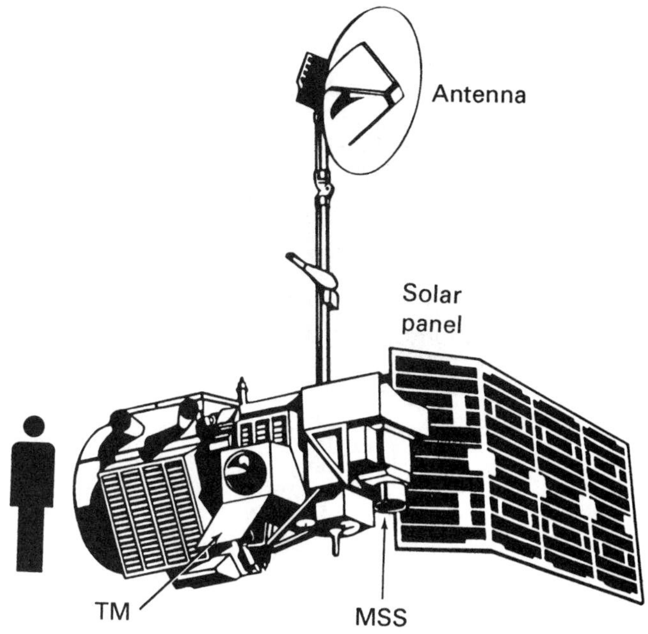

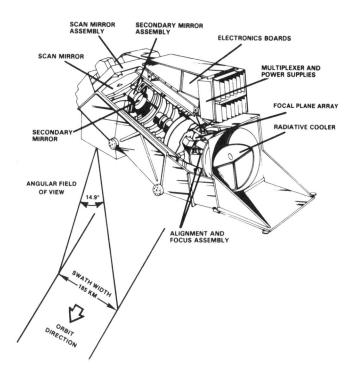

Figure 3-4 Landsats 4 and 5. The human figure (2 m high) is added for scale.

Figure 3-5 Thematic mapper imaging system.

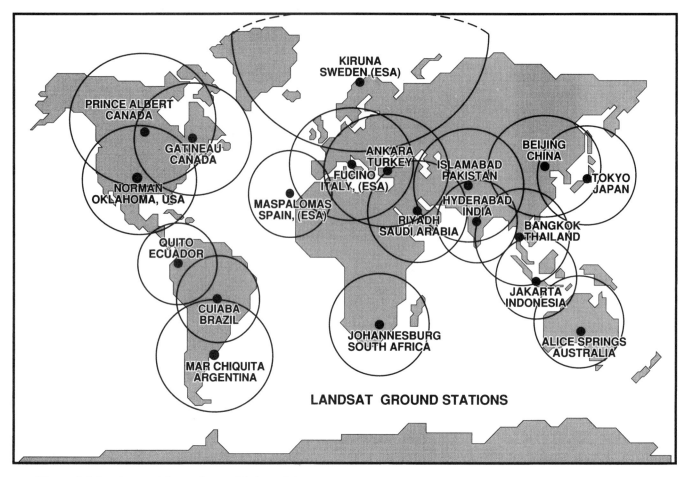

Figure 3-6 Landsat receiving stations and their receiving ranges.

Table 3-4 Landsat thematic mapper (TM) spectral bands

Band	Wavelength, μm	Characteristics
1	0.45 to 0.52	Blue-green. Maximum penetration of water, which is useful for bathymetric mapping in shallow water. Useful for distinguishing soil from vegetation and deciduous from coniferous plants.
2	0.52 to 0.60	Green. Matches green reflectance peak of vegetation, which is useful for assessing plant vigor.
3	0.63 to 0.69	Red. Matches a chlorophyll absorption band that is important for discriminating vegetation types.
4	0.76 to 0.90	Reflected IR. Useful for determining biomass content and for mapping shorelines.
5	1.55 to 1.75	Reflected IR. Indicates moisture content of soil and vegetation. Penetrates thin clouds. Provides good contrast between vegetation types.
6	10.40 to 12.50	Thermal IR. Nighttime images are useful for thermal mapping and for estimating soil moisture.
7	2.08 to 2.35	Reflected IR. Coincides with an absorption band caused by hydroxyl ions in minerals. Ratios of bands 5 and 7 are used to map hydrothermally altered rocks associated with mineral deposits.

Table 3-5 Evaluation of TM color combinations

Display colors*	Advantages	Disadvantages
1-2-3	Normal color image. Optimum for mapping shallow bathymetric features.	Lower spatial resolution due to band 1. Limited spectral diversity because no reflected IR bands are used.
2-3-4	IR color image. Moderate spatial resolution.	Limited spectral diversity.
4-5-7	Optimum for humid regions. Maximum spatial resolution.	Limited spectral diversity because no visible bands are used.
2-4-7	Optimum for temperate to arid regions. Maximum spectral diversity.	Unfamiliar color display, but interpreters quickly adapt.

*TM bands are listed in the sequence of projection colors: blue-green-red.

mapping (Chapter 5). Any three of the six visible and reflected IR bands may be combined in blue, green, and red to produce a color image. There are 120 possible color combinations, which is an excessive number for practical use. Theory and experience, however, show that a small number of color combinations are suitable for most applications. For several years after TM data became available, they were routinely produced as IR color images (bands 2, 3, and 4 combined in blue, green, and red, respectively) because that familiar combination was the standard for MSS images and matched the signatures of IR color aerial photographs. In recent years, however, we have found that other TM color combinations are more useful. The optimum band combination is determined by the terrain, climate, and nature of the interpretation project. The following examples illustrate band selection for contrasting terrains in Wyoming and Indonesia.

Images of Semiarid Terrain, Wyoming

Figure 3-7 shows the seven TM bands for the Thermopolis subscene in central Wyoming. The subscene is located in the south flank of the Bighorn Basin and includes a stretch of the Wind River and the town of Thermopolis (Figure 3-7H). Most of the area is used for ranching. Some irrigated crops are grown in the stream valleys. The Gebo and Little Sand Draw oil fields occur in the northern part of the area.

TM Color Combinations Plate 2 shows four color combinations of TM images for the Thermopolis subscene. Table 3-5 lists and compares these combinations. Plates 2A and 2B are the spectral equivalents of normal color aerial photographs and IR color photographs. Vegetation has a red signature in Plate 2B because band 4, which covers the strong vegetation response in the reflected IR region, is shown in red. Red beds of the Chugwater Formation (shown by the stippled pattern in the map of Figure 3-7H) have an orange signature in the normal color image (Plate 2A). In the IR color image (Plate 2B), Chugwater outcrops have a distinctive yellow signature that is

typical for red rocks throughout the world seen on this color combination. The IR color image has a better contrast ratio and spatial resolution than the normal color image because the blue band was not used in Plate 2B. Plate 2C uses only the reflected IR bands 4, 5, and 7 as blue, green, and red and has the best spatial resolution of all the combinations. There is little color contrast, however, between the different rock outcrops, which are monotonous shades of pale blue. Even the red beds of the Chugwater Formation are light blue and are indistinguishable from the other outcrops.

In Plate 2D the visible green band 2 is shown in blue and the IR bands 7 and 4 are combined as red and green. This combination provides the maximum range of color signatures for the rock outcrops and is optimum for interpreting geology in this semiarid area. Extensive experience in other arid and semiarid regions throughout the world confirms that the 2-4-7 color combination is optimum. Vegetation is green in this image because band 4 is shown in green. Some investigators prefer a 1-4-7 version of this combination, but I find 2-4-7 to be optimum because band 2 has less atmospheric scattering than band 1.

Image Interpretation Figure 3-8 is a geologic interpretation map of the 2-4-7 image (Plate 2D); Table 3-6 lists the formations that crop out in the Thermopolis subscene together with their ages, lithology, and signature in the 2-4-7 TM image. The image shows that the regional dip of the beds is northward. Local reversals of dip toward the south form four major anticlines. The Red Rose and Cedar Mountain anticlines are located in the southern portion of the subscene. The Gebo and Little Sand Draw anticlines in the north are oil fields that produce from the Phosphoria Formation. The cross section in Figure 3-9 shows these structural relationships. The oil fields were discovered as a result of surface mapping prior to the launch of Landsat. By studying these known oil fields we learn to recognize similar, but undrilled, structures in less well explored regions of the world.

Dip faults, which strike parallel with regional dip, are readily recognized because they offset formation contacts. Strike faults, which trend parallel with regional strike, eliminate or repeat

A. Band 1 (0.45 to 0.52 μm).

B. Band 2 (0.52 to 0.60 μm).

C. Band 3 (0.63 to 0.69 μm).

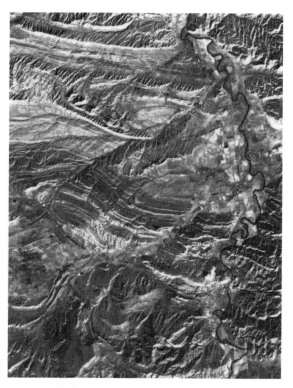

D. Band 4 (0.76 to 0.90 μm).

Figure 3-7 Landsat TM bands for the Thermopolis, Wyoming, subscene.

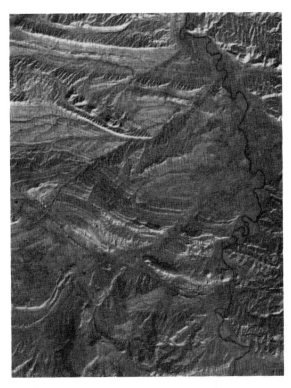

E. Band 5 (1.55 to 1.75 μm).

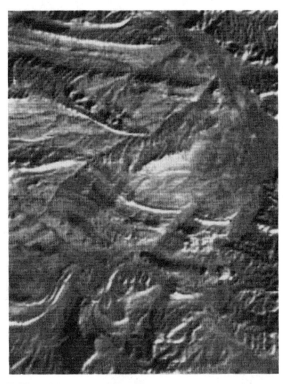

F. Band 6 (10.40 to 12.50 μm).

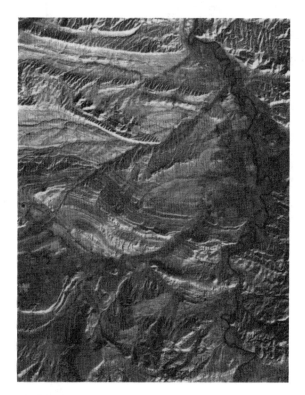

G. Band 7 (2.08 to 2.35 μm).

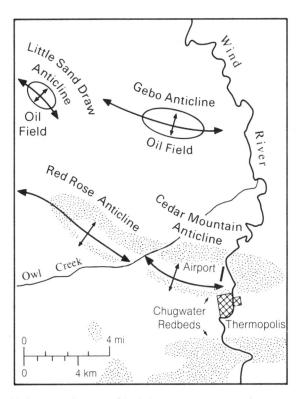

H. Interpretation map. Stippled areas are outcrops of Chugwater red beds.

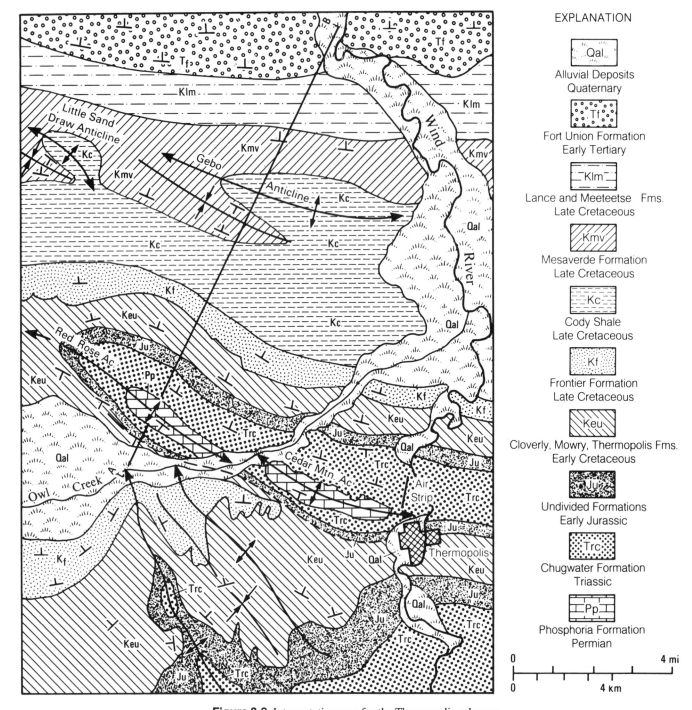

EXPLANATION

Qal
Alluvial Deposits
Quaternary

Tf
Fort Union Formation
Early Tertiary

Klm
Lance and Meeteetse Fms.
Late Cretaceous

Kmv
Mesaverde Formation
Late Cretaceous

Kc
Cody Shale
Late Cretaceous

Kf
Frontier Formation
Late Cretaceous

Keu
Cloverly, Mowry, Thermopolis Fms.
Early Cretaceous

Ju
Undivided Formations
Early Jurassic

Trc
Chugwater Formation
Triassic

Pp
Phosphoria Formation
Permian

0   4 mi

0   4 km

Figure 3-8 Interpretation map for the Thermopolis subscene.

beds and are more difficult to recognize. A strike fault along the south flank of the Red Rose and Cedar Mountain anticlines is recognized because the outcrop of the Chugwater Formation is much narrower on the south flank than on the north flank of the folds (Figures 3-8 and 3-9). The normal strike fault has cut out a portion of the Chugwater beds. The steeper dip on the south flank also contributes to the narrow outcrop.

Images of Tropical Terrain, Indonesia

Figure 3-10 shows the three visible and three reflected IR bands for the Mapia subscene in the south-central portion of Irian Jaya in Indonesia. This tropical rain-forest terrain provides a contrast with the semiarid rangeland of the Thermopolis image.

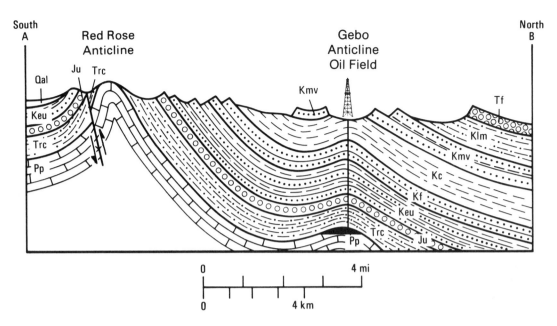

Figure 3-9 Cross section of the Thermopolis subscene. Location and formation symbols are shown in Figure 3-8.

Table 3-6 Formations in the Thermopolis TM 2-4-7 subscene

Formation	Age	Lithology	Image signature
Alluvial deposits	Quaternary	Soil in floodplains of major streams. Flat valley floors with irrigated fields.	Bright green.
Fort Union Formation	Early Tertiary	Resistant sandstone with minor shale beds. Prominent, eroded dipslopes.	Dark pink.
Meeteetsee and Lance Formations	Late Cretaceous	Nonresistant shale and sandstone. Broad valley with minor ridges.	Medium pink.
Mesaverde Formation	Late Cretaceous	Resistant sandstone with shale and coal beds. Alternating ridges and valleys.	Medium pink.
Cody Shale	Late Cretaceous	Nonresistant shale. Broad valley with minor ridges.	Light pink.
Frontier Formation	Late Cretaceous	Alternating sandstone and shale. Narrow ridges and valleys.	Dark pink.
Cloverly, Mowry, and Thermopolis Formations	Early Cretaceous	Resistant and nonresistant shale. Mapped as a single unit. Narrow ridges and valleys.	Light blue and dark pink.
Undifferentiated Formations	Early Cretaceous	Alternating sandstone and shale. Narrow ridges and valleys.	Dark pink and light blue.
Chugwater Formation	Triassic	Red sandstone and siltstone. Alternating ridges and valleys.	Yellow and orange.
Phosphoria Formation	Permian	Resistant carbonate rocks. Crops out in cores of Red Rose and Cedar Mountain anticlines.	Very light blue.

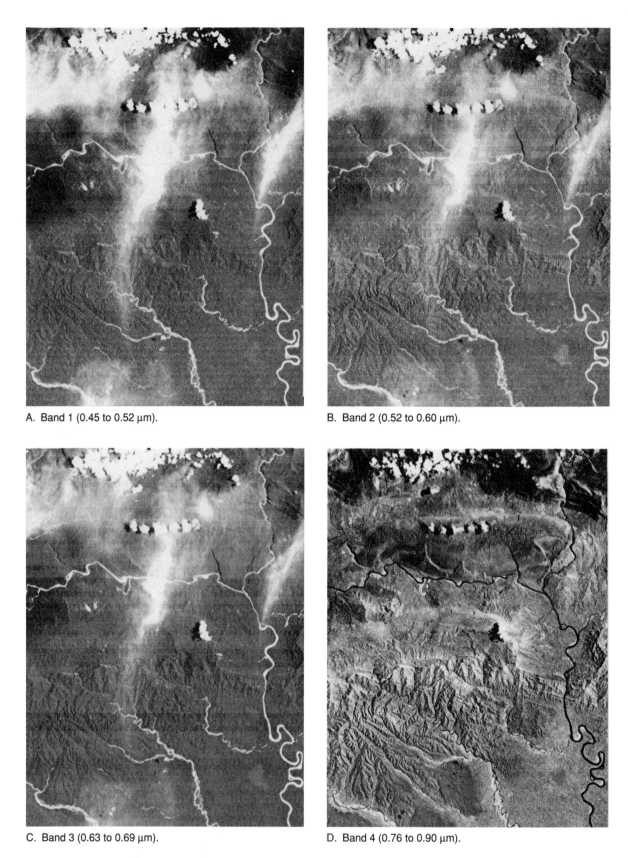

A. Band 1 (0.45 to 0.52 μm).

B. Band 2 (0.52 to 0.60 μm).

C. Band 3 (0.63 to 0.69 μm).

D. Band 4 (0.76 to 0.90 μm).

Figure 3-10 Landsat TM bands for the Mapia subscene, Irian Jaya, Indonesia.

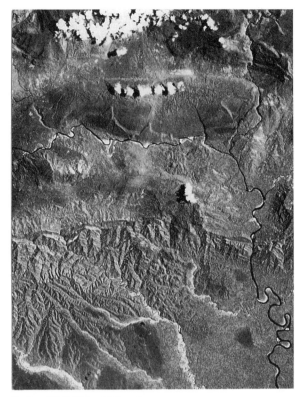

E. Band 5 (1.55 to 1.75 µm).

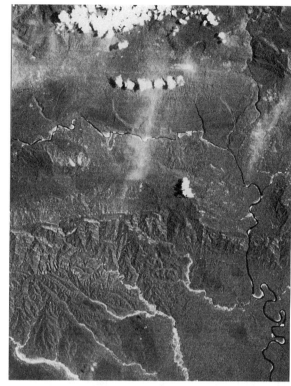

F. Band 7 (2.08 to 2.35 µm).

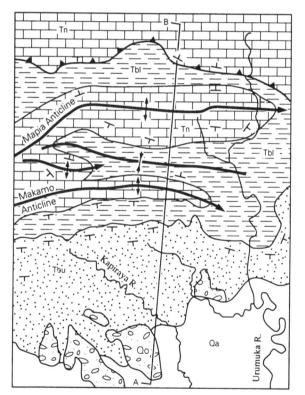

G. Geologic map.

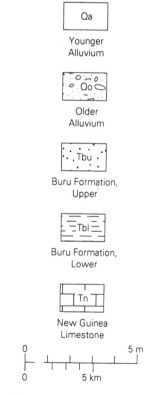

H. Explanation for map.

Qa

Younger
Alluvium

Qo

Older
Alluvium

Tbu

Buru Formation,
Upper

Tbl

Buru Formation,
Lower

Tn

New Guinea
Limestone

0   5 mi

0   5 km

Table 3-7 Formations in the Mapia TM 4-5-7 subscene

Formation	Age	Lithology	Image signature
Younger alluvium	Recent	Soil and gravel	Valleys along major drainages
Older alluvium	Recent	Gravel deposits	Eroded terraces and alluvial fans
Buru Formation	Upper Member, Early Tertiary	Sandstone and minor shale beds	Ridges and dipslopes
Buru Formation	Lower Member, Early Tertiary	Shale	Strike valleys
New Guinea Limestone	Early Tertiary	Limestone	Karst topography

TM Color Combinations The three visible bands (1, 2, 3) in Figure 3-10 have low contrast and poor spatial resolution because the high moisture content of the atmosphere strongly scatters these short wavelengths. The three reflected IR bands (4, 5, 7) have much better contrast and resolution because these longer wavelengths are less susceptible to atmospheric scattering. Plate 3 shows four color combinations for the Mapia subscene. Thanks to computer enhancement the normal color image (Plate 3A) has better color contrast than one would anticipate, based on the appearance of the black and white bands. Clouds obscure much of the image, and geologic features are difficult to discern. The 2-3-4 IR color image (Plate 3B) has more contrast and detail because band 4 (reflected IR) replaces the low-contrast band 1 image (visible blue). The clouds are greatly diminished. The red signature indicates the extensive forest cover of the region. Plate 3C is compiled from reflected IR bands 4, 5, and 7 shown in blue, green, and red. For tropical regions this combination provides optimum resolution, color contrast, reduction of clouds, and expression of geologic features. In Plate 3C vegetation has a range of color signatures, which aids interpretation, whereas in Plate 3A,B vegetation is saturated dark green or bright red. Vegetation patterns are more distinct in Plate 3C because bands 4, 5, and 7 coincide with major variations in the reflectance spectrum of vegetation, shown in Figure 3-1. Plate 3D consists of bands 2, 4, and 7 in blue, green, and red and is the second best of the four images. This 2-4-7 combination is optimum for the Thermopolis area, but in Irian Jaya the vegetation cover and humid conditions reduce its effectiveness.

Image Interpretation Figure 3-10G is the geologic interpretation map for the 4-5-7 color image of the Mapia subscene. Table 3-7 lists the formations together with their ages, lithology, and signatures on the 4-5-7 image. Despite the vegetation cover, the different formations are mappable on the image because each unit erodes to a distinctive topographic pattern. The resistant sandstones of the upper member of the Buru Formation erode to form the rugged ledge-and-slope topography in the south portion of the image (Plate 3C). The nonresistant shale of the lower member of the Buru Formation forms broad featureless strike valleys. The New Guinea Limestone crops out in the crest of the Mapia and Makamo anticlines (Figure 3-10G), where it forms broad arches. In this humid environment, solution and collapse of the limestone produce a distinctive terrain of closely spaced pits and pinnacles called *karst* topography. Karst topography is well developed on the New Guinea Limestone on the crest of Mapia and Makamo anticlines. At the small scale of Plate 3C, however, the karst pattern is somewhat difficult to recognize.

Geologic structure is interpreted for the Mapia subscene in the same manner as the Thermopolis image; however, there are significant differences in solar illumination.

	Mapia area	Thermopolis area
Sun azimuth	From ENE	From SE
Sun elevation	45°	25°

In the Mapia area, shadows and highlights are subdued because the sun has a high elevation and an azimuth nearly paral-

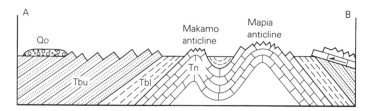

Figure 3-11 Cross section for the Mapia subscene. Location and formation symbols are shown in Figure 3-10G, H.

lel with the east-west regional strike. Despite these disadvantages, geologic structures can be interpreted. In the south portion of the subscene, resistant beds of the upper member of the Buru Formation erode to form dipslopes and antidip scarps that define regional south dips. In the central part of the subscene, the Mapia and Makamo anticlines form broad arches of the New Guinea Limestone, surrounded by lowlands of the nonresistant lower member of the Buru Formation. In the cloudy northern portion of the image the New Guinea Limestone directly overlies the lower member of Buru Formation. This relationship is interpreted as a southward-directed thrust fault. These structural relationships are shown in the north-south cross section of Figure 3-11.

ORBIT PATTERNS

Table 3-1 lists the orbital characteristics of the two generations of Landsat. In order to obtain images of the entire earth, both generations of Landsat are placed in *sun-synchronous orbits*. Figure 3-12 shows the fixed circular orbit of the second generation Landsats 4 and 5 (solid line) and the daylight hemisphere of the earth. Every 24 hours, 14.5 image swaths, shown as patterned strips in Figure 3-12, are generated. Figure 3-13 shows the southbound, daylight portion of the image swaths (185 km wide) for a 24-hour period. The northbound segment of each orbit covers the dark hemisphere. Polar areas at latitudes greater than 81° are the only regions not covered. Every 24 hours the earth's rotation shifts the image swaths westward. After 16 days the earth has been covered by 233 adjacent, sidelapping image swaths and the cycle begins again. This 16-day interval is called

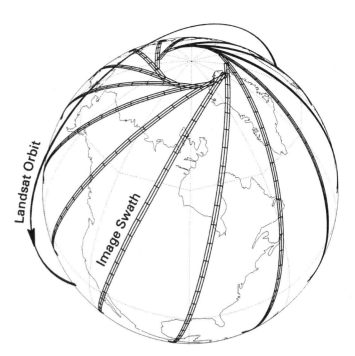

Figure 3-12 Landsat TM sun-synchronous orbit (solid circle) and image swaths (patterned lines) generated in one day. Diagram plotted with Satellite Tool Kit of Analytical Graphics Inc., King of Prussia, Pennsylvania. Courtesy D. N. Boosalis, AGI.

the *repeat cycle*. The sun-synchronous orbit pattern causes the corresponding orbits in each repeat cycle to occur at the same time. For example, every 16 days a southbound second-generation Landsat crosses Los Angeles at approximately 10 a.m. local

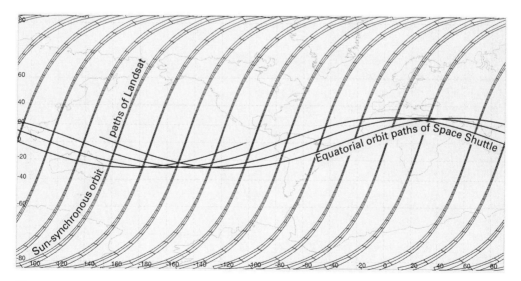

Figure 3-13 Map showing the 14.5 southbound, daytime image swaths (patterned lines) during a single day of Landsat 4 and 5. Each day the earth's rotation shifts the pattern westward; after 16 days the earth is covered and the cycle is repeated. For comparison the solid curves are equatorial orbits of a typical Space Shuttle mission. Diagram plotted with Satellite Tool Kit of Analytical Graphics Inc., King of Prussia, Pennsylvania. Courtesy D. N. Boosalis, AGI.

sun time. The midmorning schedule results in intermediate to low sun elevations, which cause highlights and shadows that enhance subtle topographic features. Sun azimuth is from the southeast for images in the Northern Hemisphere and from the northeast in the Southern Hemisphere.

Since the launch of Landsat 1 in 1972, the repeat cycles have imaged much of the earth many times, which provides several advantages:

1. Areas with persistent clouds may be imaged on the rare cloud-free days.
2. Images may be acquired at the optimum season for interpretation, as described later.
3. The repeated images record changes that have occurred since 1972, such as urbanization, deforestation, and desertification.

PATH-AND-ROW INDEX MAPS

All Landsat images are referenced to worldwide path-and-row index maps that are different for the two generations of Landsat. Figure 3-14 shows a portion of an index map for Landsats 4 and 5. The *paths* are the southbound segments of orbit paths. The paths are intersected by east-west, parallel *rows* spaced at intervals of 165 km. In Figure 3-14 the intersections are marked with circles, which are the centerpoints of the images. A TM image of Los Angeles, shown by the outline on the index map, has a centerpoint at path 41, row 36. All images of Los Angeles acquired on repeat cycles have the same centerpoint. Orbit paths for Landsats 1, 2, and 3 are more widely spaced than those for the second-generation Landsats; therefore a different set of index maps is required. Index maps are available from the EDC and are valuable aids for planning Landsat projects.

For many years Landsat images were assigned a lengthy identification number that included satellite number, image type, days since launch, and time of acquisition. Images are now identified by image type (TM or MSS), date, and path and row, which is a simpler and more useful system.

SELECTING AND ORDERING IMAGES

Landsat TM images and ordering information are available from

EOSAT Corporation
4300 Forbes Boulevard
Lanham, MD 20706
Telephone: 800-344-9933
Fax: 301-552-3762

EOSAT has approximately 2.5 million TM images from around the world and is a primary data source. A user can provide

EOSAT with the location of an area of interest, which may be identified by path and row or by latitude and longitude boundaries. The request should specify the type and format of image desired. The request may also specify a range of acquisition dates and the maximum percentage of acceptable cloud cover. The user will receive a computer printout, arranged by path and row, with the following information for each image: acquisition date, image quality, cloud cover (by quadrant), and latitude and longitude coordinates for the center and corners of the image.

A user may need an image of an area located between consecutive centerpoints on the same path. For example, in Figure 3-14 area B is located on path 41 midway between centerpoints 31 and 32. Rather than purchasing the two adjacent images, the following option is available. On the printout identify images for centerpoints 31 and 32 on path 41 that were acquired on the same date. EOSAT can then shift the northern centerpoint southward by the desired percentage, in this example 50 percent, to cover the area with a single image. Images cannot be shifted in the east-west direction.

Table 3-8 lists the formats and current prices for TM images distributed by EOSAT. The standard full-scene images (185 by 170 km) are available in the Space Oblique Mercator projection, which is described in Chapter 8. TM subscenes (100 by

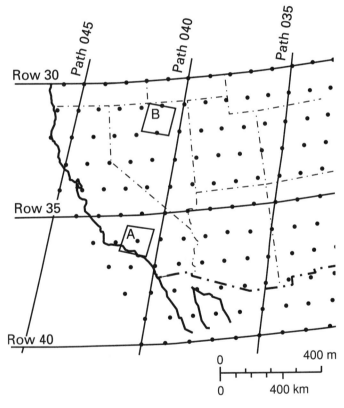

Figure 3-14 Path-and-row index map of the southwestern United States for Landsats 4 and 5. Image A at path 41, row 36 covers Los Angeles. Image B on path 40 between rows 31 and 32 is located with an optional shift of 50 percent to the south.

100 km) are available in the "map-oriented" format, which provides 18 additional map projections listed in the EOSAT *Catalog of Products and Services*. The prices in Table 3-8 are for images in which the original color transparency is created according to customer specifications. EOSAT also has an archive of transparencies for many scenes. Prints from these transparencies may be ordered for the prices listed in Table 3-8 without the charge to generate an original transparency.

TM images acquired prior to September 27, 1985, are distributed by the EROS Data Center, Sioux Falls, SD 57198. EDC also distributes all MSS images. Contact the EDC for information on prices and available images.

The Smithsonian Institution (Strain and Engle, 1992) published an annotated collection of color Landsat and other images that represents all regions of the world. Other annotated collections of MSS images have been published by Short and others (1976), Williams and Carter (1976), Short and Blair (1987), and Slaney (1981). EOSAT periodically publishes *EOSAT Notes,* which provide current information on images and application.

LANDSAT MOSAICS

The broad regional coverage of individual Landsat images can be extended by combining adjacent images into a mosaic. The east-west sidelap of adjacent orbit swaths and the north-south forward overlap of consecutive images greatly facilitate mosaic compilation. The uniform scale and minimal distortion of Landsat images also make mosaic compilation easier; anyone who has ever made mosaics of aerial photographs and attempted to match radially distorted prints will appreciate these two features. The repetition cycles have enabled the two generations of Landsat to acquire essentially cloud-free images of most of the world. Mosaics are classed as analog mosaics or digital mosaics, depending on the method used for compilation.

Analog Mosaic, Central Arabian Arch

Analog mosaics are manually compiled from prints of individual images. Areas of sidelap and overlap between adjacent images are trimmed, and the prints are assembled into the mosaic. Plate 4 is an analog mosaic of the Central Arabian Arch compiled from TM 2-4-7 prints by the Remote Sensing Research Group of Chevron. The following suggestions are useful in compiling mosaics:

1. Select images acquired on the same day along each orbit path.
2. For the entire mosaic, select images acquired during the same repeat cycle to minimize seasonal differences.
3. First assemble the images along each orbit path. Then fit the adjacent orbit paths together and distribute any mismatch along the entire length.

Table 3-8 TM image formats and prices at EOSAT

Format	Scale	Price, $
Color; full scene (185 by 170 km); path oriented*		
Positive and negative transparencies	1:1,000,000	2700
Print†	1:1,000,000	200
Print†	1:500,000	200
Print†	1:250,000	200
Color; subscene (100 by 100 km); map oriented*		
Positive and negative transparencies	1:500,000	2700
Print†	1:500,000	200
Print†	1:250,000	200
Print†	1:100,000	200

* User selects color combination.
† Must purchase color transparency first.

4. In making a large mosaic, begin with the central orbit path and work outward rather than starting at one edge.

Ryder (1981) published a black-and-white analog MSS mosaic of the United States. The images (1:1,000,000 scale) are bound in a folio and annotated with political boundaries, cities, highways, and geographic information. A number of color and black-and-white mosaics of foreign areas and individual states in the United States are available from the EDC.

Digital Mosaic, Western Pakistan

Digital mosaics are compiled by computers from digitally recorded image data. Plate 5 is a digital mosaic of southwestern Pakistan compiled from TM 2-3-4 digital data by the EDC. Compilation of digital mosaics is described in Chapter 8.

REGIONAL INTERPRETATION, SAUDI ARABIA

Regional interpretation projects are conducted at small scales and cover large areas of tens or hundreds of thousands of square kilometers. Landsat images and mosaics are well suited to these projects. The mosaic of the Central Arabian Arch (Plate 4) in the interior of Saudi Arabia was used in a regional geologic mapping project. The seven individual images of the mosaic were interpreted at a scale of 1:250,000; these maps were generalized and reduced to a scale of 1:3,000,000 to produce the map in Figure 3-15.

Figure 3-16 is a stratigraphic column of the 19 rock units that crop out in the Central Arabian Arch. For the regional map these formations were condensed into the seven map units

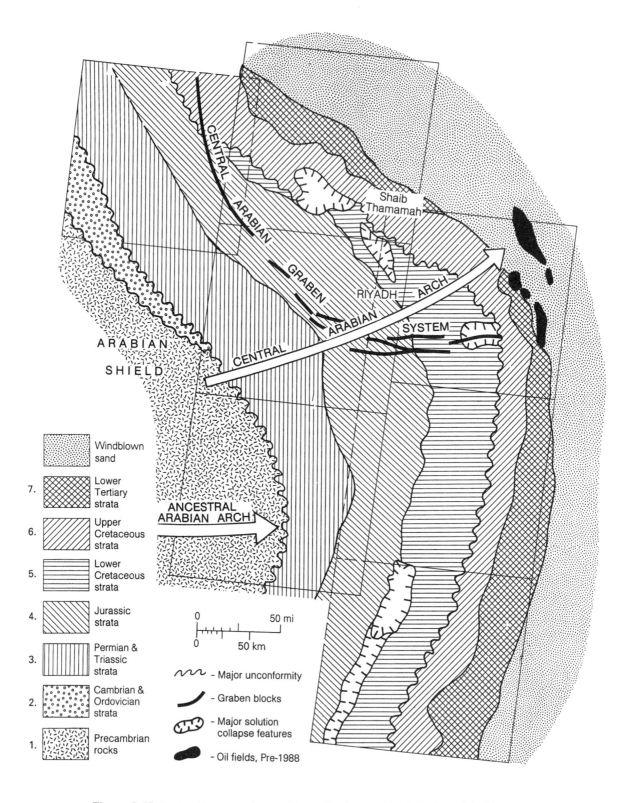

Figure 3-15 Regional interpretation map from a Landsat mosaic of the Central Arabian Arch.

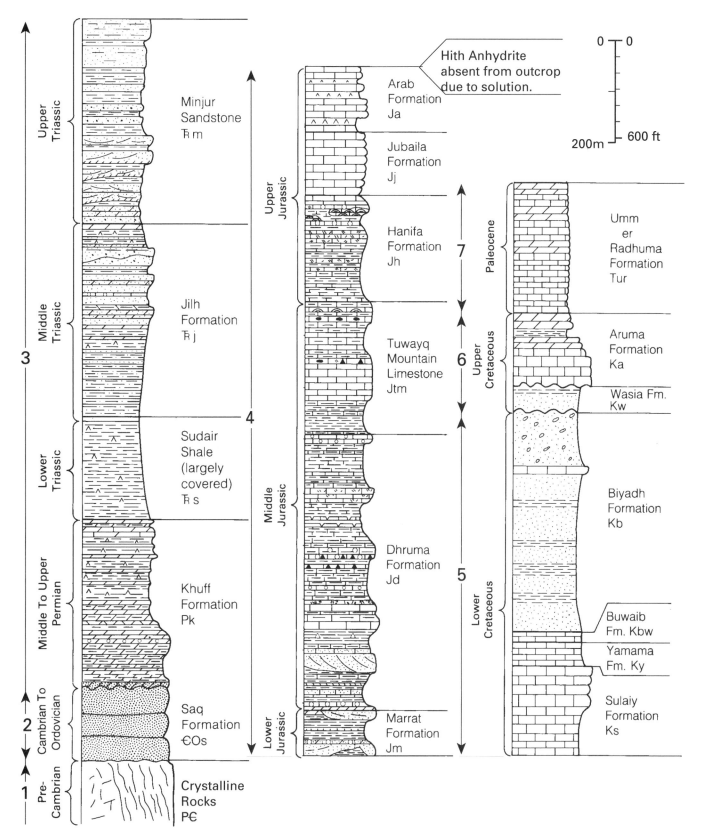

Figure 3-16 Stratigraphic column of outcrops in the Central Arabian Arch. Aggregate thickness is 3800 m. Numbers indicate mapping units used in regional mapping. Individual formations are used in detailed mapping. Compiled from Powers and others (1966).

identified by numbers in Figure 3-16. The oldest unit is the Precambrian crystalline rocks (Unit 1) of the Arabian shield that form the broad platform upon which the sedimentary units were deposited. The crystalline rocks have a dark signature on the mosaic because they are heavily coated with desert varnish. Unit 2 consists of Lower Paleozoic sandstones that crop out in the northwest part of the area and also have a dark signature. The stratified appearance of the Unit 2 sandstones distinguishes them from the underlying crystalline rocks. Unit 3 consists of a basal carbonate unit, the Khuff Formation (of Permian age), overlain by sandstones and shales of Triassic age. On the mosaic the Khuff Formation forms a prominent dipslope with a distinctive bluish purple signature. The shale intervals weather to broad strike valleys covered by windblown sand (yellow) or alluvium (light to dark blue). The Unit 3 sandstones weather to strike ridges with reddish brown signatures. Unit 4 consists largely of carbonate rocks (of Jurassic age) that form a sequence of dipslopes and antidip scarps with various shades of blue and red. Unit 5 (of Lower Cretaceous age) consists of basal carbonate strata (medium blue) overlain by sandstones and shale. The resistant sandstones form ridges with brown signatures due to desert varnish. The shales and nonresistant sandstones form valleys with light-colored signatures. Following deposition of Unit 5 the northern portion of the region was uplifted and eroded. Unit 6 was deposited upon the erosion surface. Northwest from the Central Arabian Arch (Figure 3-15) the base of Unit 6 truncates successively older beds of Unit 5 until Unit 6 rests directly upon Unit 4. This unconformable relationship is clearly shown in the mosaic.

Regional structural features, such as the Ancestral Arabian Arch, Central Arabian Arch, and Central Arabian Graben System, are interpreted on the mosaic. The regional maps provide the basis for more detailed interpretations of key areas at larger scales.

DETAILED IMAGE INTERPRETATION, SAHARAN ATLAS MOUNTAINS, ALGERIA

Subareas of TM images are readily enlarged at scales up to 1:150,000 for detailed interpretations. Plate 6 is a TM 2-4-7 subscene in the Saharan Atlas Mountains, Algeria, where lithologic units and geologic structure are well exposed and suitable for demonstrating interpretation techniques. The two major steps in geologic interpretation are to

1. define and map lithologic units, and
2. map geologic structure.

These steps are described in the following sections.

Mapping Lithology

Some areas are covered by geologic maps and reports that can be used to define lithologic units on images. No information

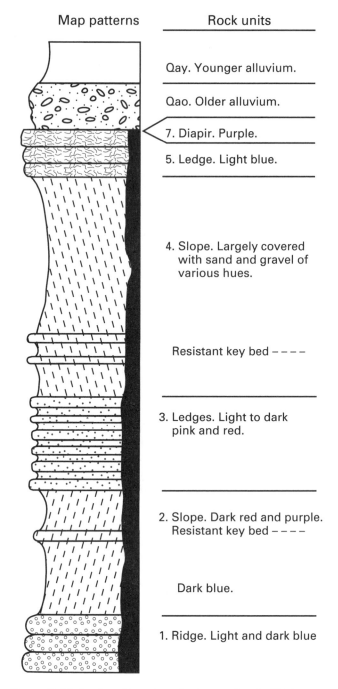

Figure 3-17 Stratigraphic column for interpreting the TM image in the Saharan Atlas Mountains, Algeria.

was available for the Saharan Atlas Mountains when this image was interpreted; therefore units were defined from the image. The first step is to identify the oldest and youngest units, which is straightforward in this well-exposed area of simple folds. Older strata crop out in the eroded crests of anticlines. Younger strata are preserved in the troughs of synclines. Figure 3-17 is a stratigraphic column showing the mapping units that were defined from the image. The profile on the left margin of the column shows topographic expression (ridges or slopes) of the units, which are identified by numbers starting with 1 for the oldest. Image signatures are given for each unit. Patterns in the column correspond to patterns in the geologic map (Figure 3-18). Unit 7 (shown in black in the column) consists of salt from a deeper layer that has penetrated upward through the strata as plugs, or diapirs, as shown in the cross section (Figure 3-19).

A large irregular area at the eastern margin of the image with a light-blue signature is older alluvium (Qoa), with a patch of Unit 5 in the center. Qoa is distinguished because it is less resistant to erosion than Unit 5. The light-blue signature of Qoa distinguishes it from the dark-blue signature of the underlying Unit 4. In the western portion of the image much of Unit 4 is covered by a thin layer of windblown sand with a yellow-orange signature that is omitted in the map (Figure 3-18).

Mapping Geologic Structure

After the lithologic units are mapped, the next step is to map geologic structure.

Attitude of Beds (Dip and Strike) from Highlights and Shadows
The first step in mapping structure is to determine attitude of the beds (dip and strike). This task is simplified with stereo images, such as the aerial photographs of the Alkali anticline in Chapter 2. TM images, however, essentially lack stereo capability and must be interpreted monoscopically. Interpretation of attitudes is facilitated by highlights and shadows caused by the low to moderate sun elevation of many TM images. This capability is demonstrated in Figure 3-20, which shows block diagrams of dipping beds. A *dipslope* is the broad, relatively gentle slope formed by the exposed surface of a dipping bed that is resistant to erosion. An *antidip scarp* is a narrow, steep escarpment formed by the eroded margin of a dipping bed. The trend of the antidip scarp shows the strike of the beds, as indicated by the long bar of the dip-and-strike symbols in Figure 3-20. The inclination of the dipslope shows the direction of dip, as indicated by the short bar of the dip-and-strike symbols. This combination of a broad dipslope with a narrow antidip scarp occurs where beds dip less than 45°, which is the common occurrence.

In Figure 3-20A the beds dip away from the sun. The dipslope is shadowed and has a broad, dark signature on an image. The antidip scarp is illuminated and has a narrow, bright signature, called a highlight, on an image. In Figure 3-20B the beds dip toward the sun. The dipslope is illuminated and has a broad highlight. The antidip scarp is shadowed and has a narrow, dark signature. These differences in width of highlights and shadows are the key to interpreting dip and strike. TM images are acquired at midmorning times. The Saharan Atlas Mountains of Algeria are located in the Northern Hemisphere; therefore the midmorning sun shines toward the northwest. This fact plus the

relationships in Figure 3-20 enable us to interpret dip and strike. In the image (Plate 6) we know that combinations of bright dipslopes and dark antidip scarps indicate beds dipping toward the south and east. Dark dipslopes and bright antidip scarps indicate beds dipping toward the north and west. For example, Locality X in Figure 3-18 is a northeast-striking syncline. The southeast limb has a narrow, bright highlight facing the sun and a parallel broad shadow; therefore these beds dip northwest. The northwest limb has a narrow, dark shadow facing away from the sun and a parallel broad shadow; therefore these beds dip southeast. These relationships between illumination and attitude are also shown in the limbs of the anticlines in the image. All the dip-and-strike symbols in Figure 3-18 were interpreted from highlights and shadows.

It is difficult to estimate the amount (vertical angle) of dip from nonstereo images. Experienced interpreters, however, can estimate categories of dip amount such as low, gentle, moderate, and steep. These criteria for shadows and highlights are applicable to any image acquired with low to intermediate illumination angles such as SPOT and radar images.

Folds and Faults
The next step is to interpret fold axes based on outcrop patterns and attitudes of beds. Anticlines have older beds in the center surrounded by successively younger beds that dip away from the center; the pattern is reversed for synclines. Faults are recognized by offsets, truncations, or repetitions of units. The large anticline in the center of Figure 3-18 is cut by a small fault that causes a minor offset of beds.

The salt diapirs shown diagrammatically in Figure 3-17 have purple signatures in the image (Plate 6). Three salt diapirs occur on the crests of anticlines where they cut across the adjacent beds. The cross section (Figure 3-19) shows a diapir along a fault that cuts an anticline.

The final step is to construct a cross section (Figure 3-19) that shows the subsurface view of the structure.

Summary
The Saharan Atlas project demonstrates the following sequence of interpretation steps:

1. Establish a sequence of mappable rock units. If published information is lacking, the sequence may be established directly from the image.
2. Determine attitudes of beds. Highlights and shadows associated with dipslopes and antidip scarps are vital clues.
3. Interpret folds and faults. Outcrop patterns and attitudes of beds are keys.
4. Prepare a cross section to accompany the interpretation map.
5. Check the interpretation in the field.

This sequence of interpretation steps is applicable to images of all types in addition to Landsat. Inexperienced interpreters often try to identify folds and faults before compiling the basic data, which leads to problems.

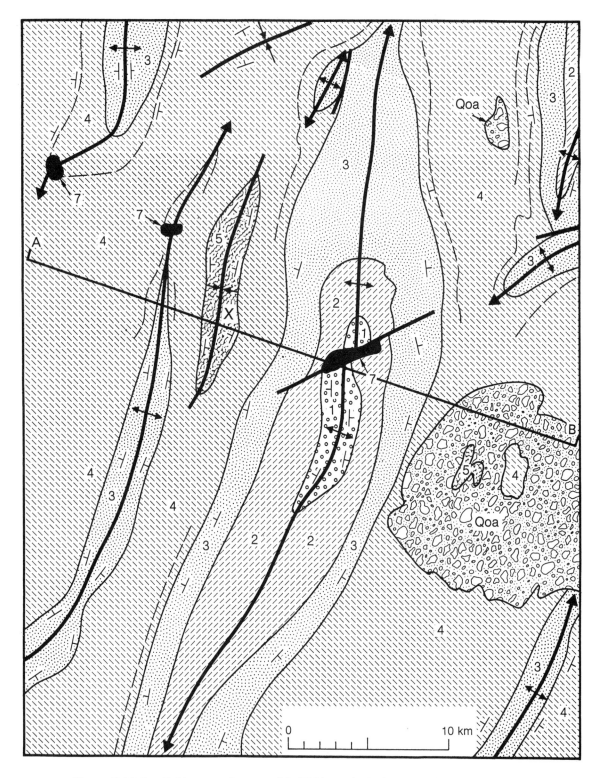

Figure 3-18 Detailed interpretation map of the TM image in the Saharan Atlas Mountains, Algeria. Numbers are keyed to Figure 3-17.

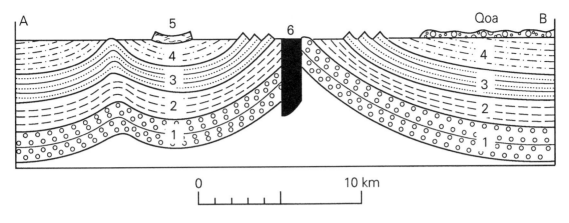

Figure 3-19 Cross section of the TM image in the Saharan Atlas Mountains, Algeria.

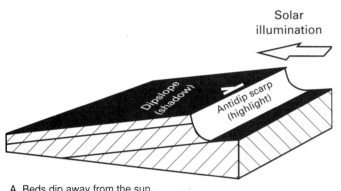

A. Beds dip away from the sun.

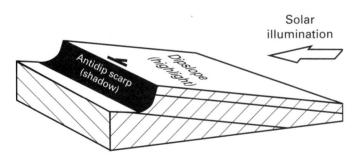

B. Beds dip toward the sun.

Figure 3-20 Dip and strike interpreted from highlights and shadows, dipslopes, and antidip scarps. Solar illumination is from the east (right).

SEASONAL INFLUENCE ON IMAGES

The repetitive coverage by Landsat can provide images that were acquired at the optimum season for interpretation. The optimum season depends on the terrain and climatic conditions and is different for various regions. In regions with seasonal rainfall patterns, images acquired in the wet season are markedly different from those acquired in the dry season. At high latitudes, there are major differences between summer and winter images.

Wet-Season and Dry-Season Images, South Africa

Plate 7A,B shows two Landsat MSS IR color images of the south flank of the Transvaal Basin; one image was acquired during the winter dry season and the other in the summer rainfall season. The area is a grass-covered plateau with ridges of resistant rock units.

Geologic Setting Grootenboer (1973) interpreted the geologic map of Figure 3-21 from the rainy-season Landsat image (Plate 7B). The northern two-thirds of the image cover the southwest flank of the Transvaal Basin where the strata dip gently north. A northeast-trending anticline separates the Transvaal Basin from the more complex Potchefstroom Basin in the southeast corner of the image. The oldest rocks in the area are granites of Archean age exposed along the axis of the anticline. The granites are overlain by the Witwatersrand quartzite and shale and the Ventersdorp andesites and sediments. The overlying Transvaal System consists of a thin basal clastic unit (Black Reef Series) overlain by massive dolomitic limestone (Dolomite Series), which is overlain by alternating quartzites and shales with some volcanic layers (Pretoria Series). Intrusive mafic sills are abundant near the top of the Pretoria Series.

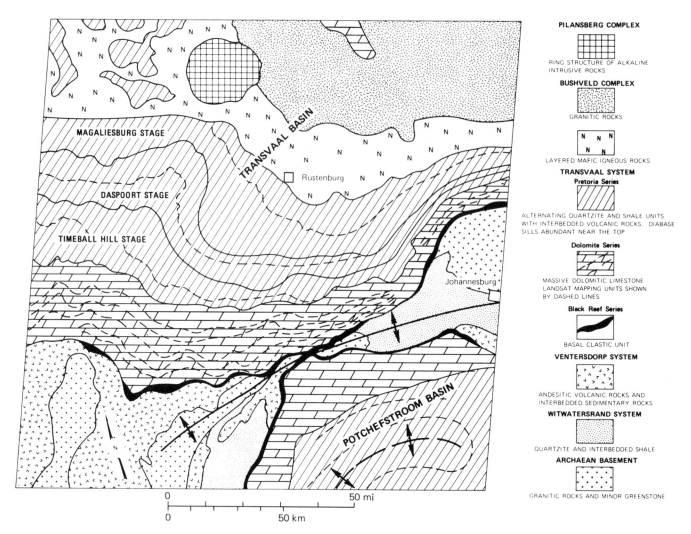

Figure 3-21 Interpretation map of an MSS image (Plate 7B) acquired during rainy season, showing south flank of the Transvaal Basin, South Africa. From Grootenboer (1973, Figure 1).

The Bushveld igneous complex occurs in the northern part of the image. The layered sequence at the base consists of mafic igneous rocks and is overlain by the Bushveld Granite. At the north edge of the image, the Bushveld Complex is intruded by the Pilansberg Complex, a circular structure of silicic intrusive rocks. Most of the area is covered by residual soil that conceals much of the bedrock. The only significant outcrops are the Pilansberg Complex, quartzites of the Witwatersrand and Transvaal Systems, and scattered exposures of the Bushveld Complex.

Comparison of Seasonal Images When the dry-season image (Plate 7A) was acquired, the area was covered with dry, brown grass, the indigenous vegetation was leafless, and the cornfields were fallow. Black patches on the image mark areas of recent burning. Slight tonal variations enable recognition of

the major stratigraphic units to a degree comparable to that on 1:1,000,000 geologic maps published prior to Landsat (Grootenboer, 1973). The wet-season image (Plate 7B) was acquired at the height of the summer rainy season, when the perennial vegetation was in full leaf. The strong color variations are directly related to bedrock lithology, particularly in the area underlain by the Transvaal System and the Bushveld Complex (Figure 3-21). Of particular interest are the seven zones of tonal variations within the outcrop of the Dolomite Series. Field checks by Grootenboer, Eriksson, and Truswell (1973) established that the four darker zones correspond to dark, chert-free dolomite and the three lighter zones to light-toned dolomite with abundant chert. During the previous 90 years of geologic investigation in the area, no such stratigraphic subdivisions had been recognized in the Dolomite Series.

Several factors contribute to the superiority of the wet-season image:

1. Windblown dust causes atmospheric haze, which severely scatters light in the dry season. During the wet season, rainfall removes dust from the air, producing a clearer atmosphere and good image contrast. The rain also washes away the surface dust layer from the outcrops.
2. Greater soil moisture enhances tonal and color differences between rock types.
3. Vegetation grows preferentially on belts of soil with higher moisture. The red stripes of vegetation in the wet-season image help delineate geologic trends.

These advantages of wet-season images have also been observed in other areas.

Winter and Summer Images, Arctic Canada

At high and moderate latitudes, winter and summer images differ in sun elevation and in snow cover. The advantages of aerial photographs acquired at low sun elevation are demonstrated in Chapter 2 and also apply to Landsat images. Figure 3-22 shows MSS band 4 (reflected IR) images of Bathurst Inlet in Arctic Canada that were acquired in the summer and late winter. On the summer image (Figure 3-22A), there is no snow; vegetation growth is vigorous, as shown by the bright signatures, and most of the small lakes have thawed, as shown by their dark signatures. Few geologic features are recognizable. Most of the lakes are only a few hundred meters in size and tend to obscure the geologic features that are thousands of meters in size. The snow cover of the winter image (Figure 3-22B) conceals the frozen lakes, thereby enhancing the appearance of geologic structures.

The relatively high sun angle (45°) of the summer image causes only minimal highlights and shadows. The low sun angle (27°) of the winter image, however, causes highlights and shadows that emphasize subtle topographic features expressing strike ridges of sedimentary rocks, folds, lineaments, faults, and igneous dikes. The geologic map (Figure 3-23) shows these features, which were interpreted from the winter image.

A major fold, outlined by strike ridges of sedimentary rocks, is surrounded by highly fractured, unstratified crystalline basement rocks. The fold is a syncline with younger strata preserved in the center. The strata are argillites, sandstones, and quartzites of Proterozoic age that overlie older crystalline rocks of Archean age. A lineae feature trending northwest across the northern part of the image is a major fault. North of the fault is a small anticline of Proterozoic strata. Also north of the fault at the east margin of the map and images is a prominent ridge that is an igneous dike. At the western boundary of the map, a distinct circular drainage anomaly 8 km in diameter marks a ring dike or an igneous plug. These geologic features are mappable only from the winter image.

This Arctic example demonstrates the advantages of snow cover and low sun angle for geologic mapping. These conditions also occur in winter images of areas at intermediate latitudes.

LINEAMENTS

In the early 1900s the American geologist William H. Hobbs (1904, 1912) recognized the existence and significance of linear geomorphic features that are the surface expression of zones of weakness or structural displacement in the crust of the earth. Hobbs defined lineaments as "the significant lines of landscape which reveal the hidden architecture of the rock basement. . . . They are character lines of the earth's physiognomy" (1912, p. 227). Over the years, additional terms have been misused as synonyms for lineament, and the resulting confusion has tended to obscure the geologic significance of lineaments. O'Leary, Friedman, and Pohn (1976) reviewed the origin and usage of the terms *linear, lineation,* and *lineament.* Their work and definitions are the basis for the following discussion.

Linear is an adjective that describes the linelike character of an object or an array of objects. Some geologists have misused the term as a noun substitute for *lineament,* which is grammatically incorrect. *Linear* is an indispensable descriptive word (linear valleys, linear escarpment, and so forth) and should be used in this sense.

Lineation is the one-dimensional fabric of internal components of a rock, such as the parallel orientation of elongate crystals in a metamorphic rock. Some geologists have used this petrographic term as a synonym for *lineament,* but this is incorrect and unacceptable.

A *lineament* is a mappable linear or curvilinear feature of a surface whose parts align in a straight or slightly curving relationship that may be the expression of a fault or other line of weakness. The surface features making up a lineament may be geomorphic (caused by relief) or tonal (caused by contrast differences). Straight stream valleys and aligned segments of valleys are typical geomorphic expressions of lineaments. A tonal lineament may be a straight boundary between areas of contrasting tone or a stripe against a background of contrasting tone. Differences in vegetation, moisture content, and soil or rock composition account for most tonal contrasts.

Although many lineaments are controlled at least in part by faults, structural displacement (faulting) is not a requirement in the definition of a lineament. On Landsat images of Precambrian shield areas, for example, long topographic lineaments are clearly defined but do not offset lithologic contacts. These and similar lineaments throughout the world are thought to represent zones of weakness in the crust. Although displacement has not occurred, the rocks may be more highly fractured and susceptible to erosion. Lineaments that coincide with lines of structural offset are called *faults.* On the Landsat image and

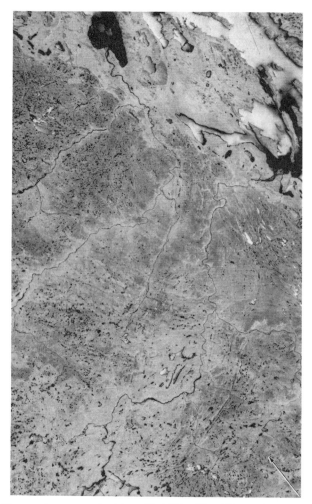 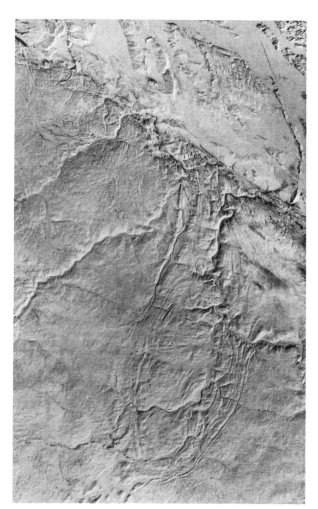

A. Summer image acquired June 18, 1973, with a 45° sun elevation.   B. Winter image acquired April 2, 1974, with a 27° sun elevation.

Figure 3-22 Landsat MSS band 4 seasonal images of Bathurst Inlet, Arctic Canada.

map of the Los Angeles region (Figures 3-2 and 3-3), the San Andreas, San Gabriel, and Garlock faults are expressed as lineaments but are called faults because they are known to be zones of structural displacement. Linear features interpreted on images are initially called lineaments; if field checking establishes the presence of structural offset, they are then designated as faults. This procedure is illustrated in the example of the Peninsular Ranges, California, later in this chapter.

There is no minimum length for lineaments, but significant crustal features are typically measured in tens or hundreds of kilometers. Lineaments are also recognized on geophysical maps by aligned highs and lows, steep contour gradients, and linear offsets of trends. Lineaments are well expressed on Landsat images because of the low sun angle, the suppression of distracting spatial details, and the regional coverage.

Analysis of Landsat Images

Scratches and other image defects may be mistaken for natural features but are identified by determining whether the questionable features appear on more than a single band of imagery. Shadows of aircraft contrails may be mistaken for tonal lineaments but are recognized by checking for the parallel white image of the contrail. Many questionable features are explained by examining several images acquired at different dates. With experience an interpreter learns to recognize linear features of cultural origin, such as roads and field boundaries.

The recommended procedure is to plot lineaments as dotted lines on the interpretation map. Field checking and reference to existing maps will identify some lineaments as faults; for these the dots are connected by solid lines on the interpretation

Figure 3-23 Interpretation map of MSS image acquired in winter, Bathurst Inlet, Arctic Canada. Dashed lines are resistant strata; other symbols are standard. Reproduced with permission from A. F. Gregory and H. D. Moore, Recent advances in geologic applications of remote sensing from space, copyright 1976, Pergamon Press, Ltd.

map. The remaining dotted lines may represent (1) previously unrecognized faults, (2) zones of fracturing with no displacement, or (3) lineaments unrelated to geologic structure.

Peninsular Ranges, California

The structure of California south of the Transverse Ranges is dominated by the southwest-trending San Andreas, San Jacinto, and Elsinore strike-slip faults. On Landsat images, however, Sabins (1973) noted prominent lineaments trending north and northeast that did not correspond to mapped faults. Lamar and Merifield (1975) interpreted the MSS band 4 image of Figure 3-24 to produce the lineament map of Figure 3-25. They then checked and evaluated the map in the field, which is a difficult task in the rugged Peninsular Ranges with limited

access and dense cover of brush. Bedrock consists of late Mesozoic plutonic rocks and roof pendants of Paleozoic and Mesozoic metamorphic rocks. Breccia and gouge zones were the main criteria for recognizing faults in the field. Displaced or terminated lithologic contacts were also used, but these features are scarce in this region. Based on fieldwork, the lineaments were assigned to three categories:

1. lineaments that correlate with previously mapped faults (two lineaments);
2. lineaments that correlate with faults that previously were unmapped or were of controversial origin (seven);
3. lineaments that do not correlate with faults or have an unknown origin (seven).

Figure 3-25 shows the lineament categories with different symbols. Two lineaments that correlate with previously mapped faults (category 1 above) require no additional explanation. Table 3-9 explains representative examples of the other two categories. The seven lineaments in category 2 are now classified as faults; characteristics of four of these are summarized in the table. The Witch Creek and Pamo Valley lineaments are examples of category 3 lineaments for which no evidence of faulting was found in the field. They are controlled by jointing and foliation directions in the bedrock or by aligned stream segments of diverse origin. Significantly, 9 of the 16 Landsat lineaments correlate with faults that were known prior to Landsat or with faults recognized on the basis of Landsat interpretation. The northwest-trending Elsinore fault (Figure 3-25) is an active fault that has moved during the past 10,000 years and caused earthquakes. The fresh appearance of the fault is evidence for the recent movement. The north- and northeast-trending lineaments and faults, however, are deeply eroded and lack recent offsets, indicating that they are not active faults. Additional criteria for recognizing active faults are given in Chapter 13.

PLATE-TECTONIC INTERPRETATIONS

The geologic discipline of plate tectonics deals with the dynamics of the lithologic plates that move over the earth. Of particular interest are the boundaries where plates diverge (spreading centers), converge (subduction zones), or shift laterally (transform faults). Triple junctions are the complex boundaries between three plates. Landsat images are ideal for interpreting plate boundaries that are exposed on the continents. Representative Landsat projects are summarized in the following sections.

Red Sea Spreading Center

The Red Sea is an active spreading center between the Arabian and Nubian plates. Earlier investigators have proposed seven

Figure 3-24 Landsat MSS band 4 image of the Peninsular Ranges, southern California. From Lamar and Merifield (1975, Figure 3). Courtesy P. M. Merifield, UCLA.

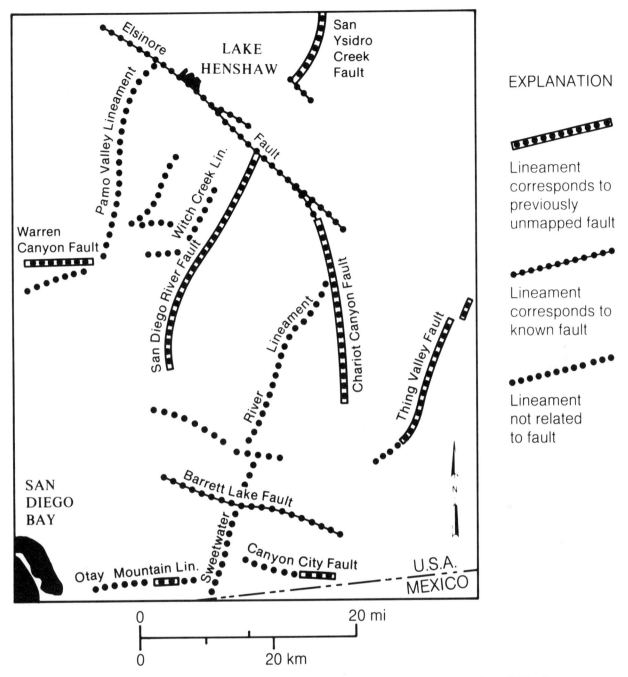

Figure 3-25 Lineaments interpreted from Landsat image of the Peninsular Ranges, southern California. From Lamar and Merifield (1975, Figure 2).

Table 3-9 Field evaluation of representative lineaments on Landsat MSS image of the Peninsular Ranges, California

Lineament	Geologic setting	Description	Remarks
Lineaments that correlate with previously unmapped or controversial faults			
San Ysidro Creek fault	Fault forms contact between granite and schist.	Fault is a zone of horizontal movement with gouge zone up to 7 m wide.	Alignment with San Diego River fault may be fortuitous.
San Diego River fault zone	Fault transects schist and plutonic rocks.	Fault is a zone of gouge and breccia, with up to 630 m of right-lateral separation.	Fault zone controls straight course of the San Diego River.
Chariot Canyon fault	Fault forms contact between schist and granitic rocks.	Fault is a broad shear zone with slickensides.	Fault is discontinuous; requires additional field mapping.
Thing Valley fault	Granite bedrock with inclusions of schist.	Fault is over 20 km long with gouge and breccia zone up to 35 m wide.	Right-lateral separation ranges from 100 m to 1 km.
Lineaments that do not correlate with faults, or have unknown origins			
Witch Creek lineament	Diorite and schist bedrock.	Aligned, straight canyons parallel with San Diego River fault zone.	Lineament may correlate with fault, but evidence is lacking.
Pamo Valley lineament	Granitic rocks.	Discontinuous, aligned straight stream segments.	No fault control.

Source: From Lamar and Merifield (1975).

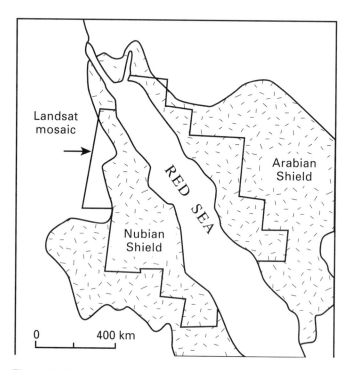

Figure 3-26 Location map showing outline of the Red Sea. From Sultan and others (1993, Figure 1).

different reconstructions to explain the divergent plate movements responsible for opening the Red Sea. These reconstructions are based on matching the coastline and geologic features on opposite sides of the Red Sea. Sultan and others (1993, Figure 2) prepared a mosaic of 23 TM images that covers the regions on both sides of the Red Sea. They used the mosaic and other data to identify and map lithologic units, mobile belts, and terrains within the Arabian and Nubian shields. Figure 3-26 shows the present configuration of the Red Sea. Figure 3-27 shows the reconstruction prior to rifting. The numbers identify features on opposite sides of the rift that were matched during the reconstruction. Sultan and others reported the following implications of their reconstruction:

1. Only a small amount of continental crust underlies the Red Sea because the restored coastlines are typically juxtaposed.
2. Only a single pole of rotation is needed, which implies that the shields were rigid plates during the rifting.
3. The reoriented coastlines are aligned with preexisting structures, which implies that rifting occurred in part along older zones of weakness.
4. Arabia has rotated 6.7° relative to Africa around a pole at 34.6°N, 18.1°E.

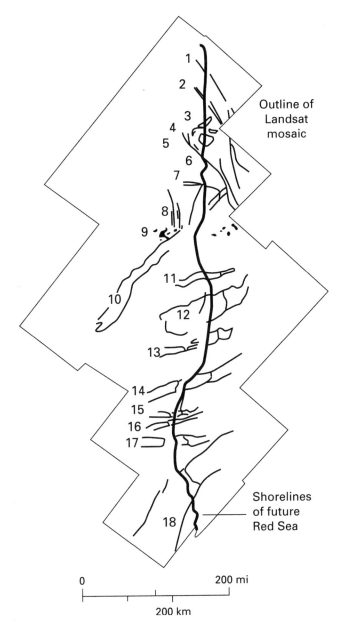

0   200 mi

200 km

Figure 3-27 Reconstruction of the Red Sea rift based on a Landsat mosaic. Numbers identify matching features in the Arabian and Nubian shields. From Sultan and others (1993, Figure 2A)

Some earlier reconstructions employed several hundred kilometers of left-lateral displacement along the Red Sea. This movement is not required by the reconstruction based on Landsat images (Figure 3-27).

Subduction Zone between the Indian and Eurasian Plates

The northward-drifting Indian plate first collided with the Eurasian plate 40 to 60 million years ago. Subsequently India

moved northward another 2000 km toward Eurasia. A major tectonic problem is to account for the vast land area that was displaced by convergence of the plates. The problem is complicated by three factors:

1. The geology of Eurasia is complex and appears to present a chaotic jumble of landforms.
2. The geology is poorly known.
3. The region has been inaccessible for conventional field mapping by outside investigators.

Molnar and Tapponnier (1977) were able to view Eurasia as a whole on Landsat images, which enabled them to recognize a simple, coherent geologic pattern attributed to the collision. Figure 3-28 shows their interpretation of structural information, much of which was previously unavailable or unknown. The large solid arrow indicates the northward movement of the Indian plate, which is shown in a stippled pattern. The authors recognized major strike-slip faults in the Eurasian plate, and for some of the faults the sense of displacement was evident. The Landsat interpretation was combined with earthquake data to determine the sense of motion of the principal faults. This analysis indicates that west-trending faults with left-lateral displacement, such as the Bolnai, Altyn Tagh, and Kang Ting faults, are predominant. Movement along the faults has displaced China eastward out of the way of India, which is reasonable because material displaced westward would encounter the resistance of the Eurasian landmass. Eastward motion (shown by large open arrows in Figure 3-28), however, is easily accommodated by the thrusting of China over the oceanic plates along the margins of the Pacific.

Chaman Transform Fault

The Chaman fault zone in eastern Afghanistan is a left-lateral, strike-slip fault zone that separates the Eurasian plate on the west from the Indian plate on the east. The Indian plate is moving northward relative to the Eurasian plate, and the Chaman fault forms a transform boundary between the plates. Plate 5 is a mosaic of TM 2-3-4 images that covers 800 km of the transform fault and associated geologic structures. Figure 3-28 shows the location of the mosaic (rectangle) and the relationship of the Chaman fault to the regional plate-tectonic features. In the area north of the mosaic (Plate 5) the transform fault merges with the east-striking Himalaya Frontal Thrust, which marks the convergence between the Indian and Eurasian plates. The broad floodplain of the Indus River is heavily vegetated and has a red signature on the mosaic. Most of the region is unvegetated, and the geology is exposed in spectacular fashion. Bannert (1992) has interpreted Landsat images (1:250,000 scale) to produce remarkably detailed maps (1:500,000 scale) of this region, which includes some of the world's most complex geology. Figure 3-29 is a generalized version of Bannert's maps that matches the small scale of the mosaic.

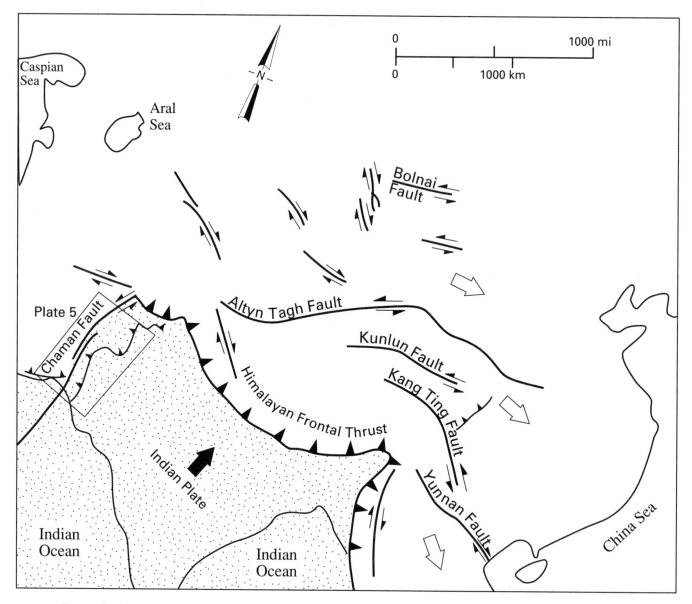

Figure 3-28 Map of the China portion of the Eurasian plate showing structures caused by collision with the Indian plate. From Molnar and Tapponnier (1977, Figure 3).

The Chaman and Ghazaband faults are the major components of the transform zone. The faults are expressed on the mosaic as lineaments along which the strata and geologic structures have been displaced for great distances. The northeast and southwest portions of the mosaic are underlain by belts of folds and thrust faults with an arcuate pattern toward the south. In addition to the regional relationships shown on the small-scale mosaic, local features are recognizable on individual Landsat images at larger scales (Bannert, 1992).

Triple Junction of Arabian, Nubian, and Somalian Plates

The Afar region, or Danakil depression, of northern Ethiopia is a triple junction at the intersection of the spreading axes of the Red Sea, Gulf of Aden, and Ethiopian rifts. The Arabian, Nubian, and Somalian plates are separating along these spreading axes. The region is inhospitable and inaccessible, but excellent Landsat images are available. Kronberg and others (1975,

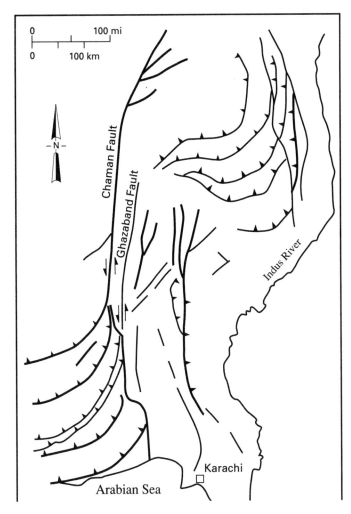

Figure 3-29 Map of the Chaman transform fault between the Indian and Eurasian plates interpreted from a Landsat mosaic. From Bannert (1992, Figure 2).

Figure 2) interpreted a mosaic of Landsat images that covers 600,000 km^2 of the region.

Figure 3-30 is an MSS image of the triple junction; Figure 3-31 is an interpretation map of the image. Lake Abbe marks the present position of the triple junction, which has migrated through time. The Gulf of Tadjura is the western end of the Gulf of Aden. Light-toned areas are recent deposits of clay, silt, and evaporite in the depressions. Basalt (darker tone) and andesite (lighter tone) of Tertiary age form the outcrops. Mount Damahali is a recent volcano with associated lava flows. Geologic structure is dominated by normal faults that form the boundaries of numerous horsts and grabens. The faults are accentuated by shadows on the image that indicate the topographic relief and the sense of throw. In the southwest part of the image, the north-trending faults belong to the Ethiopian rift system. The northwest-trending faults are paral-

lel with the Red Sea spreading axis. Prominent west-trending faults in the southeast are parallel with the Gulf of Aden spreading axis. Except for an apparent strike-slip fault northwest of Lake Asal, there is no suggestion of transform faulting. The sinuous Gawa graben in the north part of the image is a departure from the regional pattern.

COMMENTS

The Landsat program produced the following major advances in remote sensing:

1. Cloud-free images are available for most of the world with no political or security restrictions.
2. The low to intermediate sun angle enhances many subtle geologic features.
3. Long-term repetitive coverage provides images at different seasons and illumination conditions.
4. Color images are available for many of the scenes. With suitable equipment, color composites may be made for any image.
5. Synoptic coverage of each scene under uniform illumination aids recognition of major features. Mosaics extend this coverage.
6. There is negligible image distortion.
7. Images are available in digital format suitable for computer processing.
8. TM images provide regional coverage (170 by 185 km), good spatial resolution (30 m), and seven spectral bands in the visible, reflected IR, and thermal IR regions.
9. Any three of the six visible and reflected IR bands can be selected and combined to produce color images that are optimum for the climate, terrain, and objectives of the interpretation project.

In addition to the applications shown in this chapter, Landsat images are valuable for environmental monitoring, resource exploration, land-use analysis, and evaluating natural hazards (as illustrated in Chapters 9 to 13).

Another major contribution of Landsat is the impetus it has given to digital image processing (described in Chapter 8). The availability of low-cost multispectral image data in digital form has encouraged the application and development of computer methods for image processing, which are increasing the usefulness of the data for interpreters in many disciplines.

Since the first launch in 1972, Landsat has evolved from an experiment into an operational system. There has been a steady improvement in the quality and utility of the image data. Many users throughout the world now rely on Landsat images as routinely as they do on weather and communication satellites.

In early 1996 Landsat 5 was functioning normally after ten years in orbit. The intended replacement, Landsat 6, failed to

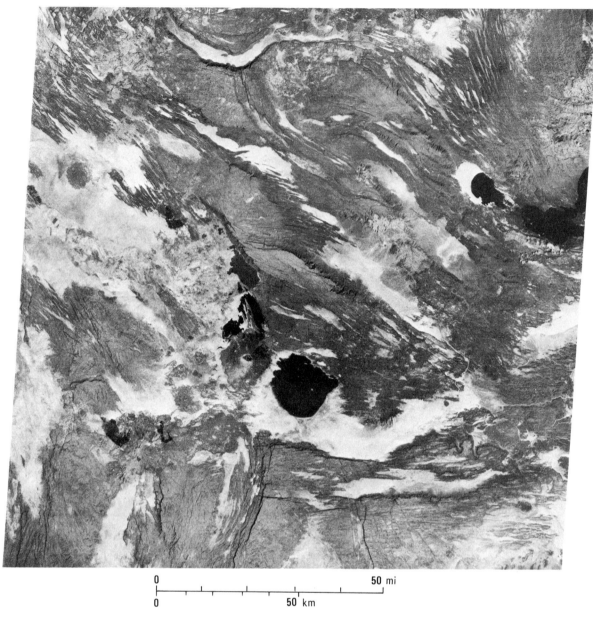

0 ——————————————— 50 mi

0 ——————————————— 50 km

Figure 3-30 MSS band 3 (reflected IR) image of the Lake Abbe region in the Afar triangle, showing features of triple junction.

reach its orbit. Current plans call for Landsat 7 to be launched in 1998, but plans are subject to budget constraints.

QUESTIONS

1. You plan to use TM images to map a strip 150 km wide spanning the border between the United States and Mexico from the Pacific Ocean to Arizona. Use Figure 3-14 to select the minimum number of images for the project. The

centerpoint of each image should be on, or very near, the border. For each path assume that coverage is available to shift any centerpoint southward. For each path select one image. P35 R38 × 30%, P36 R_____, P37 R_____, P38 R_____, P39 R_____, P40 R_____.

2. Assume that your eyes have normal spatial resolution (see Chapter 1) and that you are an astronaut traveling on the Landsat 1 satellite. What would be the dimension of the ground resolution cell you observe? How does the resolving power of your eye compare with that of MSS images?

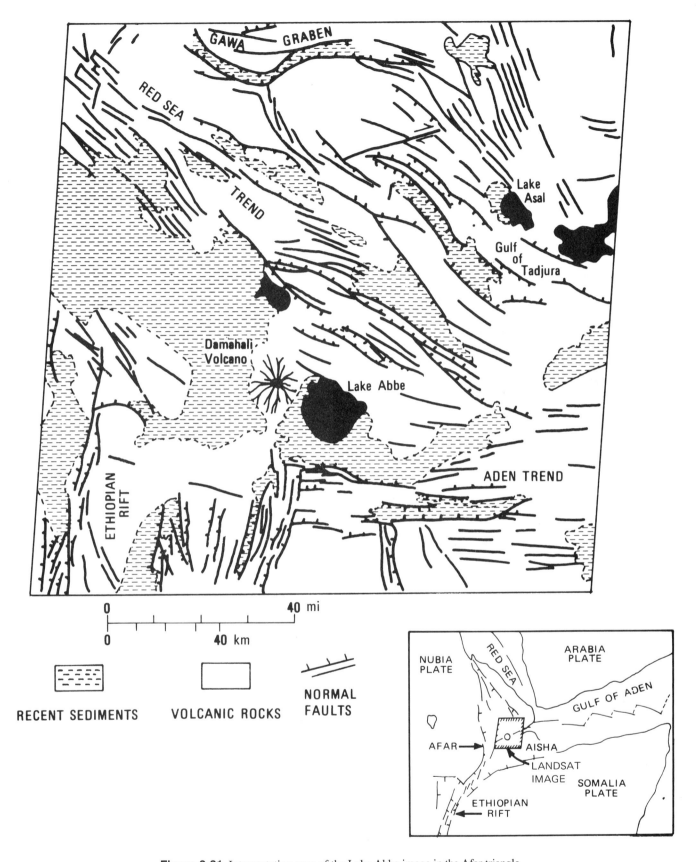

Figure 3-31 Interpretation map of the Lake Abbe image in the Afar triangle.

3. Calculate the ground resolution cell of your eyes if you were an observer on the Landsat 5 satellite. How does this resolving power compare with that of TM images?

4. Compare the spectral range of the eye with the spectral ranges of the MSS and the TM systems.

5. Based on the analyses for questions 3 and 4, discuss the relative merits of earth observations made by an astronaut versus the Landsat TM.

6. The oscillating mirror of the TM completes 14 scans each second. Calculate the dwell time for each ground resolution cell.

7. Critique and correct this statement by an image interpreter: "In the Landsat image, straight and aligned stream segments form lineations (<10 km long) and linears (>10 km long) that are undoubtedly faults, although no geologic maps are available and I have not field-checked the area."

8. The northern portion of the TM mosaic of the Central Arabian Arch (Plate 4) covers parts of two major sand seas, shown by bright orange-yellow signatures. There are two distinctly different patterns of sand dunes. Describe each of the dune patterns, including the shape, alignment, and size (of individual dunes).

REFERENCES

Bannert, D. 1992, The structural development of the Western Fold Belt, Pakistan: Geologisches Jahrbuch, Reihe B, Heft 80, Hannover, Germany.

Freden, S. C. and F. Gordon, 1983, Landsat satellites in Colwell, R. N., ed. Manual of remote sensing, second edition: ch. 12, p. 517–570, American Society for Photogrammetry and Remote Sensing, Falls Church, VA.

Gregory, A. F. and H. D. Moore, 1976, Recent advances in geologic applications of remote sensing from space: Proceedings of 24th International Astronautical Congress, p. 153–170, Pergamon Press, Oxford, England.

Grootenboer, J., 1973, The influence of seasonal factors on the recognition of surface lithologies from ERTS imagery of the western Transvaal: Third ERTS Symposium, NASA SP-351, v. 1, p. 643–655.

Grootenboer, J., K. Eriksson, and J. Truswell, 1973, Stratigraphic subdivision of the Transvaal Dolomite from ERTS imagery: Third ERTS Symposium, NASA SP-351, v. 1, p. 657–664.

Hobbs, W. H., 1904, Lineaments of the Atlantic border region: Geological Society of America Bulletin, v. 15, p. 483–506.

Hobbs, W. H., 1912, Earth features and their meaning—an introduction to geology for the student and general reader: Macmillan Publishing Co., New York, N.Y.

Kronberg, P., M. Schonfeld, R. Gunther, and P. Tsombos, 1975,

ERTS-1 data on the geology and tectonics of the Afar/Ethiopia and adjacent regions in A. Pilger and A. Rosler, eds., Proceedings of international symposium on Afar region and related rift problems: Inter-Union Commission on Geodynamics Report 14, v. I, p. 19–27, Stuttgart, Germany.

Lamar, D. L. and P. M. Merifield, 1975, Application of Skylab and ERTS imagery to fault tectonics and earthquake hazards of Peninsular Ranges, southwestern California: California Earth Science Corporation, Technical Report 75-2, Santa Monica, CA.

Molnar, P. and P. Tapponnier, 1977, The collision between India and Eurasia: Scientific American, v. 236, p. 30–41.

O'Leary, D. W., J. D. Friedman, and H. A. Pohn, 1976, Lineaments, linear, lineation—some proposed new standards for old terms: Geological Society of America Bulletin, v. 87, p. 1463–1469.

Powers, R. W., L. F. Ramirez, C. D. Redmon, and E. L. Edberg, 1966, Sedimentary geology of Saudi Arabia: U.S. Geological Survey Professional Paper 560-D.

Ryder, N. G., 1981, Ryder's standard geographic reference—the United States of America: Ryder Geosystems, Denver, CO.

Sabins, F. F., 1973, Geologic interpretation of radar and space imagery of California (abstract): American Association of Petroleum Geologists Bulletin, v. 57, p. 802.

Sabins, F. F., 1987, Remote sensing—principles and interpretation, second edition: W. H. Freeman and Co., New York, NY.

Short, N. M., P. D. Lowman, S. C. Freden, and W. A. Finch, 1976, Mission to earth, Landsat views the world: NASA Publication SP-360, U.S. Government Printing Office, stock no. 033-000-00659-4, Washington, DC.

Short, N. M. and R. W. Blair, eds., 1987, Geomorphology from space—a global overview of regional landforms: NASA Special Publication 486, Government Printing Office, stock no. 033-000-00994-1, Washington, DC.

Slaney, V. R., 1981, Landsat images of Canada—a geological appraisal: Geological Survey of Canada Paper 80-15, Ottawa, Ontario.

Strain, P. and F. Engle, 1992, Looking at earth: National Air and Space Museum, Smithsonian Institution, Washington, DC.

Sultan, M. and others, 1993, New constraints on Red Sea rifting from correlations of Arabian and Nubian Neoproterozoic outcrops: Tectonics, v. 12, p. 1303–1319.

Williams, R. S. and W. D. Carter, eds., 1976, ERTS-1, a new window on our planet: U.S. Geological Survey Professional Paper 929.

ADDITIONAL READING

McCracken, K. G. and C. E. Astley-Boden, eds., 1982, Satellite images of Australia: Harcourt, Brace, and Jovanovich Group, Sydney, Australia.

Short, N. M., 1982, The Landsat tutorial workbook: NASA Reference Publication 1078, Washington, DC.

Southworth, C. S., 1985, Characteristics and availability of data from earth-imaging satellites: U.S. Geological Survey Bulletin 1631.

CHAPTER

EARTH RESOURCE
AND ENVIRONMENTAL SATELLITES

Satellites that acquire images of the earth belong to two broad classes: earth resource satellites and environmental satellites. Landsat, described in Chapter 3, is the prototype for *earth resource satellites,* which are designed to map renewable and nonrenewable resources. In large part because of the success of Landsat, several nations have launched their own satellites, which are described in this chapter. Earth resource satellites are characterized by images with swath widths of less than 200 km and spatial resolution finer than 100 m. Each satellite requires several weeks to cover the earth. Table 4-1 lists these satellites and their characteristics.

Environmental satellites are characterized by images with swath widths of hundreds to thousands of kilometers and spa-

tial resolutions coarser than several hundred meters. The entire earth is covered daily or hourly. The images provide coverage at oceanic and continental scales and have many applications including meteorology, oceanography, and regional geography.

Figure 4-1 shows the spectral bands recorded by the major earth resource and environmental satellites, together with reflectance spectra of vegetation and rocks.

SPOT SATELLITE, FRANCE

The *SPOT-1* (Système Probatoire d'Observation de la Terre) satellite was launched by France in 1986 on an Ariane rocket

Table 4-1 Earth resource satellites with multispectral scanners

Characteristics	SPOT XS	SPOT Pan	JERS-1 OPS	IRS-1A,B LISS I	IRS-1A,B LISS II	IRS-1A,B LISS III	IRS-1C Pan	IRS-1C WiFS
Terrain coverage, km	60 by 60	60 by 60	75 by 75	148 by 148	146 by 146	141 by 141	70 by 70	810 by 810
Spatial resolution, m	20	10	20	72.5	36.3	23.5 (Vis.) 70.5 (RIR)	5.8	188
Visible bands, μm	——	——	——	0.45–0.52	0.45–0.52	——	——	——
	0.50–0.59	0.51–0.73	0.52–0.60	0.52–0.59	0.52–0.59	0.52–0.59	0.50–0.75	——
	0.61–0.68	——	0.63–0.69	0.62–0.68	0.62–0.68	0.62–0.68	——	0.62–0.68
Reflected IR bands, μm	0.79–0.89	——	0.76–0.86	0.77–0.86	0.77–0.86	0.77–0.86	——	0.77–0.86
	——	——	1.60–1.71	——	——	1.55–1.70	——	——
	——	——	2.01–2.12	——	——			
			2.13–2.25					
			2.27–2.40					

105

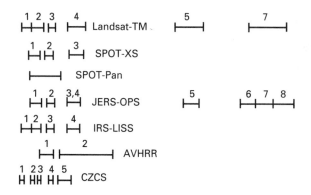

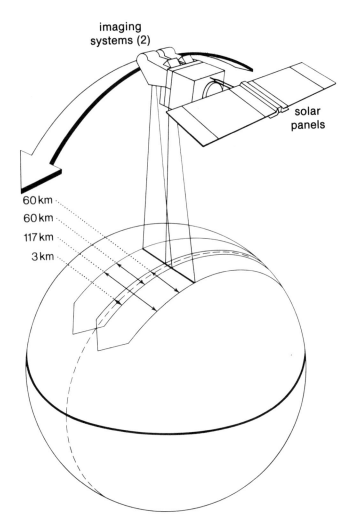

imaging systems (2)

solar panels

60 km
60 km
117 km
3 km

Figure 4-2 SPOT satellite showing the two swaths of imagery acquired by the dual imaging systems.

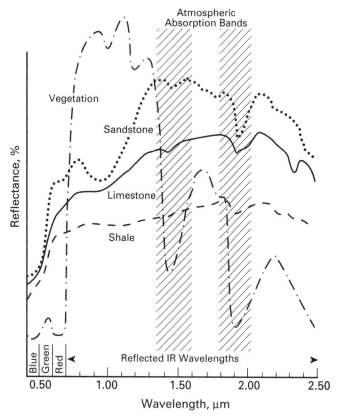

Atmospheric Absorption Bands

Vegetation

Sandstone

Limestone

Shale

Reflectance, %

Blue | Green | Red

Reflected IR Wavelengths

0.50   1.00   1.50   2.00   2.50

Wavelength, μm

Figure 4-1 Spectral bands recorded by current satellite systems, together with spectral reflectance curves.

from French Guiana. The identical SPOT-2 and SPOT-3 satellites were launched in 1988 and 1993. Like Landsat, SPOT has acquired hundreds of thousands of images in digital format that are commercially available and are used by scientists in different disciplines. Later in this section SPOT systems and images are compared with those of Landsat.

Satellite and Imaging Systems

Figure 4-2 shows the SPOT satellite, which is placed in a sun-synchronous orbit at an altitude of 832 km with a repeat cycle

of 26 days. SPOT employs a pair of along-track scanner systems such as those described in Chapter 1. Figure 4-3 is a cross section of the SPOT imaging system, which includes a tiltable mirror that can shift the field of view to either side of nadir. This is not a scanning mirror; it simply reflects light from the field of view into the optical system that focuses the image onto two sets of linear arrays of detectors (Figure 4-3). One set is a single array that acquires images in the panchromatic (pan) mode. The other set consists of three arrays of detectors that acquire images in the multispectral (XS) mode; these two modes are described in Table 4-1. *Pan images* record a single spectral band with 10-by-10-m ground resolution cells, whereas *XS images* record three spectral bands with 20-by-20-m ground resolution cells. Both modes acquire a swath of imagery that is 60 km wide. When both scanner systems are oriented in the nadir direction, there are 3 km of sidelap between the two swaths (Figure 4-2).

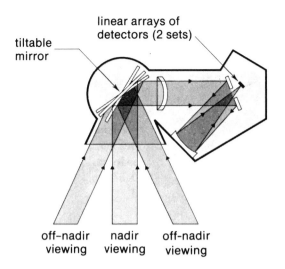

linear arrays of
detectors (2 sets)

tiltable
mirror

off-nadir
viewing

nadir
viewing

off-nadir
viewing

Figure 4-3 Cross section of a SPOT imaging system, showing the tiltable mirror.

Panchromatic (Pan) Images

Figures 4-4 and 4-5 are a pan image and location map for the Los Angeles region. The image covers 60 by 60 km with a spatial resolution of 10 m. The image was acquired by a linear array of 6000 detectors with a spectral response of 0.51 to 0.73 µm. Figure 4-6A is an enlarged subscene of the pan image that covers Los Angeles International Airport (LAX) and illustrates the high spatial resolution of the system. Figure 4-7 is a location map of the LAX subscene.

Multispectral (XS) Images

The SPOT imaging system may also be operated in the XS mode to acquire three multispectral images (green, red, and reflected IR) with ground resolution cells of 20 by 20 m. The spectral bands are listed in Table 4-1 and shown in Figure 4-1. Figure 4-6B–D shows the three XS images for the LAX subscene. Compare these images with Figure 4-6A, which demonstrates the finer spatial resolution of the pan image. Color composite XS images are prepared by projecting band 1 (green) in blue, band 2 (red) in green, and band 3 (reflected IR) in red. The resulting image is the counterpart of an IR color photograph or a Landsat IR color image.

Interpretation of an Image in Algeria

Plate 8 is a SPOT XS color composite image of the Djebel Amour area in the Saharan Atlas Mountains of Algeria. In this arid region, the rare vegetation occurs in the *wadis,* or dry streambeds, and has red signatures. Figure 4-8 is a geologic map that I interpreted from the image at a scale of 1:62,500; the map and image are greatly reduced for this book. No geo-

logic information was available when the image was interpreted. The first task, therefore, was to define rock units for mapping.

Rock Units Figure 4-9 shows the sequence of mappable units that was derived from the image, together with the symbols used to portray them on the geologic map (Figure 4-8). Characteristics of the rock units are summarized below.

1. **Recent Sand and Gravel** Deposits of recent sand and gravel occur in three forms: channel deposits, alluvial fans, and windblown sand. Channel deposits occur in the wadis and are of limited extent. Alluvial fans occur at the base of major ridges and have characteristic triangular outlines. The color signature on the image ranges from dark gray through dark blue and includes shades of blue-green, depending upon the lithology of the bedrock from which the fan material was eroded. Some dark surfaces are coated with desert varnish in this arid climate. Windblown sand occurs in the eastern and southeastern portions of the image (Plate 8). White and yellow are the dominant signatures. In this IR color image yellow hues represent red shades in the terrain. Dunes are absent and the windblown sand forms a sheet. For much of the image, the identity of the bedrock beneath the sand is inferred and shown with the appropriate rock symbol in the geologic map (Figure 4-8).

2. **Sandstone Unit** As shown in the rock column (Figure 4-9), the youngest bedrock formation consists of sandstone with minor shale interbeds. The Sandstone Unit crops out in the troughs of synclines and has a distinctive orange and grayish pink signature. The Sandstone Unit erodes to broad dipslopes with steep antidip scarps. The base of the unit is mapped at the contact between the basal sandstone ledge and the slope formed on the underlying unit.

3. **Sandstone and Shale Unit** This unit consists of an upper nonresistant shale and an underlying resistant sandstone. The shale weathers to slopes with blue signatures. The sandstone weathers to ridges with a light purple signature that is distinct from the overlying Sandstone Unit. The Sandstone and Shale Unit is relatively thin, but mapping it as a separate formation helps to define several structural features. The base of the unit is mapped at the contact between the ledge-forming sandstone and the slope of the underlying Shale Unit.

4. **Shale Unit** The Shale Unit is the most widely distributed formation in the image. It weathers to broad valleys and lowlands with a distinctive light- to medium-blue signature. Three thin resistant beds, probably sandstone, are interbedded with the Shale Unit and weather to form distinctive narrow strike ridges. In the rock column (Figure 4-9) these sandstones are designated as key beds A, B, and C. On the geologic map (Figure 4-8) each key bed is shown in a separate line pattern, which helps define folds within the Shale Unit. The base of the Shale Unit is mapped at the contact

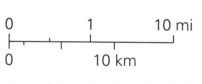

Figure 4-4 SPOT pan image of the Los Angeles region. See Figure 4-5 for location map.

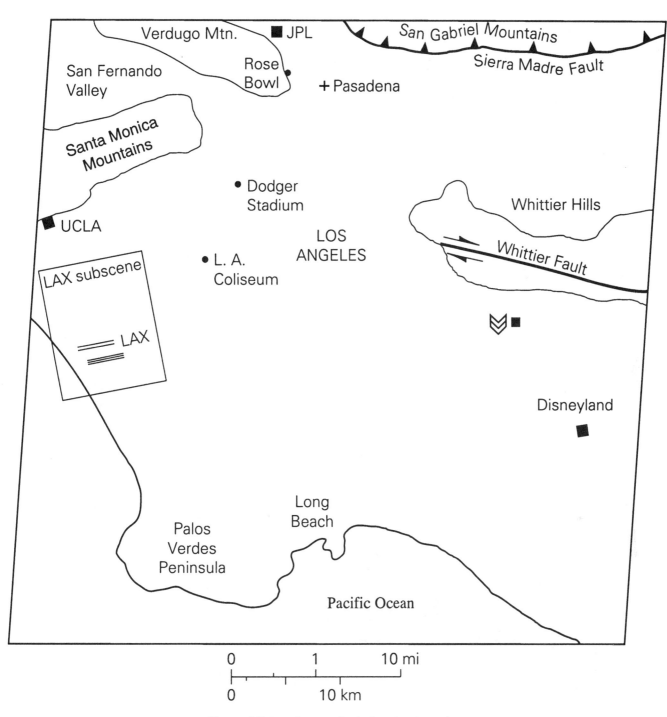

Figure 4-5 Location map for the Los Angeles region.

A. Pan image (0.51 to 0.73 μm).

B. XS band 1 (0.50 to 0.59 μm).

C. XS band 2 (0.61 to 0.68 μm).

D. XS band 3 (0.79 to 0.89 μm).

Figure 4-6 SPOT pan and XS subscenes of LAX and vicinity.

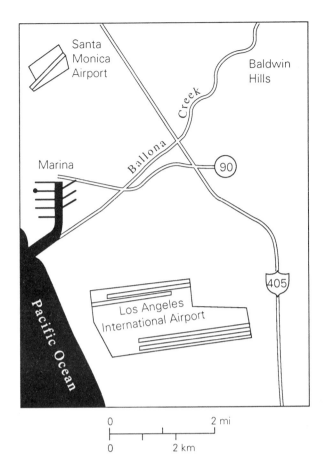

Figure 4-7 Location map for the SPOT subscenes of LAX shown in Figure 4-6.

colors. The Evaporite Unit crops out in the core of a major northeast-trending, complex anticlinal belt. In the southwest part of this belt, the Evaporite Unit is adjacent to a thick sequence of Carbonate Unit rocks. Along the strike toward the northeast, however, the Evaporite Unit cuts progressively higher in the section, and only the uppermost beds of Carbonate Unit rocks are preserved. Between the towns of El Richa and Ain Madhi (Figure 4-8), large blocks of Carbonate Unit rocks are surrounded by evaporites and appear to be roof pendants within the diapiric mass. In the northeast corner of the map, at the northeast plunge of two major anticlines, the Evaporite Unit diapirs have completely penetrated the Carbonate Unit and intrude the overlying Shale Unit. These relationships are shown diagrammatically in the rock column (Figure 4-9). The base of the Evaporite Unit is not exposed in this area.

Structural Features Structural features (dip and strike, folds, and faults) were interpreted from the XS image using the same techniques described for Landsat images in Chapter 3. The cross sections in Figure 4-10 have similar vertical and horizontal scales. The geologic structure is dominated by northeast-trending folds. In the northwest and southeast portions of the image the simple anticlines and synclines are broad and open, with gentle dips and few faults. These simple folds are separated by a complex northeast-trending belt of narrow, steeply dipping anticlines separated by strike faults that cut out the synclines. Crests of the central anticlines are extensively intruded by diapirs of the Evaporite Unit, as shown in the cross sections (Figure 4-10). These features resemble salt-cored anticlines in regions such as the Punjab Salt Range of Pakistan, Sverdrup Basin of Canadian Arctic Islands, and the Salt Valley anticlines of Utah and Colorado.

Stereo Images

The SPOT imaging systems include tiltable mirrors that enable images to be acquired as far as 475 km to the east or west of the orbit path, as shown in Figure 4-11. This off-nadir viewing capability serves two purposes:

1. The repetition cycle for images acquired in the nadir mode is 26 days. The off-nadir viewing capability, however, enables an area to be imaged more often, as shown in Figure 4-12. During a 26-day cycle, localities at the equator may be imaged 7 times; localities at 45° latitude may be imaged 11 times. This higher repetition rate is useful for observing dynamic phenomena such as floods and active volcanism; it also improves the acquisition of usable images in regions with persistent cloud cover.

2. Stereo images may be acquired by imaging an area twice from orbit paths to the east and west. Alternatively, a vertical image may be used with an image acquired from the east or west.

between the slope-forming shale and the ledges and ridges of the underlying Carbonate Unit.

5. **Carbonate Unit** The Carbonate Unit weathers to rugged ridges separated by narrow strike valleys eroded on thin interbeds of shale. Signatures range from medium blue to dark blue, with some light gray and tan beds. The shale interbeds are dark blue. The carbonate lithology is inferred from the resistant nature and dark colors of the unit. The Carbonate Unit crops out on the flanks of a major northeast-trending anticline, where it is intruded by evaporite rocks. This intrusive relationship is shown diagrammatically in the rock column (Figure 4-9). The depositional, or stratigraphic, base of the Carbonate Unit is not exposed in the area, because all contacts with the older Evaporite Unit appear to be intrusive in nature. Because of this intrusive relationship, the outcrops of the Carbonate Unit have a wide range of thickness, depending upon the amount of vertical penetration by the evaporite diapirs.

6. **Evaporite Unit** The oldest rocks in the area are the Evaporite Unit, which consists of massive, nonstratified, irregular outcrops with distinctive purple and very dark blue

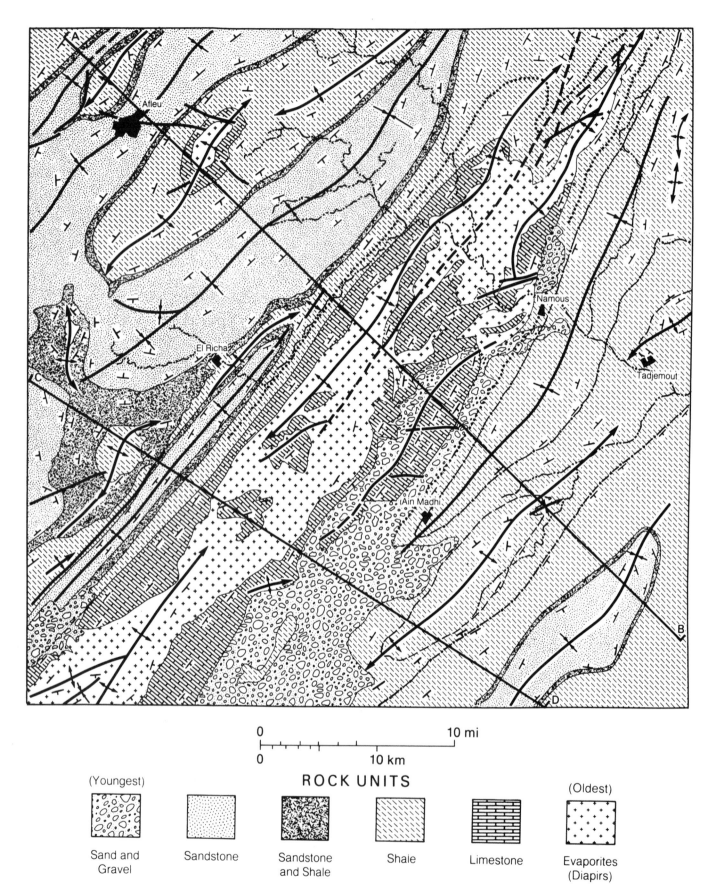

Figure 4-8 Geologic interpretation of a SPOT image, Djebel Amour area, Algeria.

ROCK UNITS

(Youngest)

Sand and Gravel

Sandstone

Sandstone and Shale

Shale

Limestone

(Oldest)

Evaporites (Diapirs)

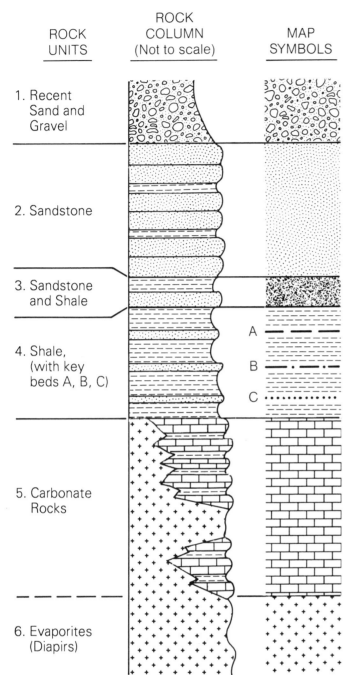

ROCK UNITS	ROCK COLUMN (Not to scale)	MAP SYMBOLS
1. Recent Sand and Gravel		
2. Sandstone		
3. Sandstone and Shale		
4. Shale, (with key beds A, B, C)	A B C	
5. Carbonate Rocks		
6. Evaporites (Diapirs)		

Figure 4-9 Rock column and mapping units for the SPOT image of the Djebel Amour area. No vertical scale.

Stereo Pair in Southern California Figure 4-13 shows the viewing geometry for an area in southern California that was imaged from the east at an incidence angle of 25.8° and from the west at 14.4°. The *incidence angle* is the angle between the vertical and a line from the satellite to the center of the image. In this example, the *image base* (the distance between the two orbit paths) is 600 km. At the SPOT height of 832 km, the base–height ratio is 0.72. The chart in Figure 2-15 shows that this ratio results in a 5× vertical exaggeration in the stereo model.

Figure 4-14 is a portion of a stereo pair of pan images that were acquired with the viewing geometry shown in Figure 4-13. By viewing this stereo pair with a stereoscope, the reader can perceive the 5× vertical exaggeration. Figure 4-15 is an interpretation map of the stereo pair that covers a syncline known as the Devils Punchbowl because it is a rugged circular depression. The Devils Punchbowl is located on the north flank of the San Gabriel Mountains, which are separated by the San Andreas fault from the Antelope Valley to the north. The San Andreas is an active fault with right-lateral, strike-slip displacement. The stereo model shows scarps, shutter ridges, and offset drainages that are diagnostic of active faults. The Punchbowl syncline is located between the San Andreas fault and the Punchbowl fault to the south. The steeply dipping beds in the flanks of the syncline are clearly seen in the stereo model. South of the Devils Punchbowl, the Punchbowl and San Jacinto faults are recognizable on the stereo pair by north-facing linear scarps and aligned northwest-trending valleys.

Topographic Map from Stereo Pair, Yemen Topographic maps are normally compiled from stereo pairs of aerial photographs. In many regions of the world, however, aerial photographs are not available and SPOT stereo pairs may be used to compile topographic maps.

For example, in the late 1980s the Chevron Corporation acquired the Gardan Block exploration concession in Yemen, where no topographic maps or aerial photographs were available. Ellis and Rossetter (1993) of Chevron Overseas Petroleum, Inc., used stereo pairs of SPOT pan images to prepare detailed, accurate topographic maps for the concession area, which covered 50 by 100 km. Figure 4-16A is a portion of one of the images. Navigation satellites provided precise information on elevation, latitude, and longitude for 11 ground localities. The ground localities were identified on the digital image files, which were then rectified to a *Universal Transverse Mercator* (UTM) map projection. The rectified stereo image data were used to calculate the elevation for each 10-by-10-m SPOT ground resolution cell. The elevation data were then used to plot a terrain model for the Gardan Block. Figure 4-16B is a portion of the model that corresponds to the SPOT image of Figure 4-16A. The model is illuminated from the east to create highlights and shadows that enhance the topography. The elevation data were then used to generate a topographic

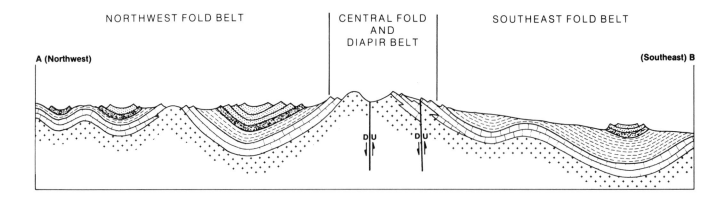

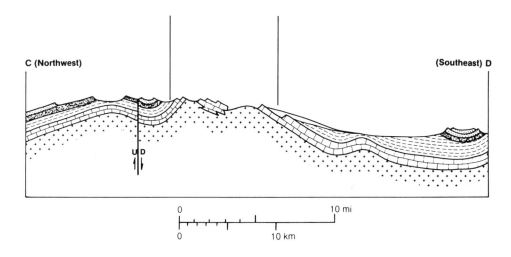

Figure 4-10 Geologic cross sections of the Djebel Amour area, Algeria. See Figure 4-8 for location of sections. Vertical and horizontal scales are approximately equal.

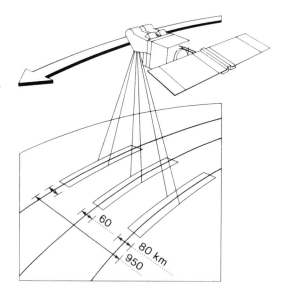

Figure 4-11 Off-nadir viewing capability of SPOT. The tiltable mirrors enable images to be acquired up to 475 km east or west of the orbit path.

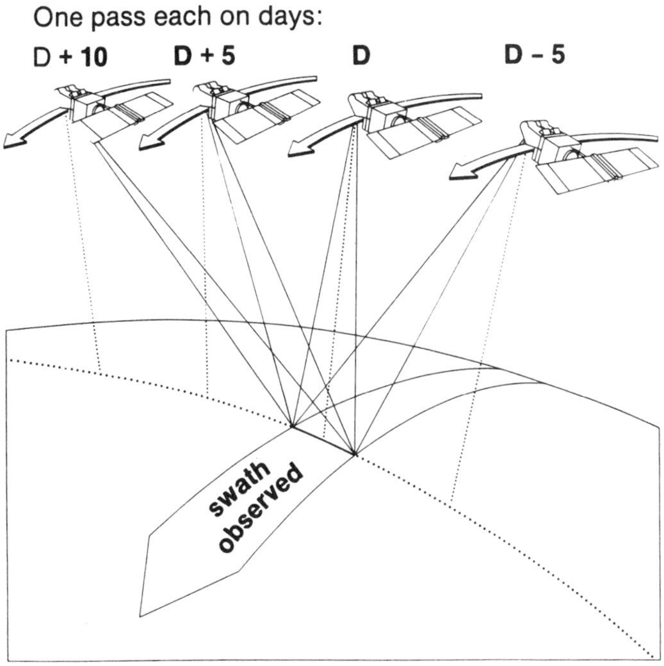

One pass each on days:

D + 10   D + 5   D   D – 5

swath observed

Figure 4-12 The off-nadir viewing capability of SPOT can acquire multiple images of a locality during a 26-day cycle. Locations at the equator can be observed 7 times; localities at latitude 45° can be observed 11 times.

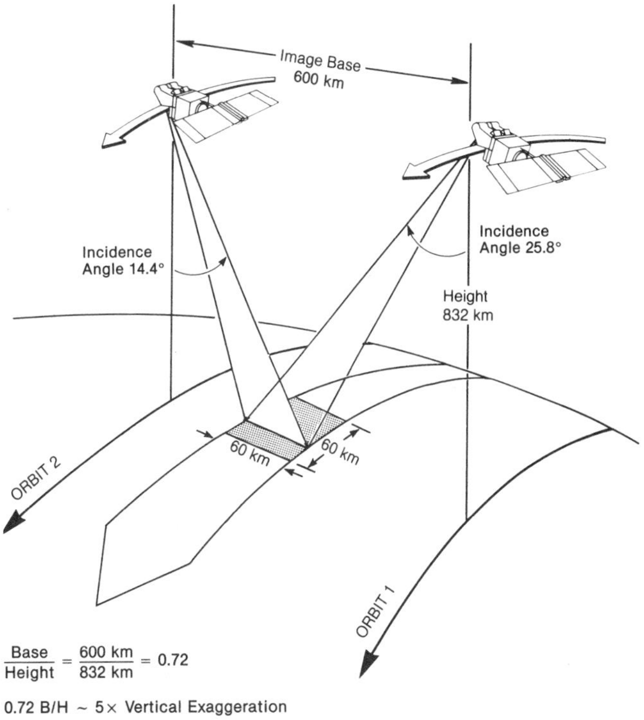

Image Base 600 km

Incidence Angle 14.4°

Incidence Angle 25.8°

Height 832 km

ORBIT 2

ORBIT 1

60 km   60 km

$$\frac{Base}{Height} = \frac{600\ km}{832\ km} = 0.72$$

0.72 B/H ~ 5× Vertical Exaggeration

Figure 4-13 Geometry of stereo images acquired by SPOT for Figure 4-14.

map (1:50,000 scale) with a contour interval of 20 m. Figure 4-17 is a small portion of the map that shows the topographic detail obtained from the stereo pair of SPOT pan images.

Orbit Pattern and Index System

SPOT operates at an altitude of 832 km in a sun-synchronous orbit pattern similar to that of Landsat. The daytime southbound orbits cross latitude 40°N at 10:00 a.m. local sun time; ground tracks are repeated at intervals of 26 days. Figure 4-2 shows the two identical imaging systems that acquire parallel swaths of images on the east and west side of the orbit track. In the nadir-viewing mode (shown in Figure 4-2) there are 3 km of sidelap between the image swaths, each of which is 60 km wide. Individual images cover 60 by 60 km of terrain.

Availability of Images and Information

SPOT images in print and digital format are available from

SPOT Image Corporation
1897 Preston White Drive
Reston, VA 22091-4368
Telephone: 703-620-2200
Fax: 703-648-1813

Users can provide this facility with the latitude and longitude coordinates of their areas of interest. SPOT Image Corporation will provide an index map and printout showing the available images and pertinent information including date of acquisition, cloud cover, and incidence angle (for determining stereo coverage). SPOT Image Corporation publishes the quarterly newsletter *SPOT Light,* which describes new developments and applications for the data. A complimentary subscription is available from the above address.

SPOT AND LANDSAT IMAGES COMPARED

SPOT and Landsat are currently the only systems that routinely provide worldwide images in the visible and reflected IR spectral regions. Therefore we will compare the images based on their suitability for remote sensing interpretation.

Terrain Coverage

Figure 4-18A compares the terrain coverage of a Landsat TM image (31,450 km^2) and a SPOT image (3600 km^2). After allowing for overlap, approximately 12 SPOT images are required to cover the area of a single TM image.

Figure 4-14 Stereo pair of SPOT pan images of the Devils Punchbowl, San Gabriel Mountains, California.

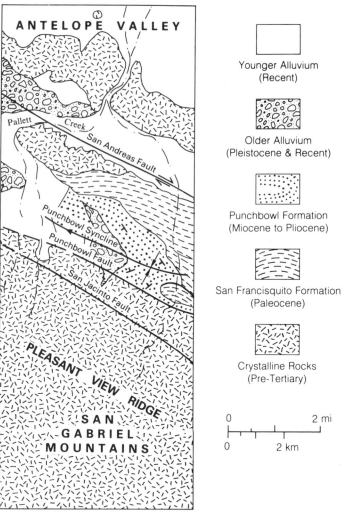

Figure 4-15 Geologic map of the Devils Punchbowl, San Gabriel Mountains, California.

Spatial Resolution

Figure 4-18B compares dimensions of ground resolution cells for SPOT pan (10 m), SPOT XS (20 m), and TM (30 m). In order to compare images, SPOT pan data were computer-processed to simulate images with larger ground resolution cells. Figure 4-19 shows the simulated SPOT and Landsat images with ground resolution cells ranging from 10 to 79 m. The images cover the town of Victorville, California (Figure 4-20) at a scale of 1:100,000, which is widely used for image interpretation. At this scale the difference between SPOT pan images (Figure 4-19A) and SPOT XS images (Figure 4-19B) is negligible. At this scale the 30-m resolution of the Landsat TM (Figure 4-19C) hampers the identification of urban features,

but most geologic features are readily interpreted. At the regional scale of 1:250,000 (not shown) the differences between 10-m and 30-m data are less perceptible.

Spectral Bands

Figure 4-1 compares the ranges of the spectral bands recorded by SPOT and Landsat. SPOT does not cover the important reflected IR region that is recorded by TM bands 5 and 7. These bands are valuable for recognizing spectral differences among different rock types. As illustrated in this chapter, the reflected IR bands of TM are preferred for preparing color images of arid or tropical regions.

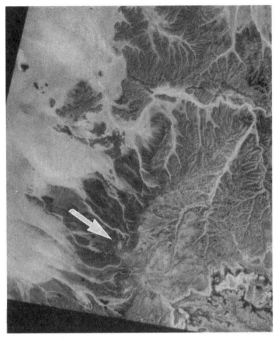

A. SPOT pan image. Arrow shows location of topographic feature called "the Hook," shown in Figure 4-17. SPOT © 1991 CNES.

B. Digital terrain model from stereo SPOT images. The model is illuminated from the east.

Figure 4-16 SPOT image and digital terrain model for a portion of the Gardan Block, Yemen. From Ellis and Rossetter (1993, Figure 9). Courtesy J. E. Ellis, Chevron Overseas Petroleum, Inc.

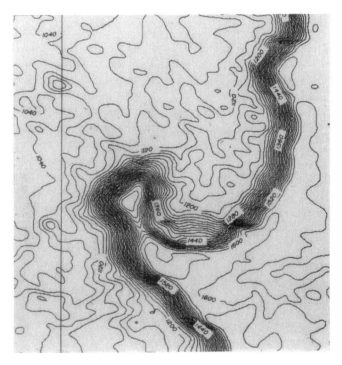

Figure 4-17 Topographic contour map of the Hook, Gardan Block, Yemen. Elevation data were derived from a SPOT stereo pair. Map is 10 km wide; contour interval is 20 m. From Ellis and Rossetter (1993, Figure 11). Courtesy J. E. Ellis, Chevron Overseas Petroleum, Inc.

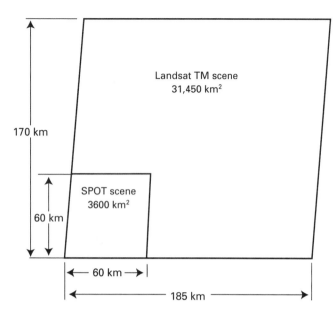

170 km

Landsat TM scene
31,450 km²

60 km

SPOT scene
3600 km²

← 60 km →

← 185 km →

A. Terrain coverage.

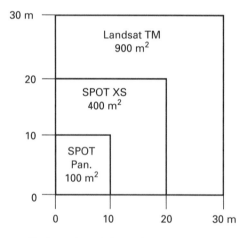

30 m

Landsat TM
900 m²

20

SPOT XS
400 m²

10

SPOT
Pan.
100 m²

0

0   10   20   30 m

B. Ground resolution cells.

Figure 4-18 Comparison of image characteristics for Landsat TM and SPOT satellites.

Stereo Images

Only SPOT provides stereo images that are valuable for many interpretation projects and for generating topographic maps.

Cost

SPOT images, in digital or print format, are almost five times as expensive per square kilometer as TM images. For many projects, however, the cost of image data is insignificant relative to other costs. In oil exploration, for example, geophysical surveys, land acquisitions, and drilling cost several orders of magnitude more than image data. The efficiency, fast response time, and information content of remote sensing images are more important than their modest costs.

Summary

The major advantages of SPOT images are their higher spatial resolution and stereo coverage. Landsat TM images also have advantages, including regional coverage, adequate spatial resolution, and optimum selection of spectral bands. For regional oil exploration projects I first interpret TM images to identify subareas with exploration potential. If these subareas have complex geology, stereo SPOT pan images are used to interpret geologic details. SPOT pan images (10-m resolution) can be digitally merged with color TM images (30-m resolution) to obtain the advantages of both data sets. This technique is described in Chapter 8.

JERS-1 SATELLITE, JAPAN

The *JERS-1* (Japanese Earth Resources Satellite) was launched February 11, 1992, from Tanegashima Space Center into a sunsynchronous orbit at an altitude of 568 km. The satellite carries both a radar imaging system (described in Chapter 7) and an optical sensor (OPS) system, described next. The name Fuyo-1 was briefly applied to the satellite but has been discontinued.

OPS Images

Nishidai (1993) described the *OPS-1 system,* which is an along-track multispectral scanner that records two image bands in the visible region and five in the reflected IR region. Table 4-1 and Figure 4-1 show the wavelength intervals of the bands. The images cover 75 by 75 km with a spatial resolution of 20 m. Figure 4-21 shows the OPS bands for the Cerros Colorados subscene in southern Argentina. Band 4 records the same wavelength range as band 3 and is used to provide stereoscopic coverage at those wavelengths. A major potential advantage of OPS is the spectral ranges of bands 6, 7, and 8. As shown in Figure 4-1, these bands subdivide the important atmospheric window from 2.01 to 2.40 μm into three narrow intervals. These bands have the potential to separate rock classes and identify types of hydrothermally altered rocks, as described in Chapter 11. SPOT records no data in this range, and the Landsat TM records only band 7, as shown in Figure 4-1.

Any three of the OPS bands may be combined in blue, green, and red to produce IR color images. Unfortunately the OPS images are marred by blurring and striping to an extent that renders them unusable for many applications. Crippen and others (1994) described these image defects. The Japanese space agency has not been able to correct the problem. No plans have been announced for launching a successor satellite to JERS-1.

Stereo Images

JERS-1 carries only one OPS imaging system. Figure 4-22 shows how this single along-track system is employed to acquire stereo pairs of images. The lens focuses the image on the focal plane of the optical system. The linear arrays of detectors for bands 1, 2, and 3 are positioned at the center of the focal plane to record images at the nadir, directly below the satellite. One linear detector array (band 4) is positioned at the edge of the focal plane to record an image from the leading edge of the field of view. The stereo base between the band 4 image and the nadir images is 156 km; the base–height ratio for a stereo pair is 0.27 (Figure 4-22), which produces a vertical exaggeration of 2.0×.

Figure 4-23 is a stereo pair of the Cerros Colorados area, Argentina. The vertical exaggeration of 2.0× is lower than for either standard aerial photographs or typical stereo pairs of SPOT images (Figure 4-14). Despite the low vertical exaggeration, however, JERS-1 stereo images are potentially useful for geologic interpretation. One advantage of the JERS-1 stereo system is that the two images are acquired simultaneously under uniform environmental and lighting conditions. SPOT stereo pairs are acquired from different orbits that may be separated in time by several months; these temporal differences may hamper interpretation of the SPOT stereo pair.

Availability of Images

OPS images are available as photographic prints or in digital format, but users should be aware of the image defects described above. For information contact

User Service Department
Remote Sensing Technology Center of Japan
Uni-Roppongi Bldg., 7-15-17, Roppongi
Minato-ku, Tokyo 106, JAPAN
Telephone: 81-3-3401-1387
Fax: 81-3-3403-1766

OPS images are indexed in a path-and-row system similar to that of Landsat.

INDIA REMOTE SENSING SATELLITE

India launched three India Remote Sensing (*IRS*) satellites between 1988 and 1995. The satellites are in sun-synchronous orbits at an altitude of 904 km with repeat cycles of 22 to 24 days. Southbound daylight orbits cross the equator at 10:25 a.m. local sun time. The satellites carry a range of imaging systems that are included in Table 4-1.

Linear Imaging Self-Scanning System

The linear imaging self-scanning (*LISS*) system is an along-track multispectral scanner that records four bands of imagery.

Table 4-1 lists the characteristics of the three versions of LISS (I, II, III) that have been deployed. LISS I and II record the same spectral bands but differ in terrain coverage and spatial resolution. LISS I records images with a swath width of 148 km and a spatial resolution of 73 m. LISS II records sidelapping pairs of images, each with a swath width of 74 km and a spatial resolution of 36.6 m. Four LISS II images cover the area of a single LISS I image.

LISS III is carried on IRS-1C, which was launched December 28, 1995. Table 4-1 lists the spectral bands of LISS III. Bands 1 to 3 have a spatial resolution of 23.5 m. Band 4 (1.55 to 1.70 µm) has a spatial resolution of 70 m. Swath width is 141 km for bands 1 to 3 and 148 km for band 4.

Panchromatic System

The panchromatic system records a single visible band (0.50 to 0.75 µm) with a spatial resolution of 5.8 m and a swath width of 70 km. The field of view can be tilted up to 26° to either side of the orbit path to acquire stereo images and to provide an image repeat cycle as short as 5 days. Figure 4-24 is a greatly enlarged portion of an IRS pan image of the outskirts of Tucson, Arizona. Closely spaced residences are resolved on this image. The IRS panchromatic system acquires the first commercial satellite images with spatial resolution finer than 10 m.

Wide Field Sensor

The wide field sensor (*WiFS*) records two images in the visible and reflected IR bands with a swath width of 810 km and a spatial resolution of 188.3 m. One application for these images is monitoring vegetation on regional and global scales, as described in Chapter 12.

Availability of Images

LISS images are indexed with a path-and-row system similar to that of Landsat. Data are available as photographic products or in digital format. For information contact

National Remote Sensing Agency
Data Center
Balangar
Hyderabad 500 037, Andhra Pradesh
India

LISS data and information are also distributed by EOSAT at the following address:

EOSAT
Customer Services
4300 Forbes Boulevard
Lanham, MD 20706
Telephone: 800-344-9933
Fax: 301-552-5476

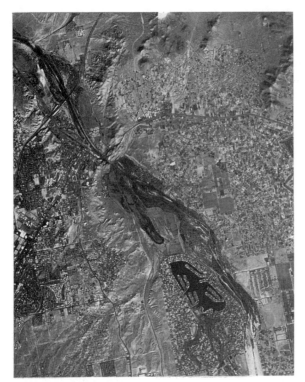

A. 10-by-10-m cell of SPOT pan.

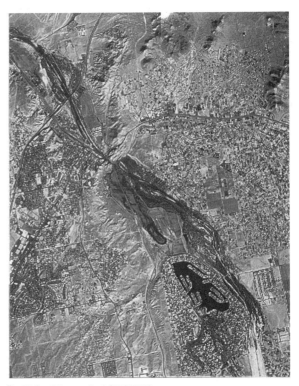

B. 20-by-20-m cell of SPOT XS.

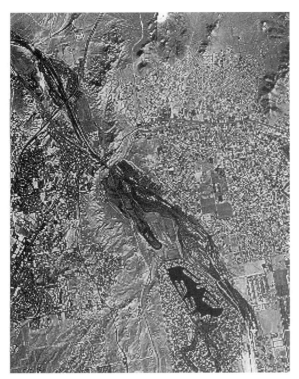

C. 30-by-30-m cell of Landsat TM.

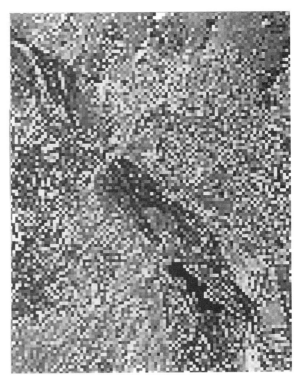

D. 79-by-79-m cell of Landsat MSS.

Figure 4-19 Ground resolution cells of SPOT and Landsat images displayed at 1:100,000 scale. The various cell sizes in B–D were produced by computer resampling of the original SPOT pan 10-m data in A. Victorville, California.

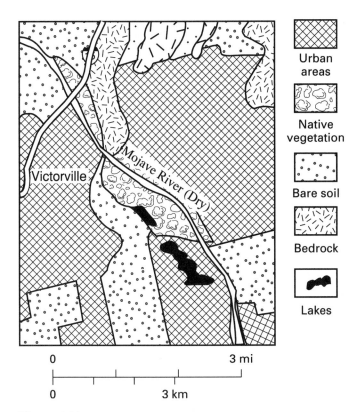

Urban areas

Native vegetation

Bare soil

Bedrock

Lakes

0   3 mi

0   3 km

Figure 4-20 Location map for Victorville, California.

Geostationary satellites travel at the same angular velocity at which the earth rotates; as a result they remain above the same point on earth at all times. A typical orbit has an altitude of almost 36,000 km and a velocity of 11,000 km · h^{-1}. Each day numerous images are acquired that cover continent-sized areas. Current geostationary satellites are listed in Table 4-2. Polar-orbiting environmental satellites are described in later sections.

Geostationary Operational Environmental Satellites

The National Oceanographic and Aeronautic Administration (NOAA) operates the Geostationary Operational Environmental Satellites (*GOES*), which are essentially "parked" above the equator. GOES East is positioned above the equator at 75° west longitude, and GOES West is positioned at 135° west longitude. Together they provide continuous coverage of North America, South America, and adjacent oceans.

The *visible and IR spin-scanning radiometer (VISSR)* on GOES acquires two bands of imagery every 30 min that cover 100° by 100° of latitude and longitude. Table 4-3 describes the VISSR and other environmental imaging systems. The visible band (0.55 to 0.70 μm) has a spatial resolution of 1 km. The full images, such as Figure 4-25, are resampled at a spatial resolution of 8 km to reduce the amount of data. At this small scale the advantages of higher spatial resolution would not be apparent. The thermal IR band (10.50 to 12.60 μm) is acquired at a spatial resolution of 8 km. The image in Figure 4-26 was acquired simultaneously with the visible band image of Figure 4-25. In GOES IR images, the temperature signatures are

Table 4-2 Currently operational environmental satellites

Satellite(s)	Nationality	Orbit	Sensor systems*
NOAA	USA	Polar	Advanced very high resolution radiometer (AVHRR)
			High resolution IR radiation sounder (HIRS)
GOES	USA	Geostationary	Visible and IR spin-scanning radiometer (VISSR)
			Visible/IR atmospheric sounder (VAS)
DMSP	USA	Polar	Operational linescan system (OLS)
			Multispectral IR radiometer (MIR)
			Special sensor microwave/imager (SSM/I)
Meteosat 3	ESA	Geostationary	Visible and IR radiometer (VIRR)
INSAT	India	Geostationary	Very high resolution radiometer (VHRR)
Meteor-Priroda	Russia	Polar	Microwave sounding units (MSU)
Meteor-3	Russia	Geostationary	Total ozone mapping spectrometer (TOMS)
			Scanning radiation budget sensor (SCARABE)

*See Table 12-2.
Source: Adapted from Greenstone and Bandeen (1993).

A. Band 1 (0.52 to 0.60 µm).

B. Band 2 (0.63 to 0.69 µm).

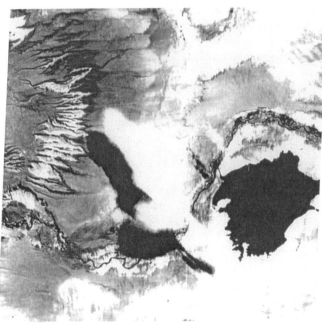

C. Bands 3 and 4 (0.76 to 0.86 µm).

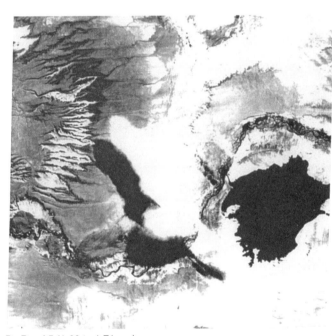

D. Band 5 (1.60 to 1.71 µm).

Figure 4.21 JERS-1 OPS images of Cerros Colorados, Argentina.
Courtesy NASDA/MITI, Tokyo, and T. Nishidai.

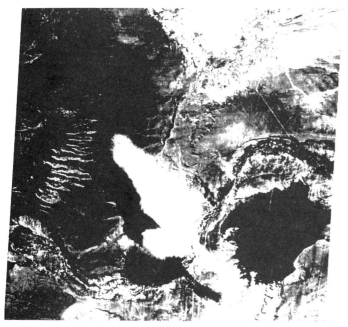

E. Band 6 (2.01 to 2.12 μm).

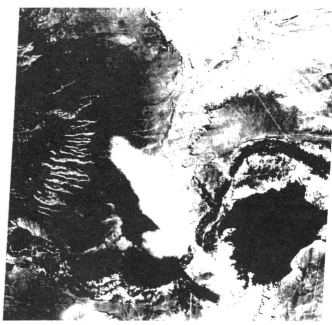

F. Band 7 (2.13 to 2.25 μm).

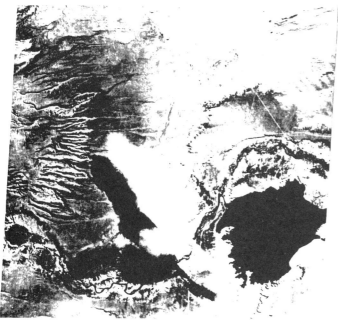

G. Band 8 (2.27 to 2.40 μm).

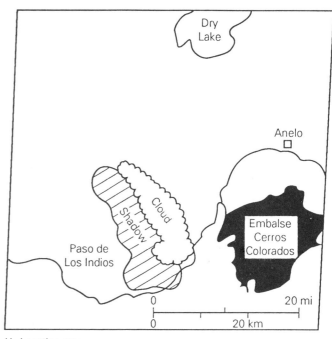

H. Location map.

Table 4-3 Remote sensing systems on environmental satellites

Sensor system[1]	Spectral bands	Spectral range, μm	Spatial resolution,[2] km	Swath width, km	Comments
AVHRR	5	0.58 to 12.5	1.1 Vis and RIR 4 TIR	2700	Vegetation; temperature
CZCS	6	0.43 to 12.50	0.825	1600	Ocean productivity
HIRS/2	20	0.7 to 15.0	17	2240	Atmospheric sounding; temperature and moisture profiles
MIR	4	9.8 to 28.3	Profile	Profile	Vertical temperature profiles
MSU	4	0.53 to 0.59 cm	109	2347	Atmospheric sounding
OLS	2	0.4 to 1.1 10.0 to 13.0	0.56 Vis 2.78 TIR	3000	Global cloud cover
SCARABE	4	0.5 to 12.5		2500	Ozone
SSM/I	4	0.35 to 1.55 cm	~55	1400	Precipitation; snow and ice cover
VAS	12	0.55 to 0.70 3.9 to 14.7	1.0 Vis 8.0 TIR	Limb-to-limb	Day/night cloud cover; atmospheric temperature and moisture content
VHRR	2	0.55 to 0.90 10.5 to 12.5	2.75 Vis 11.0 TIR	Limb-to-limb	Day/night cloud cover; temperature
VIRR	3	0.4 to 1.1 5.7 to 7.1 10.5 to 12.5	2.5 Vis 5.5 TIR 5.0 TIR	Limb-to-limb	Temperature of clouds, oceans, and land
VISSR	2	0.55 to 0.70 10.5 to 12.6	0.9 Vis 8.0 TIR	Limb-to-limb	Day/night cloud cover; temperature
TOMS	6	0.31 to 0.38	3.1	Limb-to-limb	Ozone monitoring

[1]See Table 4-2 for names of systems. CZCS (not included in Table 4-2) = coastal zone color scanner.
[2]RIR = reflected IR band; TIR = thermal IR band; Vis = visible band.
Source: Adapted from Greenstone and Bandeen (1993).

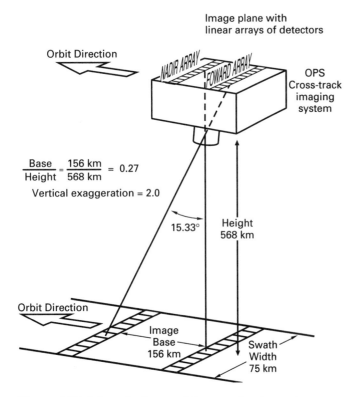

Figure 4-22 Schematic diagram of the stereo-imaging system on JERS-1.

reversed from those of other IR images in this book (Landsat TM band 6, Heat Capacity Mapping Mission (HCMM) and aircraft images). Bright tones record cool radiant temperatures and dark tones record warm temperatures. This arrangement causes clouds, which are colder than water and land, to have familiar bright signatures. In the daytime IR image of Figure 4-26, land areas are warm (darker signatures) relative to oceans (brighter signatures) for reasons given in Chapter 5.

The clouds in Figure 4-26 have a range of thermal IR signatures due to different temperatures caused by different cloud heights. The swirl of gray clouds adjacent to the west coast of South America is noticeably warmer than the nearby linear belts of white clouds. These differences are not discernible in the visible image of Figure 4-25. Another advantage of the IR band is that images are acquired both day and night. GOES data are available from

NOAA Satellite Data Services Division
Princeton Executive Square
Room 100
Washington, DC 20233

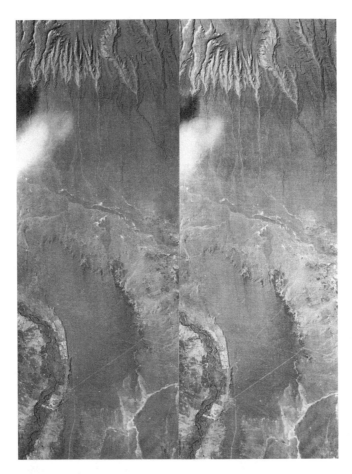

Figure 4-23 Stereo pair of the JERS-1 images, Cerros Colorados, Argentina.

Meteosat

The Meteosat geostationary satellites, operated by the European Space Agency, acquire images of Europe and Africa similar to those of GOES. Meteosat 4 is positioned at 0° longitude and provides observations every 30 minutes. The *visible and IR radiometer* (VIRR) of Meteosat acquires one image in the visible and reflected IR band (0.4 to 1.05 μm) at a 2.5-km resolution and another in the thermal IR band (10.5 to 12.5 μm) at a 5-km resolution. The images are similar to those acquired by the VISSR system on GOES. A third band in the thermal IR atmospheric absorption region records water vapor.

Defense Meteorologic Satellite Program

Satellites of the Defense Meteorologic Satellite Program (*DMSP*) acquire weather data for the U.S. Defense Department, but the data are generally available. A pair of polar-orbiting DMSP satellites cover the earth in similar fashion to the NOAA satellites. The primary imaging systems are the *operational linescan system* (OLS) and the *special sensor microwave/imager* (SSM/I), which are described in Table 4-3. Figure 4-27 shows daytime OLS images acquired in the visible band and thermal IR band. Clouds are bright in the thermal IR image because cool temperatures are shown in bright tones. Thermal IR data are available from

Figure 4-24 Enlarged portion of panchromatic image of outskirts of Tucson, Arizona. This image in the visible band (0.50 to 0.75 μm) has spatial resolution of 5.8 m. Courtesy V. L. Williams, EOSAT.

Figure 4-25 GOES East visible-band image (0.55 to 0.70 μm) of the Western Hemisphere. Courtesy NOAA Satellite Services Division.

National Geophysical Data Center
NOAA/NESDIS
325 Broadway
Boulder, CO 80303
Telephone: 303-497-6126
Fax: 303-497-6513

Snow and ice data are available from

National Snow and Ice Data Center
University of Colorado

Campus Box 449
Boulder, CO 80309
Telephone: 303-492-2378
Fax: 303-492-2468

POLAR-ORBITING NOAA ENVIRONMENTAL SATELLITES

NOAA operates a series of environmental satellites that are in sun-synchronous orbits at an altitude of 850 km. The satellites

Figure 4-26 GOES East thermal IR–band image (10.5 to 12.6 μm) of the Western Hemisphere. Dark signatures are warm radiant temperatures, and bright signatures are cool temperatures. Courtesy NOAA Satellite Services Division.

operate in pairs. The "morning" satellite crosses the equator at approximately 7:30 a.m. and 7:30 p.m. local sun time ; the "afternoon" satellite crosses at 2:30 p.m. and 2:30 a.m. Each satellite records 14 daily image swaths that are 2700 km wide and cover the entire earth twice every 24 hours. The pair of satellites cover the earth four times daily. Kidwell (1986) prepared a guide for users of data from NOAA polar-orbiting satellites. The polar-orbiting satellites are listed in Table 4-2. The NOAA satellites carry two imaging systems that are extensively employed in environmental studies: the coastal zone color scanner and the advanced very high resolution radiometer.

Coastal Zone Color Scanner

The *coastal zone color scanner* (CZCS) is a cross-track multi-spectral scanner that was carried on the Nimbus polar-orbiting satellite launched October 23, 1978. The CZCS recorded six bands of imagery in the visible, reflected IR, and thermal IR regions (listed in Table 4-4). The images have a swath width of 1600 km and spatial resolution of 825 m. Sunlight reflected from ripples on the sea surface (called *sunglint*) may be a problem. To avoid sunglint the scanner may be tilted up to 20° ahead of or behind the spacecraft (Hovis and others, 1980).

A. Visible-band image.

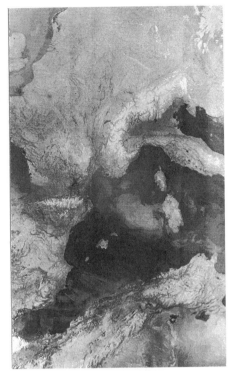

B. Thermal IR–band image. Dark signatures are warm radiant temperatures, and bright signatures are cool temperatures.

Figure 4-27 DMSP daytime OLS images of the Mediterranean region.

Table 4-4 Spectral bands of the coastal zone color scanner (CZCS)

Band	Wavelength, μm	Characteristics
1	0.43 to 0.45	Blue
2	0.51 to 0.53	Green
3	0.54 to 0.56	Green
4	0.66 to 0.68	Red
5	0.70 to 0.80	Reflected IR
6	10.50 to 12.50	Thermal IR

Figure 4-28 shows the visible and reflected IR images of a subscene covering Florida and the adjacent Gulf of Mexico and Atlantic Ocean. These subscenes are 800 km wide and show half of the full 1600-km image swath. The visible images (bands 1 to 4) show thin clouds and haze, which are reduced in the reflected IR image (band 5). Investigators have developed methods for digitally processing the images to reduce the effects of haze and atmospheric scattering. Some of these methods are described in the collection of CZCS papers edited by Barale and Schlittenhardt (1993). The images are especially useful for mapping organic productivity in the oceans, as shown in Chapter 9. CZCS data are available from

DAASC User Support Office
Code 902.2
NADSA/GSFC
Greenbelt, MD 20771
Telephone: 301-286-3209

W. A. Hovis (1984) edited an annotated and interpreted collection of CZCS images of coastal regions around the world.

Advanced Very High Resolution Radiometer

The *advanced very high resolution radiometer* (AVHRR) is a cross-track multispectral scanner that acquires images with a swath width of 2700 km. This wide swath combined with the 14 daily sun-synchronous orbits of a NOAA satellite provides complete coverage of the earth each day. Since there are two satellites, two sets of imagery are acquired, one in the morning and one in the afternoon. AVHRR acquires one visible band, one reflected IR band, and three thermal IR bands (Table 4-5), all with a spatial resolution of 1.1 km. The visible and reflected IR bands are highly effective for mapping vegetation on a global basis, as described in Chapter 12.

Figure 4-29 is a subscene of southern Louisiana that shows the three commonly used bands of AVHRR images. Band 1 (Figure 4-29A) records visible light. In the land areas, variation in gray tones relates to differences in vegetation. The

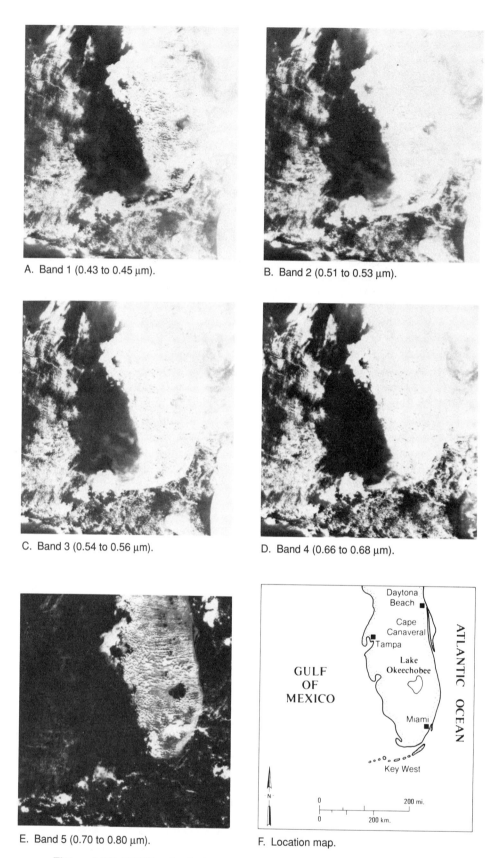

A. Band 1 (0.43 to 0.45 μm).

B. Band 2 (0.51 to 0.53 μm).

C. Band 3 (0.54 to 0.56 μm).

D. Band 4 (0.66 to 0.68 μm).

E. Band 5 (0.70 to 0.80 μm).

F. Location map.

Figure 4-28 CZCS bands of a Florida subscene. Courtesy D. K. Clark, NOAA Satellite Services Division.

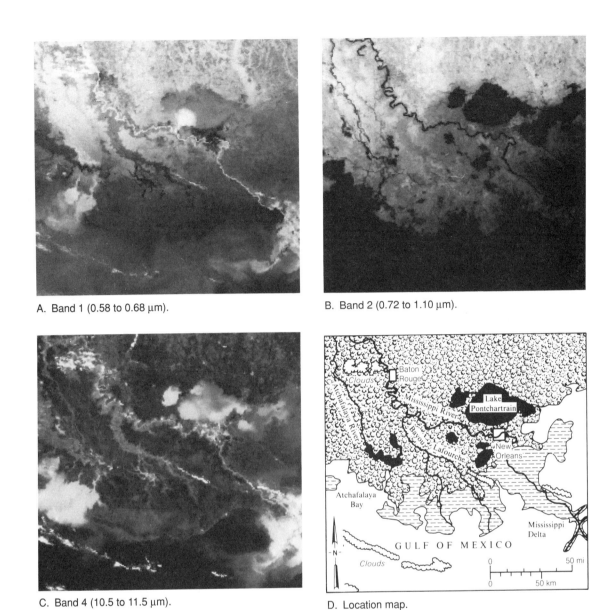

A. Band 1 (0.58 to 0.68 μm).

B. Band 2 (0.72 to 1.10 μm).

C. Band 4 (10.5 to 11.5 μm).

D. Location map.

Figure 4-29 AVHRR bands of a southern Louisiana subscene. Courtesy EROS Data Center.

TABLE 4-5 Spectral bands of the advanced very high resolution radiometer (AVHRR)

Band	Wavelength, μm	Characteristics
1	0.58 to 0.68	Red—vegetation mapping
2	0.72 to 1.10	Reflected IR—vegetation mapping
3	3.55 to 3.93	Thermal IR—hot targets such as fires and volcanoes
4	10.50 to 11.50	Thermal IR—ambient temperatures
5	11.50 to 12.50	Thermal IR—recorded only on the NOAA 7 satellite

cities of Baton Rouge and New Orleans have distinct dark signatures. In the water areas, brighter tones correlate with higher concentrations of suspended silt and mud. The Mississippi River and its discharge into the Gulf of Mexico are notably bright. Plumes of sediment-laden water discharged into Atchafalaya Bay and Lake Pontchartrain form conspicuously bright patches. Clouds are bright.

Band 2 (Figure 4-29B) records reflected IR energy. Water bodies are clearly distinguished from land by their very dark signatures, which are caused by absorption of these wavelengths. The thin clouds present in the band 1 image are readily penetrated by reflected IR energy and are not visible in band 2.

Band 4 (Figure 4-29C) records thermal IR energy. Bright signatures represent relatively cool radiant temperatures, and dark signatures are relatively warm. In this early spring season, water in the Gulf of Mexico is warm (dark tones) relative to the fresh water discharged at the Mississippi Delta and Atchafalaya Bay (bright tones). Lake Pontchartrain and adjacent lakes are also cool. The stringers of clouds are cool.

AVHRR images are well suited for studying vegetation distribution and seasonal changes on a continent-wide scale as described in Chapter 12. AVHRR data are available from the EROS Data Center (EDC). The EDC also offers a number of processed AVHRR data sets, together with ancillary data, in CD-ROM format.

A number of institutions and organizations are compiling global sets of AVHRR data with different ground resolution cells and time intervals. Butler (1994) edited a special issue of the *International Journal of Remote Sensing* with articles that describe different data sets.

ENVIRONMENTAL AND EARTH RESOURCES IMAGES COMPARED

The geostationary satellites are used primarily for meteorologic purposes. AVHRR and CZCS images from polar-orbiting satellites are the primary data sources for regional environmental studies. Images from the Landsat TM and SPOT are the primary data sources for studies of renewable and nonrenewable earth resources. Both sets of images are acquired in multispectral format from sun-synchronous orbits. The major differences are in terrain coverage and spatial resolution of the images.

Figure 4-30 illustrates these differences. North America can be covered with a dozen AVHRR images, whereas a couple of thousand TM images are required (Figure 4-30A). These differences in terrain coverage are accompanied by proportional differences in spatial resolution, as shown by the ground resolution cells in Figure 4-30B. Another difference is the repeat cycle for the images. The world is covered twice daily by AVHRR images. For the conterminous United States, experience shows that complete cloud-free coverage is obtainable at intervals of 2 weeks or less. Comparable cloud-free coverage from the 16-day repeat cycle of the TM would require at least a year. Investigators can use this comparative information to determine the optimum images for a project.

FUTURE SATELLITE SYSTEMS

Table 4-6 lists earth resource and environmental satellites that are proposed for future launches. This table is for reference purposes only, because there is no certainty that all the proposed satellites will actually be launched.

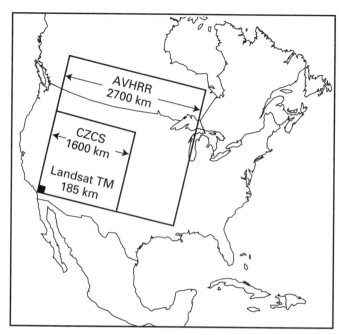

A. Terrain coverage.

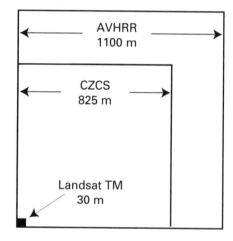

B. Ground resolution cells.

Figure 4-30 Comparison of image characteristics of two environmental imaging systems (AVHRR and CZCS) and one earth resource system (TM).

Earth Observation Satellite

NASA currently plans to launch the first Earth Observation Satellite (EOS-A) into a sun-synchronous orbit in 1998. Five imaging systems are planned. The *advanced spaceborne thermal emission and radiation radiometer* (ASTER), which is built in Japan, is the only system that will acquire images with moderate spatial resolution. ASTER is an along-track multispectral scanner that records two visible bands (spatial resolution of 15 m), seven reflected IR bands (spatial resolution of 15 to 30 m), and five thermal IR bands (spatial resolution of

Table 4-6 Proposed future satellites through 2000

Scheduled launch	Country	Organization	Mission	Spectral bands Pan[1]	Spectral bands MS[2]	Spatial resolution, m Pan	Spatial resolution, m MS	Swath width, km	Stereo
1996	USA	NASA	Lewis	1	384	5	30	13, 8	—
1996	USA	NASA	Clark	1	3	3	15	6, 30	Yes
1996	USA	Earthwatch, Inc.	Early Bird	1	3	3	15	6, 30	—
1996	Germany and Russia		PRIRODA	1	4	6	18	40, 80	Yes
1996	India	NSRA and EOSAT	IRS-1D	1	4	6	23	141	—
1996	Japan	NASDA	ADEOS	1	4	8	16	80	Yes
1997	USA	Earthwatch, Inc.	Quick Bird	1	4	1	4	6, 30	—
1997	USA	Space Imaging	Space Image	1	4	1	4	11	Yes
1997	China and Brazil	CSAT and NPE	CBERS	—	8	—	20, 80	120	Yes
1998	USA	NASA	Landsat 7	1	6	15	30	185	—
1998	USA	NASA	EOS-MODIS	2	345	250	500, 1000	2300	—
1998	USA and Japan	NASA and MITI	EOS-ASTER	—	9	—	15, 30	60	Yes
1998	USA	GDE	GDE	1	—	1	—	15	Yes
1998	USA	Orbital Sciences	OrbView-1	2	4	1, 2	8	8	Yes
1998	USA	RESOURCE 21	RESOURCE 21	—	5	—	10, 20	205	Yes
1998	Argentina	CONAE	SAC-C	—	5	—	150	315	—
1998	Russia	Russia and SAR Corp.	ALMAZ-OES	1	3, 4	2.5	4, 80	80, 300	Yes

[1]panchromatic
[2]multispectral
Source: Goldberg and Stoney (1995).

90 m). Stereo images will be acquired with a forward-looking system similar to that of JERS-1.

Space Station

The Space Shuttle will provide transportation for constructing a permanent, manned Space Station that is planned as an international venture under the direction of NASA. The Space Station is not a remote sensing project, although some imaging systems will be deployed. Because of the high cost, the future of the Space Station and its final design are uncertain.

Commercial Satellites

From its beginning, satellite remote sensing has been conducted almost exclusively by government agencies. The EOSAT Corporation and SPOT Image Corporation have evolved to operate satellites that were largely funded by governments. Beginning in 1997, however, private industry is taking the lead to build, launch, and operate several satellite remote sensing systems. A number of factors have contributed to this major change. The end of the Cold War caused governments to relax their restrictions on distribution of high-resolution images. Defense industries need nonmilitary customers for their technology. Landsat 5 launched in 1986, is an aging system. The replacement satellite, failed on launch. Because of

advances in technology, systems can be built and launched for $100 million that would have cost $1 billion 10 years ago. The advent of the "Information Age" may provide a profitable market for high-resolution images that are delivered on a timely schedule.

Table 4-6 lists characteristics of commercial imaging systems that are scheduled for launch through 1998. The systems have several common characteristics: a panchromatic imager with fine spatial resolution (1 to 3 m), a multispectral system with 3 to 4 bands and coarser resolution (4 to 15 m), and stereo capability. The images have narrow swath widths (up to 30 km). Therefore major applications will be for detailed local studies. The commercial images will supplement, but not necessarily replace, images from earth resources satellites. Fritz (1996) provides additional information on these companies and systems. There is no guarantee that all the systems will be deployed, but the successful deployment of a few satellites will be a significant advance.

COMMENTS

Earth resource satellites (SPOT, JERS-1, IRS, and Landsat) provide multispectral images with swaths less than 200 km wide and moderate spatial resolution finer than 100 m.

Because of the narrow swaths, it takes several weeks to cover the earth. Environmental satellites (GOES, Meteosat, and NOAA) provide images with swaths hundreds to thousands of kilometers wide and spatial resolutions coarser than several hundred meters. The entire earth is covered hourly or daily.

Images from earth resource satellites are employed for local and regional studies of a wide range of applications including the environment, renewable and nonrenewable resources, land use/cover, and natural hazards. Images from environmental satellites are used for regional and global studies of meteorology, climatology, oceanography, the environment, and land use/cover. Later chapters illustrate these applications of both categories of satellites.

Beginning in 1997 several corporations plan to launch remote sensing satellites and distribute image data for profit. These new commercial ventures are a welcome addition to traditional satellite operations. The commercial satellites will acquire panchromatic and multispectral images with fine spatial resolution, narrow swaths, and stereo capability.

QUESTIONS

1. View the stereo images from SPOT and JERS-1 with a stereoscope and answer the following questions. Which stereo model has greatest vertical exaggeration? Explain your answer. Which system can vary the vertical exaggeration? How is this variation accomplished? JERS-1 stereo models have one significant advantage over SPOT models. Identify and explain this advantage.

2. JERS-1 attempted to acquire three bands of imagery within the atmospheric window from 2.0 to 2.5 μm. Explain the potential advantages and applications of these bands.

3. Refer to Figure 4-30 and calculate how many TM images are required to cover: A. One CZCS image, B. One AVHRR image.

4. Figure 4-4 is a SPOT pan image of an urban scene. Plate 8 is an XS image of a geologic scene. Describe the advantages and disadvantages of both pan and XS images for interpreting both urban and geologic scenes.

5. Topographic maps are normally compiled from stereo pairs of aerial photographs. Figures 4-16 and 4-17 illustrate the use of stereo pairs of SPOT images for this application. For compiling topographic maps in inaccessible regions, compare the advantages and disadvantages of SPOT images and aerial photographs.

REFERENCES

Barale, V. and P. M. Schlittenhardt, eds., 1993, Ocean colour—theory and application in a decade of CZCS experience: Kluwer Academic Publishers, Dordrecht, Netherlands.

Butler, D. M., 1994, Preface to special issue—Global data sets for the land from the AVHRR: International Journal of Remote Sensing, v. 15, p. 3315–3639.

Crippen, R. E., J. P. Ford, R. G. Blom, and R. K. Dokka, 1994, An evaluation of Fuyo-1 (JERS-1) optical data for geologic interpretation: Environmental Research Institute of Michigan 10th Thematic Conference on Geologic Remote Sensing, Proceedings, v. II, p. 606, Ann Arbor, MI.

Ellis, J. M. and R. J. Rossetter, 1993, Remote sensing technology in support of geophysical operations: Environmental Research Institute of Michigan, Ninth Thematic Conference on Geologic Remote Sensing, Proceedings, p. 209–220, Pasadena, CA.

Fritz, L. W., 1996, The era of commercial earth observation satellites: Photogrammetric Engineering and Remote Sensing, v. 62, p. 39–45.

Goldberg, A. M. and W. Stoney, 1995, Instrument data sheets: Land Satellite Information in the Next Decade, American Society for Photogrammetry and Remote Sensing, Falls Church, VA.

Greenstone, R. and B. Bandeen, 1993, Operational and research satellites for observing the earth in the 1990s and thereafter in Gurney, R. J., J. L. Foster, and C. L. Parkinson, eds., Atlas of satellite observations related to global change, p. 449–457, Cambridge University Press, Cambridge, England.

Hovis, W. A. and others, 1980, Nimbus-7 coastal zone color scanner system description and initial imagery: Science, v. 210, p. 60–63.

Hovis, W. A., ed., 1984, Nimbus-7 CZCS coastal zone color scanner imagery for selected coastal regions: NASA Goddard Space Flight Center, Greenbelt, MD.

Kidwell, K. B., 1986, NOAA Polar Orbiter user's guide: National Oceanic and Atmospheric Administration, World Weather Building, Room 100, Washington, DC.

Nishidai, T., 1993, Early results from "Fuyo-1", Japan's earth resources satellite (JERS-1): International Journal of Remote Sensing, v. 14, p. 1825–1833.

ADDITIONAL READING

Bizzi, S., A. Arino, and P. Goryl, 1996, Operational algorithm to correct the along-track and across-track striping in the JERS-1 OPS images: International Journal of Remote Sensing, v. 17, p. 1963–1968.

Denniss, A. M., D. A. Rothery, G. Ceuleneer, and I. Amri, 1994, Lithological discrimination using Landsat and JERS-1 SWIR data in the Oman ophiolite: Environmental Research Institute of Michigan 10th Thematic Conference on Geologic Remote Sensing, Proceedings, v. II, p. 97–108, Ann Arbor, MI.

Foody, G. M. and P. J. Curran, 1994, Environmental remote sensing from regional to global scales: John Wiley & Sons, New York, NY.

Goodwin, P. B., P. D. Caldwell, and J. M. Ellis, 1991, Evaluating and updating offshore base maps using SPOT satellite images and GPS control, West Africa: Environmental Research Institute of Michigan Eighth Thematic Conference on Geologic Remote Sensing, Proceedings, p. 253–264, Ann Arbor, MI.

Kramer, H. J., 1994, Observation of the earth and its environment—survey of missions and sensors: Springer-Verlag, Berlin, Germany.

Yamaguchi, Y., S. Tsuchida, and T. Matsunaga, 1994, Evaluation of JERS-1/OPS data for lithologic mapping in Yerington, Nevada: Environmental Research Institute of Michigan 10th Thematic Conference on Geologic Remote Sensing, Proceedings, v. II, p. 91–96, Ann Arbor, MI.

CHAPTER

THERMAL INFRARED IMAGES

All matter radiates energy at thermal IR wavelengths (3 to 15 μm) both day and night. The ability to detect and record this thermal radiation as images at night takes away the cover of darkness and has obvious reconnaissance applications. For these reasons, government agencies funded the early development of thermal IR imaging technology, beginning in the 1950s. The developments were classified for security purposes. Military interpreters recognized that geologic and terrain features greatly influenced the background against which strategic targets were displayed. Word of these potential nonmilitary applications created interest in the civilian geologic community, and in the mid-1960s some manufacturers received approval to acquire images for civilian clients using the classified systems. In 1968 the U.S. government declassified systems that did not exceed certain standards for spatial resolution and temperature sensitivity. Today IR scanner systems are available for unrestricted use. The term *thermography* has been suggested as a replacement for *thermal IR imagery* but is not adopted in this book. Thermography is used for medical applications of thermal IR imagery (Heller, 1991); any change in the accepted use of this term would be confusing.

Satellite acquisition of thermal IR images began in 1960 with the U.S. meteorologic Television IR Operational Satellites (TIROS) and has continued with subsequent programs. The coarse spatial resolution of these images is optimum for monitoring cloud patterns and ocean temperatures; large terrain features are also recognizable. In 1978 the NASA Heat Capacity Mapping Mission (HCMM) obtained daytime and nighttime thermal IR images with a 600-m spatial resolution for geologic applications. In addition, the thematic mapper of Landsats 4 and 5 records a thermal IR image (band 6).

Thermal IR images generally record broad spectral bands, typically 8.0 to 14.0 μm for images from aircraft and 10.5 to 12.5 μm for images from satellites. In 1980 NASA and JPL developed the *thermal infrared multispectral scanner* (TIMS) that acquires six bands of imagery at wavelength intervals of 1.0 μm or less. Subsequently, commercial IR multispectral scanners have been developed. These multispectral images are useful for discriminating rock types based on variations of silica content.

THERMAL PROCESSES AND PROPERTIES

To interpret thermal IR images, one must understand the basic physical processes that control the interactions between thermal energy and matter, as well as the thermal properties of matter that determine the rate and intensity of the interactions.

Heat, Temperature, and Radiant Flux

Kinetic heat is the energy of particles of matter in a random motion. The random motion causes particles to collide, resulting in changes of energy state and the emission of electromagnetic radiation from the surface of materials. The internal, or kinetic, heat energy of matter is thus converted into *radiant energy*. The amount of heat is measured in calories. A *calorie* is the amount of heat required to raise the temperature of 1 g of water 1°C. *Temperature* is a measure of the concentration of heat. On the Celsius scale, 0°C and 100°C are the temperatures of melting ice and boiling water, respectively. On the Kelvin, or absolute, temperature scale, 0°K is *absolute zero,* the point

135

at which all molecular motion ceases. The Kelvin and Celsius scales correlate as follows: 0°C = 273°K, and 100°C = 373°K. The electromagnetic energy radiated from a source is called *radiant flux* (F) and is measured in watts per square centimeter (W · cm^{-2}).

The concentration of kinetic heat of a material is called the *kinetic temperature* (T_{kin}) and is measured with a thermometer placed in direct contact with the material. The concentration of the radiant flux of a body is the *radiant temperature* (T_{rad}). Radiant temperature may be measured remotely by nonimaging devices called *radiometers*. The radiant temperature of materials is always less than the kinetic temperature because of a thermal property called emissivity, which is defined later.

Heat Transfer

Heat energy is transferred from one place to another by three means:

1. *Conduction* transfers heat through a material by molecular contact. The transfer of heat through a frying pan to cook food is one example.
2. *Convection* transfers heat through the physical movement of heated matter. The circulation of heated water and air are examples of convection.
3. *Radiation* transfers heat in the form of electromagnetic waves. Heat from the sun reaches the earth by radiation. In contrast to conduction and convection, which can only transfer heat through matter, radiation can transfer heat through a vacuum.

Materials at the surface of the earth receive thermal energy primarily in the form of radiation from the sun. To a much lesser extent, heat is also conducted from the interior of the earth. There are daily and annual cyclic variations in the duration and intensity of solar energy. Energy from the interior of the earth reaches the surface primarily by conduction and is relatively constant at any locality, although there are regional variations in this heat flow. Hot springs and volcanoes are local sources of convective heat. Regional heat-flow patterns may be altered by geologic features such as salt domes and faults.

IR Region of the Electromagnetic Spectrum

This section reviews and expands some concepts introduced in Chapter 1. The IR region is that portion of the electromagnetic spectrum ranging in wavelength from 0.7 to 300 μm. The terms *near*, *short*, *middle*, and *long* have been used to subdivide the IR region but are not used here because of confusion about their boundaries and usage. The *reflected IR region* ranges from 0.7 to 3.0 μm and is dominated by reflected solar energy. Bands 4, 5, and 7 of the Landsat TM record in this region. The reflected IR region also includes the *photographic IR band* (0.7 to 0.9 μm), which may be detected directly by IR-

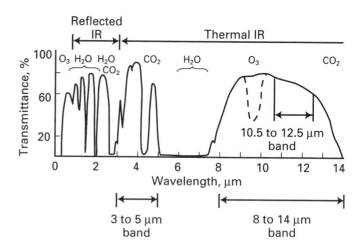

Figure 5-1 Electromagnetic spectrum showing spectral bands used in the thermal IR region. Gases that cause atmospheric absorption are indicated.

sensitive film. On IR color photographs the red signature records IR energy that is strongly reflected by vegetation and is not related to thermal radiation. IR radiation at wavelengths of 3 to 14 μm is called the *thermal IR region* (Figure 5-1). Photographic film does not detect thermal IR radiation. Special detectors and optical-mechanical scanners detect and record images in the thermal IR spectral region. The term *IR energy* mistakenly connotes heat to many people; therefore, it is important to recognize the difference between reflected IR energy and thermal IR energy.

Atmospheric Transmission The atmosphere does not transmit all wavelengths of thermal IR radiation uniformly. Carbon dioxide, ozone, and water vapor absorb energy in certain wavelength regions, called *absorption bands*. The atmosphere transmits wavelengths of 3 to 5 μm and 8 to 14 μm (Figure 5-1); these bands are called *atmospheric windows*. A number of detection devices record radiation of these wavelengths. The narrow absorption band of 9 to 10 μm (shown as a dashed curve in Figure 5-1) is caused by the ozone layer at the top of the earth's atmosphere. To avoid the effects of this absorption band, satellite thermal IR systems record wavelengths from 10.5 to 12.5 μm. Systems on aircraft, which fly beneath the ozone layer, are not affected and may record the full window from 8 to 14 μm.

Radiant Energy Peaks and Wien's Displacement Law For an object at a constant kinetic temperature, the radiant energy, or flux, varies as a function of wavelength. The *radiant energy peak* (λ_{max}) is the wavelength at which the maximum amount of energy is radiated. Figure 5-2 shows radiant energy curves for objects ranging in temperature from 300 to 700°K.

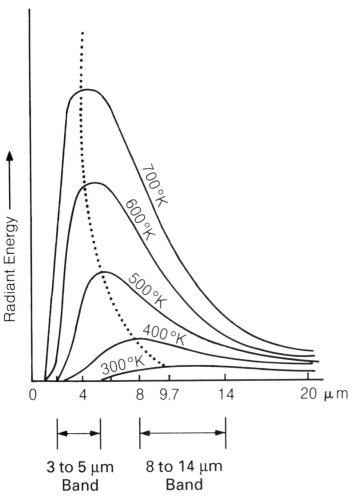

$$\lambda_{max} = \frac{2897 \; \mu m \cdot °K}{T_{rad}}$$

$$= \frac{2897 \; \mu m \cdot °K}{300°K}$$

$$= 9.7 \; \mu m$$

which is the peak for the 300°K radiant energy curve in Figure 5-2 and is the radiant energy peak of the earth. Figure 5-2 will be used to evaluate optimum wavelength bands for detecting targets at various temperatures.

Wien's displacement law also applies to hot objects that glow at visible wavelengths, such as an iron poker in a fire. As the poker heats up, the color progresses from dark red through bright red, and orange, to yellow at successively shorter wavelengths. The changing colors represent shifts in λ_{max} with increasing temperature.

Thermal Properties of Materials

Radiant energy striking the surface of a material is partly reflected, partly absorbed, and partly transmitted through the material. Therefore,

$$\text{Reflectivity + absorptivity + transmissivity = 1} \qquad (5\text{-}2)$$

Reflectivity, absorptivity, and transmissivity are determined by properties of matter and also vary with the wavelength of the incident radiant energy and with the temperature of the surface. As discussed in Chapter 2, reflectivity is expressed as *albedo* (*A*), which is the ratio of reflected energy to incident energy. For materials in which transmissivity is negligible, Equation 5-2 reduces to

$$\text{Reflectivity + absorptivity = 1} \qquad (5\text{-}3)$$

The absorbed energy causes an increase in the kinetic temperature of the material.

Blackbody Concept, Emissivity, and Radiant Temperature The concept of a blackbody is fundamental to understanding heat radiation. A *blackbody* is a theoretical material that absorbs all the radiant energy that strikes it, which means

$$\text{Absorptivity = 1} \qquad (5\text{-}4)$$

A blackbody also radiates all of its energy in a wavelength distribution pattern that is dependent only on the kinetic temperature. According to the *Stefan-Boltzmann law*, the radiant flux of a blackbody (F_b) at a kinetic temperature of T_{kin} is

$$F_b = \sigma \cdot T_{kin}^4 \qquad (5\text{-}5)$$

Figure 5-2 area and the label:

3 to 5 μm Band

8 to 14 μm Band

Figure 5-2 Spectral distribution curves of energy radiated from objects at different temperatures. Note atmospheric transmission bands at 3 to 5 μm and 8 to 14 μm. From Colwell and others (1963, Figure 2).

As temperature increases, the total amount of radiant energy (the area under each curve) increases and the radiant energy peak shifts to shorter wavelengths. The dotted line in Figure 5-2 passes through the radiant energy peaks and indicates the shift. This shift, or displacement, to shorter wavelengths with increasing temperature is described by *Wien's displacement law*, which states that

$$\lambda_{max} = \frac{2897 \; \mu m \cdot °K}{T_{rad}} \qquad (5\text{-}1)$$

where T_{rad} is radiant temperature in degrees Kelvin and 2897 μm · °K is a physical constant. The wavelength of the radiant energy peak of an object may be determined by substituting the value of T_{rad} into Equation 5-1. For example, the average radiant temperature of the earth is approximately 300°K (27°C or 80°F). Substituting this temperature into Equation 5-1 results in where σ is the *Stefan-Boltzmann constant* (5.67×10^{-12} W $\cdot$ cm$^{-2} \cdot$ °K^{-4}).

For a blackbody with a T_{kin} of 10°C (283°K), the radiant flux may be calculated from Equation 5-5 as

$$F_b = T_{kin}^4$$
$$= (5.67 \times 10^{-12} \text{ W} \cdot \text{cm}^{-2} \cdot \text{°K}^{-4}) (283°K)^4$$
$$= (5.67 \times 10^{-12} \text{ W} \cdot \text{cm}^{-2} \cdot \text{°K}^{-4}) (6.41 \times 10^9 \text{ °K}^4)$$
$$= 3.6 \times 10^{-2} \text{ W} \cdot \text{cm}^{-2}$$

A blackbody is a physical abstraction, because no material has an absorptivity of 1 and no material radiates the full amount of energy given in Equation 5-5. For real materials a property called *emissivity* (ε) has been defined as

$$\varepsilon = \frac{F_r}{F_b} \tag{5-6}$$

where F_r is radiant flux from a real material. The emissivity for a blackbody is 1, but for all real materials it is less than 1. Emissivity is wavelength-dependent, which means that the emissivity of a real material will be different when measured at different wavelengths of radiant energy.

Table 5-1 lists the emissivities of various materials in the 8-to-12-μm wavelength region, which is widely used in remote sensing. Emissivities for most materials in Table 5-1 fall within the relatively narrow range of 0.81 to 0.96. Note that water has a high emissivity and that a thin film of petroleum lowers the

Table 5-1 Emissivity of materials measured at wavelengths of 8 to 12 μm

Material	Emissivity (ε)
Granite, typical	0.815
Dunite	0.856
Obsidian	0.862
Feldspar	0.870
Granite, rough	0.898
Silicon sandstone, polished	0.909
Sand, quartz, large-grain	0.914
Dolomite, polished	0.929
Basalt, rough	0.934
Dolomite, rough	0.958
Asphalt paving	0.959
Concrete walkway	0.966
Water, with a thin film of petroleum	0.972
Water, pure	0.993

Source: From Buettner, K. J. K. and C. D. Kern, Journal of Geophysical Research, v. 70, p. 1333, 1965, copyrighted by American Geophysical Union.

emissivity, which is a significant relationship for remote sensing of oil slicks (Chapter 9).

Combining Equations 5-5 and 5-6 produces the following equation for the radiant flux of a real material:

$$F_r = \varepsilon \cdot \sigma \cdot T_{kin}^4 \tag{5-7}$$

where ε is the emissivity for that material. Emissivity is a measure of the ability of a material both to radiate and to absorb energy. Materials with high emissivities absorb large amounts of incident energy and radiate large quantities of kinetic energy. Materials with low emissivities absorb and radiate lower amounts of energy. Figure 5-3 illustrates the effect of different emissivities on the energy radiated from an aluminum block with a uniform kinetic temperature of 10°C (283°K). The portion of the block that is painted dull black has an emissivity of 0.97. The radiant flux for this material is calculated from Equation 5-7 as

$$F_r = \varepsilon \cdot \sigma \cdot T_{kin}^4$$
$$= 0.97 (5.67 \times 10^{-12} \text{ W} \cdot \text{cm}^{-2} \cdot \text{°K}^{-4}) (283°K)^4$$
$$= 3.5 \times 10^{-2} \text{ W} \cdot \text{cm}^{-2}$$

For the shiny portion of the aluminum block with an emissivity of 0.06, the radiant flux may be calculated from Equation 5-7 as

$$F_r = \varepsilon \cdot \sigma \cdot T_{kin}^4$$
$$= 0.06 (5.67 \times 10^{-12} \text{ W} \cdot \text{cm}^{-2} \cdot \text{°K}^{-4}) (283°K)^4$$
$$= 2.2 \times 10^{-3} \text{ W} \cdot \text{cm}^{-2}$$

Although the aluminum block has a uniform kinetic temperature of 283°K, the radiant flux from the surface with high emissivity is more than 10 times greater than from the surface with low emissivity.

Most thermal IR remote sensing systems record the radiant temperature (T_{rad}) of terrain rather than radiant flux. In order to determine T_{rad}, consider a blackbody and a real material that have different kinetic temperatures but the same radiant flux, so that $F_b = F_r$. For a blackbody $T_{rad} = T_{kin}$; therefore, Equation 5-5 may be written as

$$F_b = \sigma \cdot T_{rad}^4$$

This equation and Equation 5-7 may then be combined as follows:

$$F_r = \varepsilon \cdot \sigma \cdot T_{kin}^4$$
$$F_b = \sigma \cdot T_{rad}^4$$
$$F_b = F_r$$

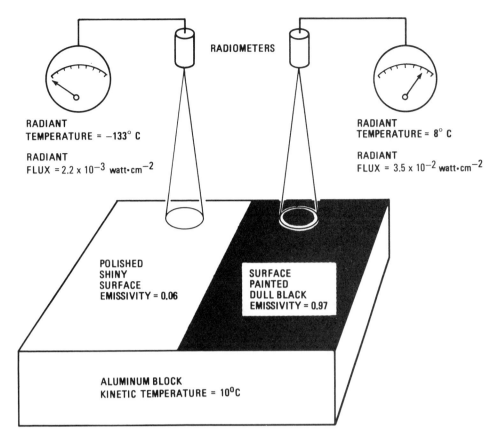

Figure 5-3 Effect of emissivities on radiant temperature. The kinetic (internal) temperature of the aluminum block is uniformly 10°C. Different emissivities cause different radiant temperatures, which are measured by radiometers.

$$\sigma\, T^4_{rad} = \varepsilon \cdot \sigma \cdot T^4_{kin}$$

$$T_{rad} = \varepsilon^{1/4} T_{kin} \qquad (5\text{-}8)$$

For a real material of known emissivity and kinetic temperature, Equation 5-8 may be used to calculate the radiant temperature. Radiant temperature is measured with radiometers. For the portion of the aluminum block in Figure 5-3 with an emissivity of 0.97, radiant temperature is calculated from Equation 5-8 as

$$T_{rad} = 0.97^{1/4} \times 283°K$$
$$= 281°K \text{ (or } 8°C)$$

For the portion of the block with an emissivity of 0.06,

$$T_{rad} = 0.06^{1/4} \times 283°K$$
$$= 140°K \text{ (or } -133°C)$$

which is 141°K lower than the radiant temperature for the portion of the aluminum block with high emissivity. An alternate

way to understand the low radiant temperature of the shiny surface involves Equation 5-3. Because the emissivity is low, the absorptivity is also low, and the reflectivity is therefore high. In the out-of-doors with no clouds, the very low temperature of outer space is reflected by the shiny aluminum surface. For this reason, metallic objects such as airplanes and metal-roofed buildings have cold radiant temperatures.

Thermal Conductivity Thermal conductivity (K) is the rate at which heat will pass through a material. It is expressed as calories per centimeter per second per degree Celsius (cal $\cdot$ cm^{-1} $\cdot$ sec^{-1} $\cdot$ °C), which is the number of calories that will pass through a 1-cm cube of the material in 1 sec when two opposite faces are maintained at a 1°C difference in temperature. Table 5-2 gives thermal conductivities for geologic materials. For any rock type the thermal conductivity may vary by up to 20 percent from the value given. Thermal conductivities of porous materials may vary by up to 200 percent depending on the nature of the substance that fills the pores. Rocks and soils are relatively poor conductors of heat. The average thermal conductivity of the materials in Table 5-2 is

TABLE 5-2 Thermal properties of geologic materials and water at 20°C

Material	Thermal conductivity (K), $cal \cdot cm^{-1} \cdot sec^{-1} \cdot °C^{-1}$	Density (ρ), $g \cdot cm^{-3}$	Thermal capacity (c), $cal \cdot g^{-1} \cdot °C^{-1}$	Thermal diffusivity (k), $cm^2 \cdot sec^{-1}$	Thermal inertia (P), $cal \cdot cm^{-2} \cdot sec^{-1/2} \cdot °C^{-1}$
1. Basalt	0.0050	2.8	0.20	0.009	0.053
2. Clay soil, moist	0.0030	1.7	0.35	0.005	0.042
3. Dolomite	0.012	2.6	0.18	0.026	0.075
4. Gabbro	0.0060	3.0	0.17	0.012	0.055
5. Granite	0.0075	2.6	0.16	0.016	0.056
6. Gravel	0.0030	2.0	0.18	0.008	0.033
7. Limestone	0.0048	2.5	0.17	0.011	0.045
8. Marble	0.0055	2.7	0.21	0.010	0.056
9. Obsidian	0.0030	2.4	0.17	0.007	0.035
10. Peridotite	0.011	3.2	0.20	0.017	0.084
11. Pumice, loose, dry	0.0006	1.0	0.16	0.004	0.009
12. Quartzite	0.012	2.7	0.17	0.026	0.074
13. Rhyolite	0.0055	2.5	0.16	0.014	0.047
14. Sandy gravel	0.0060	2.1	0.20	0.014	0.050
15. Sandy soil	0.0014	1.8	0.24	0.003	0.024
16. Sandstone, quartz	0.0120	2.5	0.19	0.013	0.075
17. Serpentine	0.0063	2.4	0.23	0.013	0.059
18. Shale	0.0042	2.3	0.17	0.008	0.041
19. Slate	0.0050	2.8	0.17	0.011	0.049
20. Syenite	0.0077	2.2	0.23	0.009	0.062
21. Tuff, welded	0.0028	1.8	0.20	0.008	0.032
22. Water	0.0013	1.0	1.01	0.001	0.036

Source: Janza and others (1975, Table 4-1).

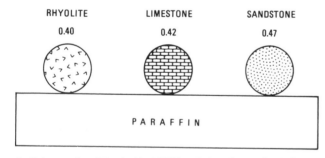

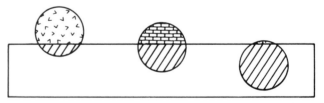

A. Spheres of rock heated to 100°C and placed on a sheet of paraffin. The value for each rock is the product of its thermal capacity (c) and density (ρ) in cal · cm^{-3} · °C^{-1}.

B. After the rocks and paraffin have reached the same temperature.

Figure 5-4 Effect of differences in the thermal capacity of various rock types.

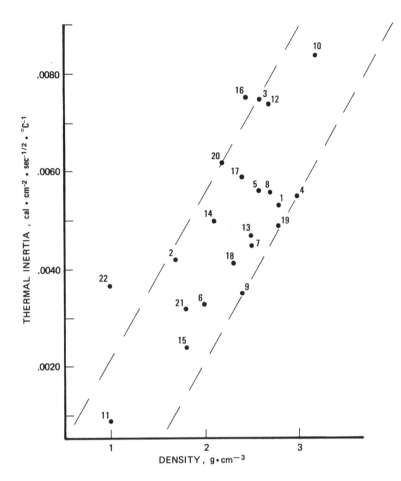

Figure 5-5 Relationship between thermal inertia and density for rocks and water. Numbers refer to the materials in Table 5-2.

0.006 cal · cm^{-1} · sec^{-1} · °C^{-1}, which is two orders of magnitude lower than the thermal conductivity of such metals as aluminum, copper, and silver.

Thermal Capacity *Thermal capacity* (*c*) is the ability of a material to store heat. Thermal capacity is the number of calories required to raise the temperature of 1 g of a material by 1°C and is expressed in calories per gram per degree Celsius (cal · g^{-1} · °C^{-1}). In Table 5-2 note that water has the highest thermal capacity of any substance. Figure 5-4 shows the difference between thermal capacity and kinetic temperature. Spheres of the same volume made from rhyolite, limestone, and sandstone are heated to a temperature of 100°C. The values for thermal capacity and density, taken from Table 5-2, are multiplied to determine the number of calories per cubic centimeter per degree centigrade that each rock stores. The rocks are assumed to have a uniform density of 2.5 g · cm^{-3}; therefore, the different values are determined solely by differences in thermal capacity. As Figure 5-4A shows, sandstone stores the greatest amount of heat and rhyolite stores the least. The

heated spheres are simultaneously placed on a sheet of paraffin. Melting ceases when the spheres and paraffin have reached a uniform temperature. As Figure 5-4B shows, the amount of melting is related to the thermal capacity of the rocks and not to their temperature.

Thermal Inertia *Thermal inertia* (*P*) is a measure of the thermal response of a material to temperature changes and is given in calories per square centimeter per second square root per degree Celsius (cal · cm^{-2} · sec$^{-1/2}$ · °C^{-1}). Thermal inertia is expressed as

$$P = (K \cdot \rho \cdot c)^{1/2} \tag{5-9}$$

where *K* is thermal conductivity, ρ is density, and *c* is thermal capacity. Of the three properties that determine thermal inertia, density is the most important. For the most part, thermal inertia increases linearly with increasing density, as shown in Figure 5-5, which plots the values for the materials listed in Table 5-2.

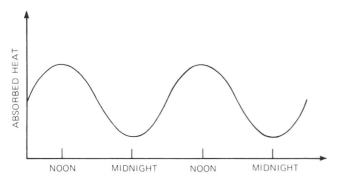

A. Solar heating cycle.

Figure 5-6 Effect of differences in thermal inertia on surface temperatures during diurnal solar cycles. Note differences in ΔT for materials with high and low thermal inertia.

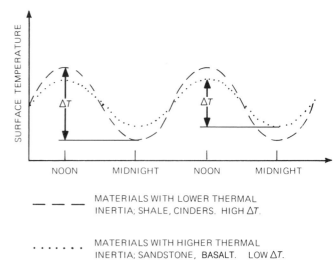

– – – – MATERIALS WITH LOWER THERMAL INERTIA; SHALE, CINDERS. HIGH ΔT.

· · · · · · MATERIALS WITH HIGHER THERMAL INERTIA; SANDSTONE, BASALT. LOW ΔT.

B. Variations in surface temperature.

Figure 5-6 illustrates the effect of differences in thermal inertia on surface temperatures. The difference between maximum and minimum temperature occurring during a diurnal solar cycle is called ΔT. Materials with low thermal inertia, such as shale and volcanic cinders, have low resistance to temperature change and have a relatively high ΔT. These materials reach a high maximum surface temperature in the daytime and a low minimum temperature at night. Materials with high thermal inertia, such as sandstone and basalt, strongly resist temperature changes and have a relatively low ΔT. These materials are relatively cool in the daytime and warm at night.

Apparent Thermal Inertia Thermal inertia cannot be measured by remote sensing methods because conductivity, density, and thermal capacity must be measured by contact methods. Maximum and minimum radiant temperature, however, can be measured from digitally recorded daytime and nighttime images. For corresponding ground resolution cells, ΔT is determined by subtracting the nighttime temperature from the daytime temperature. The fact that ΔT is low for materials with high thermal inertia and high for those with low thermal inertia (Figure 5-6) is used to determine the *apparent thermal inertia* (*ATI*) by the relationship

$$ATI = \frac{1 - A}{\Delta T} \qquad (5\text{-}10)$$

where A is the albedo in the visible band. Albedo is employed to compensate for the effects that differences in absorptivity have on radiant temperature. During the day, dark materials (with low albedo) absorb more sunlight than light materials (with high albedo). The absorbed solar energy increases kinetic temperature, which increases the radiant thermal energy.

Therefore a dark material typically has a higher ΔT than an otherwise identical material that has a light color. The term $1 - A$ corrects for some of these effects. A typical *ATI* image produced from a visible (albedo) image and daytime and nighttime thermal images is illustrated and interpreted in the later section "Heat Capacity Mapping Mission."

ATI images must be interpreted with caution because ΔT may be influenced by factors other than thermal inertia. Consider an area of uniform material, such as granite, that has high topographic relief. In a daytime IR image, the shadowed areas have a lower radiant temperature than sunlit areas. In a night-

Table 5-3 Apparent thermal inertia values measured in the field with a thermal inertia meter

Material	*ATI*[*]
1. Sandy alluvium	0.014
2. Sand, windblown	0.015
3. Rhyolite tuff	0.022
4. Clay-silt playa	0.024
5. Basalt, pahoehoe	0.039
6. Basalt, olivine	0.042
7. Basalt, aa	0.042
8. Barite	0.043
9. Andesite	0.044
10. Rhyodacite, silicified	0.048
11. Chert	0.053

[*]*ATI* values are relative to a dolomite standard with a thermal inertia of 0.984 cal $\cdot$ cm^{-2} $\cdot$ sec$^{-1/2}$ $\cdot$ °C^{-1}. Error limits are approximately ±10 percent.

Source: Kahle and others (1981, Table 4).

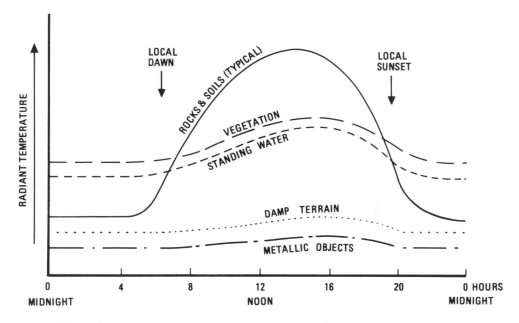

Figure 5-7 Diurnal radiant temperature curves (diagrammatic) for typical materials.

time image the sunlit and shadowed areas have similar temperatures. As a result the shadowed granite has a lower ΔT and a higher *ATI* than the same granite that is sunlit. Some *ATI* computer programs compensate for shadows by using topographic data together with information on solar elevation and azimuth. Water and vegetation typically have *ATI* values that are determined by factors other than their thermal inertias, as discussed in the later section "Influence of Water and Vegetation."

ATI may be measured with a thermal inertia meter, which employs a radiometer and standard rock samples of known emissivity and thermal inertia (Kahle and others, 1981). Electric lamps heat the target material and the standards to a uniform level, the radiometer measures radiant temperatures, and the meter then calculates the *ATI* target material. Table 5.3 lists *ATI* values determined in the field using this meter. These relative values are useful for discriminating among different materials and are not intended as absolute measures of thermal inertia.

Thermal Diffusivity The same values used to determine thermal inertia may be used to determine *thermal diffusivity* (*k*), which is given as

$$k = \frac{K}{c \cdot \rho} \tag{5-11}$$

Thermal diffusivity (given in centimeters squared per second) governs the rate at which temperature changes within a substance. More specifically, it states the ability of a substance

during a daytime period of solar heating to transfer heat from the surface to the interior and during a nighttime period of cooling to transfer stored heat to the surface.

Diurnal Temperature Variation

Figure 5-7 shows typical diurnal variations in radiant temperature. The most rapid temperature changes, shown by steep curves, occur near dawn and sunset. At the intersection of two curves (called *thermal crossover*), radiant temperatures are identical for both materials. Data on diurnal changes in radiant temperature for various materials are given in the later section "Indio Hills, California."

IR DETECTION AND IMAGING TECHNOLOGY

The pattern of radiant temperature variations of the terrain may be recorded as an image by *airborne infrared scanners.*

Airborne IR Scanners

Thermal IR images are produced by airborne scanner systems that consist of three basic components: (1) an optical-mechanical scanning system, (2) a thermal IR detector, and (3) an image recording system. As shown in Figure 5-8, the cross-track scanning system consists of an electric motor mounted in the aircraft with the rotating shaft oriented parallel with the aircraft fuselage and hence the flight direction. A front-surface

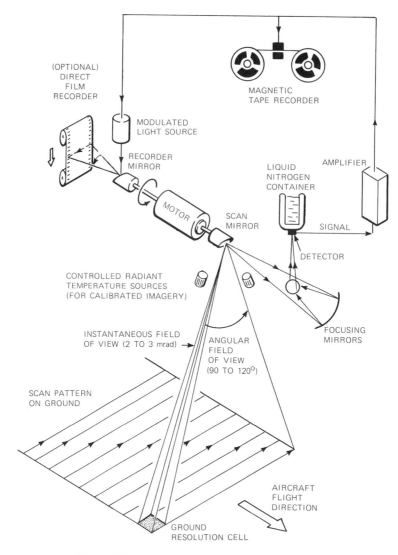

Figure 5-8 Diagram of a thermal IR scanner system.

mirror with a 45° facet mounted on the end of the shaft sweeps the terrain at a right angle to the flight path. The angular field of view is typically 90°. IR energy radiated from the terrain is focused from the scan mirror onto the detector, which converts the radiant energy into an electrical signal that varies in proportion to the intensity of the IR radiation. Detectors are cooled to reduce the electronic noise caused by the molecular vibrations of the detector crystal. In scanners such as the one in Figure 5-8, the detector is enclosed by a vacuum bottle, or *dewar*, filled with liquid nitrogen (73°K). Other detectors employ a closed-cycle mechanical cooler to maintain a low detector temperature. The scanner data are recorded on magnetic tapes that are played back onto film in the laboratory to produce images. Figure 5-8 shows an optional device for recording images directly on film during the flight.

Circular IR scanners, described in Chapter 1, are used in aircraft for nighttime navigation, reconnaissance, and target recog-

nition. The axis of rotation can be tilted forward to acquire images several kilometers ahead of the aircraft. In military aviation these are called *forward-looking IR* (FLIR) systems.

Additional information on thermal IR detectors and systems is given in Norwood and Lansing (1983). IR scanners in satellites employ the same principles as those in aircraft.

Stationary IR Scanners

Airborne scanners are designed for use in aircraft or spacecraft where the forward movement of the vehicle provides coverage along the flight path and the rotating scan mirror provides coverage at right angles to that direction. *Stationary scanners* are designed to acquire images from a fixed position. In these scanners the faceted scan mirror rotates about a vertical axis to provide coverage in the horizontal direction. Coverage in the vertical direction is provided by a plane mirror that tilts about a horizontal axis. Another system employs a television-type camera that electronically scans the scene. On both types of stationary scanners, the radiation transmitted through an IR filter is focused onto a detector that converts the radiation into an electrical signal. The resulting image is displayed in real time on a small television-type screen and may also be recorded for later analysis.

One use of stationary scanners is to record the pattern of heat radiating from the human body. This medical application is called *thermography;* the images are called *thermograms.* Tumors and impaired blood circulation are physiological disorders that are detectable on thermograms. Stationary scanners are also used to monitor industrial facilities for hot spots that may indicate potential problems. Anomalous hot spots on the exterior of industrial furnaces may be areas where the fire brick lining has eroded and failure is imminent. In electrical transmission facilities, faulty transformers and insulators have been detected by their high radiant temperatures. Railroads use IR scanners to detect overheated wheel bearings in moving trains.

CHARACTERISTICS OF IR IMAGES

On most thermal IR images, the brightest tones represent the warmest radiant temperatures, and the darkest tones represent the coolest ones (Figure 5-9). The apparent similarity of IR images to black-and-white aerial photographs results from the fact that both are displayed as gray-scale variations on film. In photography, film acts as the medium for detecting, recording, and displaying reflected energy in the 0.4-to-0.9-μm wavelength region. For thermal IR images, however, a semiconductor device detects the energy, and film serves only as a medium to display radiant temperatures.

Effects of Weather on Images

Clouds typically show the patchy warm-and-cool pattern illustrated in Figure 5-9A, where the dark signatures are relatively cool and the bright signatures are relatively warm. Scattered rain showers produce a pattern of streaks parallel with the scan lines on the image. A heavy overcast layer reduces thermal contrasts between terrain objects because of reradiation of energy between the terrain and cloud layer. Images may be acquired by flying below the cloud layer, but the resulting thermal contrast is relatively low.

Wind produces characteristic patterns of smears and streaks on images. *Wind smears* (Figure 5-9B) are parallel curved lines of alternating lighter and darker signatures that may extend over wide expanses of the image. *Wind streaks* occur downwind from obstructions on flat terrain and typically appear as the warm (bright) patterns shown in Figure 5-9C. In this example the wind is blowing from right to left across obstructions, which are clumps of trees with warm signatures. Wind velocity

A. Clouds.

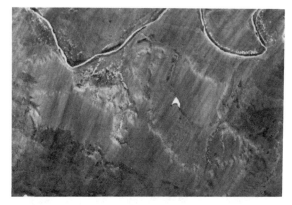

B. Surface wind smears.

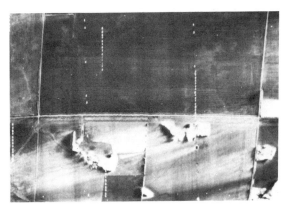

C. Surface wind streaks.

Figure 5-9 Effects of weather on thermal IR images.

is lower downwind from obstructions, which reduces the cooling effect; thus terrain in the sheltered areas is warmer than terrain exposed to the wind. Wind smears and streaks may be avoided by acquiring images only on calm nights, but, in many regions, surface winds persist for much of the year and their effects must be endured. Interpreters must be alert to avoid confusing wind-caused signatures with terrain features.

A. Visible image.

B. Thermal IR image.

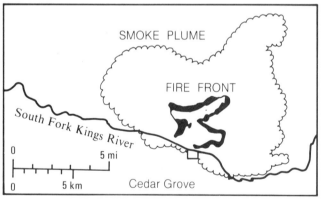

C. Location map.

Figure 5-10 Penetration of smoke by a thermal IR image. Forest fire in King's Canyon, Sequoia National Forest, California. Courtesy NASA Ames Research Center.

Penetration of Smoke Plumes

Clouds consist of tiny particles of ice or water that have the same temperature as the surrounding air. As shown in Figure 5-9B, images acquired from aircraft or satellites above cloud banks record the radiant temperature of the clouds. Energy from the earth's surface does not penetrate the clouds but is absorbed and reradiated. Smoke plumes, however, consist of ash particles and other combustion products so fine that they are readily penetrated by the relatively long wavelengths of thermal IR radiation.

Figure 5-10 shows a visible image and a thermal IR image that were acquired simultaneously during a daytime flight over a forest fire. The smoke plume completely conceals the ground in the visible image (Figure 5-10A), but terrain features are clearly visible in the IR image (Figure 5-10B), where the burning front of the fire has a bright signature. The U.S. Forest Service uses aircraft equipped with IR scanners that produce image copies in flight, which are dropped to fire fighters on the ground. These images provide information about the fire location that cannot be obtained by visual observation through the smoke plumes. IR images are also acquired after fires are extinguished in order to detect hot spots that could reignite.

Influence of Water and Vegetation

The thermal inertia of water is similar to that of soils and rocks (Table 5-2), but in the daytime, water bodies have a cooler surface temperature than soils and rocks. At night the relative surface temperatures are reversed, so that water is warmer than soils and rocks (Figure 5-7). This reversal in relative temperatures is apparent when day and night images are compared for HCMM images, which are described later in the chapter. Convection currents maintain a relatively uniform temperature at the surface of a water body. Convection does not operate to transfer heat in soils and rocks; therefore, heat from solar flux is concentrated near the surface of these solids in the daytime, causing a higher surface temperature. At night this heat radiates into the atmosphere and is not replenished by convection currents in these solid materials, causing surface temperatures to be lower than in adjacent water bodies (K. Watson, personal communication). Some images may not be annotated for the time of day at which they were acquired. The thermal signatures of water bodies are a reliable index to the time of image acquisition. If water bodies have warm signatures relative to the adjacent terrain, the image was acquired at night; relatively cool water bodies indicate daytime imagery.

As shown in the diurnal temperature curves (Figure 5-7), damp soil is cooler than dry soil, both day and night. As absorbed water evaporates, it cools the soil. Figure 5-11 clearly shows this *evaporative cooling* effect. The aerial photograph and the IR image were simultaneously acquired over an orchard of immature trees where adjacent rows were irrigated on successive days. In the photograph only the wettest, most recently irrigated soil has a discernibly darker tone than the sur-

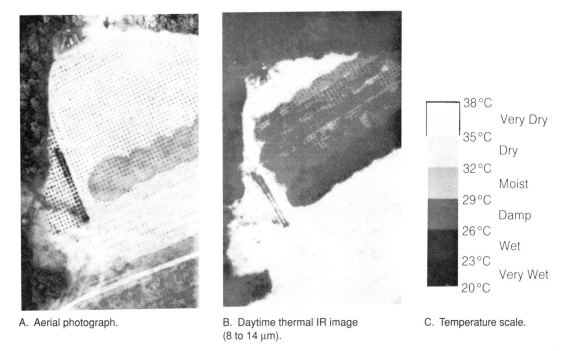

A. Aerial photograph.

B. Daytime thermal IR image (8 to 14 μm).

C. Temperature scale.

38°C — Very Dry
35°C — Dry
32°C — Moist
29°C — Damp
26°C — Wet
23°C — Very Wet
20°C

Figure 5-11 Relationship between soil moisture and radiant temperature in an irrigated orchard. Courtesy Daedalus Enterprises, Inc.

roundings. In the IR image, however, the moisture pattern is clearly visible as variations in radiant temperature. The gray scale of the image is calibrated to show radiant temperature, which in turn correlates with moisture content. Some researchers have noted that adding water to dry soil increases the thermal inertia of the soil to values comparable to those of rocks. The effect of evaporative cooling, however, dominates the radiant temperature signature of damp ground. Many geologic faults and fractures are recognizable in IR images because of evaporative cooling. Examples from California and South Africa are described later in the chapter.

Green deciduous vegetation has a cool signature on daytime images and a warm signature on nighttime images. During the day, transpiration of water vapor lowers leaf temperature, causing vegetation to have a cool signature relative to the surrounding soil. At night the insulating effect of leafy foliage and the high water content retain heat, which results in warm nighttime temperatures. The relatively high nighttime and low daytime radiant temperature of conifers, however, does not appear to be related to their water content. The composite emissivity of the needle clusters making up a whole tree approaches that of a blackbody. Dry vegetation, such as crop stubble in agricultural areas, appears warm on nighttime imagery in contrast to bare soil, which is cool. The dry vegetation insulates the ground to retain heat and causes the warm nighttime signature.

For reasons discussed earlier, water, moist soil, and vegetation have relatively low values of ΔT when their signatures are compared on daytime and nighttime IR images. When these

ΔT values are used in Equation 5-10, the resulting high *ATI* values differ significantly from the actual thermal inertias for these materials, which have low to intermediate values. Therefore one must be cautious when interpreting areas of water, moist soil, or vegetation in *ATI* images.

Temperature Calibration of Images

Most of the older IR scanners lacked temperature calibration and produced images in which the gray tones recorded relative rather than absolute radiant temperatures. These qualitative images were satisfactory for many purposes. In the early 1970s manufacturers began to equip scanners with internal temperature calibration sources that are now standard on most systems. The scanner in Figure 5-8 has temperature calibration sources mounted on either side of the angular field of view. The scanner records the radiant temperature of the first calibration source, then sweeps the terrain, and finally records the temperature of the second source. The resulting signal is recorded on magnetic tape and has the appearance shown in the upper part of Figure 5-12A. Calibration source BB1 was set at a temperature of 84°F and source BB2 was set at 102°F. These reference temperatures provide a scale for determining the temperature at any point along the magnetic tape record of the terrain temperature. During playback of the magnetic tape, the temperature range may be divided into intervals and displayed quantitatively. One playback system employs six colors plus black and white to display the temperature values. In Figure 5-12B

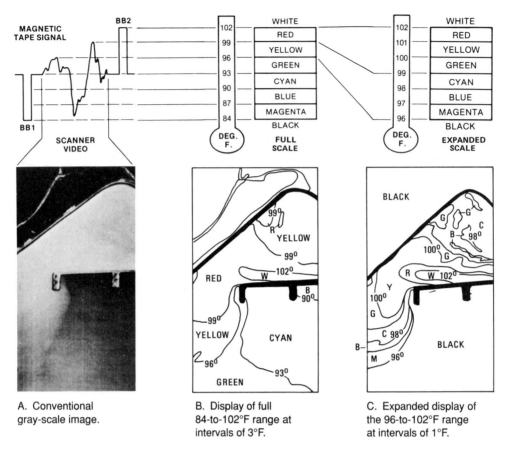

A. Conventional gray-scale image.

B. Display of full 84-to-102°F range at intervals of 3°F.

C. Expanded display of the 96-to-102°F range at intervals of 1°F.

Figure 5-12 Temperature-calibrated thermal IR image of power-plant discharge. Courtesy Daedalus Enterprises, Inc.

each color has been assigned a temperature interval of 3°F. The boundary between any two color areas is a temperature contour, or *isotherm*. Finer temperature detail may be displayed by assigning each color to a 1°F interval (Figure 5-12C). Comparison of the calibrated displays with the conventional gray-scale image (Figure 5-12A) illustrates the advantage of quantitative images. Plate 9A shows a color display of calibrated images prepared in this manner for Hawaiian volcanoes.

CONDUCTING AIRBORNE IR SURVEYS

Aircraft thermal IR surveys have not been made for most areas; therefore, airborne surveys must be conducted over an area of interest. The acquisition cost of IR images is approximately three times that of aerial photographs, although the cost per square kilometer decreases as the area of the survey increases. The flight line may be repeated at different times of day to evaluate diurnal thermal variations. Figure 5-13 shows repeated images of the Caliente and Temblor Ranges, California. For any IR survey the following factors must be considered: time of day, wavelength band, and orientation and altitude of flight lines.

Time of Day

On daytime images, topography is the dominant feature because of solar heating and shadowing. On nighttime images, however, the effects of solar heating and shadowing have been reradiated to the atmosphere and the images show thermal properties of materials. The nighttime and daytime images of the Caliente and Temblor Ranges, California, illustrate these differences (Figure 5-13). On the daytime image, the ridges and slopes that face south and east are heated by the morning sun and have bright (warm) signatures; those facing north and west are shadowed and have dark (cool) signatures. The daytime image is dominated by these topographic effects. On the nighttime image (Figure 5-13B), topographic features are largely eliminated and geologic features are emphasized, as shown by comparison with the map (Figure 5-13C). The narrow warm signatures in the Caliente Range are basalt outcrops. In the Temblor Range the bands with cool signatures are outcrops of shale and siltstone. The broad belts of warm signature are sandstone and conglomerate outcrops. These geologic features are obscure or invisible in the daytime image.

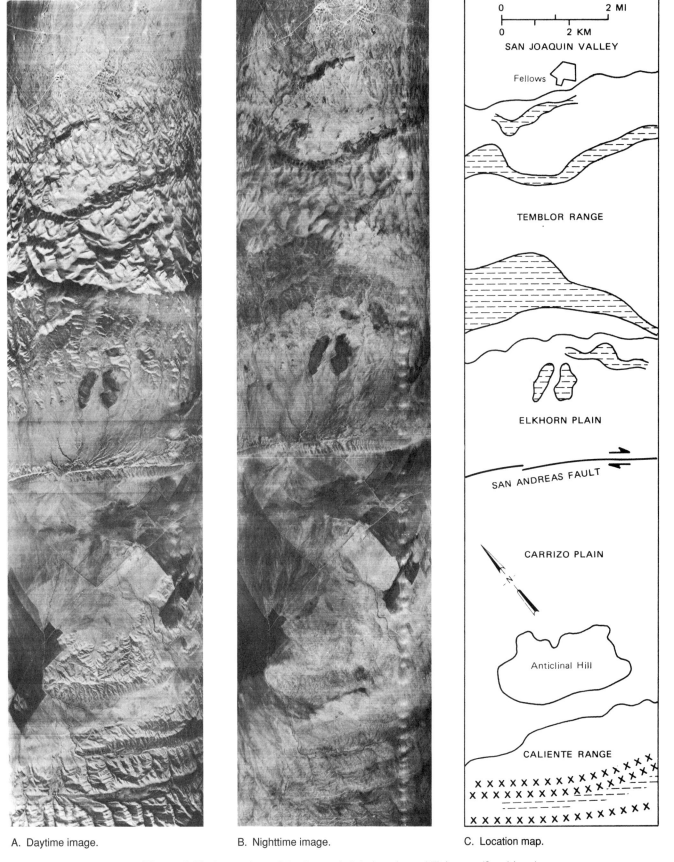

A. Daytime image. B. Nighttime image. C. Location map.

Figure 5-13 Comparison of daytime and nighttime thermal IR images (8 to 14 μm), Caliente and Temblor Ranges, California. Crosses on the map (C) are basalt outcrops; dashes are shale outcrops. From Wolfe (1971, Figures 3 and 4). Courtesy E. W. Wolfe, U.S. Geological Survey.

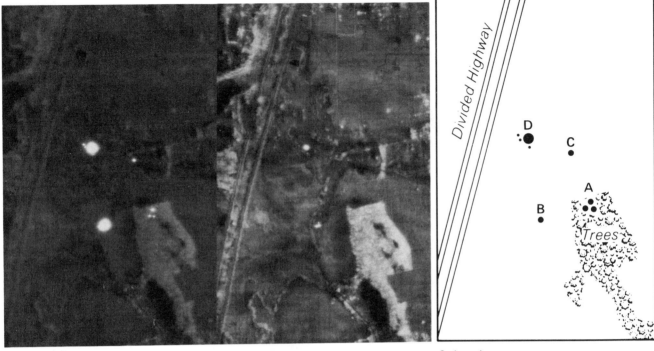

A. 3-to-5-μm band.   B. 8-to-14-μm band.   C. Location map.

Figure 5-14 Different spectral bands of nighttime thermal IR images, central Michigan. Letters on the map (C) are explained in the text. Courtesy Daedalus Enterprises, Inc.

Wavelength Bands

Thermal IR images may be acquired at wavelength bands of 3 to 5 μm and 8 to 14 μm, which are atmospheric windows (Figure 5-1). Wien's displacement law enables us to determine the temperatures at which the maximum energy will radiate for each of these bands. Figure 5-2 shows that the 3-to-5-μm band corresponds to the radiant energy peak for temperatures of 600°K and greater, which are associated with fires, lava flows, and other hot features. The 8-to-14-μm band spans the radiant energy peak for a temperature of 300°K; this is the ambient temperature of the earth, which has a radiant energy peak at 9.7 μm. Images in the 3-to-5-μm band are commonly acquired with an indium-antimonide detector; in the 8-to-14-μm band, mercury-cadmium-telluride detectors are employed. Norwood and Lansing (1983) discuss the characteristics of these and other detectors.

Figure 5-14 shows nighttime images in central Michigan acquired by a multispectral scanner that recorded images in both the 3-to-5-μm and 8-to-14-μm bands. The area on the images consists of pastures, fields, and woodlands cut by a network of roads and a few streams. In the 8-to-14-μm image (Figure 5-14B), terrain features are well expressed and have the following signatures: trees and freeways are relatively warm, fields have intermediate temperatures, and marshy areas along the left margin of the image are cool. The overall radiant temperature level of the 3-to-5-μm image (Figure 5-14A) is lower and the thermal contrasts among terrain features are much lower than in the 8-to-14-μm image. Of special interest are localities A through D in the location map (Figure 5-14C), which represent the following features:

A. Three small fires of glowing charcoal briquettes are located within a grove of trees. On the 8-to-14-μm image the warm signature of the trees effectively masks the fires, but on the 3-to-5-μm image the fires are clearly visible and the signature of the trees is subdued.
B. A large campfire in an open field is visible on both images.
C. In an open field, a pit containing a small charcoal fire is concealed beneath a pile of brush. On the 8-to-14-μm image this target could be mistaken for vegetation, but on the 3-to-5-μm image it is clearly recognizable as a hot target.
D. A large campfire and three vehicles with warm engines are located in an open field. The four targets are resolvable on the 3-to-5-μm image, but the campfire signature conceals the other targets on the 8-to-14-μm image.

This example illustrates that 8-to-14-μm images are optimum for terrain mapping, whereas 3-to-5-μm images are optimum for mapping hot targets, such as fires and volcanic eruptions.

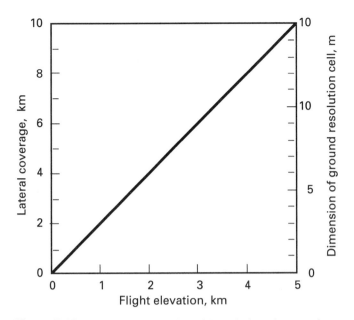

Figure 5-15 Lateral coverage and spatial resolution of scanner images as a function of flight elevation. For this scanner the angular field of view is 90° and the instantaneous field of view is 3.0 mrad.

Findlay and Cutten (1989) used a mathematical modeling approach to compare systems operating at 3 to 5 μm and 8 to 12 μm in a tropical maritime environment. They also concluded that the 8-to-12-μm band is superior for all but the hottest targets.

Orientation and Elevation of Flight Lines

When planning geologic projects, one should know the regional structural strike or tectonic "grain" of the area in order to orient flight lines in the optimum direction. If flight lines are oriented normal to the regional strike, the scan-line pattern will be parallel with the strike and may mask linear geologic features. This problem is avoided by orienting flight lines parallel with, or at an acute angle to, the regional strike. On very high quality images acquired with modern scanner systems, however, the scan lines are not a severe problem.

Flight elevation, which is height above average terrain, influences swath width and the spatial resolution of images. Figure 5-15 shows image swath width as a function of flight elevation for a scanner system with a 90° angular field of view. At an elevation of 2 km, for example, the swath width is 4 km.

Spatial Resolution and Information Content of Images

The spatial resolution of IR images is determined by the flight elevation and by the instantaneous field of view (*IFOV*) of the detector, which typically ranges from 2 to 3 mrad. At an elevation of 1000 m, 1 mrad subtends a distance of 1 m. Figure 5-15

shows the ground resolution cell as a function of elevation for a 3-mrad detector. Resolution becomes coarser toward the margins of the image because the greater viewing distance increases the size of the ground resolution cell. Spatial resolution, however, is not the sole measure of image quality. Detection of thermal patterns is determined primarily by differences in radiant temperature, which are expressed as tonal contrast on the image. Aircraft images acquired at elevations of a few kilometers provide adequate spatial resolution for most applications. For regional investigations, however, images acquired from satellite altitudes are useful, as demonstrated later in this chapter.

Ground Measurements

As described earlier, weather and surface conditions play a large role in determining terrain expression in IR images. It may be useful to collect ground information on weather conditions, soil moisture, and vegetation at the time of the IR survey. In the early days of remote sensing, the term *ground truth* was coined for these measurements, but most investigators have abandoned the term. Measurements made on the ground are no more truthful than those made remotely. Ground measurements are most practical and useful for surveys of relatively small areas that can be covered with a single flight line. If repeated flights are made, ground data on changing weather conditions and solar flux may be valuable in comparing and interpreting the images.

For larger areas ground measurements can be made at only a limited number of localities during the 3 to 4 hours required to acquire the images. Ground measurements are most valuable if they are made at localities that have anomalous image signatures. The measurements may help to explain the anomalies or may eliminate possible causes. In practice, however, it is virtually impossible to anticipate where anomalies will occur; therefore, most regional surveys omit contemporaneous ground measurements. The availability of calibrated thermal IR images also reduces the need for ground measurements.

Ground measurements to help explain image signatures may be made some time after the image flight, as in the Indio Hills example shown later in the chapter. Ground information is useful for understanding other forms of remote sensing imagery in addition to thermal IR images. The type of ground measurements will depend on the wavelength region of the airborne sensor. Dozier and Strahler (1983) describe typical ground measurements recorded in support of remote sensing surveys.

LAND USE AND LAND COVER—ANN ARBOR, MICHIGAN

The discussion of image interpretation begins with a scene familiar to all readers: a city and its surroundings (Figure 5-16). The city of Ann Arbor is located west of Detroit and includes the University of Michigan. As shown in the IR image and aerial photograph, the central city is surrounded by residential suburbs,

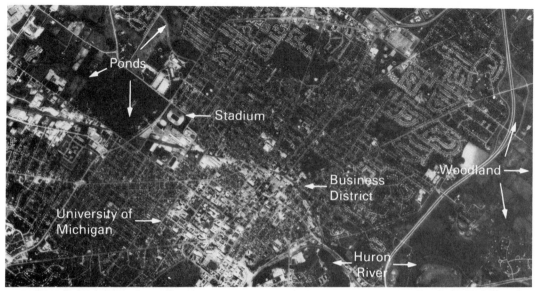

A. Aerial photograph acquired May 5, 1983.

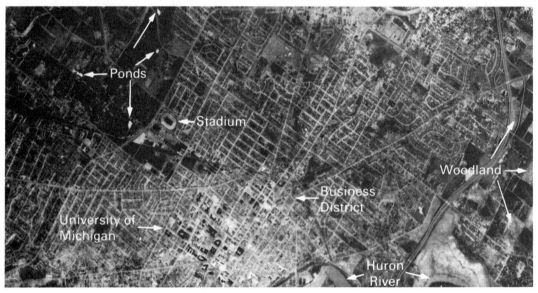

B. Thermal IR image (8 to 14 µm) acquired May 9, 1979, at 6:30 am. Courtesy Daedalus Enterprises, Inc.

Figure 5-16 Ann Arbor, Michigan.

which in turn are surrounded by open land. The daytime IR image was acquired shortly after sunrise, when radiant temperatures were rising rapidly, as seen in the diurnal curves of Figure 5-7. The road and highway network has a conspicuous warm signature because of the high thermal inertia of the relatively dense concrete and asphalt. Building roofs are cool because of the low emissivity of these materials. The inner city is noticeably warmer than the surrounding area because of the concentration of heat-generating activities. Water in the lakes and the Huron River (lower right corner of the images) is warmer than the surrounding soil, which has not yet absorbed sufficient solar energy to reach the typical high radiant temperatures of midday.

Meadows and fields are relatively cool, but wooded areas are warm. The stadium in the left central part of the scene has a warm grandstand and cool playing field, which is covered with artificial turf. IR images acquired at lower elevations with higher spatial resolution are used for heat-loss surveys of urban and industrial areas, as described in the following section.

HEAT-LOSS SURVEYS

An obvious application of thermal IR technology is to survey heated buildings, factories, and buried steam lines for anom-

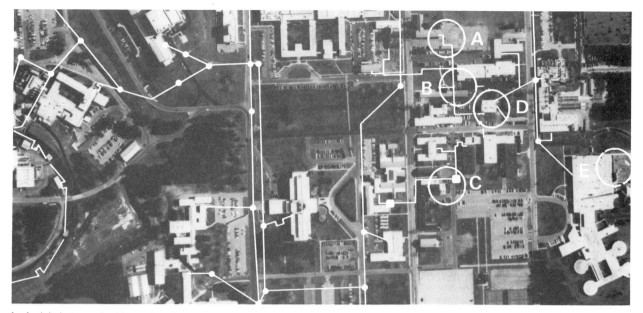

A. Aerial photograph with overlay of heating lines.

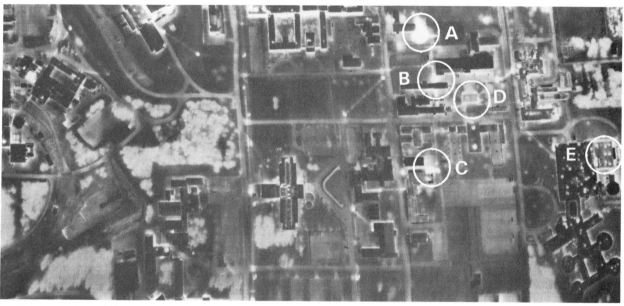

B. Night thermal IR image (8 to 14 μm).

Figure 5-17 Heat-loss survey of Brookhaven National Laboratories, Long Island, New York. Localities are explained in the text. Courtesy Daedalus Enterprises, Inc.

alous hot spots that may indicate poorly insulated roofs and steam leakage. Airborne IR surveys and image interpretations are relatively inexpensive. By locating and correcting heat losses, the fuel saved in a few months can repay the cost of a survey. In the northern and central United States, many building complexes, such as university campuses and industrial complexes, are heated by steam distributed through buried pipelines from a central generating station. As the steam heats the buildings, it condenses into hot water, which returns via

condensate lines to the steam plant, where it is used to generate more steam. Many of the pipeline systems are several decades old and have developed leaks that are difficult to detect because the lines are buried and many stretches are covered by sidewalks, streets, and parking lots.

Brookhaven National Laboratory on Long Island, New York, contracted for a heat-loss survey. Aerial photographs and nighttime thermal IR images were acquired in November 1976. The aerial photograph (Figure 5-17A) includes an overlay of buried steam and condensate lines and manholes that was provided by the Brookhaven maintenance staff. On the nighttime IR image (Figure 5-17B) trees, standing water, and pavement are relatively warm. Roofs of well-insulated buildings are cool with warm spots formed by exhaust ventilators. Sides of buildings are relatively warm because of heat radiated from windows. The buried heating lines and manholes form bright lines and spots. In some surveys the IR images revealed locations of lines for which the engineering records had been lost. Localities A, B, and C in the image and photograph are anomalous hot spots that proved to be major leaks or areas where pipe insulation had deteriorated. Localities D and E are building roofs that are significantly warmer than other roofs in the laboratory complex. Significant energy costs can be saved by improving the ceiling insulation in these buildings.

A number of other facilities have been surveyed with results similar to those at Brookhaven. Anderson and Baker (1987) conducted a heat-loss survey of central Scotland; they identified a number of facilities with high heat loss that could be eliminated to achieve considerable savings.

THE INDIO HILLS, CALIFORNIA

The Indio Hills, in the eastern part of the Coachella Valley, are a low range of deformed clastic sedimentary rocks of late Tertiary age trending southeast parallel with the San Andreas fault zone. A geologic map (Figure 5-18), an aerial photograph (Figure 5-19A), and a nighttime IR image (Figure 5-19B) cover the south end of the hills. This arid terrain, with little vegetation and well-exposed bedrock, is ideal for acquiring and analyzing IR images.

Thermal IR Signatures of Rock Types

The alluvium surrounding the hills has a relatively cool and featureless signature on the nighttime image, which is consistent with the low thermal inertia for this low-density material. Two types of bedrock are readily distinguished on the image. One type has a relatively warm and uniform signature and consists of poorly stratified, moderately to poorly consolidated conglomerate of the Ocotillo, Canebrake, and Mecca Formations

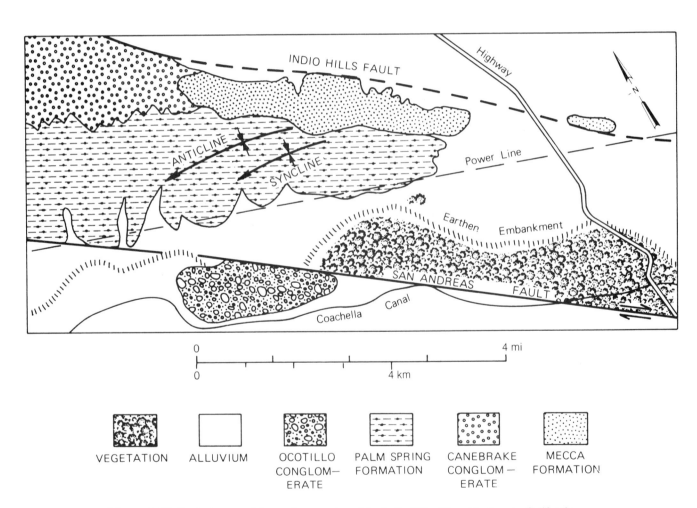

Figure 5-18 Geologic map of the southern portion of the Indio Hills, Riverside County, California.

A. Aerial photograph acquired May 5, 1953.

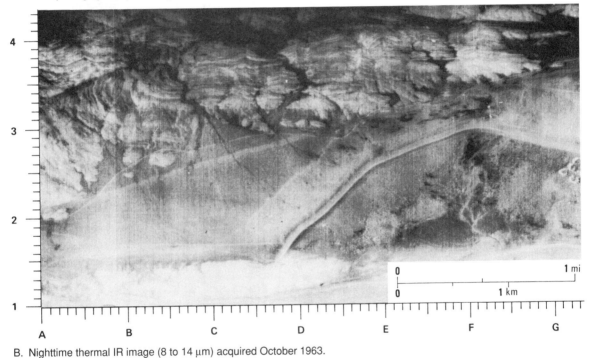

B. Nighttime thermal IR image (8 to 14 μm) acquired October 1963.

Figure 5-19 Images of the southern portion of the Indio Hills, California. From Sabins (1967, Figures 3 and 4).

(Figure 5-18). The outlying hill in the lower center of the image (location 1.3, A.8 to D.5 of Figure 5-19B) is a good example.

The other bedrock type is the Palm Spring Formation, consisting of well-stratified alternating beds of resistant conglomeratic sandstone and nonresistant siltstone up to 12 m thick that erode to form the ridge and slope topography illustrated in

Figure 5-20. The Palm Spring Formation has a distinctive pattern of alternating warm and cool bands on the nighttime IR image. Careful correlation of the image with outcrops in the field established that the warm signatures are sandstones and the cool signatures are siltstones. These signatures were later verified by radiometry studies.

Figure 5-20 Sandstones (ledges) and siltstones (slopes) of the Palm Spring Formation. Radiant temperatures of these outcrops are included in Figure 5-21.

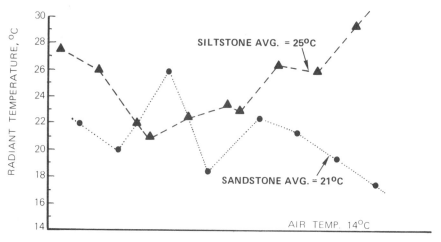

A. Daytime temperatures (8:45 a.m.). The siltstones are warmer than the sandstones, with one exception.

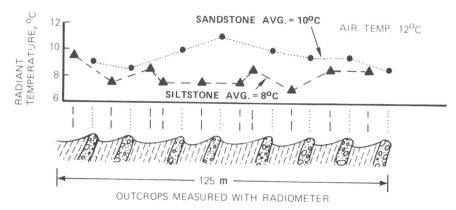

B. Nighttime temperatures. The sandstones are warmer than the siltstones, with two exceptions.

Figure 5-21 Daytime and nighttime radiant temperatures of the sandstones and siltstones of Palm Spring Formation at locality 3.4, B.7 of Figure 5-19B.

Radiometry Investigations

IR radiometers are instruments that make quantitative measurements of radiant temperature. A portable radiometer operating at 8 to 14 μm was used to measure daytime and nighttime radiant temperatures of 8 sandstone and 10 siltstone outcrops at locality 3.5, B.7 in the IR image (Figure 5-19B). Figure 5-21 shows plots of the daytime temperatures (upper diagram) and nighttime temperatures (lower diagram). The average temperature values are as follows:

	Average Temperature, °C		
Rock type	*Day*	*Night*	*ΔT*
Sandstone	21	10	11
Siltstone	25	8	17

These data show that relative to siltstones, the sandstones are cooler in daytime, warmer at night, and have a lower ΔT, which indicates that the sandstone has a higher thermal inertia than the siltstone.

In the nighttime IR image (Figure 5-19B), the sparse vegetation has a distinct warm signature and soil and alluvium are cool. In daytime IR images of similar areas (not illustrated), the signatures are reversed. In order to evaluate these relationships quantitatively, radiometry measurements were made at locality 3.1, F.3 in Figure 5-19B, where three salt-cedar trees have distinctly warmer signatures than the surrounding bare soil. Radiant temperatures were measured during a diurnal cycle for the salt cedars plus three smaller creosote bushes, and the values were averaged. For each observation period, radiant temperature measurements of six soil exposures were also averaged. Figure 5-22 plots the average values for vegetation and soil, together with air temperature readings. This diagram shows that vegetation at night is consistently warmer than soil, with a maximum temperature difference of 4°C. The temperature relationships are reversed during the day, when soil is much warmer than vegetation. Note that the thermal crossovers of the various curves occur within less than 1 hour both in the evening and morning. These diurnal temperature relationships of soil and vegetation have since been confirmed on daytime and nighttime IR images at many localities.

Folds and Faults

Geologic structures shown on the map (Figure 5-18) are well expressed on the IR image. The plunging anticlines and synclines in the Palm Spring Formation are shown by the pattern of the warm and cool signatures. The San Andreas fault borders the west side of the Indio Hills; to the south it passes along the east side of the outlier of Ocotillo Conglomerate; farther south the fault trace is concealed by alluvium and has no topographic expression. On the aerial photograph, the concealed trace of the fault is marked on the northeast side by a vegetation anomaly, which is a concentration of vegetation that abruptly terminates at the fault trace (1.6, D.2 to 1.3, G.3 in Figure 5-19A). On the IR image the fault trace is expressed by an alignment of very cool (dark) anomalies along the northeast side (1.6, E.1 to 1.6, G.3 in Figure 5-19B). The cool anomalies are not related to the vegetation distribution because vegetation is warm on the nighttime image. The cool anomalies are probably related to the barrier effect of the San Andreas fault on groundwater movement. In the spring of 1961, which was a few months before the IR survey, Cummings (1964, p. 34) observed that the water table on the east side of the fault was 15 m shallower than on the west side. The shallower water table causes a high near-surface moisture content. Near-surface moisture results in evaporative cooling, which explains the

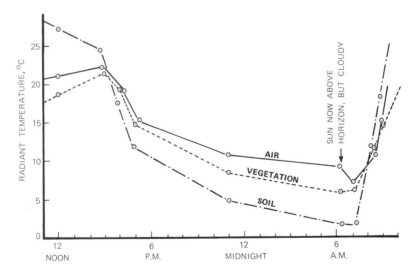

Figure 5-22 Diurnal radiant temperature curves of vegetation and soil at the Indio Hills. Locality 3.1, F.3 of Figure 5-19B.

cool anomaly on the east side of the fault. It also explains the greater density of vegetation on the east side. On IR images of the northern Indio Hills (not illustrated), the Mission Creek fault is indicated by cool anomalies where the fault is a barrier to groundwater movement. Similar thermal anomalies appear on nighttime IR images of the San Andreas fault in the Carrizo Plain 320 km to the northwest.

THE IMLER ROAD AREA, CALIFORNIA

The Imler Road area (Figures 5-23 and 5-24) on the west margin of the Imperial Valley has some similarities to the Indio Hills. Both areas are in the southern California desert and have bedrock of deformed siltstone and sandstone of late Tertiary age. There are major differences, however. The Indio Hills have rugged topography with well-exposed bedrock. The Imler Road area is a featureless plain where bedrock is partially covered by gravel and windblown sand that largely conceal geologic structures. The featureless nature of the area is shown in the aerial photograph (Figure 5-23B). The nighttime thermal IR image (Figure 5-23A), however, displays a wealth of geologic and terrain information that is explained in the following sections.

Terrain Expression

The cultivated fields in the south part of the area are irrigated by the Fillaree Canal, which has the typical warm nighttime signature of water. The very warm field north of the canal (1.8, C.1 in Figure 5-23A) was flooded with standing water when the image was acquired. The very cool fields were probably damp from recent irrigation, resulting in evaporative cooling.

Most of the area is flat desert terrain with sparse clumps of vegetation. The map (Figure 5-24) shows a number of sand dunes, stabilized by mesquite trees, that appear warm on the image and dark on the aerial photograph. The very warm, Y-shaped feature at locality 3.2, B.1 in Figure 5-23A is a thick accumulation of windblown sand lodged against an earthen embankment. Imler Road in the northern part of the area is ac-

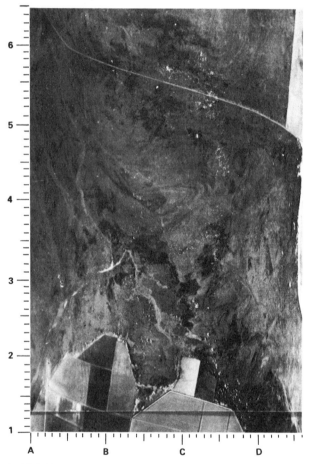

A. Nighttime thermal IR image (8 to 14 μm) acquired August 1961.

B. Aerial photograph acquired May 5, 1953.

Figure 5-23 Images of the Imler Road area, Imperial County, California. From Sabins (1969, Plate 1).

tually straight, but it appears curved because of distortion caused by the IR scanner. The road was surfaced with hard-packed sand when the image was acquired but today is paved with asphalt.

Bedrock in the area is the Borrego Formation (of Pleistocene age), which consists of brownish gray lacustrine siltstone with thin interbeds of well-cemented brown sandstone. Slabs and concretions of sandstone litter the surface where it crops out. Light-colored, nodular, thin layers in the siltstone help to define bedding trends within this monotonous sequence. On the nighttime image the siltstone is relatively cool and the sandstone is warm, corresponding to thermal signatures of similar rock types at the Indio Hills.

Anticline

An east-plunging anticline forms a conspicuous arcuate feature in the center of the IR image (4.1, C.0 in Figure 5-23A). Had this structure not been observed first on the image, one could walk across the anticline in the field without recognizing it, for there are no conspicuous lithologic or topographic patterns. After the anticline was located in the field by referring to the image, the limbs and plunge were defined by walking along the outcrops of individual beds. Structural attitudes are obscure

in the siltstone, but dips up to 45° were measured in isolated outcrops of the sandstones, and the dip-reversal across the fold axis was located.

In Figure 5-24 the anticline is mapped as solid bedrock, but there are numerous thin patches of windblown sand that cause the local gray tones on the image. The core of the anticline consists of contorted siltstone with a very cool signature. The pattern of alternating warm and cool bands outlining the anticline correlates with outcrops of sandstone and siltstone, respectively. The west end of the anticline is truncated by the southeastward projection of the Superstition Hills fault. The inferred trace of the fault is obscured by windblown sand, but siltstone outcrops in the immediate vicinity of the fault are strongly deformed, probably as a result of fault movement.

Superstition Hills Fault

This right-lateral, strike-slip fault was named for exposures in the Superstition Hills, 14 km to the northwest and projected into this area on the El Centro sheet of the *Geologic Map of California*. The fault alignment shown in Figure 5-24 differs from that on the El Centro sheet. In addition to truncating the anticline, the fault is marked in the southeast part of the image by a southeast-trending linear feature that is cooler on the east

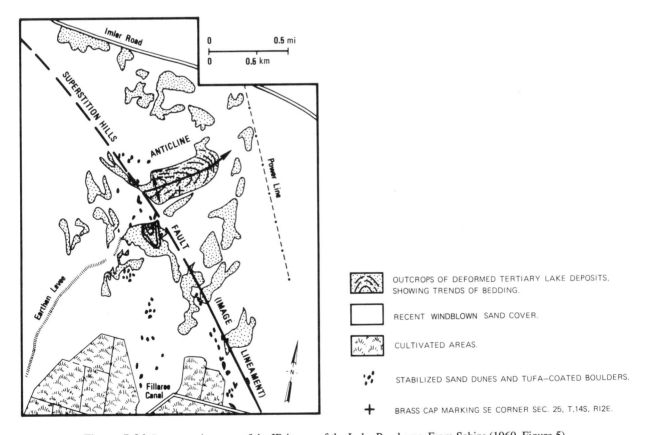

Figure 5-24 Interpretation map of the IR image of the Imler Road area. From Sabins (1969, Figure 5).

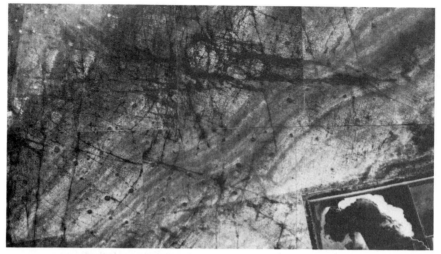

A. Nighttime thermal IR image (8 to 14 μm).

B. Aerial photograph.

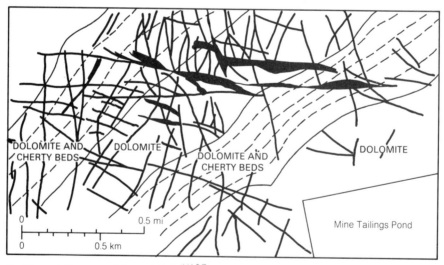

C. Interpretation map of thermal IR image.

Figure 5-25 Stilfontein area, western Transvaal, South Africa. From Warwick, Hartopp, and Viljoen (1979, Figures 8 and 9).

and warmer on the west (2.2, C.5 to 1.2, D.0 in Figure 5-23A). The trend of the linear tonal anomaly is parallel with, and about 0.2 km to the east of, the row of prominent sand dunes. On April 9, 1968, the Borrego Mountain earthquake caused surface breaks along the trace of the Superstition Hills fault that were mapped by A. A. Grantz and M. Wyss (Allen and others, 1972, Plate 2). In the area covered by Figure 5-23A, their map shows a series of breaks with less than 2.5 cm of right-lateral displacement. The trend of the breaks closely coincides with the linear anomaly on the thermal IR image (Figure 5-23A). The image, which was acquired 7 years prior to the earthquake, made visible an important structural feature that was obscure both on aerial photographs and in the field.

Comparison of the IR Image and the Aerial Photograph

The striking difference in tonal contrast and geologic information content between the IR image and the aerial photograph (Figure 5-23) is not caused by the 8-year difference in the dates when the images were acquired. This desert area is a stable environment in which natural changes occur very slowly. During annual field trips over a period of 10 years, no significant changes were noted in the area. The contrast and resolution of the aerial photograph are good and accurately record the low contrast of this area in the visible spectral region. Color and IR color aerial photographs of the anticline (not shown) are not significantly better than the black-and-white aerial photographs.

In the visible wavelengths there is little reflectance difference between the various rocks and surface materials. In the IR band, however, there are marked differences in the thermal inertia of the materials, which explains the higher information content of the IR image.

WESTERN TRANSVAAL, SOUTH AFRICA

South Africa is well suited for thermal IR surveys because of the dry climate and sparse vegetation, as shown by an example from the Stilfontein area in western Transvaal. The aerial photograph (Figure 5-25B) shows a featureless surface of low relief with a thin soil cover of 0.5 m or less and scattered trees with dark signatures. A tailings pond for a gold mine occurs in the southeast part of the image. No significant geologic information can be extracted from the photograph. The nighttime thermal IR image (Figure 5-25A), however, contains a wealth of information on geologic structure and lithology that is shown in the interpretation map (Figure 5-25C). The area is underlain by dolomite that includes a number of beds rich in chert, which is a siliceous sedimentary rock. The beds strike northeast and dip 10° to 15° to the southeast. In the IR image the dolomite has a bright (warm) signature, which is attributed to its relatively high density and high thermal inertia. The

chert-rich beds have distinctly darker (cooler) signatures caused by the lower density and lower thermal inertia of these rocks. A belt of alternating dolomite beds and chert-rich beds trends northeastward across the image and is bounded on the northwest and southeast by broad areas of dolomite with uniform warm radiant temperature.

The numerous linear features with very dark signatures are the expression of faults and joints that have been enlarged by erosion and filled with moist soil. Evaporative cooling causes the dark signatures. The two major sets of fractures trend approximately north to south and east to west. In much of the area, the bedrock is cut by closely spaced joints that produce a fine network of cool lines in the image.

The location and geology of the Stilfontein area differ from those of the Indio Hills and Imler Road areas, but in all these examples the IR image is superior to the aerial photograph for geologic interpretation. The different rock units (dolomite, chert, sandstone, siltstone, and conglomerate) have little contrast in the visible region, but are easily recognized in the IR image because of their different thermal properties. Fractures and faults in the Stilfontein area have cool signatures due to evaporative cooling of moist soil concentrated along the breaks. In the Indio Hills, evaporative cooling also marks the trace of the San Andreas fault.

MAUNA LOA, HAWAII

Mauna Loa is a classic shield volcano with a broad, convex profile and a large depression, called the Mokuaweoweo Caldera, at its summit. The calibrated nighttime IR image shown in color in Plate 9A and the aerial photomosaic (Figure 5-26A) cover the caldera and part of the Southwest Rift Zone including the pit craters, South Pit, Lua Hohonu, and Lua Hou. These features are identified on the geologic map of Figure 5-26B, which also shows the dates of the historic lava flows. Pit craters form as the surface subsides and are not primarily vents for lava. Lua Hohonu is the youngest pit crater, having formed after 1841 and partially filled with lava in 1940. The north end of the summit caldera coalesces with another pit crater, North Pit. East of this junction is Lua Poholo, a small pit crater that formed between 1874 and 1885. The cliffs bounding the summit caldera are slightly eroded fault scarps. Dashed lines in Figure 5-26B show the main faults, now covered by young lava flows, along which the floors of the caldera and the pits have subsided. Smaller faults and fractures are common. Historic lava flows shown on the geologic map originated from fissures on the floor of the summit caldera and in the Southwest Rift Zone. Lava from different flows has spilled into the pit craters. Near the fissures the lava flows are the smooth, ropy pahoehoe type, and change character downslope to the rough, fragmented aa type.

The IR image (Plate 9A) was acquired in February 1973, some 23 years after the latest eruption in June 1950. Activity

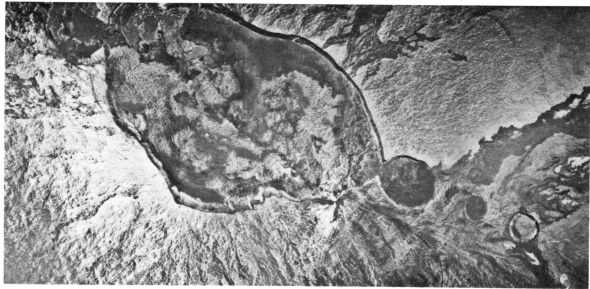

A. Aerial photograph acquired January 30, 1965.

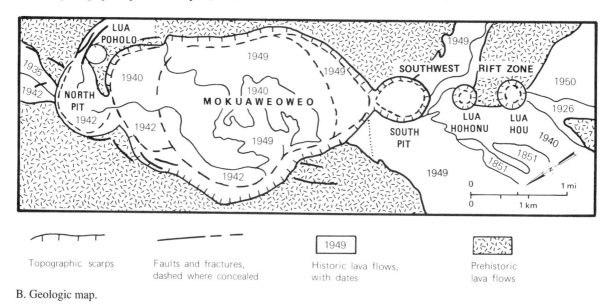

Topographic scarps

Faults and fractures, dashed where concealed

1949
Historic lava flows, with dates

Prehistoric lava flows

B. Geologic map.

Figure 5-26 Summit of Mauna Loa and upper portion of the Southwest Rift Zone, Hawaii.

resumed with the eruption of July 1975. On the thermal IR image the temperature range from –7° to +8°C has been digitally subdivided into six colors that represent 2.5° intervals. Radiant temperatures cooler than –7°C are shown in black; those warmer than 8°C are shown in white. The range –7° to +8°C was selected for color display because most features of geologic interest have radiant temperatures within this range. The low overall temperature level was caused by the 4200-m elevation of the summit of Mauna Loa. Individual lava flows cannot be distinguished on the basis of radiant temperature; the flows of the 1940s within the summit caldera have the same temperature ranges as the prehistoric flows on the flanks of the summit. The flows are more readily mapped from tonal differences on the aerial photograph (Figure 5-26A). After the flows have cooled, their radiant temperatures are determined by the albedo and thermal properties of the rocks. Geologic descriptions indicate that all the flows have similar rock composition, which suggests that the thermal properties are also similar. This similarity explains the lack of unique thermal signatures in Plate 9A for the different flows.

Some of the warmest radiant temperatures occur at the walls of the scarps bounding the summit caldera, North Pit, South Pit, and Lua Hou. The dense interior portions of lava flows that are exposed in the scarps have higher thermal inertias than the porous vesicular surfaces of the flows exposed on the flanks of the volcano and in the floors of the caldera and pit craters. The reticulate pattern of very warm signatures on the floor of the summit crater appears to be concentrated within the area of the 1940 flow (Figure 5-26B). Fumes of steam and sulfur dioxide issue from cracks in the surface of the 1940 flow (Macdonald, 1971) and are responsible for the warm reticulate pattern.

The floors of the three pit craters along the Southwest Rift Zone have different radiant temperatures. South Pit is the warmest (greater than 3°C), Lua Hohonu the coolest (less than –2°C), and Lua Hou intermediate (–2° to +0.5°C). The warmer radiant temperatures are probably caused by higher rates of convective heat transfer at South Pit and Lua Hou. The relative temperatures of the craters were confirmed in early 1975 when the summit of Mauna Loa was blanketed by 2 m of snow. Aerial photographs acquired in April 1975 (Lockwood and others, 1976, p. 12) showed that the snow had melted on the floors of South Pit and Lua Hou, while the cooler floor of Lua Hohonu was completely snow covered.

SATELLITE THERMAL IR IMAGES

Thermal IR images have been acquired during several unmanned satellite programs, including the Landsat thematic mapper (TM), the advanced very high resolution radiometer (AVHRR), and the Heat Capacity Mapping Mission (HCMM). Table 5-4 lists characteristics of these images. Thermal IR images are also acquired by environmental and meteorologic satellites of the GOES and DMSP programs, which are described in Chapter 4.

Landsat Thematic Mapper

Figure 5-27A shows wavelengths recorded by TM reflected IR bands 4, 5, and 7 and TM thermal IR band 6. The curves show the spectral distribution of energy radiated by rocks at different kinetic temperatures. As shown in Figure 5-2, at an ambient

Table 5-4 Satellite imaging systems that record radiant temperatures

System (dates)	Band	Spectral range, μm	Spatial resolution
Landsat TM (1982 to present)	5	1.55 to 1.75	30 m
	7	2.08 to 2.35	30 m
	6	10.50 to 12.50	120 m
AVHRR (1978 to present)	3	3.55 to 3.93	1.1 km
	4	10.50 to 11.50	1.1 km
	5	11.50 to 12.50	1.1 km
HCMM (1978 to 1980)	2	10.50 to 12.50	600 m

temperature of 27 °C the peak energy radiated by the earth has a wavelength of 9.7 μm. TM band 6 records wavelengths from 10.5 to 12.5 μm to avoid effects of the ozone absorption band at 9 to 10 μm. At the wavelengths of band 6, the intensity of radiant energy is very low. Two features enable band 6 detectors to record adequate energy levels: (1) a broad bandpass of 2 μm and (2) a large ground resolution cell of 120 by 120 m. At midlatitudes, the TM acquires daytime images on southbound orbit segments at 10:30 a.m. and nighttime images on northbound segments at 9:30 p.m. Path-and-row locations for the nighttime images are different from locations for daytime images. Contact EOSAT for information on any nighttime TM images that may be available.

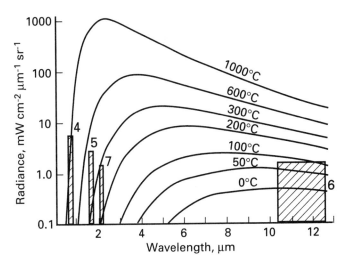

A. Landsat TM bands.

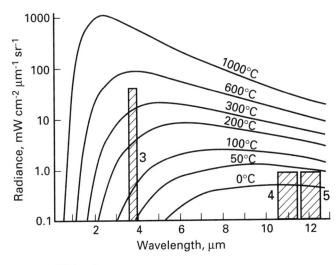

B. AVHRR bands

Figure 5-27 Bands recorded by the TM and AVHRR in the reflected IR and thermal IR regions. Curves show the radiance of rocks at various kinetic temperatures. Modified from Rothery (1989, Figure 5).

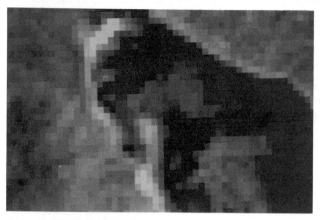

A. Band 4 (0.76 to 0.90 μm).

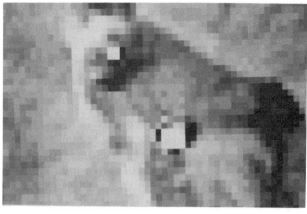

B. Band 5 (1.55 to 1.75 μm).

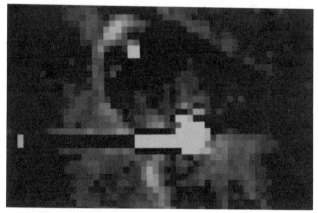

C. Band 7 (2.08 to 2.35 μm).

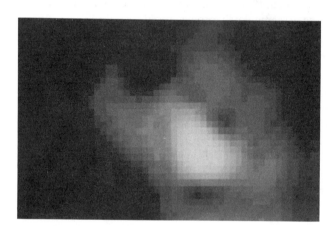

D. Band 6 (10.5 to 12.5 μm).

Figure 5-28 Subscenes of TM bands for summit of Erta 'Ale volcano, Afar region, Ethiopia. From Rothery (1989, Figure 7). Courtesy D. A. Rothery, the Open University, Milton Keynes.

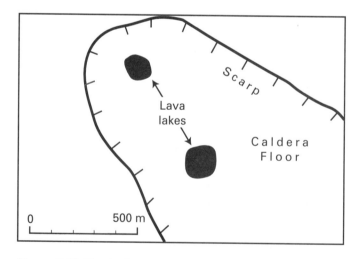

Figure 5-29 Sketch of Erta 'Ale volcano from TM images.

Hot targets with radiant temperatures of several hundred degrees centigrade radiate maximum energy in the range of 3 to 5 mm. Figure 5-27A shows that hot targets also radiate significant energy at the wavelengths recorded by TM bands 5 and 7. These bands are used to monitor active volcanoes, such as Erta 'Ale.

Erta 'Ale Volcano, Ethiopia Erta 'Ale is an active volcano in the Afar region of northern Ethiopia, or Eritrea. Figure 5-28 is a greatly enlarged subscene of the summit showing the reflected IR and thermal IR bands. Figure 5-29 is a sketch of features interpreted from the images. The band 4 image (Figure 5-28A) records reflected sunlight and shows albedo of the surface. The summit is an irregular depression, or caldera, surrounded by a steep scarp. Two lava lakes on the caldera floor have dark signatures from a thin crust of solidified lava. The lava lakes have bright signatures in the band 5 image (Figure 5-28B) caused by energy radiated at temperatures of 400°C and above. Both lava lakes are bright in the band 7 image (Figure 5-28C), indicating energy radiated at 300°C and above. Narrow bright signatures along the caldera scarps may represent heating from steam and hot gas escaping along faults.

An unusual pattern of bright and dark bands west of the southern lava lake is an artifact of the imaging process. The high radiant energy flux from the lake exceeded the maximum sensitivity level of the band 7 detectors, which is called *saturating* the detectors. This saturation is compensated by an electronic circuit called an *automatic gain control* (AGC) that reduces detector sensitivity levels to compensate for excessive temperatures. The westbound sweep of the scan mirror encountered the east edge of the hot lava lake, which saturated the band 7 detectors and resulted in anomalously bright signatures for approximately 10 pixels west of the hot lava lake. At that point the AGC responded to reduce sensitivity of the detectors, which were now imaging cooler terrain and resulted in an anomalously dark streak. This effect is also seen in TM color composite images that employ band 7. A hot target such as an active volcano, forest fire, or factory will commonly be accompanied by a color streak on the down-scan margin.

The band 6 image (Figure 5-28D) was recorded with ground resolution cells of 120 by 120 m that were resampled into picture elements representing the same 30-by-30-m cells as the other TM bands. A broad, diffuse, warm signature occupies the south portion of the caldera.

In this very hot volcanic environment the shorter wavelengths of bands 5 and 7 provided better displays of thermal features than the longer wavelengths of band 6. The finer spatial resolution of bands 5 and 7 also improved their images. Band 7 images are used to monitor the Lascar active volcano in Chile (see Chapter 13).

Appalachian Mountains and Cumberland Plateau

Figure 5-30 compares band 6 (thermal IR) images recorded on daytime and nighttime orbits over West Virginia. The images cover portions of the Appalachian Mountains in the east and the Cumberland Plateau in the west. Figure 5-30A,B shows simultaneously recorded band 6 and band 3 (visible red) images. Thin clouds in the northeast and snow cover in the northwest portions of these images produce dark (cool) signatures in the IR image and bright signatures in the visible image. In both images topography is accentuated by solar illumination from the southeast. In the visible image (Figure 5-30B) the southeast-facing slopes are highlighted and northwest-facing slopes are shadowed. In the daytime IR image, southeast-facing slopes are warm (bright) and those facing northwest are cool (dark). In the nighttime band 6 image (Figure 5-30C), the effects of differential heating and shadowing have been dissipated. In the Appalachians the ridges are warm and the adjacent valleys are cool. The ridges consist of resistant dense rocks, such as quartzite and conglomerate, which have high thermal inertia values. Valleys are eroded in less-resistant limestone and shale and are covered with soil, which has a low thermal inertia. These differences in materials and thermal inertia explain the nighttime temperature patterns. The elevation difference between ridges and valleys is only a few hundred meters; therefore, adiabatic cooling is not a significant factor. Water in the rivers and lakes has the typical cool signature on the daytime IR image and warm signature on the nighttime image.

As shown by the West Virginia example and by images from aircraft, thermal IR images in the 8-to-14-μm band must be acquired at night to be useful for most interpretations. However, most TM images are acquired in the daytime, because six of the seven TM bands only function at this time.

Advanced Very High Resolution Radiometer

The AVHRR is carried on the NOAA polar-orbiting satellites and records one visible band, one reflected IR band, and three thermal IR bands, which are listed in Table 5-4. Images of AVHRR bands are shown in Chapter 4. Figure 5-27B shows spectral ranges of the three thermal IR bands. Bands 4 and 5 together span the same spectral range as TM band 6 (Figure 5-27A). The spectral range of AVHRR band 3 matches the peak energy radiated by very hot targets. Therefore, band 3 images are widely used to monitor active volcanoes and forest fires. Examples of these applications are illustrated in Chapter 13.

Heat Capacity Mapping Mission

NASA launched HCMM specifically to record day and night thermal IR images. The system operated from 1978 to 1980. Southbound segments of orbits occurred shortly after midnight and northbound segments during early afternoon, which provided images with a maximum ΔT. Each image covers 700 by 700 km, which is approximately the area of 16 TM scenes. The thermal IR band (10.5 to 12.5 μm) has a ground resolution cell of 600 by 600 m. The visible band (0.5 to 1.1 μm) has a ground resolution cell of 500 by 500 m. An HCMM user's guide (Price, 1980) describes the system, orbit paths, and

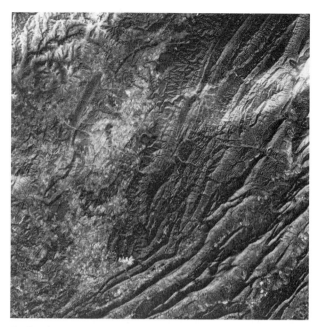

A. Daytime thermal IR image (band 6) acquired November 16, 1982.

B. Daytime visible image (band 3) acquired November 16, 1982.

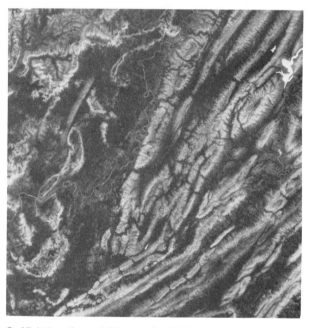

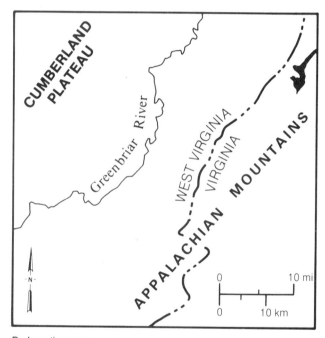

C. Nighttime thermal IR image (band 6) acquired November 8, 1982.

D. Location map.

Figure 5-30 Images acquired by the Landsat TM. Greenbriar River, West Virginia.

image format. Short and Stuart (1983) have published numerous examples and descriptions of HCMM images.

Daytime and Nighttime Images of California and Nevada

Figure 5-31 is an HCMM subscene that extends from San Francisco Bay northeast to Pyramid Lake, Nevada, and covers portions of the Coast Ranges, Central Valley, Sierra Nevada, and Basin Range. The map in Figure 5-32 shows these features. Figure 5-31A,B shows daytime visible and thermal IR images recorded in the daytime; Figure 5-31C is a nighttime thermal image. On the IR images, bright signatures are warm and dark signatures are cool. Water is cooler than its surroundings in the day and warmer at night. This temperature reversal is obvious at San Francisco Bay, Lake Tahoe, and Pyramid Lake. In the nighttime image, water in San Francisco Bay is warmer than the Pacific Ocean. An outgoing tide moves a plume of warmer bay water west through the Golden Gate into the Pacific Ocean, where it is carried south by the California Current.

In the daytime IR image major topographic features are shown in two ways:

1. Large ridges in the Coast Ranges and canyons in the Sierra Nevada have warm and cool signatures caused by solar heating and shadowing.
2. The higher portions of the Sierra Nevada have a cool signature of regional extent that is caused by *adiabatic cooling*, which is the phenomenon that causes air temperature to decrease by 6.5°C per 1000 m of altitude increase.

At night the effects of topography are dissipated and thermal signatures are dominated by moisture content, vegetation, and thermal inertia.

ATI Image of the San Rafael Swell, Utah

Kahle and others (1981) used daytime and nighttime HCMM images of the San Rafael Swell in central Utah (Figure 5-33) to prepare an image showing apparent thermal inertia (*ATI*). Digital data for the daytime and nighttime images were geometrically registered. For each ground resolution cell, the nighttime radiant temperature was subtracted from the daytime temperature to determine ΔT. Albedo values from the daytime visible image were used to calculate the term $1 - A$ for each cell. Equation 5-10 was then used to determine *ATI*. The *ATI* values were converted into an image (Figure 5-33D) in which dark tones display low values of *ATI* and bright tones display high values.

The San Rafael Swell is a low domal feature bounded by the Wasatch Plateau on the northwest, the Buckhorn Plateau on the northeast, and the San Rafael Desert of windblown sand on the southeast. The swell is semiarid with little vegetation, and bedrock is well exposed except where it is covered by sand and gravel. Areas of vegetation occur along the east slope of the Wasatch Plateau. Additional vegetation occurs at the higher el-

evations of the plateaus. Standing water occurs in three small reservoirs in the northwest part of the area.

The interpretation map (Figure 5-34) was prepared by delineating areas of high, intermediate, and low values of *ATI* from Figure 5-33D. Published maps were used to identify the terrain features that cause the *ATI* signatures. Water, vegetation, and plateau terrain have the following signatures and explanations:

Water (high *ATI* value) Has a nearly constant day and night temperature, which for reasons described earlier, results in a very low ΔT and a corresponding high *ATI* value.

Vegetation (high to intermediate *ATI* values) Has a low ΔT, relative to soil, as shown in the earlier example from the Indio Hills. Therefore, vegetation should have a high *ATI* value, but most of the vegetation along the eastern slope of the Wasatch Plateau has only a moderate *ATI* value. The Landsat image shows that much of this vegetation occurs along drainage channels separated by areas of bare soil. Each 600-by-600-m ground resolution cell of HCMM includes areas of vegetation and soil, which results in intermediate values of *ATI*.

Plateau terrain (high and intermediate *ATI* values) Because of their high elevation, the Wasatch and Buckhorn Plateaus are relatively cool both day and night. These areas also support a moderate vegetation cover. Both elevation and vegetation contribute to relatively low ΔT and high *ATI* values. In the Wasatch Plateau the ridges have high *ATI* values and the valleys have intermediate values.

Erosion of the San Rafael Swell anticline has exposed sedimentary rocks ranging in age from late Cretaceous to late Paleozoic. The rocks are sandstone, shale or siltstone, and minor limestone. Two bodies of mafic igneous rock occur at the southwest plunge of the arch. Alternating outcrops of sandstone and shale surround the swell. The large ground resolution cell of HCMM does not resolve these details, and only some broad zones with contrasting signatures are visible. In the *ATI* image these signatures are light gray (intermediate *ATI* values) and very dark gray (low *ATI* values). The following correlation of *ATI* signatures with rock units was made by comparing the image with a geologic map of the region:

Sandstone (intermediate *ATI* values) The zones of intermediate *ATI* values correlate with sandstones, which have relatively high thermal inertias.

Shale and siltstone (low *ATI* values) The zones of low *ATI* values correlate with shale and siltstone, which have low thermal inertia values.

Mafic igneous rocks (intermediate *ATI* values) The two exposures of igneous rock have intermediate signatures that correspond to published values of thermal inertia for these rocks.

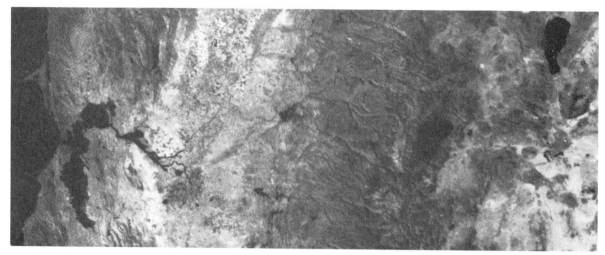

A. Daytime visible image acquired September 20, 1978.

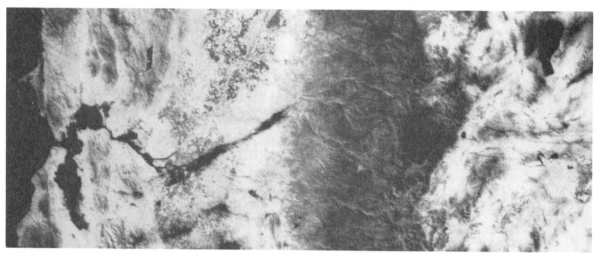

B. Daytime thermal IR image acquired September 20, 1978.

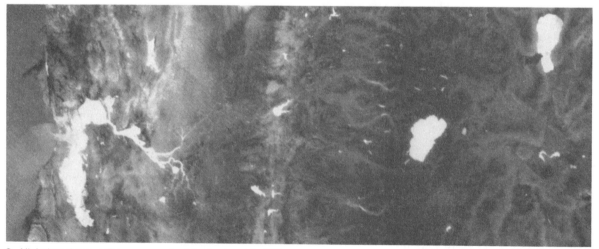

C. Nighttime thermal IR image acquired September 19, 1978.

Figure 5-31 HCMM subscenes of central California and Nevada.

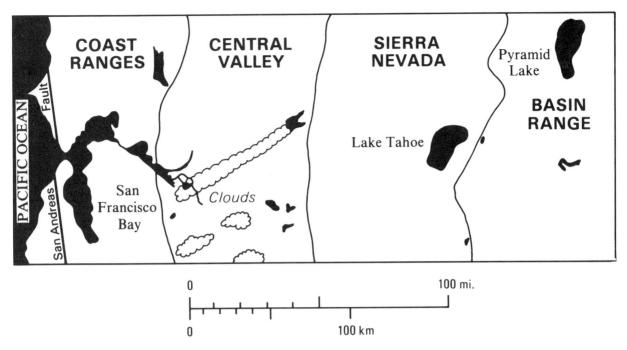

Figure 5-32 Location map for HCMM images in Figure 5-31.

Windblown sand (low *ATI* values) The windblown sand of the San Rafael Desert has a dark signature in the *ATI* image. The very low density of this unconsolidated sand explains the low thermal inertia.

This example from Utah demonstrates that, despite the coarse spatial resolution, HCMM image data may be processed to express thermal characteristics of terrain and geologic features. Cassinis and others (1984) prepared and evaluated *ATI* images from HCMM data of Sardinia, Italy. Despite problems of atmosphere, topography, surface moisture, and vegetation, these authors were able to interpret geologic features on the *ATI* images.

THERMAL IR SPECTRA

Thermal IR spectrometers are used to record spectra of energy emitted from materials in the 8-to-14-μm band. Figure 5-35 shows spectra for igneous rocks with silica contents ranging from 50 to 71 percent. The spectra show a broad emission minimum called the *Reststrahlen band*. In silicate rocks this minimum is due to stretching vibrations between silicon and oxygen atoms bonded in the silicate crystal lattice. The position and depth of the emission minimum are related to the crystal structure of the constituent minerals and are especially sensitive to the quartz content of the rocks. The spectra in Figure 5-35 are arranged in the order of decreasing silica (and quartz)

content, which is shown for each rock. An arrow marks the geometric center of each emission minimum, which is not necessarily the position of least emission. For the leucogranite with maximum silica the arrow is located at a wavelength of slightly less than 9.0 μm. For the anorthosite with minimum silica the arrow is located at a wavelength of greater than 10.0 μm. Figure 5-36 is a plot of silica content as a function of the minimum emission wavelength for igneous rocks similar to those in Figure 5-35. The plot shows a linear relationship between decreasing silica and increasing wavelength for the emission minimum.

Figure 5-37 shows spectra of additional silicate rocks, together with clay minerals and carbonate rocks. The silicate rocks show the typical emissivity minimum at wavelengths of 8 to 12 μm. The silicate spectra are arranged in order of decreasing silica content from quartzite through basalt, and show the shift toward longer wavelengths of the emissivity minimum. In the clay minerals (kaolinite and montmorillonite), spectral features in the 8-to-14-μm region are attributed to various Si-O-Si and Si-O stretching vibrations and to an Al-O-H bending mode. The spectral features in montmorillonite are less distinct than in kaolinite because the numerous exchangeable cations and water molecules in the montmorillonite structure allow many different vibrations.

Spectra of the carbonate rocks (limestone and dolomite) in Figure 5-37 have a major emissivity minimum at 6 to 8 μm due to internal vibrations in the carbonate anion. This wavelength region, however, coincides with an atmospheric absorption

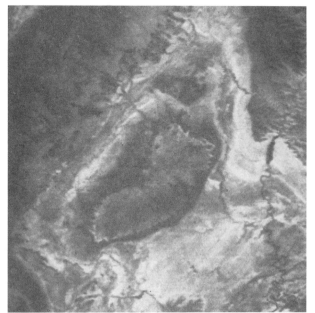

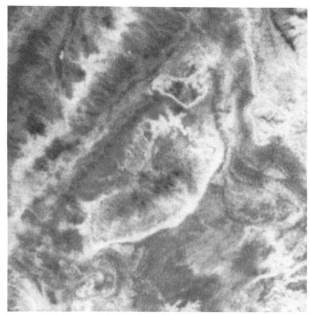

A. Daytime thermal IR image (10.5 to 12.5 μm) acquired August 28, 1978.

B. Nighttime thermal IR image (10.5 to 12.5 μm) acquired August 27, 1978.

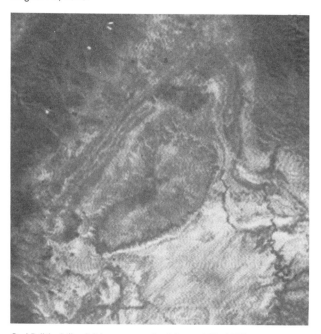

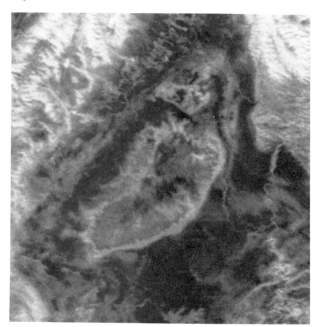

C. Visible (albedo) image acquired August 28, 1978.

D. Apparent thermal inertia image.

Figure 5-33 Enlarged HCMM images of the San Rafael Swell, Utah. From Kahle and others (1981). Courtesy A. B. Kahle, Jet Propulsion Laboratory.

band and is not usable in remote sensing of the earth. The spectra of pure carbonate rocks are featureless in the region of 8 to 12 μm, but the spectrum of limestone with clay and quartz has an emissivity minimum near 9.0 μm caused by clay.

The IR spectra described here were measured in laboratories. Hoover and Kahle (1986) have described a portable system for recording thermal IR spectra in the field.

THERMAL IR MULTISPECTRAL SCANNER

In order to obtain images of the spectral information in the thermal IR region, NASA and JPL developed the thermal infrared multispectral scanner (TIMS). Kahle and Goetz (1983) described TIMS, which records six spectral bands of imagery at wavelength intervals of 8.19 to 8.55, 8.58 to 8.96, 9.01 to 9.26,

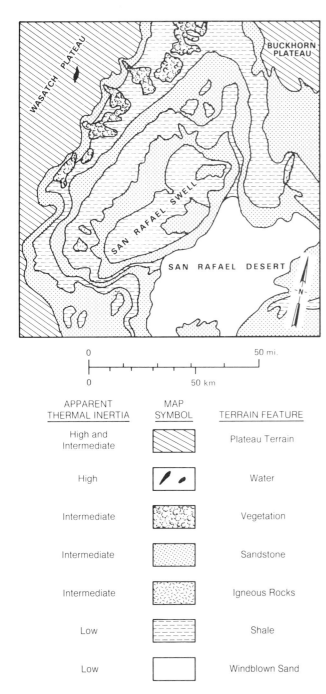

Figure 5-34 Interpretation map of the *ATI* image (Figure 5-33D) of the San Rafael Swell, Utah.

APPARENT THERMAL INERTIA	MAP SYMBOL	TERRAIN FEATURE
High and Intermediate		Plateau Terrain
High		Water
Intermediate		Vegetation
Intermediate		Sandstone
Intermediate		Igneous Rocks
Low		Shale
Low		Windblown Sand

9.65 to 10.15, 10.34 to 11.14, and 11.29 to 11.56 μm; Figure 5-37 shows the position of these bands and the spectra of rocks. The TIMS has an *IFOV* of 2.5 mrad and an angular field of view of 80°. The TIMS is an experimental system, and the data are not commercially available. Kahle and Abbott (1986), Abbott (1990, 1991) and Realmuto (1992) published the results of a series of workshops for TIMS investigators.

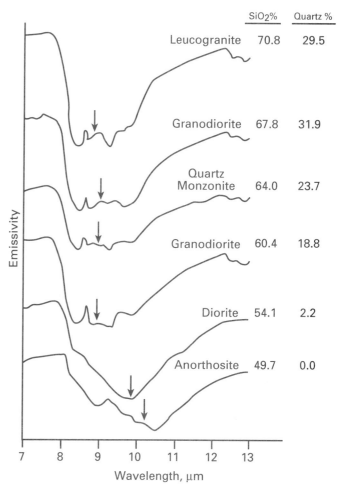

Figure 5-35 Emissivity spectra of igneous rocks with different silica and quartz contents. Arrows show centers of absorption bands. Spectra are offset vertically. From Sabine, Realmuto, and Taranik (1994, Figure 3, revised).

Plate 9B,C shows two color composite images of daytime TIMS bands 1, 2, and 3 of the Cuprite Hills in west-central Nevada. Figure 5-38 is a geologic map of the area, which includes the western portion of the Cuprite mining district. S. J. Hook of JPL provided the images and the following interpretation. Plate 9B shows radiant temperatures of the three bands. Recall that radiant temperature is the product of emissivity and kinetic temperature, as determined by Equation 5-8. Kinetic temperature dominates the signature in each band. The radiant temperature image (Plate 9B) shows topography because of differences in solar heating and shadowing. Areas of color variation represent materials with the maximum variation in emissivity from band to band.

Plate 9C shows emissivity information that was extracted for each band by Hook and others (1992). Topography is absent because temperature information is excluded. Lithologic information is emphasized in this image. The Thirsty Canyon

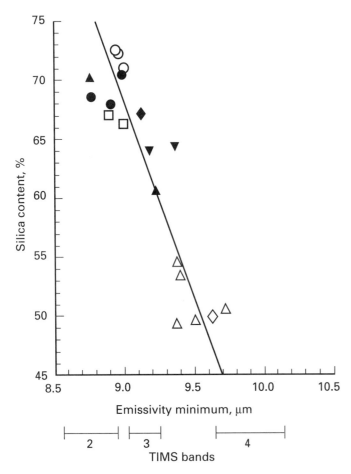

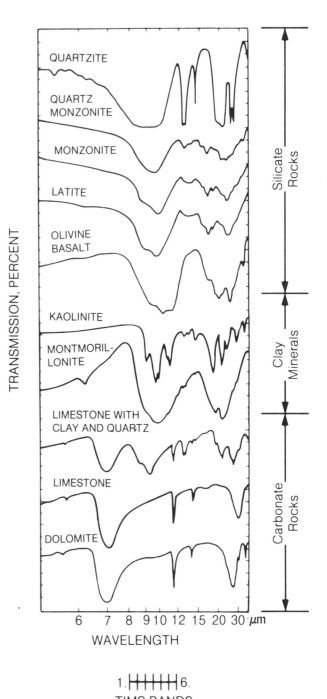

Figure 5-36 Plot showing the relationship between emissivity minima and silica content. Vertical axis shows the wavelength of emissivity minima (indicated by arrows in Figure 5-35). Horizontal axis shows the percentage of silica. Symbols represent various types of igneous rocks. From Sabine, Realmuto, and Taranik (1994, Figure 4).

Figure 5-37 Thermal IR spectra of rocks and minerals. Spectra are offset vertically. From Kahle (1984, Figure 4).

Tuff in the Cuprite mining district has a distinctive violet signature. The Mule Spring Limestone is bright blue. Metamorphosed sedimentary rocks of the Harkless Formation are various shades of orange. The Siebert Tuff is dark blue and purple. Alluvial deposits have the same signatures as the outcrops from which they were eroded.

Kahle and others (1988) used TIMS images to differentiate young volcanic flows in Hawaii. The differences in spectral emittance were attributed to different weathering characteristics, rather than compositional differences, among the flows.

COMMENTS

Thermal IR images record the radiant temperature of materials. Radiant temperature is determined by a material's kinetic temperature and by its emissivity, which is a measure of its ability to radiate and to absorb thermal energy. The diurnal temperature range (ΔT) is a function of thermal inertia, which is directly related to density of materials. Thermal IR images are useful for many applications, including

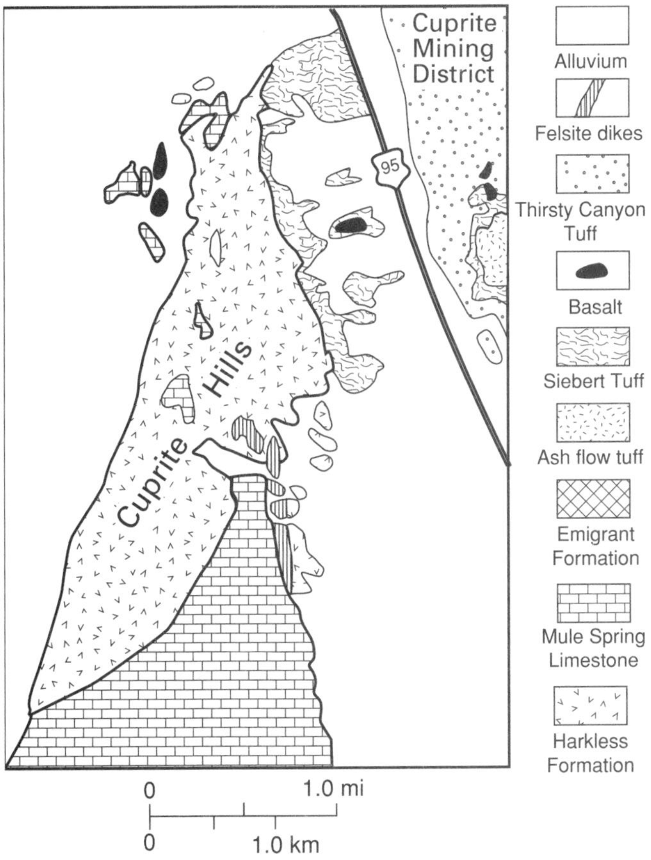
Cuprite Mining District
95
Cuprite Hills
Alluvium
Felsite dikes
Thirsty Canyon Tuff
Basalt
Siebert Tuff
Ash flow tuff
Emigrant Formation
Mule Spring Limestone
Harkless Formation
0
1.0 mi
0
1.0 km

Figure 5-38 Interpretation map of TIMS images of the Cuprite Hills, Nevada. From Hook and others (1992, Figure 5).

1. **Differentiating materials** Denser materials, such as basalt and sandstone, have higher thermal inertias than less dense materials, such as cinders and siltstone. On nighttime thermal IR images, the rocks with higher thermal inertia values have warmer signatures.
2. **Mapping surface moisture** Damp ground has a cool signature on thermal IR images that is caused by evaporative cooling.
3. **Monitoring active volcanoes and subsurface coal fires** Additional examples are shown in Chapter 13.
4. **Monitoring forest fires** Thermal IR wavelengths penetrate smoke and show hot spots.
5. **Mapping environmental features** Chapter 12 shows images of sea ice, oil films, thermal plumes, and current patterns in water bodies.

Thermal IR images are affected by environmental factors that the interpreter must consider:

1. **Clouds and surface winds** These produce confusing thermal patterns.
2. **Time of day** Daytime images record the differential solar heating and shadowing of topographic features. Geologic and other interpretations require nighttime images.
3. **Surface moisture and dense vegetation** These effectively mask other features on thermal IR images; therefore, thermal IR images are most useful in arid and semiarid terrain.

The HCMM experiment demonstrated the value of low-resolution IR images acquired at night from satellites. Band 6 of the TM has demonstrated the value of images with medium spatial resolution, but additional nighttime data should be acquired. The TIMS aircraft system records images at narrow spectral bands in the 8-to-14-μm atmospheric window. When they are suitably processed, these images discriminate different rock types, primarily on the basis of variations in silica content.

QUESTIONS

1. An iron poker in a fire glows dull red, which corresponds to a visible wavelength of 0.65 μm and represents λ_{max}. Use Wein's displacement law (Equation 5-1) to calculate the radiant temperature of the iron.
2. The filament in a light bulb is heated to a radiant temperature of 7000°K. What is the wavelength at which the maximum energy radiates from the filament?
3. Calculate the radiant flux (F_b) for a blackbody with a kinetic temperature of 21°C.
4. Calculate radiant flux (F_r) from a block of rough dolomite (see Table 5-1) with a kinetic temperature (T_{kin}) of 15°C.
5. Calculate radiant temperature (T_{rad}) of a block of rough dolomite (see Table 5-1) with a kinetic temperature (T_{kin}) of 13°C.
6. Calculate the thermal inertia of a rock with the following properties: a thermal conductivity (K) of 0.005 cal · cm^{-1} · sec^{-1} · °C^{-1}; a density (ρ) of 2.7 g · cm^{-3}; and a thermal capacity (c) of 0.19 cal · g^{-1} · °C^{-1}.
7. You are interpreting daytime and nighttime HCMM thermal IR images and a visible image of an area. A rock outcrop has a daytime radiant temperature of 20°C and a nighttime temperature of 10°C. The albedo from the visible image is 0.50. Calculate the apparent thermal inertia (*ATI*) for this rock.
8. Describe the appearance of clouds and of smoke plumes on thermal IR images. Explain the difference in appearance.
9. List the signature (warm or cool) of the following targets on daytime and nighttime IR images: dry soil, damp soil, and standing water. Explain these signatures.
10. In arid terrain such as southern California and South Africa, thermal IR images commonly portray more geologic information than do aerial photographs. Explain this advantage of IR images.

11. Compare the TIMS images in Plate 9 with the geologic map (Figure 5-38). Which image is most useful for differentiating geologic units? Explain your answer.

REFERENCES

Abbott, E., 1990, Proceedings of the second Thermal Infrared Multispectral Scanner (TIMS) workshop: Jet Propulsion Laboratory Publication 90-56, Pasadena, CA.

Abbott, E., 1991, Proceedings of the third Thermal Infrared Multispectral Scanner (TIMS) workshop: Jet Propulsion Laboratory Publication 91-29, Pasadena, CA.

Allen, C. R., M. Wyss, J. N. Brune, A. Grantz, and R. E. Wallace, 1972, Displacements on the Imperial, Superstition Hills and San Andreas faults triggered by the Borrego Mountain earthquake in The Borrego Mountain earthquake of April 9, 1968: U.S. Geological Survey Professional Paper 787, p. 87–104.

Anderson, J. M. and D. Baker, 1987, The thermographic survey of central Scotland: International Journal of Remote Sensing, v. 8, p. 779–788.

Buettner, K. J. K. and C. D. Kern, 1965, Determination of infrared emissivities of terrestrial surfaces: Journal of Geophysical Research, v. 70, p. 1329–1337.

Cassinis, R. and others, 1984, Thermal inertia of rocks—an HCMM experiment on Sardinia, Italy: International Journal of Remote Sensing, v. 5, p. 79–94.

Colwell, R. N., W. Brewer, G. Landis, P. Langley, J. Morgan, J. Rinker, J. M. Robinson, and A. L. Sorem, 1963, Basic matter and energy relationships involved in remote reconnaissance: Photogrammetric Engineering, v. 29, p. 761–799.

Cummings, J. R., 1964, Coachella Valley investigation: California Department of Water Resources, Bulletin 108.

Dozier, J. and A. H. Strahler, 1983, Ground investigations in support of remote sensing in Colwell, R. N., ed., Manual of remote sensing, second edition: ch. 23, p. 969–989, American Society for Photogrammetry and Remote Sensing, Falls Church, VA.

Findlay, G. A. and D. R. Cutten, 1989, Comparison of performance of 3 to 5 and 8 to 12 μm infrared systems: Applied Optics, v. 28, p. 5029–5037.

Heller, D. J., 1991, IR imaging takes on long-standing military medical problems: Advanced Imaging, July, p. 48–76.

Hook, S. J., A. R. Gabell, A. A. Green, and P. S. Kealy, 1992, A comparison of techniques for extracting emissivity information from thermal infrared data for geologic studies: Remote Sensing of the Environment, v. 42, p. 123–135.

Hoover, G. and A. B. Kahle, 1986, A portable spectrometer for use from 5 to 15 μm: Jet Propulsion Laboratory Publication 86-19, Pasadena, CA.

Janza, F. J. and others, 1975, Interaction mechanisms in Reeves, R. G., ed., Manual of remote sensing, first edition: ch. 4, p. 75–179, American Society for Photogrammetry and Remote Sensing, Falls Church, VA.

Kahle, A. B., 1984, Measuring spectra of arid lands in El-Baz, F., ed., Deserts and arid lands: ch. 11, p. 195–217, Martinus Nijhoff Publishers, The Hague, Netherlands.

Kahle, A. B. and E. Abbott, 1986, The TIMS data users workshop: Jet Propulsion Laboratory Publication 86-38, Pasadena, CA.

Kahle, A. B. and A. F. H. Goetz, 1983, Mineralogic information from a new airborne thermal infrared multispectral scanner: Science, v. 222, p. 24–27.

Kahle, A. B., J. P. Schieldge, M. J. Abrams, R. E. Alley, and C. J. LeVine, 1981, Geologic applications of thermal inertia imaging using HCMM data: Jet Propulsion Laboratory Publication 81-55, Pasadena, CA.

Kahle, A. B., A. R. Gillespie, E. A. Abbott, M. J. Abrams, R. E. Walker, and G. Hoover, 1988, Relative dating of Hawaiian lava flows using multispectral thermal infrared images—a new tool for geologic mapping of young volcanic terranes: Journal of Geophysical Research, v. 93, p. 15,239–15,251.

Lockwood, J. P., R. Y. Koyanagi, R. I. Tilling, R. T. Holcomb, and D. W. Peterson, 1976, Mauna Loa threatening: Geotimes, v. 21, p. 12–15.

Macdonald, G. A., 1971, Geologic map of the Mauna Loa Quadrangle, Hawaii: U.S. Geological Survey Geologic Quadrangle Map GQ-897.

Norwood, V. T. and J. C. Lansing, 1983, Electro-optical imaging systems in Colwell, R. N., ed., Manual of remote sensing, second edition: ch. 8, p. 335–367, American Society for Photogrammetry and Remote Sensing, Falls Church, VA.

Price, J. C., 1980, Heat Capacity Mapping Mission (HCMM) data users handbook for Applications Explorer Mission (AEM): NASA Goddard Space Flight Center, Greenbelt, MD.

Realmuto, V. J., ed., 1992, TIMS workshop in Summaries of the third annual JPL airborne geoscience workshop June 1–5, 1992: Jet Propulsion Laboratory Publication 92-14, v. 2, Pasadena, CA.

Rothery, D. A., 1989, Volcano monitoring by satellite: Geology Today, v. 5, p. 128–132.

Sabine, C., V. J. Realmuto, and J. V. Taranik, 1994, Quantitative estimation of granitoid composition from thermal infrared multispectral scanner (TIMS) data, Desolation Wilderness, northern Sierra Nevada, California: Journal of Geophysical Research, v. 99, p. 4261–4271.

Sabins, F. F., 1967, Infrared imagery and geologic aspects: Photogrammetric Engineering and Remote Sensing, v. 29, p. 83–87.

Sabins, F. F., 1969, Thermal infrared imagery and its application to structural mapping in southern California: Geological Society of America Bulletin, v. 80, p. 397–404.

Salisbury, J. W. and D. M. D'Aria, 1992A, Emissivity of terrestrial materials in the 8 to 14 μm atmospheric window: Remote Sensing of the Environment, v. 42, p. 83–106.

Salisbury, J. W. and D. M. D'Aria, 1992B, Emissivity of terrestrial materials in the 3 to 5 μm atmospheric window: Remote Sensing of the Environment, v. 47, p. 345–361.

Salisbury, J. W., L. S. Walter, and D. D'Aria, 1988, Mid-infrared (2.5 to 13.5 μm) spectra of igneous rocks: U.S. Geological Survey Open-File Report 88-686.

Short, N. M. and L. M. Stuart, 1983, The Heat Capacity Mapping Mission (HCMM) anthology: NASA SP 465, U.S. Government Printing Office, Washington, DC.

Warwick, D., P. G. Hartopp, and R. P. Viljoen, 1979, Application of the thermal infrared linescanning technique to engineering geological mapping in South Africa: Quarterly Journal of Engineering Geology, v. 12, p. 159–179.

Watson, K., 1971, Geophysical aspects of remote sensing: Proceedings of the International Workshop on Earth Resources Survey Systems, NASA SP 283, v. 2, p. 409–428.

Wolfe, E. W., 1971, Thermal IR for geology: Photogrammetric Engineering and Remote Sensing, v. 37, p. 43–52.

ADDITIONAL READING

Becker, F. and Z. L. Li, 1995, Surface temperature and emissivity at various scales—definition, measurement, and related problems: Remote Sensing Reviews, v. 12, p. 225–253.

Estes, J. E., E. J. Hajic, and L. R. Tinney, 1983, Fundamentals of image analysis—analysis of visible and thermal infrared data *in* Colwell, R. N., ed., Manual of remote sensing, second edition: ch. 24, p. 987–1124, American Society for Photogrammetry and Remote Sensing, Falls Church, VA.

Hook, S. J., K. E. Karlstrom, C. F. Miller, K. J. W. McCaffrey, 1994, Mapping the Piute Mountains, California, with thermal infrared multispectral scanner images: Journal of Geophysical Research, v. 99, p. 15,605–15,622.

Kahle, A. B., 1980, Surface thermal properties *in* Siegal, B. S. and A. R. Gillespie, eds., Remote sensing in geology: ch. 8, p. 257–273, John Wiley & Sons, New York, NY.

Norman, J. M. and F. Becker, 1995, Terminology in thermal infrared remote sensing of natural surfaces: Remote Sensing Reviews, v. 12, p. 159–173.

Prata, A. J., 1995, Thermal remote sensing of land surfaces temperature from satellites—current status and future prospects: Remote Sensing Reviews, v. 12, p. 175–224.

Sabins, F. F., 1980, Interpretation of thermal infrared images *in* Siegal, B. S. and A. R. Gillespie, eds., Remote sensing in geology: ch. 9, p. 275–295, John Wiley & Sons, New York, NY.

CHAPTER

RADAR TECHNOLOGY
AND TERRAIN INTERACTIONS

Radar is an *active* remote sensing system because it provides its own source of energy. The system "illuminates" the terrain with electromagnetic energy, detects the energy returning from the terrain (called *radar return*), and records the return energy as an image. *Passive* remote sensing systems, such as photography and thermal IR, detect the available energy reflected or radiated from the terrain. Radar systems operate independently of lighting conditions and are largely independent of weather. In addition, the terrain can be "illuminated" by employing the optimum viewing geometry to enhance features of interest.

Radar is the acronym for "radio detection and ranging"; it operates in the microwave band of the electromagnetic spectrum, ranging from millimeters to meters in wavelength. The reflection of radio waves from objects was noted in the late 1800s and early 1900s. Definitive investigations of radar began in the 1920s in the United States and Great Britain for the detection of ships and aircraft. Radar was developed during World War II for navigation and target location and used the familiar rotating antenna and circular *cathode-ray-tube* (CRT) display. The continuous-strip mapping capability of *side-looking airborne radar* (SLAR) was developed in the 1950s to acquire reconnaissance images without the necessity of flying over politically unfriendly regions. Fischer (1975) has prepared a comprehensive history of radar development. Moore (1983) and Moore and others (1983) give details of radar theory and practice.

This chapter describes the technology and terrain interactions that produce radar images. The chapter is illustrated with images acquired from both aircraft and satellites. Chapter 7 describes satellite radar systems and images.

RADAR SYSTEMS

Imaging radars are complex systems that are continually evolving. For example, phased array radars are replacing the conventional antenna radars described in this chapter. Nevertheless, a general description of principles and components is sufficient to understand how images are acquired.

Components

Figure 6-1 shows the components of a radar imaging system. The timing pulses from the pulse-generating device serve two purposes: (1) they control the bursts of energy from the transmitter, and (2) they synchronize the recording of successive energy returns to the antenna. The pulses of electromagnetic energy from the transmitter are of specific wavelength and duration, or pulse length. (Pulse length is further defined later in the chapter.)

The same antenna transmits the radar pulse and receives the return from the terrain. An electronic switch, or *duplexer*, prevents interference between transmitted and received pulse by

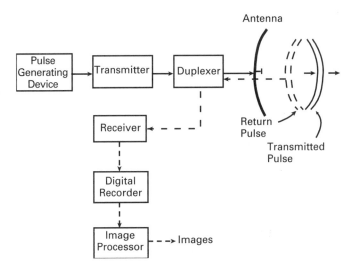

Figure 6-1 Block diagram of a side-looking radar system.

Figure 6-2 Radar survey aircraft. The radar antenna is housed in the pod beneath the fuselage. Courtesy Intera Technologies, Ltd.

terrain. A *receiver* amplifies the weak energy waves collected by the antenna. At the same time it preserves the variations in intensity of the returning pulse. The receiver also records the timing of the returning pulse, which determines the position of terrain features on the image. The return pulse is recorded on a digital medium for computer processing into images.

Imaging Systems

Figure 6-2 shows a radar survey aircraft. The antenna is housed in a pod mounted with its long axis parallel with the aircraft fuselage. Pulses of energy transmitted from the antenna illuminate strips of terrain in the *look direction* (also called the *range direction*). The look direction is oriented normal to the *azimuth direction* (aircraft flight direction). Figure 6-3 illustrates such a strip of terrain and the shape and timing of the energy pulse that it returns to the antenna. The return pulse is displayed as a function of two-way travel time on the horizontal axis. The shortest times are shown to the right, at the *near range*, which is the closest distance to the aircraft flight path. The longest travel times are shown at the *far range*. Travel time is converted to distance by multiplying by the

blocking the receiver circuit during transmission and blocking the transmitter circuit during reception. The *antenna* is a reflector that focuses the pulse of energy into the desired form for transmission and also collects the energy returned from the

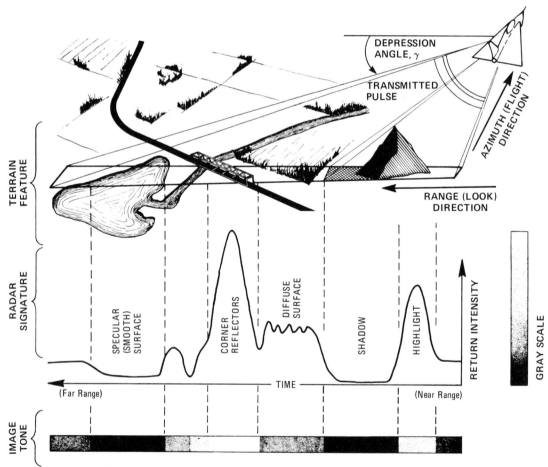

Figure 6-3 Terrain returns and image signatures for a pulse of radar energy.

Table 6-1 Typical terrain features and signatures on radar images

Image signature	Image tone	Terrain feature	Cause of signature
Highlights	Bright	Steep slopes and scarps facing *toward* antenna	Much energy is reflected back to antenna.
Shadows	Very dark	Steep slopes and scarps facing *away* from antenna	No energy reaches terrain; hence there is no return to antenna.
Diffuse surfaces	Intermediate	Vegetation	Vegetation scatters energy in many directions, including returns to antenna.
Corner reflectors	Very bright	Bridges and cities	Intersecting planar surfaces strongly reflect energy toward antenna.
Specular surfaces	Very dark	Calm water, pavement, dry lake beds	Smooth, horizontal surfaces totally reflect energy, with angle of reflectance opposite to angle of incidence.

speed of electromagnetic radiation (3×10^8 m · sec^{-1}). The amplitude of the returned pulse is a complex function of the interaction between the terrain and the transmitted pulse.

The digital record of the radar flight line is plotted onto an image film record in the following manner. The long dimension of the film roll represents the flight direction. A radar scan line is recorded across the film, beginning with the near-range data and ending with the far-range data. The film is advanced so that successive scan lines are plotted side by side. The analog display of a scan line shown in Figure 6-3 is plotted on the image by assigning the darkest tones of a gray scale to returns of the lowest intensity and the brightest tones to returns of the highest intensity. Figure 6-4A is an image that was recorded and plotted in this fashion.

In addition to being an active system, radar differs from other remote sensing systems such as cameras and optical-mechanical scanners by recording electromagnetic energy as a function of time rather than angular distance. Time is much more precisely measured and recorded than angular distance; hence radar images can be acquired at longer ranges with higher resolution than images from other remote sensing systems. Atmospheric absorption and scattering are negligible except at the very shortest radar wavelengths.

Table 6-1 lists the terrain features illustrated in Figure 6-3 together with their signatures and tones in a radar image. The table also summarizes the causes of the signatures and tones. The following section illustrates and describes the appearance of these terrain features on a typical radar image.

Typical Images

Figure 6-4 is an aircraft radar image and map of Weiss Lake and vicinity in northeastern Alabama that includes examples of the terrain features shown diagrammatically in Figure 6-3. The image was acquired with the look direction toward the west, which is toward the left margin of the image; thus the look di-

rection has the same orientation for both Figures 6-3 and 6-4A. The east-facing slopes of the mountains and ridges in Figure 6-4A face the radar look direction and form bright signatures, or *highlights*, caused by the strong radar returns. The west-facing slopes are oriented away from the radar look direction and form dark *shadows* because no energy reached these areas.

The flat terrain adjacent to Weiss Lake and the Coosa River is covered with various types of crops and native vegetation which have *diffuse* signatures with intermediate gray tones. The bridge southwest of Cedar Bluff (Figure 6-4B) has a bright signature caused by *corner reflectors*, formed by intersecting planar structures. Energy encountering corner reflectors is strongly reflected toward the antenna. The towns of Cedar Bluff and Centre likewise have very bright signatures because of abundant corner reflectors. The calm waters of Weiss Lake and the Coosa River are *specular*, or smooth, surfaces that reflect all incident energy such that the angle of reflection is equal but opposite to the angle of incidence. Hence no energy returns to the antenna, and a very dark signature results.

From the preceding descriptions it is obvious that tone alone (bright, intermediate, or dark) is insufficient to identify terrain features on radar images. Topographic scarps facing away from the antenna and specular surfaces both have dark tones but are completely different features. Radar signatures, however, are determined not only by tone but also by size, shape, texture, and associations of the image feature. The size and shape of specular features are different from radar shadows. Radar shadows are generally associated with highlights, which are absent from specular features.

Radar Wavelengths

Table 6-2 lists the various radar wavelengths, or bands, and their corresponding frequencies. The bands are designated by letters that were assigned randomly rather than in order of

A. Radar image.

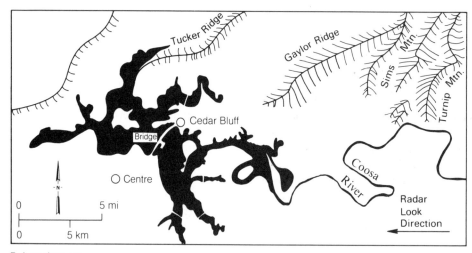

B. Location map.

Figure 6-4 Radar image and map of Weiss Lake and vicinity, northeastern Alabama.

Table 6-2 Radar wavelengths and frequencies used in remote sensing

Band designation*	Wavelength (λ), cm	Frequency (ν), GHz
K	0.8 to 2.4	40.0 to 12.5
X (3.0 cm)	2.4 to 3.8	12.5 to 8.0
C (6 cm)	3.8 to 7.5	8.0 to 4.0
S (8.0 cm, 12.6 cm)	7.5 to 15.0	4.0 to 2.0
L (23.5 cm, 25.0 cm)	15.0 to 30.0	2.0 to 1.0
P (68 cm)	30.0 to 100.0	1.0 to 0.3

*Wavelengths commonly used in imaging radars are shown in parentheses.

increasing wavelength. These random designations are a relict of World War II security policies to avoid mentioning any wavelengths. Frequency is a more fundamental property of electromagnetic radiation than is wavelength. As radiation passes through media of different densities, frequency remains constant, whereas velocity and wavelength change. Most interpreters, however, can visualize wavelengths more readily than frequencies; also, wavelengths are used to describe the visible and infrared spectral regions. Therefore, wavelengths are used here to designate various radar systems. Equation 1-1 (Chapter 1) enables any frequency (ν) to be converted into wavelength (λ) in the following manner:

$$c = \lambda \cdot \nu \tag{1-1}$$

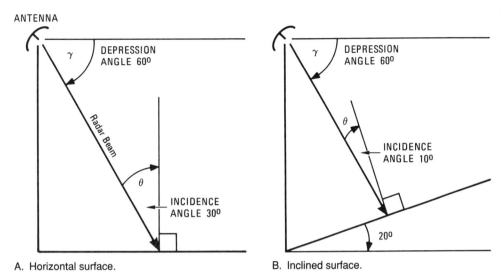

A. Horizontal surface.   B. Inclined surface.

Figure 6-5 Depression angle and incidence angle.

$$\lambda = \frac{3 \times 10^8 \ m \cdot sec^{-1}}{v}$$

where *c* is the speed of electromagnetic radiation. A convenient version of Equation 1-1 for converting radar frequencies into wavelength equivalents is

$$\lambda \ (in \ centimeters) = \frac{30}{v \ (in \ gigahertz)} \qquad (6\text{-}1)$$

In the 1960s and early 1970s, the first unclassified airborne radar was a Ka-band system that was discontinued by the mid-1970s. Modern commercial aircraft radar systems operate at X-band wavelengths. Aircraft radar experiments are also conducted at various other wavelengths.

Depression Angle

Spatial resolution in both the range (look) direction and azimuth (flight) direction is determined by the engineering characteristics of the radar system. In addition to wavelength, a key characteristic is the *depression angle* (γ), defined as the angle between a horizontal plane and a beam from the antenna to a target on the ground (Figure 6-5A). The depression angle is steeper at the near-range side of an image strip and shallower at the far-range side. The average depression angle of an image is measured for a beam to the midline of an image strip. An alternate geometric term is *incidence angle* (θ), defined as the angle between a radar beam and a line perpendicular to the surface. For a horizontal surface, θ is the complement of γ (Figure 6-5A), but for an inclined surface there is no correlation between the two angles (Figure 6-5B). The incidence angle more correctly describes the relationship between a radar beam and a surface than does depression angle; however, in actual practice, the surface is usually assumed to be horizontal and the incidence angle is taken as the complement of the depression angle. Several other terms have been applied to this complementary angle, including *look angle* and *off-nadir angle*. This book uses *depression angle* to describe radar viewing geometry.

Slant-Range and Ground-Range Images

Side-scanning radar systems acquire images as *slant-range displays* that are geometrically distorted as a function of the depression angle. The cross section in Figure 6-6A shows that locations on the ground (ground range) are projected into the slant-range display, which is the plane from the antenna to the far range of the image swath. As a result, features in the near-range portion of the swath are compressed relative to features in the far range. For example, Features A and B are of equal size, but on the slant-range display image A′ is much smaller than image B′. Figure 6-6B is a slant-range image that illustrates this distortion.

All modern radar systems automatically project the images into ground-range displays using the transformation

$$G = H \left(\frac{1}{\sin^2 \gamma} - 1 \right)^{1/2} \qquad (6\text{-}2)$$

where

G = ground-range distance

H = height of the antenna

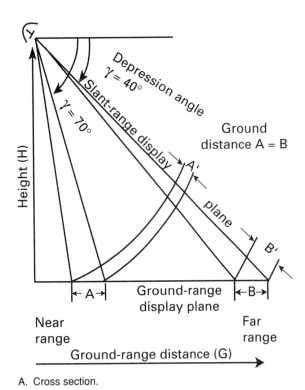

A. Cross section.

Figure 6-6C is the same image, now transformed into a ground-range display. This transformation assumes a horizontal ground surface and does not correct for the topographic variations discussed in the following section.

Spatial Resolution

The spatial resolution of a radar image is determined by the dimensions of the ground resolution cell, which are controlled by the combination of range resolution and azimuth resolution.

Range Resolution *Range resolution* (R_r), or resolution in the radar look direction, is determined by the depression angle and by the pulse length. *Pulse length* (τ) is the duration of the transmitted pulse and is measured in microseconds (μsec, or 10^{-6} sec); it is converted from time into distance by multiplying by the speed of electromagnetic radiation ($c = 3 \times 10^8$ m · sec^{-1}). Range resolution is determined by the relationship

$$R_r = \frac{\tau \cdot c}{2 \cos \gamma} \tag{6-3}$$

B. Slant-range display.

C. Ground-range display.

Figure 6-6 Restoration of slant-range display into ground-range display. Aircraft L-band image in central Illinois. Image covers an area of 10 by 10 km. Courtesy J. P. Ford, Jet Propulsion Laboratory.

Figure 6-7 shows this relationship graphically. Targets A and B, spaced 20 m apart, are imaged with a depression angle of 50° and a pulse length of 0.1 μsec. For these targets range resolution is calculated from equation (6-3) as

$$R_r = \frac{(0.1 \times 10^{-6} \text{ sec})(3 \times 10^8 \text{ m} \cdot \text{sec}^{-1})}{2 \cos 50°}$$

$$= \frac{30 \text{ m}}{2 \times 0.64}$$

$$= 23.4 \text{ m}$$

Targets A and B are not resolved because they are closer together (20 m) than the range resolution distance. In other words, they are within a single ground resolution cell and cannot be separated on the image.

In Figure 6-7 targets C and D are also spaced 20 m apart but are imaged with a depression angle of 35°. For this depression angle, range resolution is calculated as 18 m. Therefore targets C and D are resolved because they are more widely spaced than the ground resolution cell.

One method of improving range resolution is to shorten the pulse length, but this reduces the total amount of energy in each transmitted pulse. The energy and pulse length cannot be reduced below the level required to produce a sufficiently strong return from the terrain. Electronic techniques have been developed for shortening the apparent pulse length while providing adequate signal strength.

Azimuth Resolution Azimuth resolution (R_a), or resolution in the azimuth direction, is determined by the width of the terrain strip illuminated by the radar beam. To be resolved, targets must be separated in the azimuth direction by a distance greater than the beam width as measured on the ground. As shown in Figure 6-8, the fan-shaped beam is narrower in the near range than in the far range, causing azimuth resolution to be smaller in the near-range portion of the image. *Angular beam width* is directly proportional to the wavelength of the transmitted energy; therefore, azimuth resolution is higher for shorter wavelengths, but the short wavelengths lack the desirable weather penetration capability. Angular beam width is inversely proportional to *antenna length*; therefore, resolution improves with longer antennas, but there are practical limitations to the maximum antenna length.

The equation for azimuth resolution (R_a) is

$$R_a = \frac{0.7 \cdot S \cdot \lambda}{D} \tag{6-4}$$

where S is the slant-range distance and D is the antenna length. For a typical X-band system ($\lambda = 3.0$ cm; $D = 500$ cm) with a slant-range distance (S_{near} in Figure 6-7) of 8 km, R_a is calculated from Equation 6-4 as

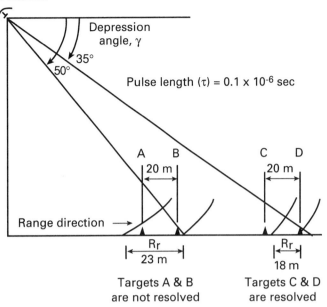

Antenna

Depression angle, γ

35°

50°

Pulse length (τ) = 0.1 x 10⁻⁶ sec

A B

20 m

C D

20 m

Range direction →

Rr
23 m

Rr
18 m

Targets A & B
are not resolved

Targets C & D
are resolved

Figure 6-7 Range resolution for different depression angles.

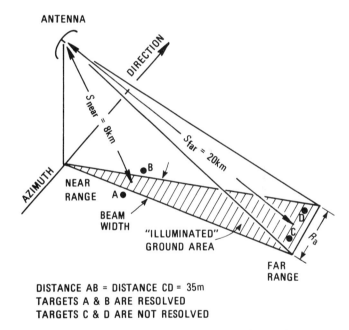

ANTENNA

DIRECTION

S_{near} = 8km

S_{far} = 20km

AZIMUTH

NEAR RANGE

•B

A•

BEAM WIDTH

"ILLUMINATED" GROUND AREA

D

C

R_a

FAR RANGE

DISTANCE AB = DISTANCE CD = 35m
TARGETS A & B ARE RESOLVED
TARGETS C & D ARE NOT RESOLVED

Figure 6-8 Azimuth resolution and beam width for a real-aperture system. From Barr (1969).

$$R_a = \frac{0.7(8 \text{ km} \times 3.0 \text{ cm})}{500 \text{ cm}}$$

$$= 33.6 \text{ m}$$

Therefore, targets in the near range, such as A and B in Figure 6-8, must be separated by approximately 35 m to be resolved. At the far-range position the slant-range distance (S_{far} in Figure

6-8) is 20 km, and R_a is calculated as 84 m; thus targets C and D, also separated by 35 m, are not resolved. Synthetic-aperture radar systems achieve much finer range resolutions than given in Equation 6-3, as described in the following section.

Real-Aperture and Synthetic-Aperture Systems

The two basic systems are real-aperture radar and synthetic-aperture radar, which differ primarily in the method used to achieve resolution in the azimuth direction. *Real-aperture radar* (the "brute force" system) uses an antenna of the maximum practical length to produce a narrow angular beam width in the azimuth direction, as illustrated in Figure 6-8.

The *synthetic-aperture radar* (SAR) employs a small antenna that transmits a relatively broad beam (Figure 6-9A). The Doppler principle and data-processing techniques are employed to synthesize the azimuth resolution of a very narrow beam. Using the familiar example of sound, the Doppler principle states that the frequency (pitch) of the sound heard differs from the frequency of the vibrating source whenever the listener and the source are in motion relative to one another. The rise and drop in pitch of the siren as an ambulance approaches and recedes is a familiar example. This principle is applicable to all harmonic wave motion, including the microwaves employed in radar systems.

Figure 6-9A illustrates the apparent motion of a target through the successive radar beams from points A to C as a consequence of the forward motion of the aircraft. The curve shows changes in Doppler frequency during the elapsed time between targets A, B, and C. The frequency of the energy pulse returning from the targets increases from a minimum at A to a maximum at B normal to the aircraft. As the target recedes from B to C, the frequency decreases.

A digital system records the amplitude and phase history of the returns from each target as the repeated radar beams pass across the target from A to C. The digital record is computer-processed to produce an image. The record of Doppler frequency changes enables each target to be resolved on the image as though it had been observed with an antenna of length L, as shown in Figure 6-9B. This synthetically lengthened antenna produces the effect of a very narrow beam with constant width in the azimuth direction, shown by the shaded area in Figure 6-9B. Comparing this narrow, constant beam with the fan-shaped beam of a real-aperture system (Figure 6-8) demonstrates the advantage of synthetic-aperture radar, especially for satellite systems. At the hundreds-of-kilometers-range distances of satellites, a real-aperture beam is so wide that azimuth resolution is in hundreds of meters. Synthetic-aperture systems carried on satellites, described in Chapter 7, have an azimuth resolution of a few tens of meters. For both real-aperture and synthetic-aperture systems, resolution in the range direction is determined by pulse length and depression angle (Figure 6-7). All modern aircraft and satellite radars operate in the SAR mode.

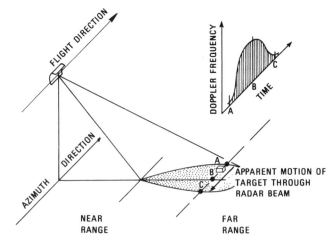

A. Shift in Doppler frequency caused by relative motion of target through radar beam.

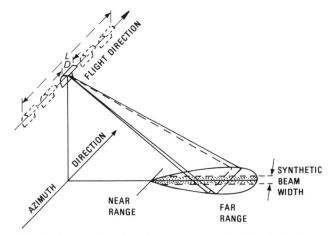

B. Azimuth resolution of synthetic-aperture radar. The physical antenna length *D* is synthetically lengthened to *L*.

Figure 6-9 Synthetic-aperture radar system. From Craib (1972, Figures 3 and 5).

CHARACTERISTICS OF RADAR IMAGES

The nature of radar illumination causes specific geometric characteristics in the images that include shadows and highlights, slant-range distortion, and image displacement. The geometric characteristics provide stereo models and the capability to compile topographic maps.

Highlights and Shadows

Figure 6-3 illustrated how topographic features interact with a radar beam to produce highlights and shadows. Figure 6-10A is an image of the White Mountains in eastern California that was acquired with a look direction toward the west (bottom

A. Aircraft radar image acquired with a depression angle of 17°.

B. Landsat TM band 4 image acquired with a solar elevation of 40°.

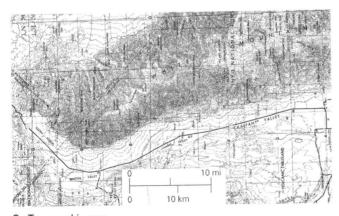

C. Topographic map.

Figure 6-10 Radar shadows and highlights, White Mountains, California.

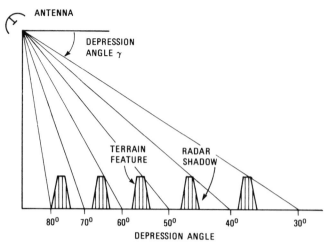

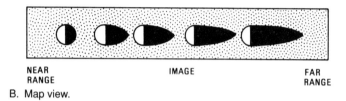

A. Cross-section view.

B. Map view.

Figure 6-11 Radar illumination and shadows at different depression angles.

margin of image) and a depression angle of 17°. This gentle depression angle results in the pronounced highlights on the east flank of the range and shadows on the west flank. The shadow records the profile of the range. Even in the low-relief terrain west of the mountains (lower portion of Figure 6-10A) subtle topographic features are clearly expressed by highlights and shadows. For comparison, Figure 6-10B is a Landsat TM image acquired with a sun elevation of 40°, which results in minimal highlights and shadows. The bright and dark signatures are largely the expression of differences in albedo of the surface. Figure 6-10C is a topographic contour map that shows variations in relief of the area.

The cross section in Figure 6-11A shows that the depression angle is steeper in the near range and shallower in the far range; therefore, terrain features of constant elevation have proportionally longer shadows in the far-range portion of an image, as illustrated in the map view of Figure 6-11B. For terrain with low relief, it is desirable to acquire radar images with a small depression angle in order to produce maximum highlights and shadows. In terrain with high relief, an intermediate depression angle is desirable, because the extensive shadows caused by a low depression angle may obscure much of the image. Images such as aerial photographs and Landsat are acquired with solar illumination which has a constant depression angle; thus, shadow lengths are directly proportional to topo-

graphic relief throughout these images. When interpreting a radar image one must recall that depression angles and shadow lengths are not constant throughout the image.

Look Direction and Terrain Features

Many natural and cultural linear features of the terrain, such as fractures or roads, have a strong *preferred orientation* and are expressed as lines on images. The geometric relationship between the preferred orientation and the radar look direction strongly influences the signatures of linear features.

Geologic Features, Venezuela Lineaments, faults, and outcrops of layered rocks are natural linear features that may be enhanced or subdued in radar images depending on their orientation relative to the look direction. Features trending normal or at an acute angle to the look direction are enhanced by highlights and shadows. Features trending parallel with the look direction produce no highlights or shadows and are suppressed in an image. These relationships are illustrated by an area in Venezuela (Figure 6-12) where separate X-band images (3-cm wavelength) were acquired with look directions toward the south and west. The geologic map (Figure 6-12C) shows a high mesa capped by relatively horizontal beds of resistant quartzite. The mesa is surrounded by lowlands of folded metamorphosed sedimentary rock that clearly express the bedding trends.

In the large fold in the eastern part of the area, one limb trends north and the second limb trends northwest. The north-trending limb is enhanced by the west look direction (Figure 6-12B). The northwest-striking limb trends at an acute angle to either look direction and is equally enhanced in both images. In the western part of the area, a major lineament (probably a fault) strikes south through the metamorphic terrain and cuts the west portion of the mesa where it forms a linear valley. The lineament is strongly enhanced by the west look direction (Figure 6-12B) and subdued by the south look direction (Figure 6-12A). Other lineaments in the mesa that trend generally east or west of north are enhanced or subdued in the same fashion on the two images. In the southeast part of the area, a number of parallel lineaments trend slightly north of east in the metamorphic terrain; these are clearly visible with the south look direction but are not recognizable with the west look direction. The west-trending, parallel light and dark bands in Figure 6-12B are minor defects in the image. If this area were to be imaged with only a single look direction, an orientation toward the southwest or northeast would enhance most of the linear features.

Cultural Features, New Orleans, Louisiana In urban areas a strong preferred orientation is formed by street patterns and by buildings with their walls aligned parallel and normal to the streets. Radar energy with a look direction that is normal or parallel to a street direction will thus be oriented perpendic-

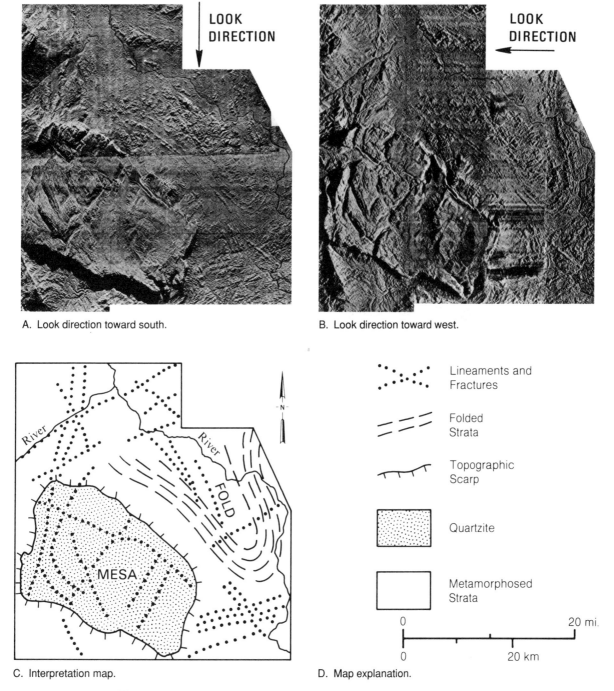

A. Look direction toward south.

B. Look direction toward west.

C. Interpretation map.

D. Map explanation.

Lineaments and Fractures

Folded Strata

Topographic Scarp

Quartzite

Metamorphosed Strata

Figure 6-12 Expression of linear geologic features on images with different look directions. Aircraft X-band images in southeast Venezuela.

ular to many walls, resulting in a strong return and a bright signature. A lock direction that is oblique to a street direction will be oriented obliquely to walls, resulting in a weak return and an intermediate signature.

New Orleans, Louisiana, is well suited for evaluating the relationship between the radar look direction and the orientation of cultural features. Figure 6-13A was acquired with the look direction toward the northwest. Figure 6-13B was acquired with the look direction toward the northeast. Figure 6-13C is an aerial photograph. Figure 6-14 is a map showing land cover and land use. Calm water in the lakes and in the Mississippi River is dark in all three images of Figure 6-13; bright spots on

A. Radar image with look direction toward northwest.

B. Radar image with look direction toward northeast.

C   Aerial photograph.

Figure 6-13 Land-use signatures on images acquired with different look directions. L-band images of New Orleans, Louisiana. Courtesy J. P. Ford, Jet Propulsion Laboratory.

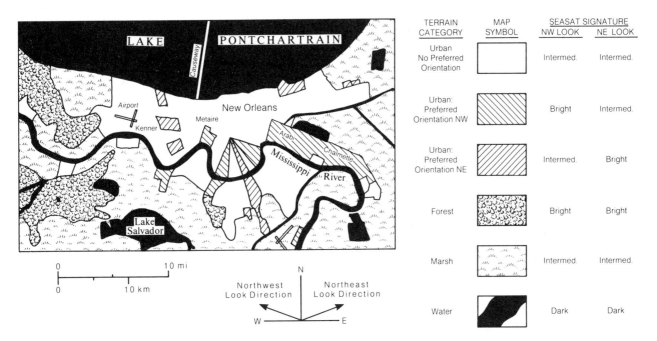

Figure 6-14 Map showing preferred orientation of cultural features for radar images of New Orleans, Louisiana.

the river are ships and barges. Forested areas are bright in both radar images because the trees have no preferred orientation and produce strong volume scattering. Much of the area is covered with marsh vegetation, which has a medium-gray signature on both radar images.

The aerial photograph (Figure 6-13C) shows a wide diversity of street patterns and orientations in the area. The more recently developed areas around the shore of Lake Ponchartrain and the towns of Arabi and Chalmette (Figure 6-14) have rectangular street grids. The older part of New Orleans, which is located within a bend of the Mississippi River, has a radial street pattern that was laid out to provide access to the river.

In the newer urban areas, the orientation of streets and building walls is oblique to both the northwest and northeast radar look directions, which results in dark to intermediate signatures in both images (Figure 6-13A,B). In the map (Figure 6-14) these areas are shown with a blank pattern ("No Preferred Orientation" category). The map also shows the orientation of the two radar look directions. Of special interest are urban areas that are bright on one image but dark or intermediate on the other. On the map these areas are patterned to indicate whether they are bright on the northwest look direction or on the northeast look direction. The radial street pattern in the bend of the river produces these different returns, which are known as *cardinal point effects*. Figure 6-13A (northwest look direction) shows two wedges with bright signatures in this radial street pattern. The bright wedge on the western side is caused by the radial streets oriented normal to the northwest look direction. Parallel streets oriented normal to the northwest look direction

cause the bright wedge in the eastern part. In Figure 6-13B (northeast look direction) there is a single bright wedge from the radial streets in the central part of the bend. The street directions are used for convenient reference; actually the buildings are responsible for the bright signatures in both images. In the Arabi-Chalmette area, streets are oriented parallel and perpendicular to the look direction, and this produces the bright signature in Figure 6-13A and the dark signature in Figure 6-13B. Similar explanations apply to the other northwest and northeast areas indicated on the map (Figure 6-14). This relationship between street patterns, radar look direction, and image signatures has been observed elsewhere (Bryan, 1979).

Significance of the Look Direction

The X-band images of Venezuela and the L-band images of New Orleans demonstrate that, irrespective of radar wavelength, there is an important relationship between radar look direction, image signatures, and orientation of linear features in the terrain. Linear geologic features such as faults, outcrops, and lineaments that are oriented at a normal or oblique angle to the radar look direction are enhanced by highlights and shadows in the image. Features oriented parallel with the look direction are suppressed and are difficult to recognize in the image. The importance of radar look direction is comparable to the importance of sun azimuth in aerial photographs and in Landsat images acquired at low sun angles. A radar survey should be oriented with the look direction at a normal or acute angle relative to the known geologic trends of the area.

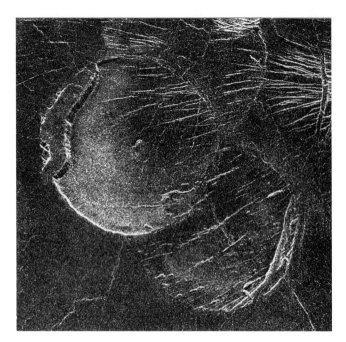

Figure 6-15 Topographic inversion caused by orientation of the radar look direction. Are these circular features depressions (calderas) or uplifts (domes)? Magellan radar image of the Guinevere Planitia lowland, Venus. The look direction is toward the upper margin at a depression angle of 51°. From Ford and others (1993, Figure 9-18). Courtesy J. P. Ford, Jet Propulsion Laboratory.

In urban areas the orientation of streets and buildings relative to the look direction also influences the signatures. In addition, linear metal features, such as fences and railroads, have bright signatures where they are oriented normal to look directions but are not recognizable where they are oriented parallel with the look direction. Interpreters of radar images of geologic or urban scenes must be aware of these geometric relationships.

Topographic Inversion

Ramachandran (1988) noted that the human visual experience of the world is based on two-dimensional images: namely, the flat patterns of varying light intensity and color that fall on the single layer of cells in the retina of the eye (Figure 1-9). Yet our brains come to perceive solidity and depth. A number of cues about depth are available in the retinal image: shading, perspective, occlusion of one object by another, and stereoscopic vision. The brain utilizes these cues to recover the three-dimensional shapes of objects. Experiments show that our visual system assumes that an image is illuminated by only one light source and that the light comes from above, that is, from the upper margin toward the lower margin of the image. This assumption gives us a strong bias when we view images

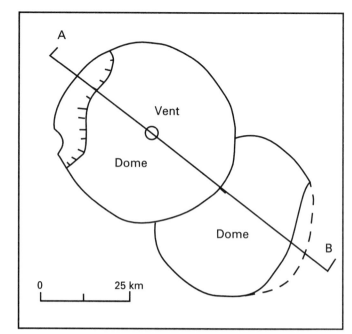

A. Map.

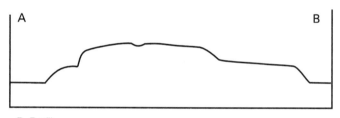

B. Profile.

Figure 6-16 Interpretation map and profile of rotated Magellan image of circular features on Guinevere Planitia, Venus (Figure 6-15).

with pronounced shadows, such as low-sun-angle photographs and radar images.

This bias is demonstrated by the radar image shown in Figure 6-15. Do you perceive these circular features as uplifts (domes) or as depressions (calderas)? Most viewers see them as depressions. For example, 24 students in my UCLA class viewed the image; 19 students (79 percent) perceived the features as craters, while 5 students (21 percent) perceived domes. The image is shown with the illumination and radar shadows oriented toward the upper margin, which is the reverse of our assumption. The circular features are actually uplifts (domes), as shown in the map and cross section of Figure 6-16. Rotate the image (Figure 6-15) 180° so that look direction and shadows are oriented toward the lower margin. The features should now appear in their correct perspective as domes. This phenomenon is called *topographic inversion* and is common to all images with pronounced highlights and shadows. Interpreters should determine the illumination direction and orient the image with illumination and shadows toward the lower margin.

Radar and Landsat Images Compared, Indonesia

Figure 6-17 covers the Mapia anticline in the southern portion of Irian Jaya, which is the Indonesian province in the western part of the island of New Guinea. Figure 6-17A is a TM subscene acquired with a 45° solar elevation, which results in minimal shadows. Figure 6-17B is a radar image acquired with a depression angle of 17°. On the images and geologic map (Figure 6-18), north is toward the lower margin, in order to direct the radar look direction and shadows toward the lower margin. The solar direction is toward the right margin (west).

In the upper portion of the images resistant sandstones of the upper member of the Buru Formation dip south and form prominent dipslopes and antidip scarps. Nonresistant shale of the lower member forms a broad valley. The resistant New Guinea Limestone forms the crest of the Mapia anticline. In the northern (lower) portion of the image a thrust plate of New Guinea Limestone overrides the upper member of the Buru Formation and rests on the lower member. These geologic features are relatively obscure on the TM image, which has only minor shadows oriented westward, parallel with the geologic strike. The features are pronounced in the radar image, which has prominent shadows that are favorably oriented normal to the geologic strike.

The cross sections in Figure 6-19 show the relationship between vertical illumination angle and geologic structure for both images. The constant 45° solar elevation of the TM image produces minimal highlights and shadows (Figure 6-19A). The 17° radar depression angle produces strong highlights and shadows (Figure 6-19B). Beds dipping toward the antenna have broad highlights on the dipslopes and narrow shadows on the antidip scarps. These radar highlights and shadows are used to interpret dip-and-strike orientation in the same manner that solar highlights and shadows were used in the TM image of the Saharan Atlas Mountains in Chapter 3. The north look direction of the radar image is ideally oriented to enhance the west-trending Mapia anticline. An east or west look direction would have obscured the structural pattern.

This example demonstrates that in tropical regions radar is a valuable mapping system, not only because it penetrates cloud cover, but because it also provides optimal illumination geometry.

Image Foreshortening

The curvature of a transmitted radar pulse causes the top of a tall vertical feature to reflect energy in advance of its base, which results in displacement of the top toward the near range on the image. This distortion is called *foreshortening* or *image layover*. Figure 6-20A shows an image of Mount Saint Helens, acquired prior to the eruption, that is an extreme example of foreshortening. The cross section (Figure 6-20A) shows that

the wavefront encounters the mountaintop at A in advance of the base B. Therefore, the return from the top is received in advance of the return from the base, which foreshortens the mountain (Figure 6-20B). On the image (Figure 6-20C) the foreshortened mountain appears to lean toward the near-range direction. Comparing the image to the topographic map (Figure 6-20D) illustrates the extent of foreshortening on the image. Foreshortening cannot be readily corrected because Equation 6-3 assumes a horizontal ground surface.

The amount of foreshortening is influenced by the following factors:

1. **Height of targets** Higher features are foreshortened more than lower features.
2. **Radar depression angle** Features imaged with steep (large) depression angles are foreshortened more than those acquired with shallow (small) depression angles.
3. **Location of targets** Features located in the near-range portion of the image swath are foreshortened more than comparable features located in the far range because the depression angle is steeper in the near range.

Foreshortening can be minimized by acquiring images at shallow depression angles, but the resulting radar shadows may be excessive for terrain with high topographic relief. The systematic geometric distortion introduced by foreshortening enables sidelapping images to be viewed in stereo, as described in the following section.

Stereo Images

Aircraft radar surveys are normally flown with a flight line spacing that provides 60 percent sidelap between adjacent swaths of imagery. Figure 6-21 shows the flight lines and image swaths for a portion of a survey in Papua New Guinea (PNG). Each image swath is 45 km wide; the area of sidelap (37 km wide) is shown by the cross-hatched pattern in the plan view (Figure 6-21A). The imaging geometry for the sidelapping swaths is shown in the profile view (Figure 6-21B). The left image of the stereo pair is from the far-range portion of swath 1 and was acquired with a gentler depression angle (11°). The right image is from the near-range portion of swath 2 and was acquired with a steeper depression angle (20°). The different depression angles result in different amounts of foreshortening, which enables the overlapping swaths to be viewed in stereo. This configuration is called *same-side stereo* because the look direction is toward the same side of the image swaths, which is toward the north in this example. *Opposite-side stereo* images are acquired with flight lines on opposite sides of the two swaths, which results in opposite look directions. Opposite-side stereo is not desirable because shadows are oriented in opposite directions on the two images and more terrain is obscured by shadows. Stereo images may also be acquired by repeating a flight line at two different altitudes, which results in two different depression angles for the same

A. Landsat TM band 4 image with a solar elevation angle of 45°.

B. X-band radar image with a depression angle of 17°.

Figure 6-17 Radar and Landsat images compared, Mapia anticline, Irian Jaya, Indonesia.

terrain. This method is rarely employed. Plaut (1993) gives details of the geometry of radar stereo images.

Figure 6-22 is a stereo pair of the Mananda anticline in the Papuan Fold Belt of the PNG portion of the island of New Guinea. The overlapping images were acquired with the geometry shown in Figure 6-21. The left image of the pair (Figure 6-22A) is a far-range image recorded with a very shallow depression angle that results in relatively long shadows. The right image (Figure 6-22B) is a near-range image recorded with a steeper depression angle that results in relatively short shadows. When viewed with a stereoscope the stereo model has a vertical exaggeration estimated between 1.5× and 2.0×. Figure

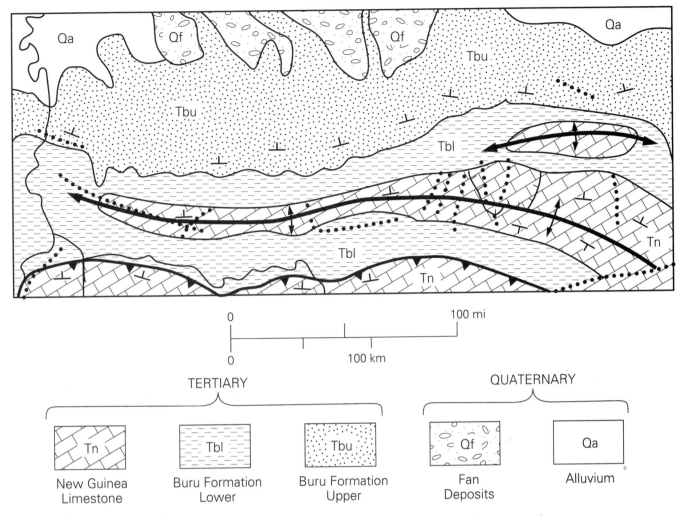

TERTIARY

QUATERNARY

Tn	Tbl	Tbu	Qf	Qa
New Guinea Limestone	Buru Formation Lower	Buru Formation Upper	Fan Deposits	Alluvium

Figure 6-18 Geologic map of Mapia anticline (Figure 6-17).

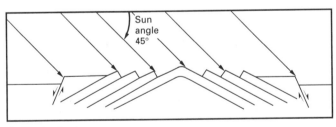

A. Illumination by sun.

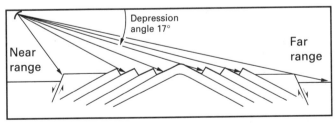

B. Illumination by radar.

Figure 6-19 Generalized cross sections comparing illumination geometry for Landsat and radar images.

6-23 is a geologic map interpreted from the stereo pair. In the northwest corner (upper right) of the stereo model are the eroded cones of Mount Sisa and an unnamed volcano, together with associated lava flows. The Mananda anticline was folded so recently that it has not yet been breached by erosion and forms a distinct topographic ridge capped by the Darai Limestone. Tropical weathering conditions have dissolved parts of the limestone to produce the distinctive pitted texture known as *karst topography*. The western portion of the model is underlain by Orubadi Sandstone, which weathers to a ledge-and-slope topography. The older Darai Limestone has been thrust southward over the younger Orubadi Sandstone. The outcrop of the thrust fault is marked by the strong, narrow, curvilinear highlight along the south (left) flank of the Ma-nanda anticline (Figures 6-22 and 6-23). The anticline was folded by the same horizontal compression that caused the thrusting. A number of steep fault scarps are shown by linear highlights.

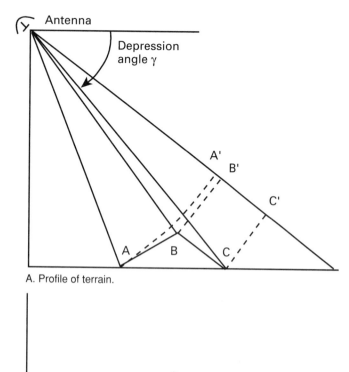

A. Profile of terrain.

B. Profile of terrain with layover.

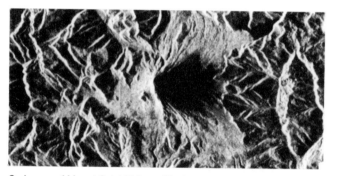

C. Image of Mount Saint Helens, Washington, showing foreshortening.

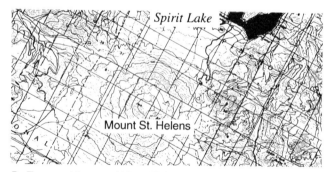

D. Topographic map of Mount Saint Helens.

Figure 6-20 Foreshortening of topographic features on radar images.

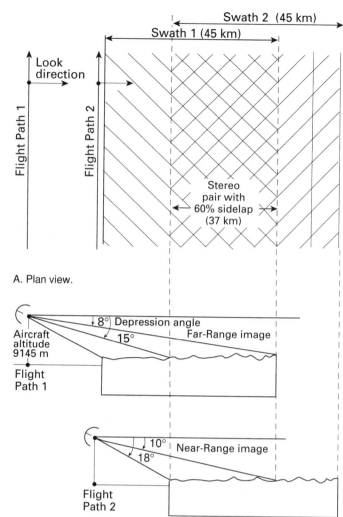

A. Plan view.

B. Profile view.

Figure 6-21 Geometry of sidelapping radar swaths for stereo viewing.

Topographic Maps Derived from Stereo Images

In a stereo pair of images, the overlapping portions are acquired at different depression angles (Figure 6-21B). Therefore, a topographic feature such as a mountain peak is foreshortened to different degrees in the two images. This difference in foreshortening may be used to compute the elevation of the feature. The elevation data are then used to produce contour maps. Figure 6-24 is a topographic map compiled from stereo radar images of a portion of the Canadian Rocky Mountains. The actual elevations of contours are determined from ground localities (called *benchmarks*) where elevations have been established by conventional surveys. Radar energy does not penetrate vegetation, as discussed in Chapter 7; therefore, contour elevations in forested terrain, such as the Rocky Mountains, give the height at tree canopy rather than at ground level.

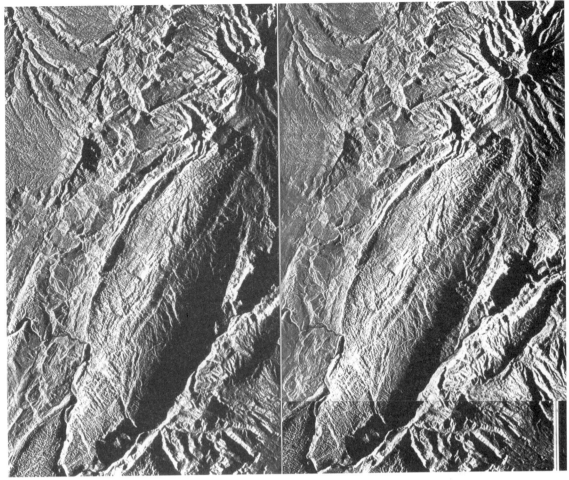

A. Left image.

B. Right image.

Figure 6-22 Stereo pair of radar images of Mananda anticline, Papuan Fold Belt, Papua New Guinea. Courtesy J. M. Ellis. Chevron Overseas Petroleum, Inc.

The ability to derive contour maps from radar images is especially valuable in cloudy regions that lack aerial photographs. In 1988 a group of international oil companies needed topographic maps in the Congo, Africa, in order to conduct exploration projects. A contractor flew over the area with sidelapping radar swaths, which were compiled into topographic maps. A contour interval of 10 m was used in this coastal plain region with low relief. Ellis and Richmond (1991) describe the survey and map preparation; they report that the maps have an 85 percent reliability factor. Mercer and others (1991) describe a similar radar mapping project in Irian Jaya, which is also cloud covered. The 1:50,000 scale maps of this rugged region have contour intervals of 50 or 100 m, depending upon relief of the terrain. The horizontal accuracy is 20 m, and vertical accuracy is 30 to 40 m.

Radar Mosaics

Radar swaths are commonly several hundred kilometers long and range from 20 to 60 km in width. In addition to providing stereo models, the sidelapping swaths may be combined into a mosaic that provides more extensive coverage. The acquisition of images and the construction of a mosaic are exacting tasks. Aircraft navigation must be precise to provide proper sidelap of image strips. Altitude and speed must be controlled to provide uniform scale on the images. All image swaths are acquired with the same look direction; otherwise shadows and highlights will be reversed in different parts of the mosaic.

Figure 6-25A shows the plan view for the same image swaths with 60 percent sidelap that were used for the stereo pairs shown in Figure 6-21. The central 27.5-km-wide strip

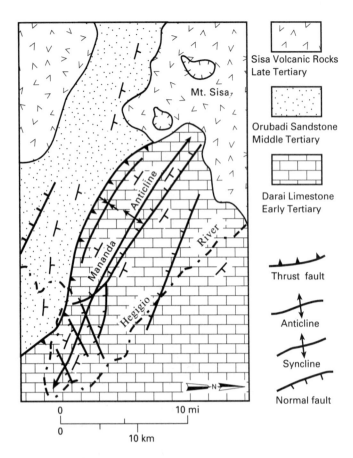

Figure 6-23 Interpretation map of radar stereo pair of Mananda anticline (Figure 6-22).

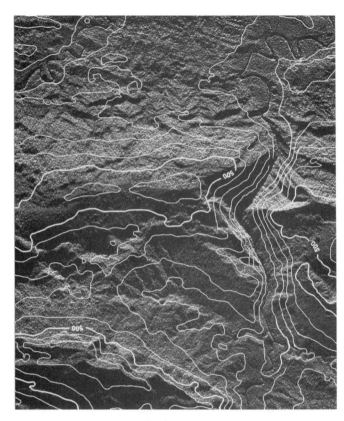

Figure 6-24 Topographic contour map derived from a stereo pair of radar images. Contours (interval 100 m) are superimposed on the image. Canadian Rocky Mountains, Alberta, Canada. Courtesy Intera Technologies, Ltd.

from each swath is selected for the mosaic. The profile view (Figure 6-25B) shows that this selection provides relatively uniform depression angles for all image strips. For strip 1 the depression angles range from 12° to 20°. For strip 2 the depression angles range from 10° to 18°. As a result highlights and shadows are relatively uniform throughout the mosaic. Figure 6-26 is a portion of a mosaic of the Central Range in PNG that was produced in this manner. This mosaic represents the technology of the mid-1980s, when prints of individual image strips were manually trimmed and spliced into a mosaic. Today most mosaics are compiled directly from digital data, using the techniques described in Chapter 8.

Radar mosaics of portions of the United States including Alaska (scale 1:250,000) are available from the EROS Data Center (EDC), which can provide index maps and price lists. Wynn and others (1993) published a radar mosaic of the Guiana shield in Venezuela.

RADAR RETURN AND IMAGE SIGNATURES

Stronger radar returns produce brighter signatures on an image than do weaker returns, as shown diagrammatically in Figure 6-3 and in the images of this chapter. The intensity of the radar return, for both aircraft and satellite systems, is determined by the following properties:

1. Radar system properties
   Wavelength
   Depression angle
   Polarization
2. Terrain properties
   Dielectric properties (including water content)
   Surface roughness
   Feature orientation

The following sections discuss these properties in more detail.

Dielectric Properties and Water Content

The *dielectric constant* is an electrical property that influences the interaction between matter and electromagnetic energy, especially at radar wavelengths. At radar wavelengths the dielectric constant of dry rocks and soils ranges from about 3 to 8, while water has a value of 80. Therefore, as moisture content

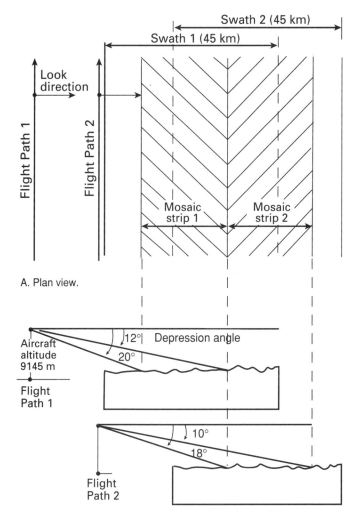

Swath 2 (45 km)

Swath 1 (45 km)

Look direction

Flight Path 1

Flight Path 2

Mosaic strip 1

Mosaic strip 2

A. Plan view.

Aircraft altitude 9145 m

12° | Depression angle

20°

Flight Path 1

10°

18°

Flight Path 2

B. Profile view.

Figure 6-25 Geometry of sidelapping image swaths for compiling a radar mosaic.

of a material increases, the dielectric constant can reach values of 25 or more, as shown for sand and clay in Figure 6-27.

Backscatter coefficient is a quantitative measure of the intensity of energy returned to the antenna. Figure 6-28 plots the relationship between backscatter coefficient and moisture content for fields of corn, bare soil, and milo, a grain resembling millet. As moisture increases, the backscatter coefficient also increases, which in turn produces brighter image signatures.

The experimental relationships between moisture content and image tone are illustrated by an image in Iowa (Figure 6-29) where a rainstorm moved northeastward across an agricultural region 12 hours before the image was acquired. The map of rainfall measurements shows that the bright signature of the western portion of the image correlates with areas that

received up to 0.35 in. (0.89 cm) of rain, whereas the dark eastern area was dry. The bright streaks in the otherwise dark eastern portion of the image are attributed to small rainstorms that moved northeastward in advance of the main weather front. Similar signatures of damp and dry agricultural areas occur in images acquired at other dates and localities (Ulaby, Brisco, and Dobson, 1983).

At a test area in Oklahoma, soil moisture was measured at a number of sites at the same time a radar image was acquired (Blanchard and others, 1981). For each site the scattering coefficient was plotted as a function of soil moisture. The resulting graph (Figure 6-30) shows that radar backscatter, or image brightness, increases linearly with increasing soil moisture. The test sites plotted in Figure 6-30 include bare soil, alfalfa, and milo. Backscatter measurements were also made for cut and standing cornfields, which showed no correlation with soil moisture. Apparently both the cut and standing corn effectively masked the underlying soil from any interaction with the radar beam. Increasing soil moisture reduces the penetration of radar energy beneath the surface.

These image signatures and experimental data suggest that radar images may be used to estimate soil moisture, which would be valuable information for hydrology and agronomy. Radar backscatter, however, is also strongly influenced by other characteristics of the scene, such as surface roughness, which is discussed later in this chapter. A number of researchers are working at sorting out the effects of these various characteristics of a scene.

The preceding comments refer to absorbed water. Standing water (fresh or salty) is highly reflective of radar energy and has dark or bright signatures depending upon whether the surface is calm or agitated.

Surface Roughness

Surface roughness is measured in centimeters and is determined by textural features comparable in size to the radar wavelength, such as leaves and twigs of vegetation and sand, gravel, and cobble particles. Surface roughness strongly influences the strength of radar returns and is distinct from *topographic relief*, which is measured in meters and hundreds of meters. Topographic relief features include hills, mountains, valleys, and canyons that are expressed on images by highlights and shadows. The average surface roughness within a ground resolution cell determines the intensity of the return for that cell. Ground resolution cells are typically 10 by 10 m for airborne systems and several times larger for satellite systems. Surface roughness is a composite measure of the vertical and horizontal dimensions and spacing of the small-scale features, together with the geometry of the individual features (leaves, twigs, sand, and gravel particles). Because of the complex geometry of most natural surfaces, it is difficult to characterize them mathematically, particularly for the large area of a resolution cell. For most surfaces the *vertical relief*, or average

Figure 6-26 Radar mosaic in Papua New Guinea.

height of surface irregularities, is an adequate approximation of surface relief. Surfaces may be grouped into the following three roughness categories:

1. A *smooth surface* reflects all the incident radar energy with the angle of reflection equal and opposite to the angle of incidence (Snell's law).

2. A *rough surface* diffusely scatters the incident energy at all angles. The rays of scattered energy may be thought of as enclosed within a hemisphere, the center of which is located at the point where the incident wave encounters the surface.

3. A surface of *intermediate roughness* reflects a portion of the incident energy and diffusely scatters a portion.

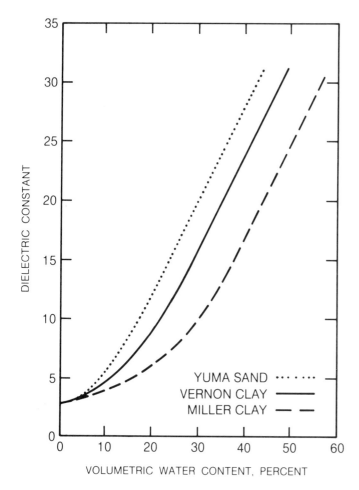

Figure 6-27 Variation of dielectric constant (at 27-cm wavelength) as a function of moisture content in sand and clay. From Wang and Schmugge (1980, Figure 3).

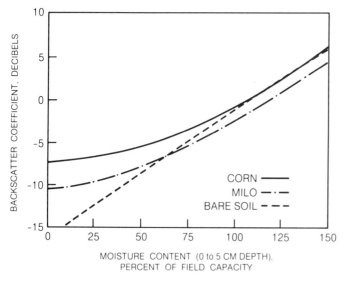

Figure 6-28 Variation of backscatter (at 6.7-cm wavelength) as a function of moisture content. Milo is a grain resembling millet. From Ulaby, Aslam, and Dobson (1982, Figures 4 and 7).

A. Radar image.

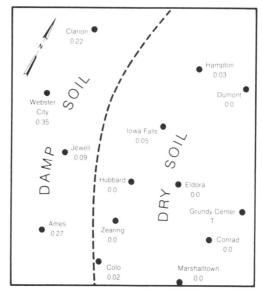

B. Map showing rain-gauge data.

Figure 6-29 L-band image and map of rain-gauge data for Ames, Iowa, and vicinity acquired 12 hours after a rainfall. From Ulaby, Briscoe, and Dobson (1983, Figure 2). A, courtesy J. P. Ford, Jet Propulsion Laboratory.

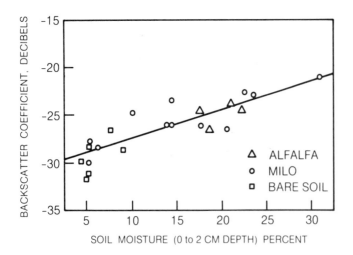

Figure 6-30 Variation in backscatter coefficient (L band) as a function of soil moisture for individual fields in a test site at Guymon, Oklahoma. From Blanchard and others (1981).

The roughness of a surface return is determined by the relationship of surface relief, at the scale of centimeters, to radar wavelength and to the depression angle of the antenna.

Roughness Criteria The *Rayleigh criterion* considers a surface to be smooth if

$$h < \frac{\lambda}{8 \sin \gamma} \qquad (6\text{-}5)$$

where

h = vertical relief

λ = radar wavelength

γ = depression angle

Both h and λ are given in the same units, usually centimeters. For a system with a wavelength of 23.5 cm and a depression angle of 70°, the surface relief below which the surface will appear smooth is determined by substituting into Equation 6-5:

$$h < \frac{23.5 \text{ cm}}{8 \sin 70°}$$

$$< \frac{23.5 \text{ cm}}{8 \times 0.94}$$

$$< 3.1 \text{ cm}$$

Therefore, a vertical relief of 3.1 cm is the theoretical boundary between smooth and rough surfaces for the given wave-

length and depression angle. Peake and Oliver (1971) modified the Rayleigh criterion by defining upper and lower values of h for surfaces of intermediate roughness. By their *smooth criterion*, a surface is smooth if

$$h < \frac{\lambda}{25 \sin \gamma} \qquad (6\text{-}6)$$

Substituting for a system in which $\lambda = 23.5$ cm and $\gamma = 70°$ results in

$$h < \frac{23.5 \text{ cm}}{25 \sin 70°}$$

$$< \frac{23.5 \text{ cm}}{25 \times 0.94}$$

$$< 1.0 \text{ cm}$$

Therefore, a vertical relief of 1.0 cm is the boundary between smooth surfaces and surfaces of intermediate roughness for the given wavelength and depression angle.

Peake and Oliver (1971) also derived a *rough criterion* that considers a surface to be rough if

$$h > \frac{\lambda}{4.4 \sin \gamma} \qquad (6\text{-}7)$$

Substituting for a system in which $\lambda = 23.5$ cm and $\gamma = 70°$ results in

$$h > \frac{23.5 \text{ cm}}{4.4 \sin 70°}$$

$$> \frac{23.5 \text{ cm}}{4.4 \times 0.94}$$

$$> 5.7 \text{ cm}$$

Therefore, a vertical relief of 5.7 cm is the boundary between intermediate surfaces and rough surfaces for the given radar wavelength and depression angle. Note that the value determined earlier from the Rayleigh criterion ($h < 3.1$ cm) is intermediate between those derived for the smooth criterion and the rough criterion.

Figure 6-31 illustrates the interaction between a transmitted pulse of energy and surfaces with smooth, intermediate, and rough relief. The smooth surface (Figure 6-31A) reflects all the energy, and no energy is returned to the antenna, which results in a dark signature. The surface with intermediate roughness (Figure 6-31B) reflects part of the energy and scatters the remainder; the backscattered component results in an intermediate signature. The rough surface (Figure 6-31C) diffusely scat-

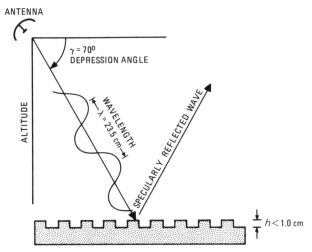

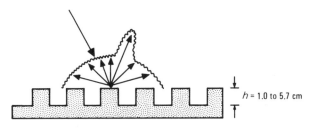

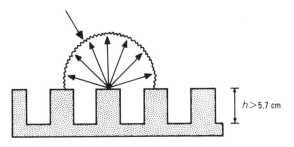

C. Rough surface, strong return.

Figure 6-31 Models of surface roughness criteria and return intensity for radar images at 23.5-cm wavelength.

ters all the energy, causing a relatively strong backscattered component, which produces a bright signature.

Equations 6-5 and 6-6 were used to calculate the values of vertical relief (*h*) that define smooth, intermediate, and rough surfaces for three radar systems listed in Table 6-3. This information is shown in a different fashion in Table 6-4, where the *roughness response* for different values of *h* is given for different radar wavelengths. For example, a surface with a vertical relief of 1.40 cm, typical of fine gravel, appears rough on X-band images, intermediate on C-band images, and smooth on L-band images. The image signature of this surface will be bright, medium gray, and dark at these wavelengths. Table 6-4

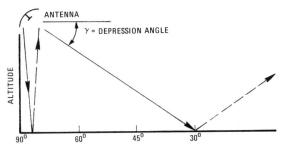

A. Smooth surface with specular reflection.

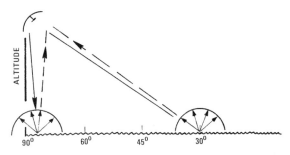

B. Rough surface with diffuse scattering.

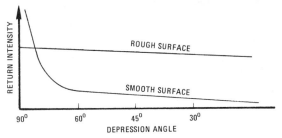

C. Return intensity as a function of depression angle.

Figure 6-32 Radar return from smooth and rough surfaces as a function of depression angle.

also illustrates the advantage of acquiring radar images at more than one wavelength for terrain analysis. By comparing the image signatures at two or more wavelengths, one can estimate the surface roughness more accurately than by looking at the image signature of a single wavelength. Images acquired at different wavelengths of the Copper Canyon alluvial fan are compared later in this chapter.

Depression Angle and Surface Roughness As discussed earlier, the depression angle (γ) affects the smooth and rough criteria. Figure 6-32A shows that at low to intermediate depression angles the specular reflection from a smooth surface returns little or no energy to the antenna. At very high depression angles (80° to 90°), however, the specularly reflected wave may be received by the antenna and produce a strong return. A rough surface produces diffuse scattering of relatively uniform strong intensity for a wide range of depression angles (Figure 6-32B), which results in strong returns at all depression angles. Figure 6-32C compares the relative return inten-

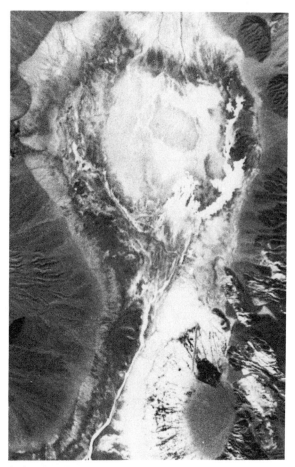

A. Landsat TM band 3 (red) image.

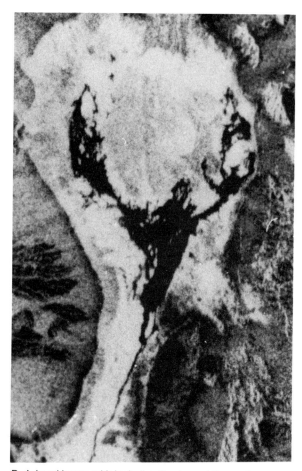

B. L-band image with look direction toward the east.

Figure 6-33 Radar and Landsat images of Cottonball Basin, Death Valley, California. From Sabins (1984, Figure 5).

Table 6-3 Surface roughness categories for representative radar systems. Depression angle (γ) is 40° for all systems.

Roughness category	X band λ = 3 cm	C band λ = 6 cm	L band λ = 23.5 cm
Smooth, cm	$h < 0.19$	$h < 0.37$	$h < 1.46$
Intermediate, cm	$h = 0.19$ to 1.06	$h = 0.37$ to 2.12	$h = 1.46$ to 8.35
Rough, cm	$h > 1.06$	$h > 2.12$	$h > 8.35$

Table 6-4 Roughness response for different values of vertical relief (h) at different radar wavelengths. Depression angle (γ) is 40° for all systems.

Vertical relief (h), cm	X-band image (λ = 3 cm)	C-band image (λ = 6 cm)	L-band image (λ = 23.5 cm)
0.10	Smooth	Smooth	Smooth
0.40	Intermediate	Smooth	Smooth
1.40	Rough	Intermediate	Smooth
5.00	Rough	Rough	Intermediate
10.00	Rough	Rough	Rough

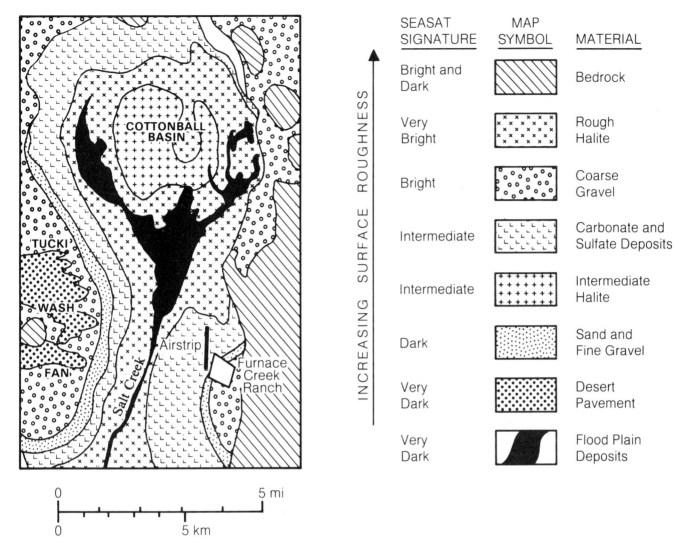

SEASAT SIGNATURE	MAP SYMBOL	MATERIAL
Bright and Dark		Bedrock
Very Bright		Rough Halite
Bright		Coarse Gravel
Intermediate		Carbonate and Sulfate Deposits
Intermediate		Intermediate Halite
Dark		Sand and Fine Gravel
Very Dark		Desert Pavement
Very Dark		Flood Plain Deposits

INCREASING SURFACE ROUGHNESS

Figure 6-34 Map showing distribution of surface materials at Cottonball Basin. Based on geologic map by Hunt and Mabey (1966, Plate 1).

sity for smooth and rough surfaces at different depression angles. The relatively uniform return from the rough surface decreases somewhat at low depression angles because of the greater two-way travel distance. Smooth surfaces produce strong returns at depression angles near vertical, but little or no return at lower angles.

Radar Signatures and Surface Roughness at Death Valley

The relationship between radar backscatter and surface roughness was initially based on theoretical analyses supplemented by laboratory studies of artificial surfaces. Subsequently, field studies in Death Valley established the relationship between

radar roughness criteria and natural surfaces with different degrees of roughness. Key papers are by Schaber, Berlin, and Brown (1976) and Schaber, Berlin, and Pitrone (1975). Cottonball Basin, at the north end of Death Valley, is a dessicated salt lake with a variety of saline deposits on the old lake floor. Figure 6-33A is a Landsat TM image of Cottonball Basin; Figure 6-34 is a map showing the deposits that have a wide range of surface roughness values. The gravel surfaces of the alluvial fans bordering the basin provide additional degrees of surface roughness. The diversity of surface materials makes this an excellent site for correlating radar signatures with materials of different degrees of surface roughness. Figure 6-33B is an L-band radar image of the basin; Table 6-5 lists the roughness categories calculated for the image. Table 6-5 also lists

Table 6-5 Signatures of materials on an L-band image ($\gamma = °$) of Cottonball Basin, Death Valley, California

Image signature	Roughness category	Roughness criteria, cm	Materials
Dark	Smooth	$h < 1.0$	Desert pavement, sand and fine gravel, floodplain deposits
Intermediate	Intermediate	$h = 1.0$ to 5.7	Intermediate halite, carbonate, and sulfate deposits
Bright	Rough	$h > 5.7$	Coarse gravel, rough halite

the materials with surface relief that correspond to roughness criteria. Figure 6-35 shows field photographs of the materials, which are described in the following sections.

Coarse Gravel (h = 12.0 cm) Cobbles and boulders eroded from the mountains surrounding Death Valley form alluvial fans on the east and west margins of Cottonball Basin (Figure 6-34A). Tucki Wash Fan in the southwest part of the map (Figure 6-34) is a good example. The gravel is deposited by intermittent streams and slope wash during the infrequent rainstorms in the mountains. The gravel consists of a wide range of rock types that are determined by the lithology of their bedrock source. In the radar image (Figure 6-33B) the rough gravel has a bright signature that contrasts with the dark signatures of desert pavement and sand that also occur in the fan.

Rough Halite (h = 29.0 cm) Rough halite (Figure 6-35B) is the roughest of all materials in the Cottonball Basin. At the end of the glacial period a salt lake that filled Death Valley evaporated, depositing a layer of halite (rock salt). The salt crystals have partially dissolved and recrystallized, causing stresses that break the salt into jumbled slabs a meter or more in diameter. The slabs are covered with sharp pinnacles formed by solution during the infrequent rains. This material has the brightest signature in the radar image. In the Landsat image (Figure 6-33A) the signature of rough halite ranges from bright to dark, depending on the amount of silt contained in the salt.

Intermediate Halite (h = 6.0 cm) Halite deposits in the lower elevations of the basin are periodically flooded, a process that reduces surface relief to a hummocky surface of intermediate roughness (Figure 6-35C). This surface has an intermediate gray signature in the radar image and a similar gray signature in the Landsat image.

Carbonate and Sulfate Deposits (h = 6.0 cm) Carbonate and sulfate deposits (Figure 6-35D) form a belt around the margin of the basin, where they are mixed with sand. Periodic wetting and drying produces a puffy surface with intermediate relief that has a gray signature in the radar image. In the TM image the carbonate and sulfate deposits have a light gray tone.

Desert Pavement (h = 1.0 cm) Desert pavement (Figure 6-35E) forms on alluvial fans where older gravel deposits have been subjected to prolonged weathering, which disintegrates the surface layer of cobbles and boulders into slabs and chips. These fragments form a smooth surface that resembles a tile mosaic. Areas of desert pavement are common at the Tucki Wash Fan and other fans on the west side of Death Valley. Much of the desert pavement is coated with desert varnish, which causes a dark signature in the TM image that contrasts with the brighter signature of the younger deposits of coarse gravel. In the radar image the smooth desert pavement has a distinctive dark signature.

Sand and Fine Gravel (h = 1.0 cm) A narrow belt of sand and fine gravel (not illustrated in Figure 6-35) occurs at the margin of the Tucki Wash Fan and forms a narrow dark band in the radar image. The signature is consistent with the relatively fine grain size and low relief of this material. The belt of sand and fine gravel has a distinctive gray signature in the TM image.

Floodplain Deposits (h = 0.2 cm) The ephemeral streams in Cottonball Basin have formed floodplains of silt (Figure 6-35F) that are saturated with brine and are extensively coated with thin salt crusts. These floodplains are the smoothest surfaces in the basin and have distinctive dark signatures in the radar image. In the TM image the tone ranges from bright to medium gray depending on the thickness of the white salt crust and the degree of water saturation.

Bedrock Bedrock is exposed in the eastern portion of the radar image (Figure 6-33B). Relief is measured in tens of meters. Radar signatures of bedrock are dominated by topographic highlights and shadows, rather than by degrees of surface roughness.

Conclusions The Cottonball Basin example demonstrates that calculated roughness criteria correlate with the surface relief of materials in the field and with signatures on a radar image. This close correlation between radar signature and surface relief points out another important fact, namely, that radar signatures alone cannot be used to identify the composition of materials. For example, coarse gravel and rough halite have completely different compositions, but both have similar bright radar signatures. Floodplain deposits, sand, desert pavement, and standing water are very different materials, but all have dark signatures because of their smooth surfaces. These materials can be distinguished in radar images because of their

A. Coarse gravel, h = 12.0 cm.

B. Rough halite, h = 29.0 cm.

C. Intermediate halite, h = 6.0 cm.

D. Carbonate and sulfate deposits, h = 6.0 cm.

E. Desert pavement, h = 1.0 cm.

F. Floodplain deposits, h = 0.2 cm.

Figure 6-35 Oblique ground photographs of materials at Cottonball Basin, with values for typical vertical relief. Areas are approximately 50 cm wide.

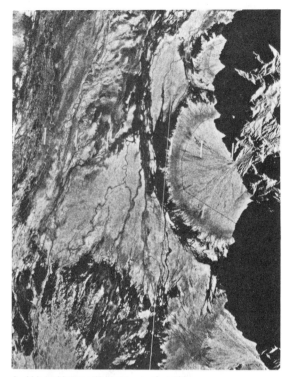

A. X-band image (3.0-cm wavelength).

B. L-band image (23.5-cm wavelength).

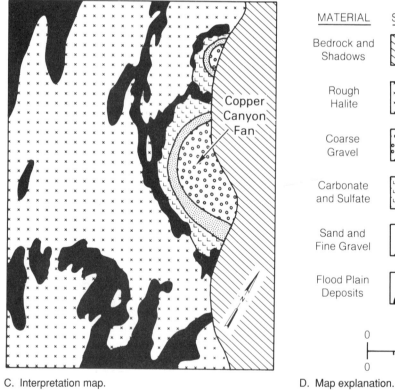

C. Interpretation map.

D. Map explanation.

MATERIAL	SYMBOL	RADAR SIGNATURES X-BAND	L-BAND
Bedrock and Shadows		Bright and Dark	Bright and Dark
Rough Halite		Very Bright	Very Bright
Coarse Gravel		Bright	Intermediate
Carbonate and Sulfate		Bright	Intermediate
Sand and Fine Gravel		Intermediate	Dark
Flood Plain Deposits		Dark	Dark

0   4 mi

0   4 km

Figure 6-36 X-band and L-band images of the Copper Canyon alluvial fan and vicinity, Death Valley. From Sabins (1984, Figures 9, 10).

distribution and associations, but not solely on the basis of their radar signature.

The correlations between roughness and radar signature demonstrated at Cottonball Basin have been confirmed on radar images covering all of Death Valley and other regions.

X-Band and L-Band Images Compared

The Copper Canyon alluvial fan on the east side of Death Valley is an instructive site for comparing the signatures of images acquired at different wavelengths. Figure 6-36A is an X-band image (3-cm wavelength), and Figure 6-36B is an L-band image (23.5-cm wavelength). The comparison is subjective because the two images were acquired and processed separately and were not calibrated to a known standard. Figure 6-36C shows the distribution of materials at the Copper Canyon Fan that are similar to those at Cottonball Basin. The X-band image (Figure 6-36A) has a smooth criterion of 0.3 cm and a rough criterion of 1.6 cm. The L-band image (Figure 6-36B) has a smooth criterion of 1.0 cm, and a rough criterion of 5.7 cm. The X-band image (5-m resolution) has finer spatial detail than the L-band image (25-m resolution).

Rough halite exceeds the rough criterion of both images and has bright signatures. Floodplain deposits have less relief than either of the smooth criteria and is dark in both images. The gravel of the Copper Canyon Fan is finer grained than that at the Tucki Wash Fan; its relief of 5 to 6 cm is rough for the X-band image (bright signature) but intermediate for the L-band image (gray signature). The belt of sand and fine gravel near the margin of the fan (Figure 6-36C) has a vertical relief of 1 cm, which is intermediate for the X-band image (gray signature) and smooth for the L-band image (dark signature). A belt of carbonate and sulfate deposits, with a vertical relief of 6 cm, occurs between the sand and the floodplain deposits. These deposits are bright in the X-band image and intermediate in the L-band image. This example demonstrates the relationship of radar wavelength to roughness of materials and to their radar signatures. Images acquired at different wavelengths provide better definitions of surface relief. Consider the gravel at the Copper Canyon Fan, which has a relief greater than 1.6 cm in the X-band image and 1 to 5.7 cm in the L-band image. The combination of images provides an estimated roughness range of 1.6 to 5.7 cm for this gravel.

POLARIZATION

The electric field vector of the transmitted energy pulse may be polarized (or vibrating) in either the vertical or horizontal plane (Figure 6-37). On striking the terrain, most of the energy returning to the antenna usually has the same polarization as the transmitted pulse. This energy is recorded as *parallel-polarized* (or like-polarized) imagery and is designated HH (*horizontal transmit, horizontal return*) or VV (*vertical transmit,*

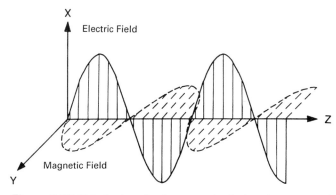

Figure 6-37 Polarization of radar energy (and other electromagnetic energy).

vertical return). A portion of the returning energy has been *depolarized* by the terrain surface and vibrates in various directions. The mechanisms responsible for depolarization are not definitely known, but the most widely accepted theory attributes it to multiple reflections at the surface. This theory is supported by the fact that depolarization effects are much stronger from vegetation than from bare ground. Leaves, twigs, and branches are believed to cause the multiple reflections responsible for depolarization. Some experimental radar systems have a second antenna element that receives the depolarized energy vibrating at right angles to the plane of the transmitted pulse. The resulting imagery is termed *cross-polarized* and may be either HV (*horizontal transmit, vertical return*) or VH (*vertical transmit, horizontal return*). Most mapping radar systems operate in the HH mode because this mode produces the strongest return signals.

Parallel-Polarized and Cross-Polarized Images

Figure 6-38 is a set of HH, HV, VV, and VH images (25-cm wavelength) that were simultaneously acquired of Furnace Creek Ranch and the vicinity in Death Valley. The cross-polarized images were digitally processed to increase their brightness to match the parallel-polarized images. Figure 6-39 is an aerial photograph and map of the Furnace Creek Ranch area. Despite the different polarizations, all four radar images are very similar. Smooth surfaces (airport runway, floodplain, sand, and fine gravel) have dark signatures on all the images. Rough surfaces (rough halite) have bright signatures on all images. Surfaces with intermediate relief (carbonate and sulfate deposits) have intermediate signatures. Bedrock has similar highlights and shadows on all four images. For all of these materials the backscattering processes are apparently similar in both the parallel- and cross-polarized modes.

The only apparent differences in polarization signatures are caused by vegetation, which consists of small mesquite trees and shrubs of creosote bush. Distribution of the sparse plant community is shown in the aerial photograph and map (Figure 6-39). Native vegetation is concentrated in the dry washes that radiate across the carbonate and sulfate deposits south and

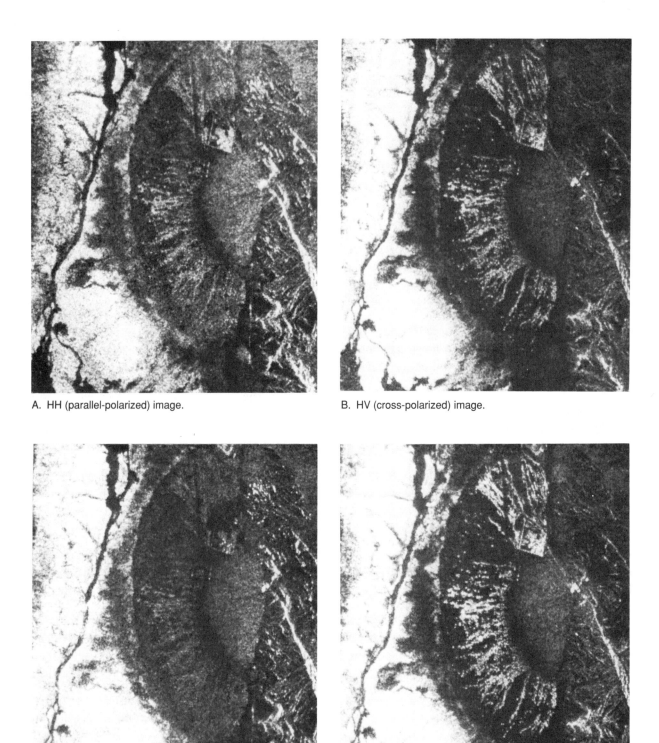

A. HH (parallel-polarized) image.

B. HV (cross-polarized) image.

C. VV (parallel-polarized) image.

D. VH (cross-polarized) image.

Figure 6-38 Parallel-polarized and cross-polarized L-band images of Furnace Creek Ranch and vicinity, Death Valley, California. Courtesy Jet Propulsion Laboratory.

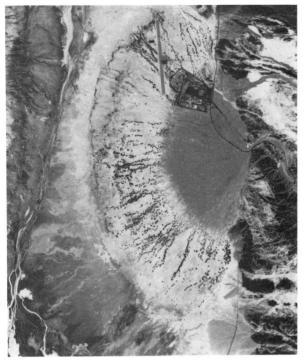

A. Aerial photograph.

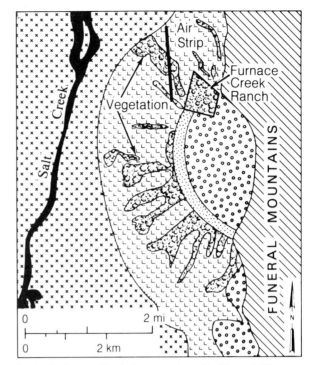

B. Geologic map. Symbols are explained in Figure 6-36.

Figure 6-39 Aerial photograph and map of Furnace Creek Ranch and vicinity, Death Valley, California. Vegetation forms dark stripes on the margin of the alluvial fan.

west of Furnace Creek Ranch (Figure 6-39B). In the photograph (Figure 6-39A) native vegetation forms dark stripes that contrast with the bright tone of the carbonate and sulfate deposits. Furnace Creek Ranch is enclosed by a windbreak of tamarisk trees and inside includes a date palm grove, a golf course, and scattered tamarisk trees.

In the parallel-polarized HH and VV images (Figure 6-38A,C), vegetation is indistinguishable from the background in the VV mode and is only faintly brighter than the background in the HH mode. In both cross-polarized HV and VH images (Figure 6-38B,D), however, the cultivated vegetation at Furnace Creek Ranch and the native vegetation are distinctly brighter than the background. The twigs and branches cause multiple reflections of incident radar energy (sometimes referred to as *volume scattering*), which depolarizes a significant proportion of the energy. For a horizontally transmitted pulse of energy, the major return from the scene is HH-polarized, but returns from vegetation will also have a significant HV-polarized signal. Unvegetated terrain, however, does not cause multiple reflections and does not depolarize the incident energy; thus the HV return from bare surfaces is weak relative to the HV return from vegetation. The same relationships apply to the VV and VH images (Figure 6-38C,D) and account for the bright signature of vegetation in the VH image. This correla-

tion between vegetation and relatively bright signatures on cross-polarized images has been observed in numerous other images. Donovan, Evans, and Held (1985) have published multiple-polarized images of several test sites.

Polarimetric Images

Digitally recorded parallel-polarized and cross-polarized images (Figure 6-38) can be processed to produce images of intermediate degrees of polarization difference between extremes, such as from HH to HV. Polarimetric images record intermediate degrees of polarization difference. Ulaby and Elachi (1990) edited a volume that describes and illustrates these images. Currently, polarimetric images are recorded by experimental aircraft systems and are not generally available. Polarimetric systems may be deployed on future satellite missions.

INTERFEROMETRY

Interferometry is the field of physics that deals with the interaction between superimposed wave trains. Figure 6-40 shows two waves traveling from left to right that are identical in wavelength and amplitude; however, the crest of wave 1 arrives

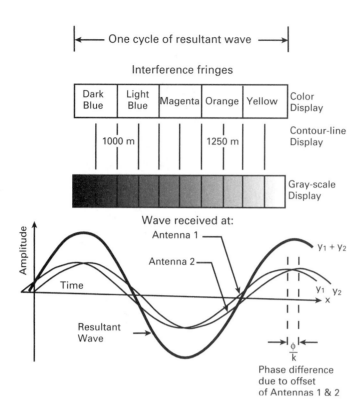

One cycle of resultant wave

Interference fringes

Dark Blue	Light Blue	Magenta	Orange	Yellow	Color Display

Contour-line Display 1000 m 1250 m

Gray-scale Display

Wave received at:
Antenna 1
Antenna 2

Amplitude
Time
Resultant Wave

$y_1 + y_2$
y_1 y_2
x

$\frac{\phi}{k}$

Phase difference due to offset of Antennas 1 & 2

Figure 6-40 Diagram showing origin of interference fringes from two received radar waves that are not in phase.

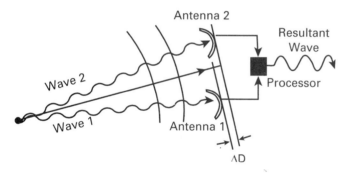

Antenna 2
Resultant Wave
Wave 2
Processor
Wave 1
Antenna 1
ΔD

Figure 6-41 Radar interferometer system. Only one antenna transmits the radar wave. The return waves are received by two antennas that are spaced at different distances from the target; therefore, the two return waves are not in phase. Interference results when the two return waves are combined by the processor.

A. Radar image used in interferogram.

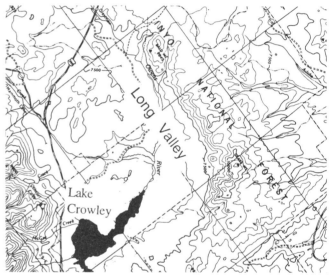

B. Conventional topographic map from aerial photographs.

Figure 6-42 Radar image and map, Long Valley, California.

The resultant wave may be displayed as an image by assigning a gray scale or a color spectrum to each cycle (from crest to crest), as shown in the upper portion of Figure 6-40. Each cycle of the color spectrum or gray scale is called an *interference fringe,* and the image is an *interferogram.*

Figure 6-41 shows how radar interferograms are produced. Two antennas are mounted in an aircraft or spacecraft so that antenna 1 is offset slightly closer to the earth's surface than antenna 2, with the difference shown as ΔD in Figure 6-41. One antenna transmits a pulse of energy, but the return waves are received by both antennas. Because of the offset, a wave crest is received at antenna 1 before the equivalent wave crest is re-

at position P (located at right margin of figure) shortly before the equivalent crest of wave 2. The difference in arrival time is a measure of the *phase difference* between the two waves. Superposing and combining wave 1 plus wave 2 produces the *resultant wave* (shown by the heavy curve in Figure 6-40), which is a record of the interference between waves 1 and 2.

ceived at antenna 2, which means there is a phase difference between the two return waves. The digital processor superposes the two waves to produce the resultant wave (Figure 6-41). The resultant wave may then be displayed as an interferogram with color spectra representing each interference cycle. The number of color cycles, also called *color fringes*, is a measure of topographic relief. Dixon (1995) edited a volume that describes the generation of radar interferograms and their applications for detecting surface changes. Zebker and others (1994) describe the accuracy of topographic maps generated from interferograms of satellite radar data.

Plate 10A is an interferogram of Long Valley, California, and vicinity. Figure 6-42A is one of the two images that were combined to produce the interferogram. For comparison, Figure 6-42B is from a published contour map that shows the topography. Each color fringe in the interferogram represents a range of elevations. The low-relief terrain of the valleys has few or no fringes. Terrain with high relief in the lower left portion of the interferogram (Plate 10A) has numerous closely spaced fringes. Plate 10B is a topographic contour map that was produced from the interferogram. The map colors are not interferometer fringes; they indicate ranges of elevation from blue (lowest) to pink (highest). In the future this technique should enable us to produce maps of cloudy regions where aerial photographs are difficult to obtain. The section on earthquakes and active faults in Chapter 13 includes an interferogram that records topographic displacements caused by a recent earthquake.

COMMENTS

Because radar is an active system that supplies its own illumination at long wavelengths, images can be acquired at night and through cloud cover. The illumination direction can be oriented to enhance particular linear features of the terrain, such as street patterns, fractures, and faults. Interaction between materials and radar energy is determined by the wavelength, depression angle, and polarization of the radar beam and by the dielectric properties, surface roughness, and orientation of the material. Rough surfaces produce stronger radar backscatter, which is recorded as brighter tones on the images. The oblique angle of radar illumination causes highlights and shadows that enhance subtle terrain features such as geologic structure. Topographic maps are generated from sidelapping images or from radar interferograms.

QUESTIONS

1. A typical C-band radar system operates at a frequency of 5 GHz. What is the wavelength of this radar?
2. Calculate the range resolution for a radar system with a pulse length of 0.2 μsec and a depression angle of 30°.
3. For a real-aperture, X-band system with an antenna length of 300 cm, calculate the azimuth resolution at a slant-range distance of 15 km.
4. What is the principal difference between real-aperture and synthetic-aperture radar systems?
5. Summarize the differences (aside from wavelength) between radar stereo images and stereo pairs of aerial photographs.
6. Define and discuss the optimum depression angle (relatively steep or relatively shallow) for imaging terrain (a) with low relief, such as coastal plains; and (b) high relief, such as mountain chains.
7. Prepare a version of Table 6-3 for aircraft systems at X-, C-, and L-band wavelengths, all with a depression angle of 30°.
8. Predict the dielectric constant for Yuma sand, Vernon clay, and Miller clay at 20 percent and at 40 percent volumetric water content.
9. Assume L-band images are available for the Guymon, Oklahoma, test site. Use Figure 6-30 to predict the signature (bright, intermediate, or dark) of milo fields with soil moistures of: (a) 5 percent, and (b) 30 percent.

REFERENCES

Barr, D. J., 1969, Use of side-looking airborne radar (SLAR) imagery for engineering studies: U.S. Army Engineer Topographic Laboratories, Technical Report 46-TR, Fort Belvoir, VA.

Blanchard, B. J., A. J. Blanchard, S. Theis, W. D. Rosenthal, and C. L. Jones, 1981, Seasat SAR response from water resources parameter: U.S. Department of Commerce/NOAA Contract 78-4332, Final Report 3891, Washington, DC.

Bryan, M. L., 1979, The effect of radar azimuth angle on cultural data: Photogrammetric Engineering and Remote Sensing, v. 45, p. 1097–1107.

Craib, K. B., 1972, Synthetic-aperture SLAR systems and their application for regional resources analysis *in* Sahrokhi, F., ed., Remote sensing of earth resources, v. 1, p. 152–178, University of Tennessee Space Institute, Tullahoma, TN.

Dixon, T. H., 1995, SAR interferometry and surface change detection: University of Miami, Rosensteil School of Marine and Atmospheric Science, Technical Report TR 95-003, Miami, FL.

Donovan, N., D. Evans, and D. Held, eds., 1985, NASA/JPL aircraft SAR workshop proceedings: Jet Propulsion Laboratory Publication 85-39, Pasadena, CA.

Ellis, J. M. and D. A. Richmond, 1991, Mapping the coastal plain of the Congo with airborne digital radar: Environmental Research Institute of Michigan, Eighth Thematic Conference on Geologic Remote Sensing, Proceedings, p. 1–15, Ann Arbor, MI.

Fischer, W. A., ed., 1975, History of remote sensing *in* Reeves, R. G., ed., Manual of remote sensing, first edition: ch. 2, p. 27–50, American Society of Photogrammetry, Falls Church, VA.

Ford, J. P. and others, 1993, Guide to Magellan radar image interpretation: Jet Propulsion Laboratory Publication 93-24, Pasadena, CA.

Hunt, C. B. and D. R. Mabey, 1966, Stratigraphy and structure of Death Valley, California: U.S. Geological Survey Professional Paper 494-A.

Mercer, J. B. and others, 1991, STARMAP processing of SAR imagery for petroleum exploration in Irian Jaya: Environmental Research Institute of Michigan, Eighth Thematic Conference on Geologic Remote Sensing, Proceedings, Ann Arbor, MI.

Moore, R. K., 1983, Radar fundamentals and scatterometers *in* Colwell, R. N., ed., Manual of remote sensing, second edition: ch. 9, p. 369–427, American Society of Photogrammetry, Falls Church, VA.

Moore, R. K., L. J. Chastant, L. Porcello, and J. Stevenson, 1983, Imaging radar systems *in* Colwell, R. N., ed., Manual of remote sensing, second edition: ch. 10, p. 429–474, American Society of Photogrammetry, Falls Church, VA.

Peake, W. H. and T. L. Oliver, 1971, The response of terrestrial surfaces at microwave frequencies: Ohio State University Electroscience Laboratory, 2440-7, Technical Report AFAL-TR-70-301, Columbus, OH.

Plaut, J. J., 1993, Stereo imaging *in* Ford, J. P. and others, Guide to Magellan image interpretation: Jet Propulsion Laboratory Publication 93-24, Pasadena, CA.

Ramachandran, V. S., 1988, Perceiving shape from shading: Scientific American, August, p. 76–83.

Sabins, F. F., 1983, Remote sensing laboratory manual, second edition: Remote Sensing Enterprises, La Habra, CA.

Sabins, F. F., 1984, Geologic mapping of Death Valley from thematic mapper, thermal infrared, and radar images: Proceedings of Third Thematic Conference, Remote Sensing for Exploration Geology, p. 139–152, Environmental Research Institute of Michigan, Ann Arbor, MI.

Schaber, G. G., G. L. Berlin, and W. E. Brown, 1976, Variations in surface roughness within Death Valley, California—geologic evaluation of 25-cm wavelength radar images: Geological Society of America Bulletin, v. 87, p. 29–41.

Schaber, G. G., G. L. Berlin, and D. J. Pitrone, 1975, Selection of remote sensing techniques—surface roughness information from 3-cm wavelength SLAR images: American Society of Photogrammetry, Proceedings of 42nd Annual Meeting, p. 103–117, Washington, DC.

Ulaby, F. T., A. Aslam, and M. C. Dobson, 1982, Effects of vegetation cover on radar sensitivity to soil moisture: IEEE Transactions on Geoscience and Remote Sensing, v. GE-20, p. 476–481.

Ulaby, F. T., B. Brisco, and M. C. Dobson, 1983, Improved spatial mapping of rainfall events with spaceborne SAR imagery: IEEE Transactions on Geoscience and Remote Sensing, v. GE-21, p. 118–121.

Ulaby, F. T. and C. Elachi, 1990, Radar polarimetry for geoscience applications: Artech House, Norwood, MA.

Wang, J. R. and T. J. Schmugge, 1980, An empirical model for the complex dielectric permittivity of soils as a function of water content: IEEE Transactions on Geoscience and Remote Sensing, v. GE-18, p. 288–295.

Wynn, J. C., F. Gray, and N. J. Page, 1993, Geology and mineral resource assessment of the Venezuelan Guiana shield: U.S. Geological Survey Bulletin 2062.

Zebker, H. A., C. L. Werner, P. A. Rosen, and S. Hensley, 1994, Accuracy of topographic maps derived from ERS-1 interferometric radar: IEEE Transactions on Geoscience and Remote Sensing, v. 32, p. 823–836.

ADDITIONAL READING

Alberti, G., S. Esposito, and S. Ponte, 1996, Three-dimensional digital elevation model of Mt. Vesuvius from NSA/JPL TOPSAR: International Journal of Remote Sensing, v. 17, p. 1797–1801.

Elachi, C., 1987, Solid surface sensing—microwave and radio frequencies *in* Introduction to the physics and techniques of remote sensing: ch. 6, p. 161–241, John Wiley & Sons, New York, NY.

Evans, D. and A. Freeman, 1995, Future directions for synthetic aperture radar *in* Land satellite information in the next decade: American Society for Photogrammetry and Remote Sensing, Falls Church, VA.

Gens, R. and J. L. Van Genderen, 1996, SAR interferometry—issues, techniques, applications: International Journal of Remote Sensing, v. 17, p. 1803–1835.

Henderson, F. M. and A. J. Lewis, eds., 1996, Principles and applications of imaging radar in the geosciences: Manual of remote sensing, third edition, American Society for Photogrammetry and Remote Sensing, Falls Church, VA.

Lowman, P. D., 1994, Radar geology of the Canadian shield—a 10-year review: Canadian Journal of Remote Sensing, v. 20, p. 198–209.

Society of Photo-optical Instrumentation Engineers, 1993, Radar polarimetry: SPIE Proceedings, v. 1748, Bellingham, WA.

Zebker, H. A., T. G. Farr, R. O.Salazar, and T. H. Dixon, 1994, Mapping the world's topography using radar interferometry: IEEE Proceedings, v. 82, p. 1774–1786.

CHAPTER

SATELLITE RADAR SYSTEMS AND IMAGES

The development of synthetic-aperture radar systems enabled us to acquire images from satellites at distances of hundreds of kilometers with good spatial resolution. Table 7-1 lists the characteristics of the unclassified radar satellite systems that have been launched, beginning with the NASA Seasat in 1978. The description of radar technology and terrain interactions in Chapter 6 applies equally to satellite systems and images.

SEASAT

As the name implies, Seasat was designed primarily to investigate oceanic phenomena (such as roughness, current patterns, and sea-ice conditions), but the images were also valuable for terrain observations. The unmanned satellite was launched in June 1978 and prematurely ceased operation in October 1978 because of a major electrical failure. Table 7-1 lists characteristics of the radar system. Image data were telemetered to earth receiving stations and processed to produce images. The satellite was in a polar orbit that covered the earth. Because of the limited number of receiving stations, however, images were recorded only for North America and western Europe. Ford and others (1980) prepared an atlas of representative Seasat images.

Santa Barbara Coast, California

Figure 7-1A is a Seasat image of the Santa Barbara coast acquired with a look direction toward the northeast. The steep depression angle (70° average) of Seasat causes the pronounced foreshortening of mountain ridges toward the west (left margin of the image). The extensive bright patches in the Pacific Ocean are rough water and contrast with the dark signature of calm water. Linear dark patterns in the ocean in the western part of the image are streaks of calm water caused by slicks from submarine oil seeps, as shown in the location map (Figure 7-1C). Chapter 9 explains radar signatures of oil slicks. Metal structures of oil production platforms cause small, bright signatures that are detectable where the adjacent water is calm and has a dark signature. Ships have similar bright signatures. In areas of rough water, however, platforms and ships are virtually undetectable amid the bright background. The cities of Ventura and Santa Barbara along the coast have bright signatures. The calm surfaces of Lake Cachuma and Casitas Lake have distinctive dark signatures.

San Andreas Fault, California

Figure 7-2A is a Seasat image of the San Andreas fault in the Durmid Hills, which are a broad, gentle arch of arid terrain along the eastern shore of the Salton Sea in southern California. The grainy appearance of the Seasat image is due to its enlargement from the small-scale original image. An enlarged satellite photograph (Figure 7-2B) is included for comparison, together with an interpretation map (Figure 7-2C) of the Seasat image. Rough water of the Salton Sea has a bright signature in the Seasat image. In the Seasat image a tonal lineament extends the length of the Durmid Hills. The lineament is formed by the contact between bright tones to the southwest and dark tones to the northeast. Equations 6-5 and 6-6 indicate that Seasat has a smooth criterion of 1.0 cm and a rough criterion of 5.7 cm. Therefore, the terrain with dark signatures should be a smooth surface ($h < 1.0$ cm), and the terrain with bright signatures should be a rough surface ($h > 5.7$ cm).

Table 7-1 Satellite radar systems (unclassified)

	Seasat	SIR-A	SIR-B	SIR-C	ERS-1	JERS-1	Almaz	Magellan	Radarsat
Launch date	1978	1981	1984	1994	1991	1992	1991	1989	1995
Wavelength, cm	23.5 (L)	23.5 (L)	23.5 (L)	3.0 (X) 6.0 (C) 24.0 (L)	5.7 (C)	23.5 (L)	9.6 (S)	12.6 (S)	5.7 (C)
Depression angle	70°	40°	30 to 75°	Variable	67°	55°	40 to 58°	65°	40 to 70°
Spatial resolution, m	25	38	25	Variable	218	18	15	100	Variable
Polarization	HH	HH	HH	Multiple	VV	HH	HH	HH	HH
Swath width	100	50	40	30 to 60	100	75	50 to 100	20	50 to 100
Altitude, km	790	250	225	225	785	568	350	290 to 2000	800
Latitude covered	72°N to 72°S	50°N to 50°S	58°N to 58°S	57°N to 57°S	Polar	Polar	Polar	Polar	Polar
Nationality	USA	USA	USA	USA	Europe	Japan	Russia	USA	Canada

R. G. Blom and C. Elachi (both from JPL) and I visited the Durmid Hills to (1) compare the actual terrain roughness with predicted values of vertical relief and (2) determine the significance of the Seasat lineament. Figure 7-3A is a field photograph of the terrain with bright signatures on the Seasat image. This terrain is underlain by the Borrego Formation (Pleistocene), which consists of poorly consolidated siltstone with beds and concretions of hard, resistant sandstone. The infrequent rains erode the soft siltstone to form a surface littered with gravel and boulders having a relief of 10 to 20 cm. Figure 7-3B is terrain with the dark Seasat signatures that is covered with sand and silt deposited in Lake Cahuila, which is now extinct. The roughness of this fine-grained material is less than the 1.0-cm limit for smooth surfaces. The linear contact between the Cahuila deposits and the Borrego outcrops corresponds to the tonal lineament of the Seasat image and to the trace of the San Andreas fault. The fault is obscure in the satellite photograph (Figure 7-2B) because the terrain on opposite sides of the fault has little contrast in the visible and reflected IR regions.

SIR-A MISSION

The NASA Space Shuttle, which was described in Chapter 2, has carried three *Shuttle imaging radar* (SIR) experiments that are listed in Table 7-1. Because of communication and recovery requirements, all Shuttle missions are placed in near-equatorial orbits and do not provide the global coverage of polar-orbiting satellites. Figure 3-13 compares the orbit pattern of a typical Shuttle mission with the pattern of Landsat TM. The radar antenna is folded and carried in the cargo bay. After

reaching orbit, the Shuttle is inverted to orient the cargo bay toward the earth, and the antenna is deployed to acquire images. The first radar mission (SIR-A) was conducted in November 1981. Figure 7-4 shows the configuration of the SIR-A system, which acquired 50-km-wide image swaths with an average depression angle of 40°. Cimino and Elachi (1982) published a collection of SIR-A images. Ford, Cimino, and Elachi (1983) published additional examples and an index map of the 10 million km² of imagery that is confined to the midlatitudes covered by the SIR-A orbits.

Figure 7-1B is a SIR-A image of the Santa Barbara coast of California that corresponds to the Seasat image of the same area (Figure 7-1A). The look direction was toward the northeast for the Seasat image and toward the north for the SIR-A image, which explains the different orientation of shadows and highlights. The moderate depression angle (40°) of SIR-A causes less topographic foreshortening than in Seasat images; therefore, terrain features are more readily interpreted in SIR-A images. In the SIR-A image the Pacific Ocean lacks the expression of currents and *sea state* (variations in roughness) that are clearly expressed in the Seasat image. Some of this difference may be due to differences in sea state in 1978 (when the Seasat image was acquired) and 1981 (when the SIR-A image was acquired); however, most of the difference is attributed to the steeper depression angle of Seasat, which emphasizes roughness differences in the ocean.

SIR-A IMAGES OF TROPICAL TERRAIN, INDONESIA

The SIR-A mission acquired images of portions of Indonesia. Persistent cloud cover in this tropical region has hampered

A. Seasat image acquired 1978.

B. SIR-A image acquired November 1981.

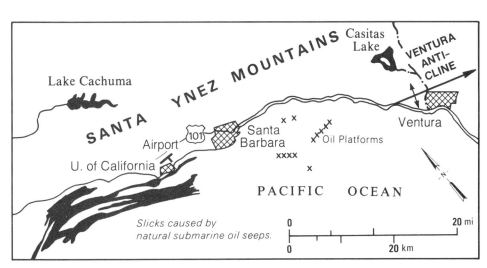

C. Location map.

Figure 7-1 Seasat and SIR-A images of Santa Barbara coast, California.

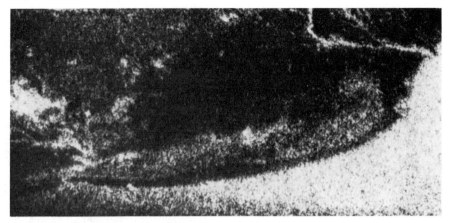

A. Seasat radar image.

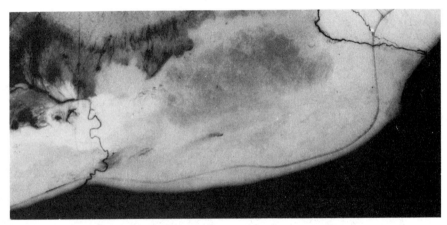

B. Satellite photograph.

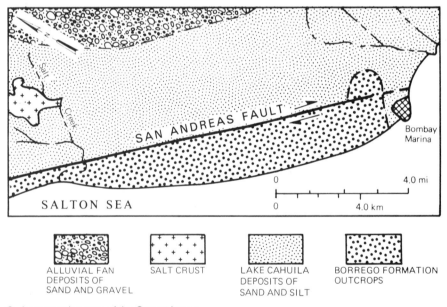

C. Interpretation map of the Seasat image.

Figure 7-2 Seasat and Skylab images of Durmid Hills, southern California. From Sabins, Blom, and Elachi (1980, Figures 5 and 9).

A. Outcrops of the Borrego Formation on the southwest side of the San Andreas fault. This rough surface has a bright signature on the Seasat image.

B. Silt and clay deposits on northeast side of San Andreas fault. This smooth surface has a dark signature on the Seasat image.

Figure 7-3 Terrain features associated with tonal lineament on the Seasat image of Durmid Hills (Figure 7-2). From Sabins, Blom, and Elachi (1980, Figures 6 and 7).

acquisition of aerial photographs and Landsat images, but the SIR-A images are of excellent quality. I interpreted the SIR-A images in five steps (Sabins, 1983):

1. Produced base maps by tracing shorelines, drainage patterns, and the sparse cultural features (roads, cities, and airports) from the images. The contrasting signatures of water and vegetation made this a straightforward task.
2. Defined terrain categories that are recognizable throughout the region, and mapped their distribution.

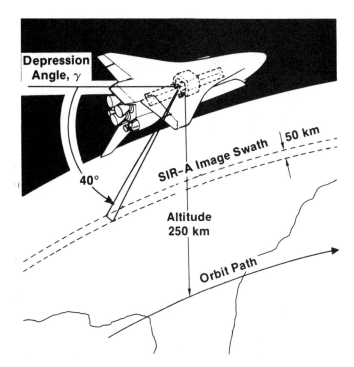

Figure 7-4 Shuttle imaging radar (SIR-A) system.

3. Mapped geologic structures including dips and strikes, faults, folds, and lineaments.
4. Constructed cross sections.
5. Evaluated the completed maps by comparing them with existing geologic maps.

The remainder of this section describes steps 2 through 5.

Terrain Categories

Much of Indonesia is densely forested, but the forest conforms to the terrain because the surface of the tree canopy is parallel with the underlying topography. Topography is controlled by geologic structure and by the erosional characteristics of the underlying rocks. The six terrain categories shown in Figure 7-5 were recognized on the basis of their expression on the SIR-A images. These categories are not restricted to images of Indonesia but are recognizable on radar images of forested regions throughout the world.

Carbonate Terrain In humid environments, solution and collapse of carbonate rocks produce karst topography, which is readily recognized by the distinctive pitted surface (Figure 7-5A). In Indonesia, carbonate terrain generally occurs as uplands surrounded by lowlands eroded from less resistant rocks. Because of the relatively coarse spatial resolution of SIR-A images, small patches of karst terrain may not be recognizable.

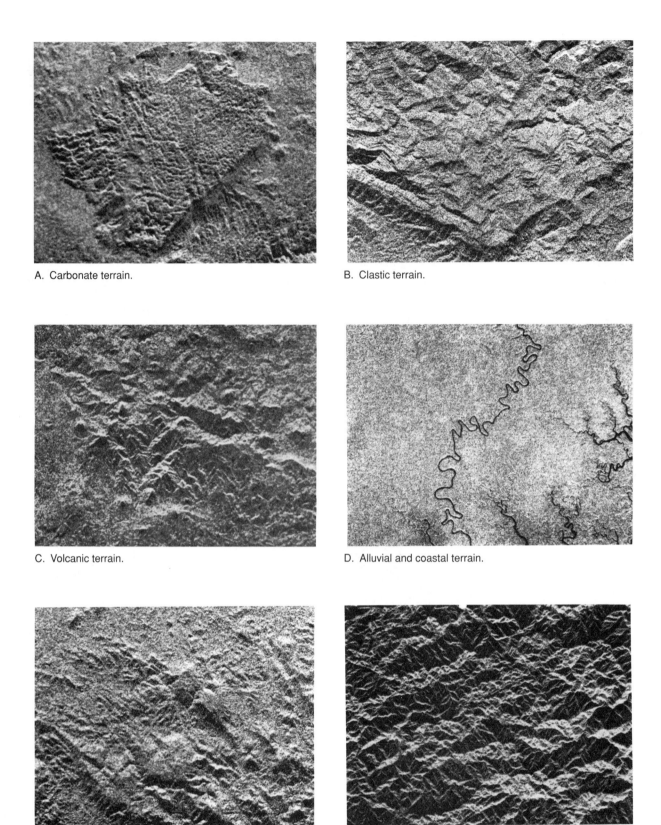

A. Carbonate terrain.

B. Clastic terrain.

C. Volcanic terrain.

D. Alluvial and coastal terrain.

E. Mélange terrain.

F. Metamorphic terrain.

Figure 7-5 SIR-A images of typical Indonesian terrain categories. Areas covered are 28 km wide (left to right). Look direction is toward the lower margin of each image. From Sabins (1983, Figure 4).

Also, faulting and stream erosion may obscure the expression of karst topography.

Clastic Terrain Terrain formed on clastic sedimentary rocks, primarily sandstone and shale, is recognized by the stratification that forms asymmetric ridges, called *cuestas* and *hogbacks*, where the rocks are dipping, as seen in Figure 7-5B. Flat-lying clastic rocks form mesas, terraces, and associated erosional scarps. The lack of karst topography generally distinguishes clastic terrain from carbonate terrain in humid regions.

Volcanic Terrain Young volcanic rocks form irregular flows associated with cinder cones or eroded volcanic necks (Figure 7-5C). Because of erosion and deformation, older volcanic terrains lack these distinctive features and so cannot be recognized on the images without additional information.

Alluvial and Coastal Terrain This category is characterized by low relief, a uniform bright signature of heavily vegetated floodplains, and numerous estuaries and meandering streams (Figure 7-5D). Despite the generally featureless nature of this terrain, careful interpretation of the images often reveals subtle lineaments and drainage anomalies that may be expressions of geologic structure.

Mélange Terrain *Mélange* refers to rocks formed in subduction zones as a mixture of clastic sediments and oceanic crustal and mantle rocks. Lenticular rock fragments of a wide range of sizes, up to kilometers in length, are enclosed in a matrix of clay. Erosion of these rocks produces an irregular, rounded terrain with unsystematic drainage patterns (Figure 7-5E). At the scale of the radar images, neither stratification nor individual rock fragments are detectable

Metamorphic Terrain Sedimentary rocks that have been metamorphosed to slate, quartzite, and schist occur in portions of Irian Jaya (Figure 7-5E). The original stratification is no longer recognizable. The strongly dissected metamorphic terrain has high relief and angular ridges that distinguish it from the lower relief and rounded appearance of mélange terrain. Foliation trends are not discernible in Indonesian metamorphic terrain at the scale of SIR-A images.

Undifferentiated Bedrock There are areas where several terrain types are juxtaposed in such a complex manner that individual categories are not recognizable at the scale and resolution of SIR-A images. The category "undifferentiated bedrock" is used for these areas.

Regional Interpretation

The second step in the interpretation process was to recognize the terrain categories and map their distribution on regional SIR-A images. The image in Figure 7-6A is located in the northwest portion of the Vogelkop region at the west end of Irian Jaya. The image was acquired with the radar look direction toward the northeast, which causes shadows to extend from the lower margin of the image toward the upper margin. This orientation may cause topographic inversion for some viewers, which may be corrected by rotating the image 180° so the shadows extend toward the lower margin of the image.

The geologic interpretation map (Figure 7-6B) shows the distribution of the five terrain categories that occur in this SIR-A image. The central part of the area, called the Kemum block, consists of metamorphic rocks with typical rugged terrain. Toward the south (lower right corner of Figure 7-6A,B) the metamorphic rocks are overlain by clastic strata, which in turn are overlain by carbonate rocks that form a broad expanse of karst terrain. This vertical sequence of rocks is shown at the southeast end (right side) of the cross section (Figure 7-6C). In the northwest part of the image (left side), the Tamrau Mountains consist of a variety of rocks and are mapped as undifferentiated bedrock. Belts of clastic strata occur on the north and south sides of the Tamrau Mountains.

The third step was to interpret geologic structure. Strikes and dips are inferred from the attitudes of dipslopes and anti-dip scarps formed by stratified rocks. Lineaments are mapped with a dot pattern and belong to two major types. The most common type is formed by straight streams and aligned segments of streams. These lineaments are especially abundant in the metamorphic terrain of the Kemum block. The second type of lineament category consists of straight escarpments that are expressed in the image as linear highlights or shadows depending on their orientation relative to the radar look direction. Additional structural information is gained from the outcrop pattern of rock types, which may indicate the plunge of folds and the sense of offset along faults.

The fourth step was to construct cross section A–B (Figure 7-6C), which shows subsurface relationships of the structures.

The fifth step was to compare the SIR-A interpretation map (Figure 7-6B) with recently published geologic maps of the area. Distribution of the terrain types conforms to the patterns in the published maps, and the attitudes of the beds are in agreement. The radar lineaments were compared with mapped faults; where the two correspond, a solid line was added to the dot patterns of the lineaments (Figure 7-6B). Many lineaments correlate with mapped faults; others may represent faults or extensions of faults that were not recognized in the field because of dense forest cover and inaccessible terrain. For example, one prominent lineament correlates with faults at each end; the lineament probably indicates a long, continuous fault that was not mapped in the field. The Tamrau Mountains are bounded on the north and south by two major west-trending lineaments that are formed by prominent linear valleys and stream segments. The northern lineament correlates with the Koor fault zone (Figure 7-6B). The lineament on the south side of the Tamrau Mountains extends for 80 km across the image and marks the trace of the Sorong fault. This fault is a major

A. SIR-A image. Look direction toward upper margin of image.

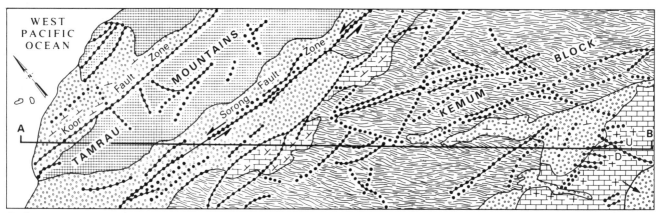

B. Geologic interpretation map.

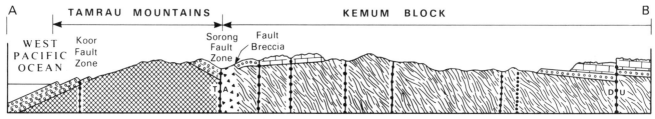

C. Cross section along line A–B.

TERRAIN CATEGORIES

Alluvial and
coastal plains

Carbonate
rocks

Clastic
rocks

Metamorphic
rocks

Undifferentiated
bedrock

LINEAMENTS

Correlate with
mapped faults

Possible
faults

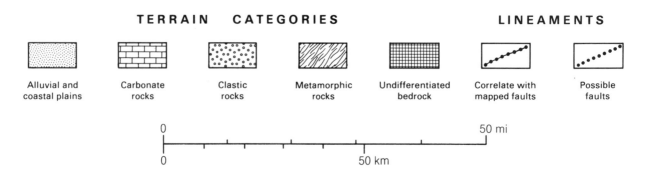

Figure 7-6 SIR-A image and interpretation in northwest Vogelkop region, Irian Jaya, Indonesia. From Sabins (1983, Figures 10 and 11).

tectonic feature in the Vogelkop region and is described later in the chapter in the section "Strike-Slip Faults."

Comparing SIR-A maps with published geologic maps confirms the validity of the radar interpretations. This example from the Vogelkop region is a limited subset of the SIR-A interpretation maps by Sabins (1983), which cover portions of the mainland of Irian Jaya and of Kalimantan. In 1983, published geologic maps were lacking for Kalimantan, and the SIR-A maps provided the most detailed information.

Detailed Interpretation

The small-scale SIR-A image and interpretation map of the Vogelkop region show the regional use of these data but lack the details of larger-scale versions of the image. Figure 7-7 shows representative geologic features at a uniform scale. Figure 7-8 shows the geologic interpretation for each image.

Strike and Dip Strike-and-dip information can be interpreted in carbonate and clastic terrains and in some volcanic terrains where flow surfaces are well expressed. The radar signature of dipping layers depends upon the relationship between the dip direction and the radar look direction. In the image (Figure 7-7A) and the map (Figure 7-8A), the strata are dipping generally toward the top of the image, which is toward the radar antenna. The dipslopes therefore have bright signatures, and the antidip scarps, which are in the radar shadow, have dark signatures. Where beds dip away from the radar antenna, the antidip scarps are bright and the dipslopes are dark, as seen in the flanks of the folds in Figure 7-7C.

Thrust Faults Thrust faults are difficult to interpret from radar and other remote sensing images for the following reasons:

1. The planes of thrust faults are commonly parallel or nearly parallel with bedding planes of associated strata. Therefore, most thrust faults do not cause the discordant geometric relationships that are associated with many normal and strike-slip faults.
2. Thrust faults are commonly recognized by anomalous rock relationships such as older beds over younger beds, repetition of beds, and omission of beds. These relationships are difficult to recognize on images without the aid of field data.

Despite these difficulties, thrust faults were interpreted from the SIR-A image of the Paniai Lake region, mainland Irian Jaya, where several imbricate thrust plates occur. Figures 7-7B and 7-8B show thrust plates that dip gently toward the lower margin of the image and are terminated updip at eroded antidip scarps. The trend of these scarps is locally discordant with the trend of the underlying rocks. At places such as the right margin of Figure 7-7B, an upper thrust plate overrides and truncates an underlying plate. Elsewhere in the Paniai region, thrust faults are recognized by the repetition of belts of distinctive karst topography that alternate with belts of clastic terrain.

Folds, Moderately Eroded Folds may be recognized by attitudes of beds, outcrop patterns, and topographic expression. In this example of moderately eroded folds taken from Kalimantan, the strike-and-dip attitudes are readily interpreted using the criteria described earlier. Outcrop patterns are also diagnostic in Figure 7-7C, where the youngest beds (carbonate rocks with characteristic karst topography) are preserved as mesas in the troughs of synclines but are eroded from the crest of the anticline. These erosion patterns commonly result in *topographic reversal*, in which topographic elevations (mesas) correspond to structural depressions (synclines). In the upper right portion of Figures 7-7C and 7-8C, note that the resistant clastic beds at the crest of the anticline have been breached to expose older strata.

Folds, Deeply Eroded Deep erosion may produce a nearly planar surface, where plunging folds are marked by subdued parallel ridges with arcuate patterns that are formed by outcrops of resistant strata (Figure 7-7D). The noses and axes of these folds are readily mapped (Figure 7-8D). Anticlines cannot be distinguished from synclines, however, because dip attitudes cannot be interpreted from the subdued ridges.

Lineaments In the SIR-A images of Indonesia, lineaments are expressed as scarps, linear valleys, and aligned valleys. In the upper part of the image (Figure 7-7E) and map (Figure 7-8E), two linear scarps form the boundary between mélange terrain and alluvial terrain and may be the expression of faults. In this remote area along the Kindjau River in Kalimantan, however, there are no geologic maps with which to evaluate the SIR-A interpretations. Not all topographic scarps are linear; those with an irregular trend, such as in the upper left part of Figure 7-7E, are not classed as lineaments.

Linear valleys are a lineament category formed by a single straight or curvilinear channel, of which there are several examples in Figure 7-7E. Two or more separate valleys may be aligned end to end to form a lineament. Many linear valleys and aligned valleys follow faults and fracture zones that are preferentially eroded zones of weakness.

Strike-Slip Faults The image (Figure 7-7F) and map (Figure 7-8F) cover a segment of the Sorong fault enlarged from the regional SIR-A image in Figure 7-6. The prominent lineament is formed by a linear valley that has the following characteristics of a strike-slip fault: (1) aligned notches; (2) shutter ridges; (3) linear terraces; and (4) offset stream channels, which in this example indicate left-lateral fault displacement. The preservation of these tectonic features in this region of heavy rainfall and rapid erosion indicates that the Sorong fault is active; that is, movement has occurred in the past 10,000 years. Strike-slip

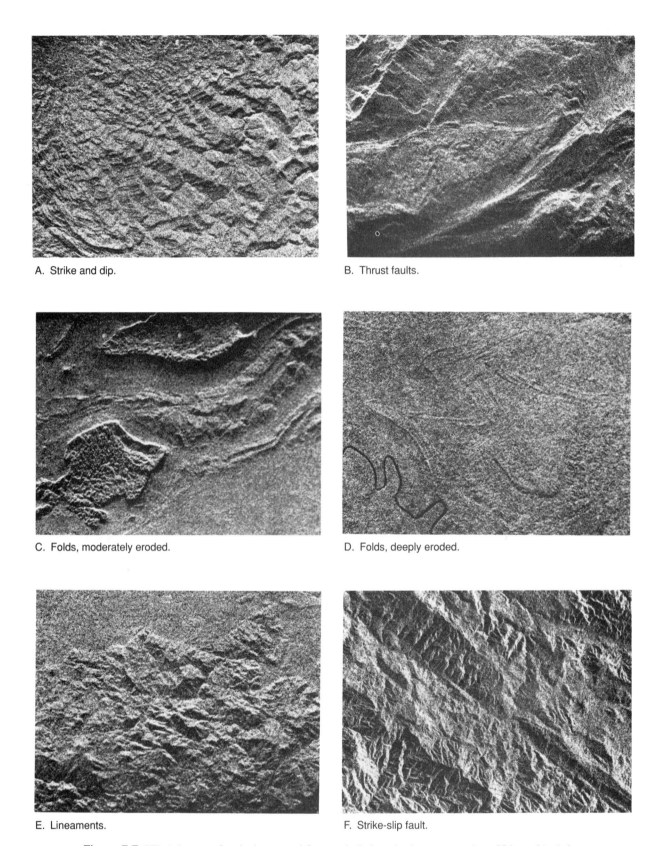

A. Strike and dip.

B. Thrust faults.

C. Folds, moderately eroded.

D. Folds, deeply eroded.

E. Lineaments.

F. Strike-slip fault.

Figure 7-7 SIR-A images of typical structural features in Indonesia. Areas covered are 28 km wide (left to right). Look direction is toward the lower margin of each image. From Sabins (1983, Figure 5)

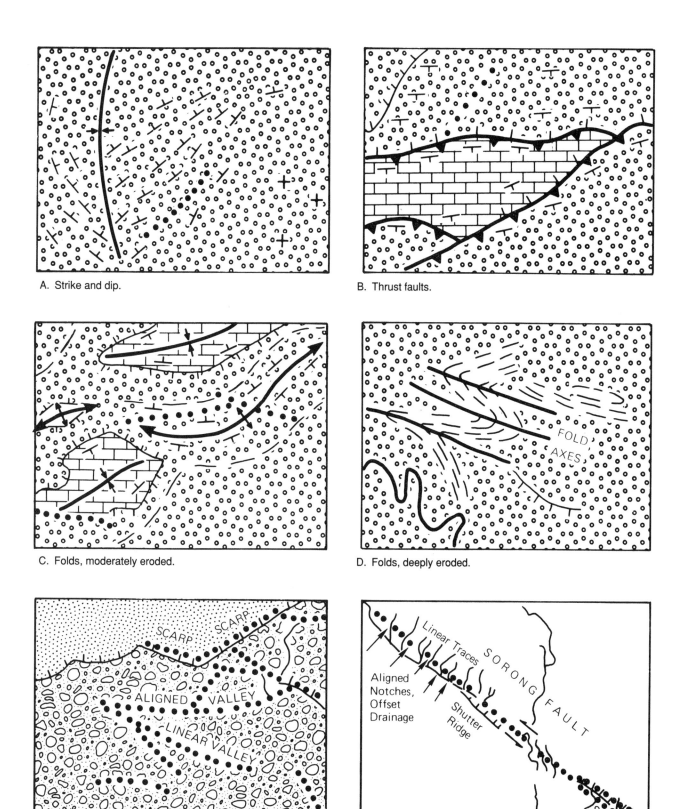

A. Strike and dip.

B. Thrust faults.

C. Folds, moderately eroded.

D. Folds, deeply eroded.

E. Lineaments.

F. Strike-slip fault.

Figure 7-8 Interpretation maps of structural features (shown in Figure 7-7); areas covered are 28 km wide (left to right). From Sabins (1983, Figure 6).

faults are also recognizable on images by the lateral offset of geologic units on opposite sides of the fault.

Summary

Despite persistent cloud cover, SIR-A acquired images of portions of Indonesia from which geomorphology, rock types, and geologic structure were interpreted. The following section explains why such interpretations were possible despite the dense vegetation cover. Where geologic maps of Irian Jaya were available for comparison, many SIR-A interpretations were confirmed and others suggested locations for additional fieldwork. In Kalimantan, geologic maps are lacking and thus the SIR-A interpretations provide valuable new information.

RADAR INTERACTION WITH VEGETATED TERRAIN

In radar images of forested terrain, such as Indonesia, one can interpret detailed geologic information. This observation may lead to the incorrect conclusion that the image portrays the ground surface beneath the forest canopy. In fact, radar signatures of vegetated terrain are dominated by interactions within the vegetation cover, rather than with the ground surface.

L-band radar systems operate at the maximum wavelength currently available for imaging radars and have the greatest potential for penetrating foliage. However, the rough criterion for SIR-A is 8.4 cm and for Seasat is 5.7 cm. The relief of most vegetated surfaces greatly exceeds these values, and radar theory states that such surfaces will strongly scatter the incident radar energy. Growing vegetation has a high water content, which increases the dielectric constant and the radar reflectivity, as discussed in Chapter 6. Thus radar theory predicts that because of the rough surfaces and high moisture content, vegetation will scatter and reflect incident radar energy and little or no energy will be transmitted.

The interaction with vegetation is illustrated by SIR-A and Seasat images of agricultural areas in the Great Plains (Figure 7-9) where centerpoint irrigation is practiced. Clusters of bright circles are fields, predominantly wheat, with small dark interstices caused by smooth, bare soil. Differences in brightness of the fields are caused by differences in the growth and harvest status of the crops. The relevant point is that if the L-band radar energy had penetrated the vegetation, the images would show only the dark signature of the underlying soil. It would be impossible to recognize crops and variations in crops. Since these images demonstrate the inability of radar to penetrate crops less than 1 m in height, it is even less likely that radar could penetrate a forest canopy, such as in Indonesia.

Figure 7-10 explains how features of the ground surface are recognizable in images of forested terrain. Figure 7-10A shows two faults and their associated topographic scarps in bedrock covered by forest. The trees are of relatively uniform height and the top surface of the canopy is controlled by the bedrock surface. The radar system produces an image of the canopy (Figure 7-10B) in which the underlying geology is enhanced by two mechanisms:

A. SIR-A image in northwest Texas, November 1981.

B. Seasat image in southwest Kansas, September 1978.

Figure 7-9 Radar signatures of agricultural fields on enlarged satellite radar images. Areas covered are 8 km wide (left to right).

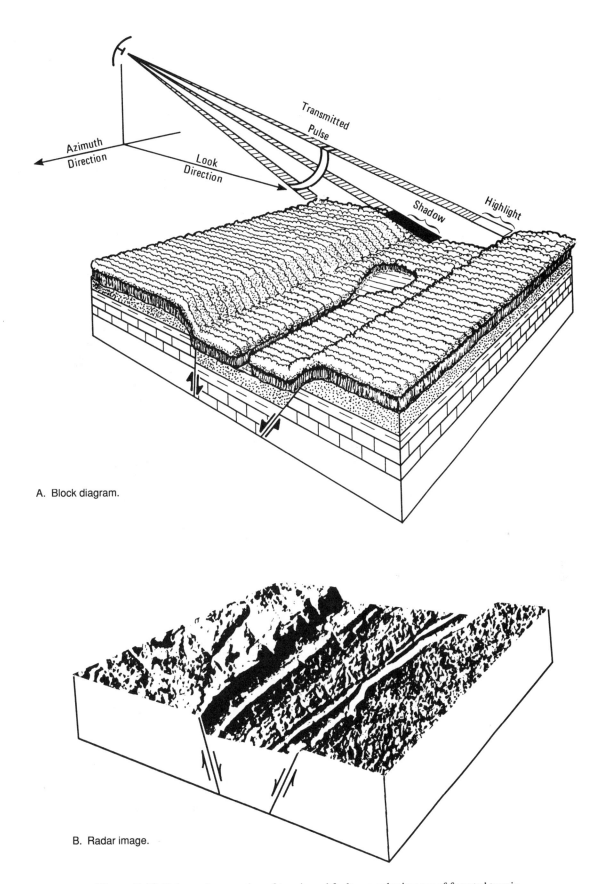

A. Block diagram.

B. Radar image.

Figure 7-10 Enhanced expression of terrain and faults on radar images of forested terrain.

A. Landsat MSS band 2 image.

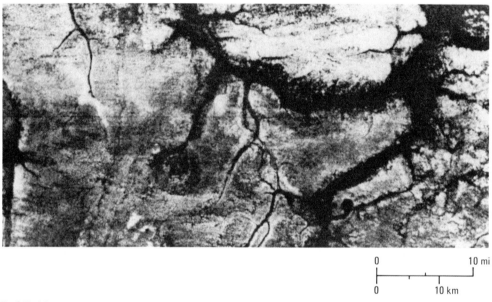

0   10 mi

0   10 km

B. SIR-A image.

Figure 7-11 SIR-A and Landsat images of the Selima sand sheet, northwest Sudan. From McCauley and others (1982). Courtesy G. G. Schaber, U.S. Geological Survey.

1. The inclined illumination produces highlights and shadows that emphasize topographic lineaments associated with the faults. Other structural features, such as folds and dipping strata, are similarly enhanced.
2. Because of the relatively large ground resolution cells (10 to 15 m for aircraft radar, 25 m for Seasat, and 38 m for SIR-A), individual trees are not resolved, which improves the topographic expression of geologic features with dimensions of hundreds and thousands of meters. In other words, the large resolution cell of radar acts as a filter to remove the high-frequency spatial detail of vegetation "noise," thereby enhancing the lower frequency geologic "signals."

In addition to these factors, radar is not affected by the cloud cover that is associated with most forested regions, especially in the tropics. For these reasons, radar is an excellent system for mapping in forested areas.

SIR-A IMAGES OF DESERT TERRAIN, EASTERN SAHARA

Over the years, many aircraft and Seasat images have been acquired of sand-covered desert areas, principally in North America. Sheets of sand and fine gravel have dark signatures

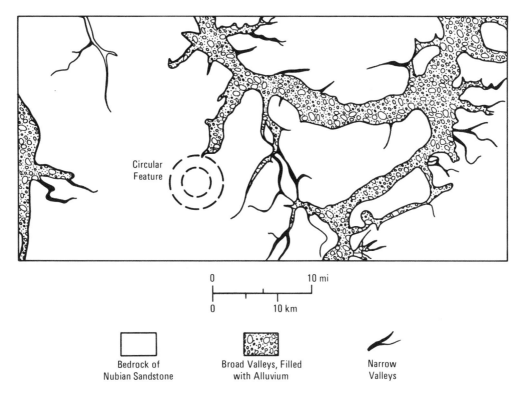

Circular
Feature

0   10 mi

0   10 km

Bedrock of
Nubian Sandstone

Broad Valleys, Filled
with Alluvium

Narrow
Valleys

Figure 7-12 Interpretation map of SIR-A image of the Selima sand sheet
(Figure 7-11B).

because of their smooth surfaces, and sand dunes have highlights and shadows. The radar signatures are clearly determined by the surface roughness and landforms. Because of this
experience, investigators were startled by SIR-A images of the
Selima sand sheet of the eastern Sahara in Egypt and the
Sudan, for reasons described below.

Figure 7-11A is a Landsat MSS image of a portion of the
Selima sand sheet that shows a featureless surface of windblown sand. Based on this signature in a visible spectral band,
one would predict the sand sheet to have a uniform dark signature on SIR-A images. Instead the SIR-A image (Figure 7-
11B) shows details of ancient drainage patterns eroded into the
bedrock that underlies the sand sheet. The SIR-A radar energy
has clearly penetrated the sand sheet and returned from the
bedrock surface. McCauley and others (1982) investigated the
images of the Selima sand sheet in the field. The sand sheet is
so devoid of features that satellite navigation systems were
used in field vehicles to locate specific sites on the SIR-A images. A number of pits were excavated in the sand and established the following:

1. The bright areas in the SIR-A images are the rough, eroded
   surface of the Nubian Sandstone.
2. The dark signatures indicate extinct drainage channels filled
   in with pebbly alluvium.

These relationships are shown in the interpretation map
(Figure 7-12). The diagrammatic cross section (Figure 7-13)
shows that some of the transmitted radar energy is reflected by
the smooth sand surface. However, much of the energy penetrates the sand and is refracted to a steeper depression angle as
it enters this denser medium. The energy that encounters the
rough bedrock surface is strongly backscattered, and a relatively large proportion returns to the antenna to produce a
bright signature. The transmitted energy that encounters the
smooth surface of a channel is specularly reflected (Figure
7-13), and no energy is returned to the antenna, which produces the dark signature of the channels.

RADAR INTERACTION WITH SAND-COVERED
TERRAIN

The radar penetration of the Selima sand sheet is due to the hyperarid environment, where rainfall occurs at intervals of 30 to
50 years and the sun evaporates any absorbed moisture. Field
excavations in the eastern Sahara (Schaber and others, 1986)
showed that 1.5 m was the maximum sand thickness through
which images of the substrate could be recorded. Most of the
modern sand sheet in the eastern Sahara Desert is "transparent" to L-band radar.

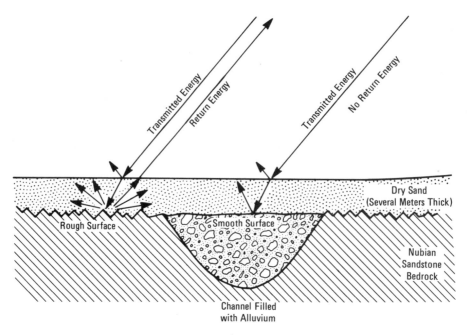

Figure 7-13 Cross section showing the interaction between SIR-A energy and hyperarid terrain in northwest Sudan. From information provided by G. G. Schaber, U.S. Geological Survey.

Radar images of other desert regions have been examined for evidence of surface penetration, but only a few additional examples have been reported. Blom, Crippen, and Elachi (1984), detected a sand-covered igneous dike on a Seasat image of an arid valley in the Mojave Desert, California. Berlin and others (1986) interpreted SIR-A and SIR-B images of the Nafud sand sea in northern Saudi Arabia. In this hyperarid area they reported that bedrock was imaged through sand up to 1.24 m thick. The general lack of radar penetration of other deserts is due to the presence of subsurface moisture, since most deserts have rainfall every few years. The water soaks into the sand and later evaporates from a thin surface layer that insulates the deeper sand. In many deserts the sand has appreciable moisture at depths of less than a meter.

LAND USE AND LAND COVER, IMPERIAL VALLEY, CALIFORNIA

Land use and land cover may be interpreted from radar images, as shown by the SIR-A image of part of the Imperial Valley in southern California and Mexico (Figure 7-14A). The map (Figure 7-14B) shows the various categories of land use and land cover that have been interpreted from the image. The Imperial Valley is a broad, flat depression with elevations below sea level; its fertile soil and warm climate support year-round irrigated agriculture. The rectangular field patterns in

California reflect the township and range system of landownership in the United States. Mexico does not use this system, so the pattern is irregular. On both sides of the border the agriculture cropland consists predominantly of cotton, sugar beets, alfalfa, and vegetables, all with bright signatures. Fallow fields have dark signatures. Towns, which are readily recognized by their bright signatures, belong to the category "urban, mixed." At greater magnifications of the image, street patterns are recognizable. The agriculture area is bordered on the east and west by desert, which is classified as "barren land, sand, and gravel." There is no evidence of radar penetration of the sand in this area, which receives 10 cm or more of rain annually. San Felipe Creek and other ephemeral streams in the desert have bright signatures caused by gravel and native vegetation along the stream courses. The mountains and hills of bedrock that project through the sand and gravel are classified as "barren land, exposed rock." These bedrock outcrops are identified by their bright signature and rugged topography.

SIR-B MISSION

A second Shuttle radar experiment (SIR-B) was conducted in October 1984 with the antenna modified to change the depression angle during the mission within a range from 30° to 75°. Multiple images of certain areas were acquired at different depression angles for two purposes: (1) to evaluate

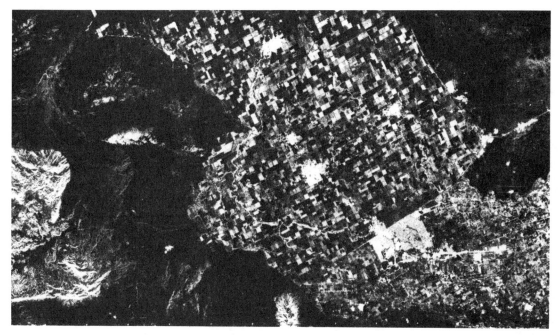

A. SIR-A image.

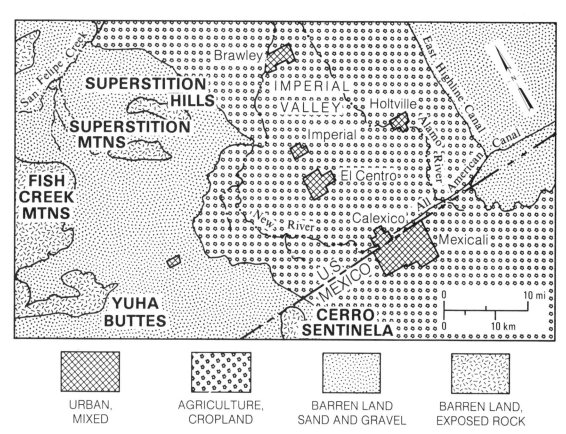

B. Map showing land-use and land-cover categories interpreted from SIR-A image.

Figure 7-14 Land use and land cover interpreted from a SIR-A image of the Imperial Valley region, southern California and northern Mexico.

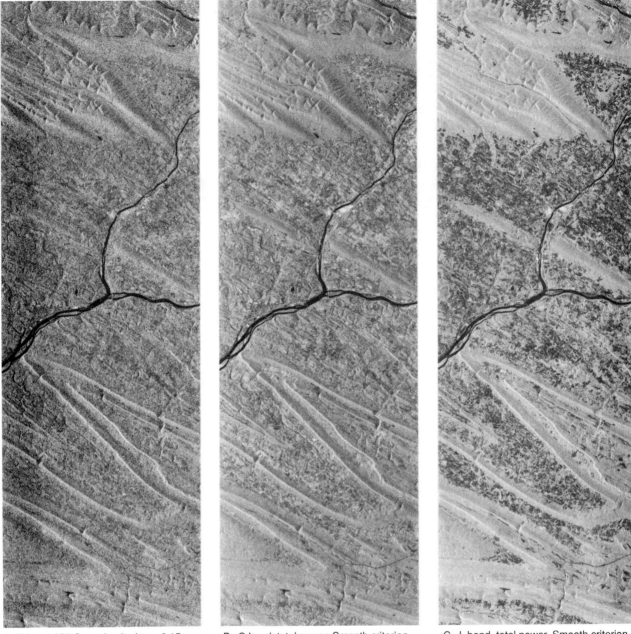

A. X band, VV. Smooth criterion = 0.15 cm; rough criterion = 0.88 cm.

B. C band, total power. Smooth criterion = 0.31 cm; rough criterion = 1.76 cm.

C. L band, total power. Smooth criterion = 1.23 cm; rough criterion = 7.06 cm.

Figure 7-15 SIR-C images of Mahantango Mountain area, Pennsylvania, acquired at a 51° depression angle. Courtesy A. England, Jet Propulsion Laboratory.

changes in radar return as a function of depression angle for different terrain types, and (2) to produce stereo images based on the parallax provided by differences in viewing geometry.

Cimino, Holt, and Richardson (1988) provide details of the mission and an index to the images that were acquired. Ford and others (1986) published an annotated collection of repre-

sentative SIR-B images. The *International Journal of Remote Sensing* (May 1988, vol. 9, no. 5) published a collection of 14 papers on different aspects of the mission. Ford (1988) edited a series of SIR-B research papers. Ford and Sabins (1986) interpreted SIR-B image strips across the island of Borneo and recognized categories of terrain and structure similar to those on earlier SIR-A images of Indonesia.

SIR-C MISSION

The SIR-C mission was carried on two Shuttle flights in 1994 and was a major technologic advance over earlier missions. Three antennas recorded X (3.0 cm), C (6.0 cm), and L (24.0 cm) bands of imagery at multiple polarizations and variable depression angles (Table 7-1).

Representative Images

Figure 7-15 shows images at the three wavelengths recorded by SIR-C, together with their smooth and rough criteria, for Mahantango Mountain and vicinity in the Appalachian Mountains of Pennsylvania (Figure 7-16). The land cover is a mixture of agriculture and woodlands. The region is underlain by folded sedimentary rocks that erode to form sinuous ridges of resistant quartzite and conglomerate. The structural attitudes (strike and dip) of the asymmetric gentle dipslopes and steep antidip scarps are recognized by using the highlight-and-shadow method that was described for Landsat images (Chapter 3). The SIR-C images in Figure 7-15 were illuminated with a look direction from the west (left margin). Line Mountain has a narrow highlight (antidip scarp) on the west and a parallel broad shadow (dipslope) on the east, which indicates east-dipping beds. At Little Mountain the same beds have a reversed pattern of highlights and shadows, indicating west-dipping beds. These relationships show that Line Mountain and Little Mountain are opposing limbs of a syncline (Figure 7-16). The same illumination criteria show that Line Mountain and Mahantango Mountain are opposing limbs of an anticline.

The digitally recorded SIR-C images are coregistered and can be combined in colors in the same manner as multispectral images such as Landsat. Plate 10C shows a color image of Mahantango scene that combines the L-, C-, and X-band images in red, green, and blue. Plate 10D is an image of Mount Pinatubo, an active volcano in the Philippines. The crater at the crest was formed by the eruption in June 1991. The red signature on the higher slopes is ash. The broad, dark signatures along major drainages are not water, but are mudflows of ash and water from heavy rains that followed the eruption. Evans and others (1994) published a collection of multicolor SIR-C images.

Figure 7-17 compares SIR-C images of central Java acquired at C-band and L-band wavelengths. Figure 7-18 identifies the major towns and terrain features, which are dominated by active and dormant volcanoes. The smooth and rough criteria have been calculated for each wavelength: the L-band wavelength is four times that of the C band; therefore, the L-band criteria (smooth = 1.3 cm, rough = 7.6 cm) are four times the C-band criteria (smooth = 0.3 cm, rough = 1.9 cm). These roughness differences are expressed in the images (Figure 7-17) where the C-band image is brighter than the L-band image. In the western (lower) portion of the images, fields with

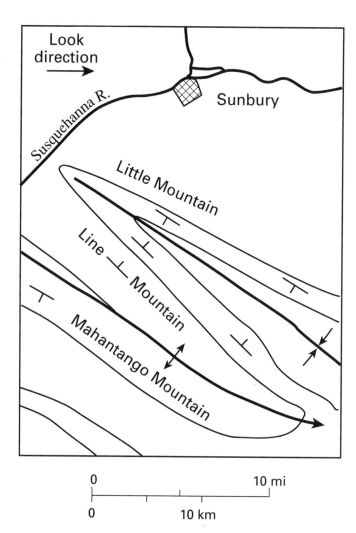

Figure 7-16 Location map for Mahantango Mountain, Pennsylvania.

standing rice crops (bright signature) and fallow or flooded fields (dark) are more readily distinguished on the L-band image. In the eastern (upper) portion of each image, however, this distinction is more apparent in the C-band image.

Sources of Data

The Jet Propulsion Laboratory (JPL) has set up a World Wide Web site with the following address:

http://southport.jpl.nasa.gov/

The site contains information on the radar program at JPL, radar images and results from the two SIR-C missions, sample

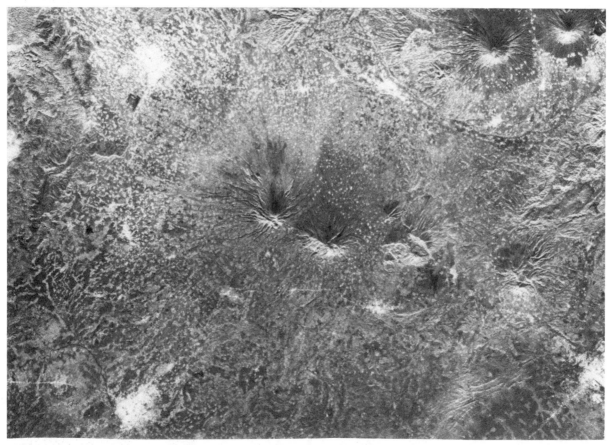

A. C band (6 cm). Smooth criterion = 0.3 cm; rough criterion = 1.9 cm.

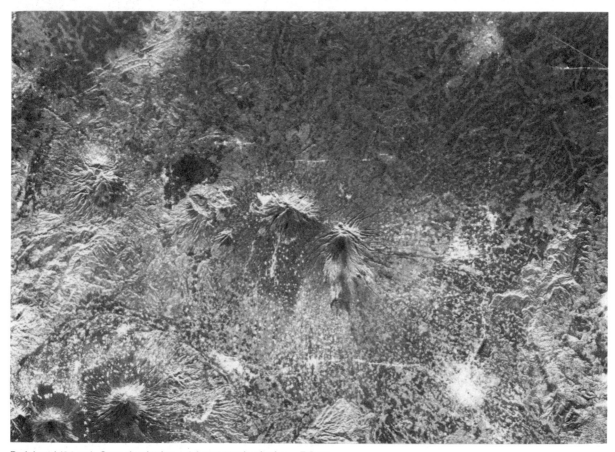

B. L band (24 cm). Smooth criterion = 1.3 cm; rough criterion = 7.6 cm.

Figure 7-17 SIR-C images of central Java, Indonesia, acquired with a 46° depression angle.

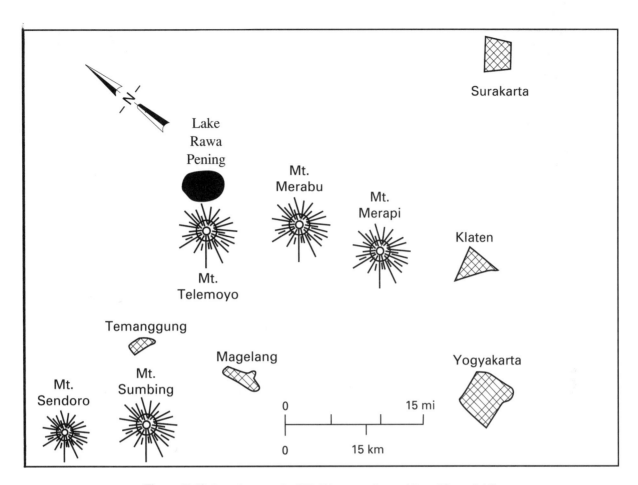

Figure 7-18 Location map for SIR-C images of central Java (Figure 7-17).

data and videos, and information on how to obtain radar images and software for displaying images. This site includes links to other Internet sites that distribute radar images, including the EROS Data Center (EDC), which has SIR-C survey (low-resolution) images on-line at the following address:

http://sun1.cr.usgs.gov/landdaac/sir-c/sir-c.html/

This Internet access enables users to preview low-resolution versions of SIR-C images on their terminals.

The actual digital data are distributed by the EDC, Sioux Falls, SD 57198. Data are available in two formats:

1. **Survey data** are uncalibrated, low-resolution (50-m ground resolution cell) images of entire data takes. A standard data take contains 4 min of acquisition time, which is an image swath 1600 km long. Survey data are provided on CD-ROMs, each of which contains up to 12 data takes. Survey images are useful for (a) regional interpretations, (b) determining extent of coverage, and (c) selecting areas for precision products, which are described next.
2. **Precision product data** cover subareas (which measure full-swath width by 50 or 100 km long) with all the polar-

izations that were recorded. The images in Figure 7-15 are precision products. The data are calibrated and at full resolution (10 to 45 m). Digital data for precision images are provided on 8-mm tapes.

Contact the EDC for additional information.

Syracuse University provides a SIR-C Teachers Guide at the following address:

http://ericir.syr.edu/NASA/nasa/htm1

JERS-1

The Japanese JERS-1 satellite carries a radar imaging system that is summarized in Table 7-1. Figure 7-19 is an image of the Tsugaru Peninsula, Japan, that covers 75 by 75 km. Water in the ocean, lakes, and rivers has the typical dark signature. Urban areas are very bright. The vegetated terrain with moderate to high relief has subdued shadows and highlights because of the steep depression angle. Terrain with low relief along the river valley ranges from bright to dark, depending upon the nature of the

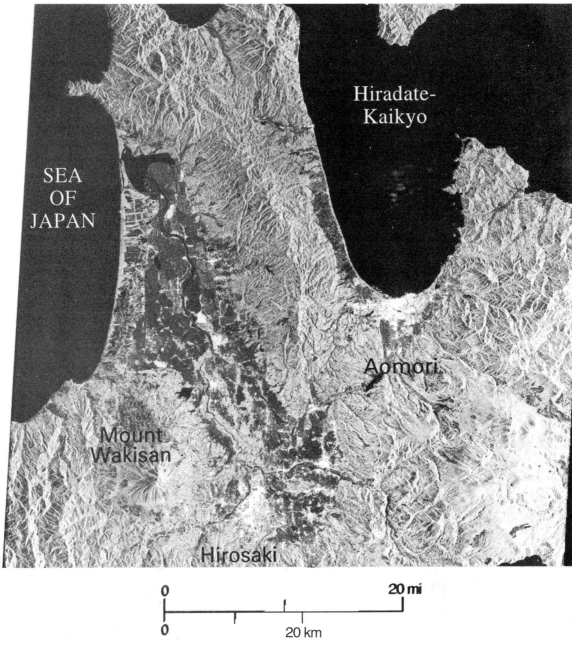

Figure 7-19 JERS-1 radar image of the Tsugaru Peninsula, Japan. Courtesy T. Nishidai, ERS-DAC.

land cover. The 55° depression angle of JERS-1 is intermediate between those of SIR-A (40°) and ERS-1 (67°). Therefore, the foreshortening on JERS-1 images is intermediate between that for SIR-A and ERS-1 images.

JERS-1 images are available as black-and-white prints or as digital data. U.S. users can obtain information and place orders at EOSAT at the following on-line address:

http://www.eosat.com

Other users can contact

Remote Sensing Technology Center of Japan (RESTEC)
Attn.: User Service Department
Uni-Roponggi Bldg., 7-15-17, Ropongi
Minato-ku, Tokyo 106, Japan
Telephone: 81-3-3401-1387
Fax: 81-3-3403-1766
http://hdsn.eoc.nasda.go.jp/

Figure 7-20 Figure 7-20 ERS-1 radar image of the St. Lawrence River, Canada.

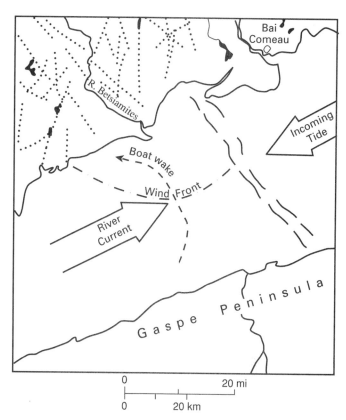

Figure 7-21 Interpretation map of the St. Lawrence River (Figure 7-20). Dotted lines are lineaments.

ERS-1

ERS-1, the European Remote-sensing Satellite, acquires images with VV polarization and a steep depression angle (67°) to emphasize oceanographic features (Table 7-1). The European Space Agency (1993) published a user handbook that provides details of the system. Figures 7-20 and 7-21 are an ERS-1 image and an interpretation map, respectively, of the St. Lawrence River and vicinity in eastern Canada. Currents and wind conditions in the river are clearly shown by roughness signatures. In the eastern portion of the river a pair of bright bands extends from shore to shore and separates two distinctly different surface patterns. The bright bands are interpreted as rougher water at the leading edge of a tidal bore that is advancing up the river from the ocean a short distance to the east. The low terrain north of the river is underlain by crystalline rocks (of Precambrian age) that are intensely fractured and faulted. The resulting topographic lineaments are well expressed on the image. The Gaspé Peninsula, on the south side of the river, is underlain by gently dipping sedimentary rocks (of lower Paleozoic age) with few structural features. The image shows patterns of agriculture and Pleistocene glaciation.

The St. Lawrence image demonstrates the value of ERS-1 images for interpreting features of water and terrain with low relief. In terrain with moderate to high relief, however, the steep depression angle results in severe topographic layover.

Figure 7-22A,B compares ERS-1 and JERS-1 images of Mount Fuji, Japan, with the look direction from the north. Figure 7-23C is a topographic map showing the symmetric conic shape of the mountain. The very steep depression angle (67°) of ERS-1 causes severe layover of the mountain. The steep depression angle (55°) of JERS-1 causes less severe layover of Mount Fuji. On both images the topographic ridges surrounding Mount Fuji appear as dipslopes dipping to the south. Because of this distortion, ERS-1 and JERS-1 images are of limited use for interpreting terrain with any degree of relief. Lalonde, Posehn, and Sabins (1993) compared ERS-1 images with aircraft radar images acquired with gentle depression angles and concluded that the aircraft images were superior for geologic interpretation.

ERS-1 image prints and digital data are distributed from three centers. U.S. users can contact EOSAT. Users in Europe, North Africa, and Middle East should contact

Euroimage ERS Customer Services
ESRN
Via Galileo Galilei
00044 Frascatti, Italy
Telephone: 39-6-94180-478
Fax: 39-6-94180-510

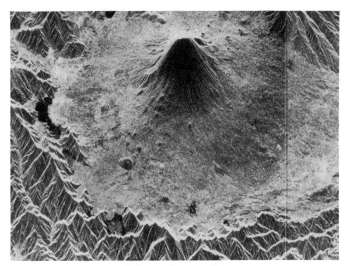

A. ERS-1 image acquired with a 67° depression angle.

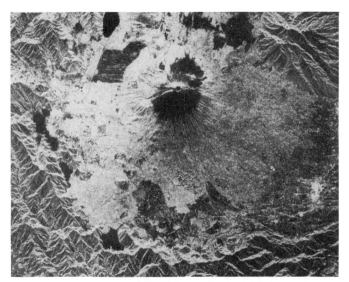

B. JERS-1 image acquired with a 55° depression angle.

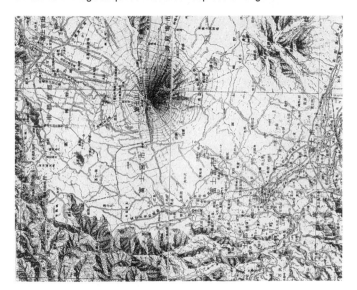

C. Topographic map.

Figure 7-22 Comparison of ERS-1 and JERS-1 images of Mount Fujiyama, Japan, showing the effects of different depression angles.

Users in Canada should contact

Radarsat International ERS Ordere Desk
851 Shell Road, Suite 200
Richmond, BC, V6X 2W2
Canada
Telephone: 604-244-0400
Fax: 604-244-0404

Users in other locations should contact

SPOT Image ERS Order Desk
5 rue des Satellites
B.P. 4359
F-31030 Toulouse Cedex
France
Telephone: 33-62-19-41-46
Fax: 33-62-19-40-55

ALMAZ-1

The Russian Almaz-1 satellite (launched March 31, 1991) records images on holographic film that is converted into image film. Individual images cover 40 by 40 km. Figure 7-23 is an Almaz image, acquired with a 50.4° depression angle, of the Hedinia anticline in the Papuan Fold and Thrust Belt on the island of New Guinea. The look direction is toward the upper margin of the image, which may cause topographic inversion for some viewers. For comparison, an aircraft radar image and map of the Hedinia region is shown in Chapter 10. Almaz images are available from various sources, including

Hughes STX Corporation
Satellite Mapping Technologies
4400 Forbes Boulevard
Lanham, MD 20706
Telephone: 301-794-5020
Fax: 301-306-0963

RADARSAT

Radarsat, Canada's first earth observation satellite, was launched November 4, 1995. Table 7-1 lists characteristics of the system. The satellite completes 14 orbits daily in a sun-synchronous pattern that covers the earth (north of latitude 80°S) every 6 days. Latitudes north of 48°N are covered every 4 days; those north of 70°N are covered daily. One objective is to monitor sea-ice conditions in the Arctic Ocean during periods of winter darkness and seasonal cloud cover.

Figure 7-24A shows the geometry of the system that images a swath 500 km wide. A wide range of image formats is

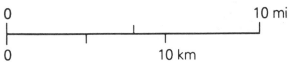

Figure 7-23 Almaz image acquired with a depression angle of 50.4° and the look direction toward upper margin of image. Hedinia anticline in the Papuan Fold and Thrust Belt, island of New Guinea. Courtesy W. A. Kennedy, Hughes STX Corporation.

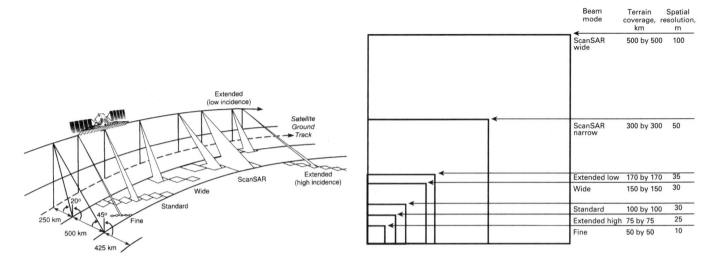

A. Imaging modes and formats.

B. Terrain coverage of image formats.

Figure 7-24 Imaging modes and terrain coverage of Radarsat. From Radarsat International (1995).

Table 7-2 Characteristics of Radarsat image formats

Beam mode	Beam position	Incidence angle range, degrees	Terrain coverage, km	Spatial resolution, m
Fine	F1	37 to 40	50 by 50	10
	F2	39 to 42		
	F3	41 to 44		
	F4	43 to 46		
	F5	45 to 48		
Standard	S1	20 to 27	100 by 100	30
	S2	24 to 31		
	S3	30 to 37		
	S4	34 to 40		
	S5	36 to 42		
	S6	41 to 46		
	S7	45 to 49		
Wide	W1	20 to 31	165 by 165	30
	W2	31 to 39	150 by 150	
	W3	39 to 45	130 by 130	
ScanSAR (narrow)	SN1	20 to 40	300 by 300	50
	SN2	31 to 46		
ScanSAR (wide)	SW1	20 to 50	500 by 500	100
Extended high incidence	H1	49 to 52	75 by 75	25
	H2	50 to 53		
	H3	52 to 55		
	H4	54 to 57		
	H5	56 to 58		
	H6	57 to 59		
Extended low incidence	L1	10 to 23	170 by 170	35

Agricultural Region

Eldorado International Airport

BOGOTA

Folded and Faulted Sedimentary Rocks

Embalse del Muna

0   10 mi
|——|——|——|——|——|——|——|——|——|——|
0   10 km

Figure 7-25 Radarsat image of Bogota, Colombia. The image was acquired in the Fine mode (F5) with a spatial resolution of 8 m at an incidence angle of 46° (depression angle of 44°). Look direction is toward the west. This image was acquired during Radarsat's commissioning phase. Not to be used for analysis or interpretation. ©Canadian Space Agency 1996. Image received by the Canada Centre for Remote Sensing. Processed and distributed by Radarsat International. Courtesy C. Aspden, Radarsat International.

available from within the swath. Table 7-2 lists characteristics of the formats. Figure 7-24B shows terrain coverage, which ranges from the ScanSAR wide format (full swath width, 500 by 500 km) to the Fine format (50 by 50 km). Other formats are available beyond the near-range and far-range limits of the primary image swath.

Figure 7-25 is an image of Bogota, Colombia, acquired in the Fine mode (F5) with a spatial resolution of 8 m and a depression angle of 44°. The look direction is toward the west. The urban complex of Bogota in the eastern portion of the image has a medium-to-bright signature caused by the buildings. Local areas with very bright signatures are attributed to streets and buildings being oriented to produce maximum radar returns, as illustrated on images of New Orleans (Figure 6-12). Streets and the airport have dark signatures because of their smooth surfaces. The flat terrain west of Bogota is intensely cultivated. Fields are sparated by bright lines caused by fences and tree rows. Lakes have bright to intermediate signatures which indicate rough water due to strong winds. Much of the area is underlain by folded and faulted sedimentary rocks. Shadow and highlight criteria can be used to interpret dip and strike of the beds, as illustrated for SIR-A images in Figure 7-7A.

Images are available as digital data or as prints at a range of scales. Radarsat International (1995) has published a guide to their products and services. The guide and additional information are available from

Radarsat International ERS Order Desk
3851 Shell Road, Suite 200
Richmond, BC, V6X 2W2
Canada
Telephone: 604-244-0400
Fax: 604-244-0404
http://radarsat.space.gc.ca/

MAGELLAN MISSION TO VENUS

Of all planets in our solar system, Venus most closely resembles the orbital and physical properties of the earth, as summarized in Table 7-3. Venus is spoken of as the "sister planet" of the earth. The major differences are the high atmospheric pressure and temperature of Venus, which cause major differences in the geology. Knowledge of the geology of Venus can aid in understanding the origin of the earth, in part because Venus has no concealing cover of water or vegetation cover and almost no erosion. The thick cloud cover (mostly CO_2) concealed the surface of Venus until the advent of planetary radar systems.

Radar images of Venus were acquired from the earth and from satellites from the 1960s to the early 1980s. Despite their coarse spatial resolution (a few km to a few tens of km), these images showed a complex planetary surface and helped justify the Magellan mission.

Table 7-3 Orbital and physical properties of the earth and Venus

Property	*Venus*	*Earth*
Mean distance from sun, (earth = 1)	0.723	1.0
Mean distance from sun, 10^6 km	108.2	149.6
Orbital period, days	224.7	365.25
Mean orbital velocity, km · sec^{-1}	35.05	29.79
Mass, kg	4.871×10^{24}	5.976×10^{24}
Mass (earth = 1)	0.815	1.0
Equatorial radius, km	6051.92	6378
Equatorial radius (earth = 1)	0.949	1.0
Mean density, gm · cm^{-1}	5.24	5.52
Atmospheric pressure, bars	98	1
Surface temperature, °C	450	15

Source: Ford and others (1989, Appendix A).

Magellan Radar System

The Magellan satellite (Figure 7-26) was launched May 4, 1989, from Cape Canaveral on-board the Space Shuttle Atlantis, from which it was sent on its mission to Venus. Magellan began transmitting images on September 15, 1990. Ford and others (1993) provide details of the imaging system, orbit pattern, and typical images. The polar orbit of Magellan has a minimum altitude of 290 km and a maximum of 2000 km at polar latitudes.

The imaging radar has the following characteristics:

Wavelength	12.6 cm
Polarization	HH
Spatial resolution:	
Along-track	120 m
Slant-range, effective	88 m
Depression angle, nominal	65°
Roughness criteria (at $\gamma = 65°$):	
Rough criterion	3.2 cm
Smooth criterion	0.6 cm

Magellan also carried a radar altimeter for topographic mapping and a radiometer for measuring emissivity at radar wavelengths.

Images were acquired from a total of 4225 orbits from September 1990 through September 1992. Over 98 percent of Venus was imaged. Many areas were viewed more than once with different look directions and depression angles. The image recorded on a typical orbit is 20 km wide and 17,000 km

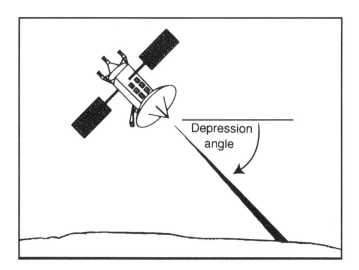

Figure 7-26 Magellan satellite. Depression angle varies systematically from ~70° near the poles to ~40° at 40°N latitude.

long. This narrow format is not optimum for interpretation; therefore, several adjacent images are combined into a strip with a broader swath width. Many strips were combined to produce the mosaic of the entire planet shown in Figure 7-27. Incidentally, no comparable radar mosaic exists for planet earth.

Ford and others (1993) published and interpreted a representative collection of Magellan images. The April 12, 1991, issue of *Science* (vol. 252, no. 5003) contained a number of papers on preliminary results of the mission. The August 25 and October 25, 1992, special issues of the *Journal of Geophysical Research* (vol. 97, nos. E8 and E10) were devoted to detailed papers on various aspects of the mission. Because of the volume of imagery and its recent availability, many of the interpretations are preliminary and subject to later modification.

Tectonic Features

Magellan's complete coverage of Venus has enabled investigators to interpret tectonic features on both local and regional scales. Solomon and others (1992) prepared an overview of Venus tectonics that forms the basis for the following summary. They noted several major similarities and differences between tectonics on Venus and on the earth.

Similarities:

1. Tectonic features of both compressional and tensional natures are present on both planets.
2. Many features (faults, folds, rifts, and others) are similar on both planets.

Differences:

1. Horizontal displacements on Venus are much smaller than on the earth.
2. Venus has some features (called complex ridged terrains) with no analogs on earth.
3. Plate tectonism does not occur on Venus, as shown by the

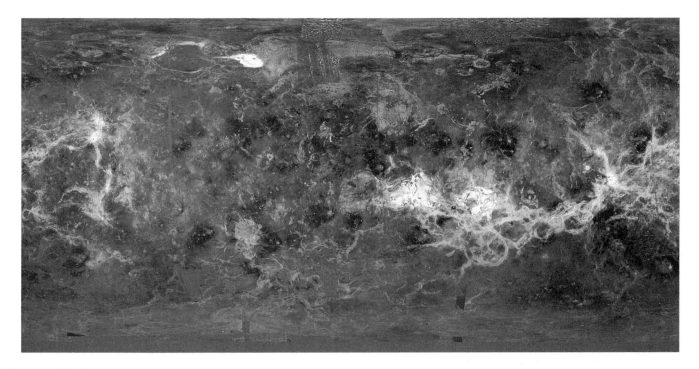

Figure 7-27 Global radar mosaic of Venus. From Ford and others (1993, Figure 2-2). Courtesy J. P. Ford, JPL (Ret.).

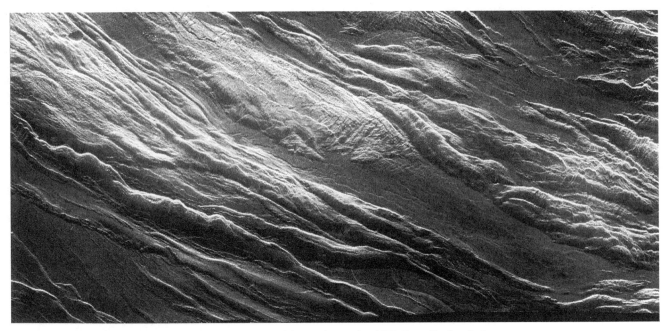

A. Image with look direction toward lower margin (east) and depression angle of 66°. From Ford and others (1993, Figure 8-4). Courtesy J. P. Ford, JPL (Ret.).

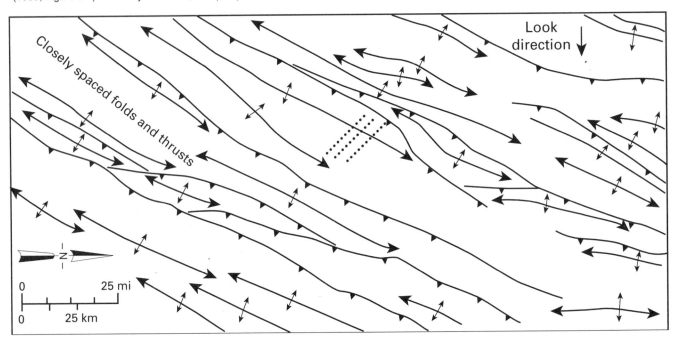

Look direction

Closely spaced folds and thrusts

0   25 mi

0   25 km

B. Interpretation map.

Figure 7-28 Magellan image and interpretation map of the fold belt Akna Montes, Venus. The steep depression angle exaggerates the asymmetric appearance of the folds with steep northwest flanks and gentle southeast flanks.

absence of the following characteristic plate-tectonic features:

a. spreading centers,
b. long transform faults with major strike-slip displacements,
c. collision zones marked by long narrow belts of folded mountains.

Despite the absence of plate tectonism, a wide variety of tectonic features occur on Venus and are described below.

Fold Belts Figure 7-28A shows the fold belt Akna Montes. The image is oriented with the look direction (and shadows) extending toward the lower margin, thereby avoiding topographic

A. Color multispectral image, San Pablo Bay, California.

B. AVIRIS hyperspectral image, Cuprite mining district, Nevada.

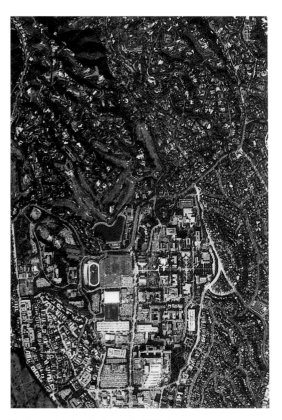

C. Normal color photograph, UCLA campus.

D. IR color photograph, UCLA campus.

Plate 1　Multispectral images and aerial photographs.
Image B, from Hook and others (1991). Courtesy S. J. Hook, Jet Propulsion
Laboratory (JPL). See "References Cited in Color Plate Captions," which fol-
lows the Glossary.

A. Normal color. Bands 1-2-3 = BGR.

B. IR color. Bands 2-3-4 = BGR.

C. All IR color. Bands 4-5-7 = BGR.

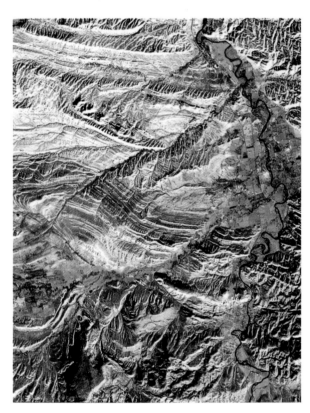

D. IR plus visible color. Bands 2-4-7 = BGR.

Plate 2   Color combinations of Landsat TM bands,
Thermopolis, Wyoming.

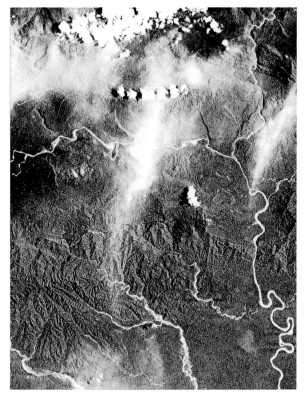

A. Normal color. Bands 1-2-3 = BGR.

B. IR color. Bands 2-3-4 = BGR.

C. All IR color. Bands 4-5-7 = BGR.

D. IR plus visible color. Bands 2-4-7 = BGR.

Plate 3   Color combinations of Landsat TM bands,
Mapia region of Irian Jaya, Indonesia.

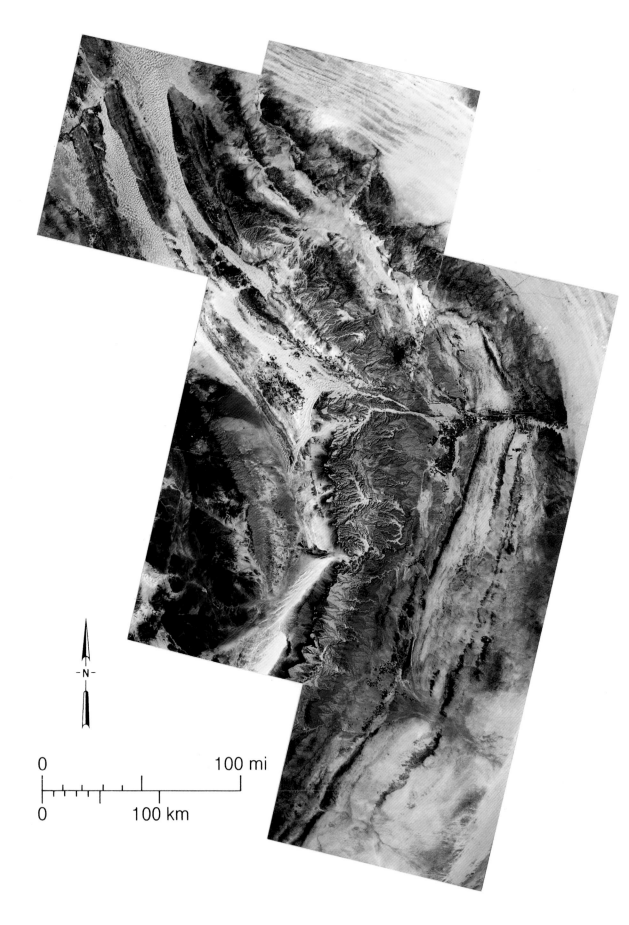

Plate 4 Analog mosaic of the Central Arabian Arch compiled from Landsat TM 2-4-7 prints.

Plate 5 Digital mosaic of western Pakistan compiled from
Landsat TM 2-3-4 digital data. Courtesy EROS Data Center.

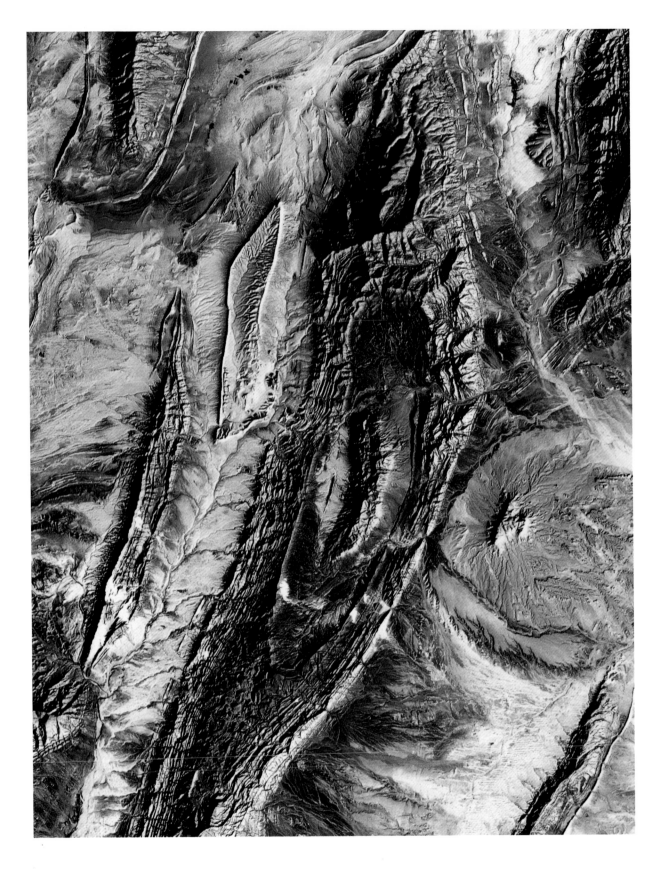

Plate 6 Landsat TM 2-4-7 subscene, Saharan Atlas
Mountains, Algeria.

A. Dry-season MSS image of Transvaal Basin, South Africa.

B. Wet-season MSS image of Transvaal Basin, South Africa.

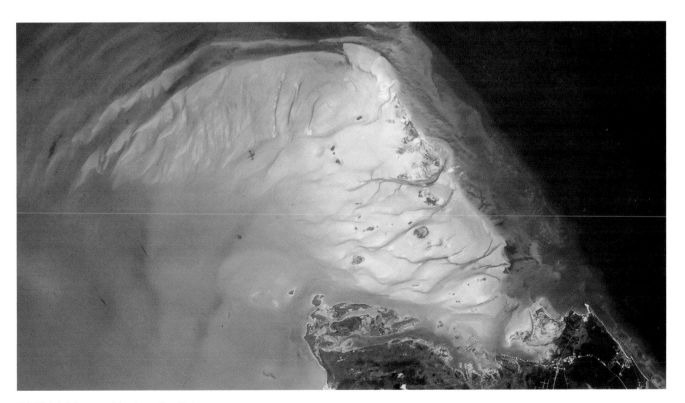

C. TM 1-2-3 image of Joulters Key, Bahamas.

Plate 7 Landsat images. Image C, from Harris and Kowalik (1994, Figure 37).

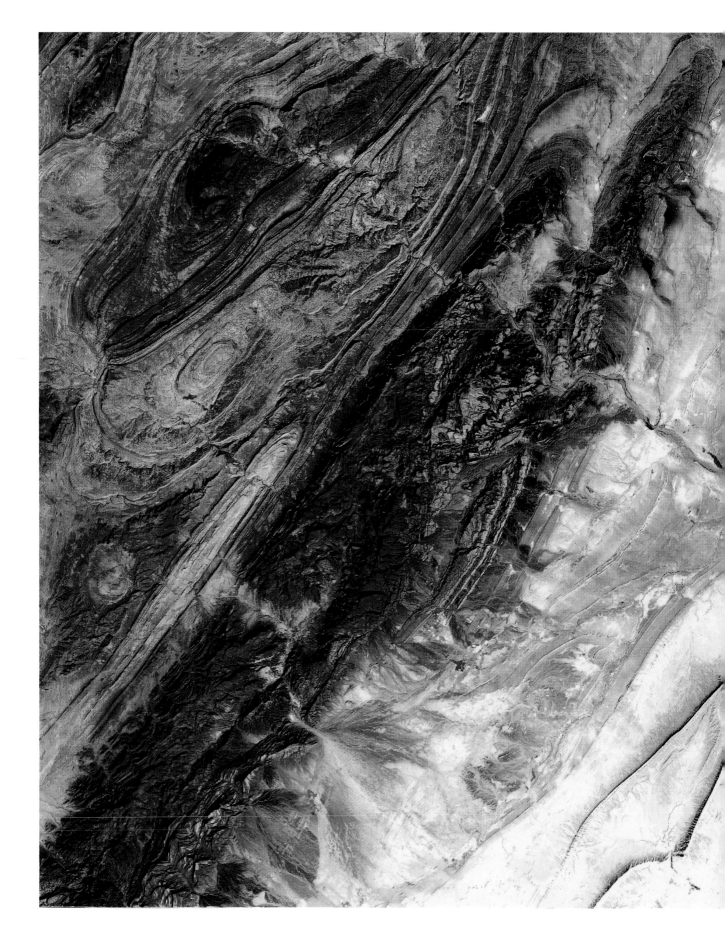

Plate 8 SPOT XS IR color image, Djebel Amour, Algeria.

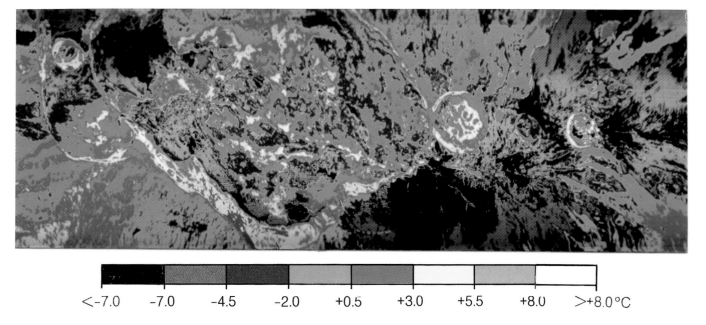

<-7.0	-7.0	-4.5	-2.0	+0.5	+3.0	+5.5	+8.0	>+8.0°C

A. Mauna Loa, Hawaii.

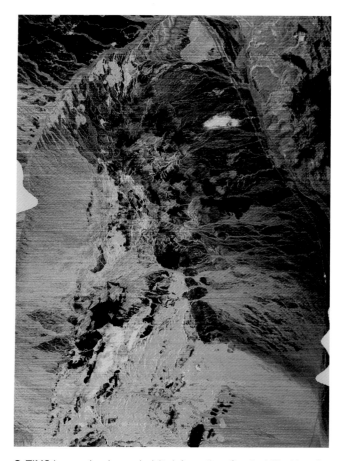

B. TIMS image showing kinetic temperature and emissivity, Cuprite Hills, Nevada.

C. TIMS image showing emissivity information, Cuprite Hills, Nevada.

Plate 9 Thermal IR images. Image A, courtesy Daedalus Enterprises, Inc. Images B and C, courtesy S. J. Hook, JPL.

A. Interferogram, Long Valley, California.

B. Topographic map from interferogram, Long Valley, California.

C. SIR-C color composite image, Mahantango Mountain, Pennsylvania. Bands L-C-X = RGB.

D. SIR-C color composite image, Mount Pinatubo, Philippines.

Plate 10 Radar images. Courtesy A. Freeman and E. O'Leary, JPL.

A. Before IHS. TM 1-2-3 image, Thermopolis, Wyoming.

B. After IHS. TM 1-2-3 image, Thermopolis, Wyoming.

C. Before IHS. TM 2-4-7 image, western Bolivia.

D. After IHS. TM 2-4-7 image, western Bolivia.

Plate 11   Images before and after enhancing saturation with IHS transformation.

A. TM 2-3-4 image (30-m resolution), Los Angeles, California.

B. Combined TM and SPOT pan images, Los Angeles, California.

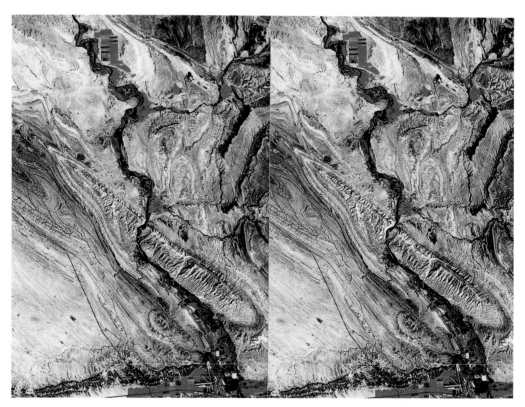

C. Synthetic stereo pair created by merging a TM 2-3-4 image
with digital elevation data, Sheep Mountain, Wyoming.

Plate 12 Digitally merged images.

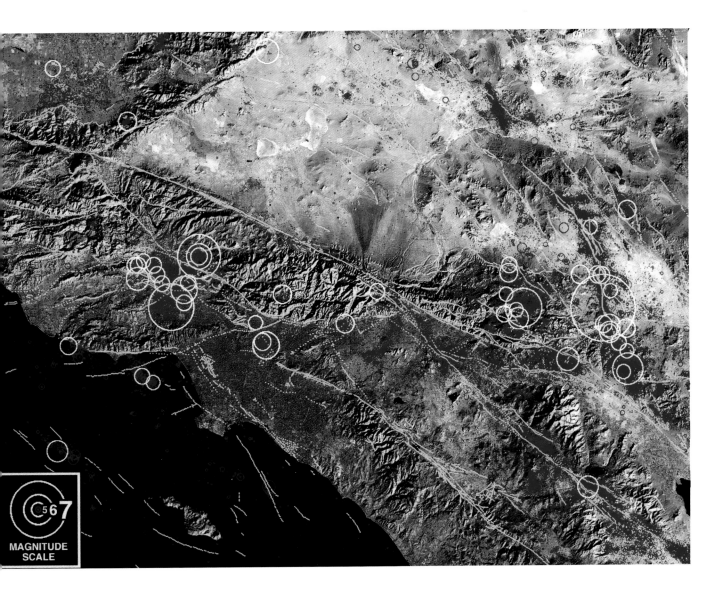

Plate 13 Digital mosaic of TM images of southern California. Band 7 is shown in red, band 4 in green, and the average of bands 1 and 2 in blue. Yellow lines are traces of active faults. White and red circles and dots are magnitude of earthquakes recorded from 1970 to 1995 by the Southern California Seismic Network. Courtesy R. E. Crippen, JPL.

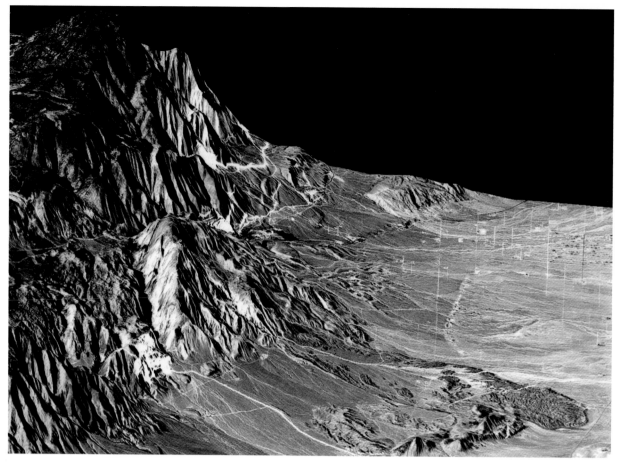

A. Composite of TM 2-4-7 image, SPOT pan image, and digital elevation data. Blackhawk landslide, California. Courtesy R. E. Crippen, JPL.

B. TM principal-component color image, Thermopolis, Wyoming. PC 2 = red, PC 3 = green, PC 4 = blue.

C. TM ratio color image, Thermopolis, Wyoming. Ratio 3/1 = red, 5/7 = green, 3/5 = blue.

Plate 14 Digitally processed images.

SYMBOL	CLASS	PERCENT
	Redbeds	8.4
	Sandstone	48.3
	Shale	18.9
	Agriculture	16.2
	Native vegetation	5.2
	Water and shadows	1.9
	Unclassified	1.1

A. Supervised-classification map and explanation.

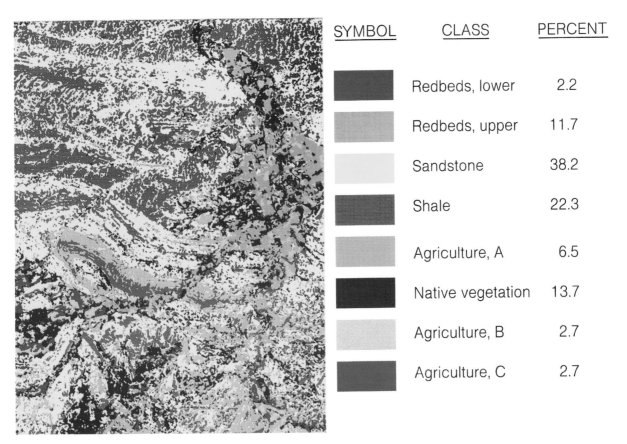

SYMBOL	CLASS	PERCENT
	Redbeds, lower	2.2
	Redbeds, upper	11.7
	Sandstone	38.2
	Shale	22.3
	Agriculture, A	6.5
	Native vegetation	13.7
	Agriculture, B	2.7
	Agriculture, C	2.7

B. Unsupervised-classification map and explanation.

Plate 15   Digital classifications of TM image, Thermopolis, Wyoming.

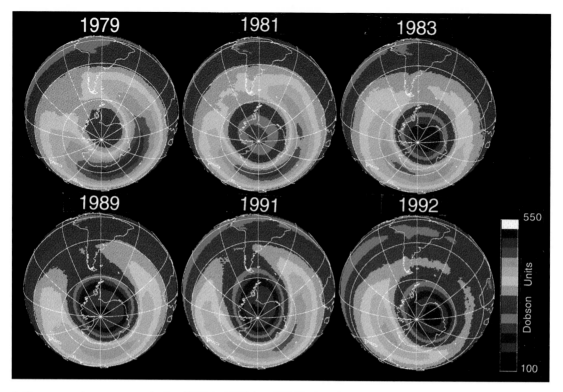

A. Mean annual ozone concentration (1979 to 1992) from TOMS data. The ozone hole is the expanding area of low values at the South Pole. After Schoeberl (1993, Figure 5). Courtesy E. Beach, Ozone Processing Team, NASA Goddard Space Flight Center.

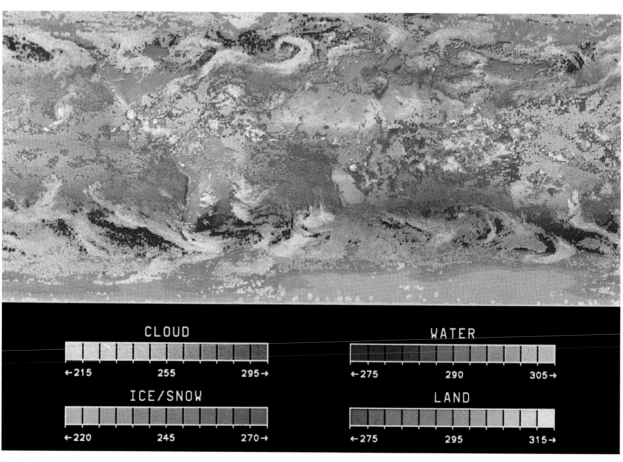

B. Global temperature map compiled from ISCCP data for July 4, 1983 at 1500 GMT. From Rossow (1993, Figure 1). Courtesy W. B. Rossow, NASA Goddard Institute for Space Studies.

Plate 16   Environmental remote sensing images.

inversion. The steep depression angle (66°) causes the folds to appear highly asymmetric with gentle east flanks and very steep west flanks. In reality the folds should be more symmetric in cross section. Figure 7-28B is an interpretation map of the fold belt. Thrust faults are interpreted where fold axes are truncated along their strike by through-going trends. The pattern of folds and thrust faults on the Venus image resembles that on the aircraft radar images of the Papuan Fold and Thrust Belt in Papua New Guinea that are described in Chapter 10.

Rift Valley Systems Extensive rift valley systems occur on both Venus and the earth. Figure 7-29A covers a portion of Devana Chasma, which is located on the crest of a major topographic rise. The extensive normal faults trend northward with steep scarps that face east (dark shadows) or west (bright highlights). The orientation of scarps indicates the downthrown side of each fault, which is shown by tick marks in the interpretation map (Figure 7-29B). The impact crater Balch is cut by faults and extended by about 10 km (Solomon and others, 1992, Figure 5). Devana Chasma has up to 6 km of vertical relief over distances of 200 km.

Complex Ridged Terrains (Tesserae) *Complex ridged terrains,* also called *Tesserae,* are plateau regions characterized by at least three intersecting sets of closely spaced linear features that may include folds, thrust faults, graben, and strike-slip faults. Aside from volcanic plains, complex ridged terrains are the most extensive tectonic feature on Venus. In the north portion of Figure 7-29A a complex ridged terrain is cut by the Devana Chasma rift system.

Volcanic Features

Volcanoes on the earth are classified according to the eruptive process, which ranges from quiet outpourings of lava that form shield volcanoes (such as Hawaii) to explosive events that form stratovolcanoes (such as Mount St. Helens). These end members have distinctive landforms and associated volcanic deposits that are readily recognized on images. The terrestrial classification is not applicable to Venus, where the explosive category is absent. The high atmospheric pressure of Venus prevents the violent expulsion of gases that causes explosive volcanism on the earth.

Classes of Volcanic Features Head and others (1992) interpreted over 1660 volcanic landforms and deposits from the Magellan images and developed a classification based on morphology. Figures 7-30 and 7-31 show images and maps of six major classes. Characteristics of the classes are summarized below, based on Head and others (1992).

Shields and shield fields (Figures 7-30A, 7-31A) are concentrations of shields that are typically between 100 to 200 km in diameter and lack associated flows and deposits.

Shields are small volcanoes (< 20 km in diameter) that range from low and shieldlike to domelike features commonly surmounted by a pit. Individual shield volcanoes are among the most abundant feature on Venus. Figures 7-30A and 7-31A show a shield field located within a corona, which is a volcanic feature described in a following section. Shield fields result from concentrations of numerous individual volcanic vents and indicate a larger magma source at depth, as shown by their association with coronae and large shields.

Large volcanoes (Figures 7-30B, 7-31B) are large centers (> 100 km in diameter) with radial lava flows and topographic highs. The example in Figure 7-30B and 7-31B is 520 km in diameter and lacks the concentric and radial fractures associated with many other large volcanoes. Bright (rough) and dark (smooth) lava flows radiate and widen with distance from the central vent region near the summit. The vent region is darker and more featureless than the flanks.

Intermediate volcanoes (not illustrated) are smaller versions of the "large volcanoes" class and include volcanic centers between 20 and 100 km in diameter characterized by radial lava flows, fracture patterns, and a topographic high. Head and others (1992) describe several categories of intermediate volcanoes.

Calderas (Figures 7-30C and 7-31C) are circular to elongated depressions surrounded by closely spaced concentric fractures. Figures 7-30C and 7-31C show the typical smooth, dark floor caused by a late filling of lava. Lava flows of intermediate roughness radiate from the rim of the caldera. Radial fractures are rare; the west-trending fractures are not related to the caldera. Calderas resemble meteorite impact craters, which are described in a following section. Calderas are distinguished from impact craters by the following:

1. Presence of radial lava flows
2. Presence of detailed concentric structure
3. Absence of hummocky raised rim
4. Absence of mottled or speckled ejecta deposits.

Calderas are also common on the earth in places where the central depression has subsided as a result of underlying magma being removed.

Coronae (Figures 7-30D and 7-31D) have the following characteristics:

1. A central domal feature with fractures and small shield volcanoes
2. Concentric fractures
3. Peripheral moats or troughs flooded with dark lava
4. Associated smaller domes, such as the feature on the west flank

A. Image with look direction toward lower margin and depression angle of 66°. From Ford and others (1993, Figure 8-1C). Courtesy J. P. Ford, JPL (Ret.).

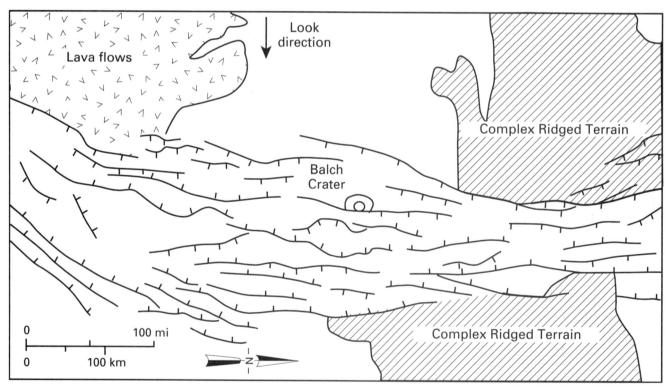

B. Interpretation map.

Figure 7-29 Magellan image and map of the rift valley and complex ridged terrain at Devana Chasma, Venus.

Coronae are distinguished from large volcanoes by the prevalence of concentric fracture patterns and the absence of radial patterns.

Arachnoids (Figures 7-30E and 7-31E) are characterized by the association of two patterns of fractures and ridges: a concentric pattern is concentrated around the center and is intersected by a more extensive radial pattern. The features are called arachnoids because they resemble multilegged spiders sitting on webs of interconnected bright fractures. Arachnoids are distinguished from coronae, which are larger and lack the strong radial patterns.

Novae (Figures 7-30F and 7-31F) are prominent starburst or stellate patterns of radiating fractures, lacking concentric patterns. The term *nova* is based on the starburst appearance. The stellate pattern is generally centered on a broad domal high (> 100 km in diameter) within radial flow patterns. The radial fractures form boundaries of downdropped graben blocks. Volcanic features, such as shields and flows, are rare within arachnoids as compared with coronae. The scarcity of these features, together with their domal topography, suggests that arachnoids are due to uplift and fracturing.

Lava flows (Figure 7-32) coalesce and overlap to form extensive volcanic plains that cover more than 80 percent of Venus. Figure 7-32 shows a spectacular flow sequence approximately 1000 km long and 300 km wide. Based on overlapping relationships, the flows are dated from 1 (oldest) to 4 (youngest). The flows erupted from faults caused by regional rifting. The flow field has an area of 180,000 km^2 and covers older fractured and faulted terrain. Fifty-three major flow fields have been identified and indicate voluminous outpourings of lava over relatively short periods. The flows have resurfaced major portions of Venus and covered older impact craters.

Lava channels (Figure 7-33) occur in lava plains, have sinuous patterns, and range from 0.5 to 2.0 km in width. Figure 7-33 is a prominent channel 2 km wide with channel cutoffs and local branching patterns that flowed down a regional topographic gradient. Most of the 50 lava channels are less than 400 km long, but a few are much longer.

Venus and Earth Volcanoes Compared Figure 7-34 compares the distribution of Venusian volcanoes and active volcanoes on the earth. On the earth (Figure 7-34B), volcanoes are concentrated in narrow belts that coincide with subduction zones where crustal plates are colliding. The Rim of Fire surrounding the Pacific Ocean formed in this manner. The aligned volcanoes in East Africa occur along a rift zone where plates are moving apart. This linear pattern is completely lacking on Venus (Figure 7-34A), where volcanoes are widely distributed. This difference indicates that plate movements do not occur on Venus, or they occur at rates too slow to influence volcanism.

One explanation is that the high surface temperature (450°C) makes the Venus crust too weak to behave as rigid plates.

Impact Features

All of the terrestrial planets have been bombarded by bolides, which have profoundly affected their surfaces. On the earth the extinction of dinosaurs at the end of the Cretaceous period is attributed to climatic changes caused by a giant impact. Most of the impact features on the earth, however, have been obliterated or obscured by erosion, deposition, and tectonism. Venus and the other planets have more complete cratering records that tell us a great deal about their histories and that of our solar system.

Schaber and others (1992) and Strom, Schaber, and Dawson (1994) interpreted Venus impact features from the Magellan images; their work is the basis for this summary. Crater morphologies on Venus are similar to those on other planets. There are, however, two major differences:

1. There are no Venus craters smaller than 1.5 km in diameter and only 135 less than 5 km. Indeed, comparison with other planets shows that Venus has a major "shortage" of craters less than 35 km in diameter. Friction in the dense atmosphere destroys the smaller bolides capable of forming the missing craters.
2. Unlike other planets, Venus craters are largely unaltered by later volcanism and tectonism; the implications are explained in the later section "Resurfacing Event."

Figure 7-35 shows images and Figure 7-36 shows maps of representative craters. Radar images of craters show both morphology (expressed as highlights and shadows) and materials (expressed as smooth to rough signatures).

Morphology and Materials of Impact Craters Figure 7-37 shows the morphology and materials of impact craters. Table 7-4 describes the materials and their radar signatures.

Lava plains are the older volcanic terrain that has been impacted by the crater-forming bolides. Lava plains generally have dark radar signatures that indicate smooth surfaces. Portions of the plains are faulted and fractured to form scarps that have bright linear signatures. In Figures 7-35A,D the plains were fractured prior to the impacts. In Figure 7-34F fracturing occurred after the impacts.

The **crater** is the circular to irregularly shaped depression formed by the impact and the associated explosion. A range of features occurs within craters, as shown in the map and cross section of Figure 7-37. The **crater floor** is the level plain within the crater that has signatures ranging from dark to bright. Dark floors (Figure 7-35A) consist of impact melt, or later volcanic fill, with a smooth surface. Bright floors (Figure 7-35E) consist of rough shocked rock fragments.

(text continues on page 252)

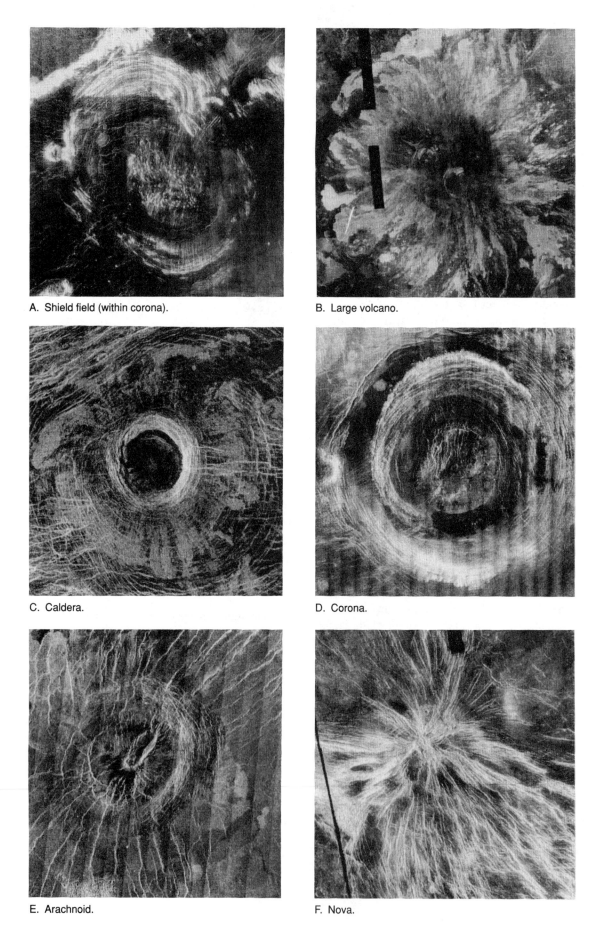

A. Shield field (within corona).

B. Large volcano.

C. Caldera.

D. Corona.

E. Arachnoid.

F. Nova.

Figure 7-30 Magellan images of volcanic features on Venus. From Head and others (1992). Courtesy J. W. Head, Brown University.

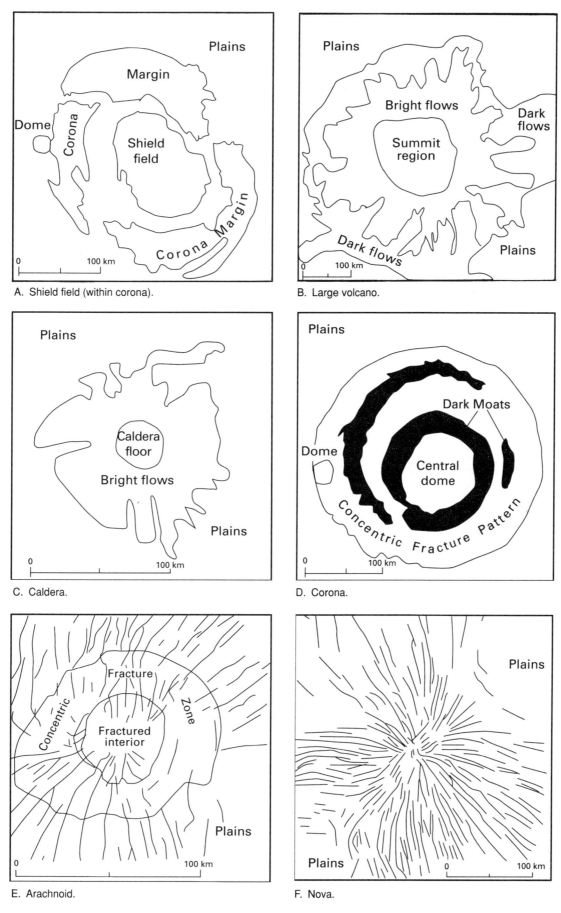

A. Shield field (within corona).

B. Large volcano.

C. Caldera.

D. Corona.

E. Arachnoid.

F. Nova.

Figure 7-31 Maps of volcanic features shown in Figure 7-30. North is toward upper margins. From Head and others (1992).

A. Image with look direction toward lower margin.

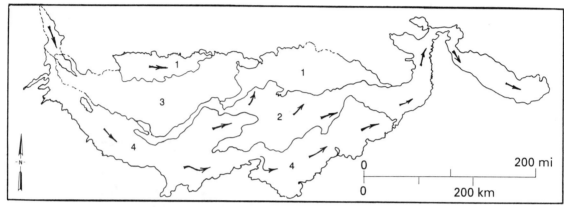

B. Map showing relative ages of flows (1 = oldest) based on overlapping relationships. Arrows show direction of flow.

Figure 7-32 Magellan image and map of extensive lava flows erupted from regional rifts on Venus. From Head and others (1992 , Figure 9a). Courtesy J. W. Head, Brown University.

Table 7-4 Characteristics of materials associated with craters

Material	*Description*	*Radar signature*
Lava plain	Lava flows that have been impacted by bolides. Many plains are flat and featureless; others are faulted and folded to various degrees.	Dark to intermediate, depending upon roughness. Bright linear features are fault scarps.
Rim	Ejecta deposited up to three crater radii from the center. Coarse, hummocky material near crater, becoming somewhat finer outward.	Bright near rim, becoming intermediate outward, with bright patches.
Wall	Rough, terraced deposits on steep inner slopes. Slump blocks and talus are common.	Bright to intermediate. Annular zone surrounding crater floor.
Floor	Level to smooth interior plains ranging from smooth to hummocky surfaces. Materials include impact melt, shocked rock, fallback, or volcanic fill.	Bright to dark. Circular to irregular outline.
Peaks and rings	Peaks and ridges within the crater that rise above the floor materials. Uplifted shocked, crushed, and sheared rock.	Very bright with characteristic morphology.
Outflow	Fluidized material formed by impact that resembles a lava flow. At some larger craters, outflows extend for hundreds of kilometers.	Bright to intermediate with flowlike outline.

Source: Compiled from Schaber and others (1992).

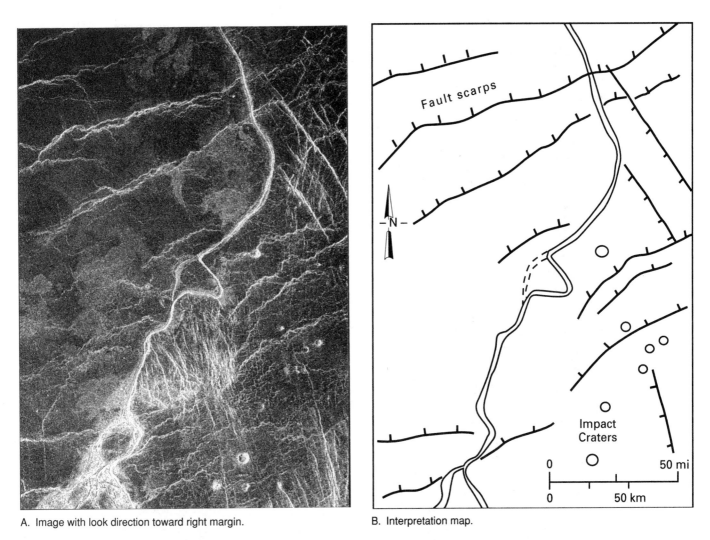

A. Image with look direction toward right margin.

B. Interpretation map.

Figure 7-33 Magellan image and map of a lava channel on Venus. From Head and others (1992). Courtesy J. W. Head, Brown University

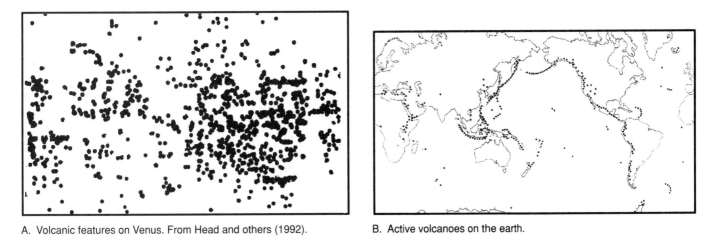

A. Volcanic features on Venus. From Head and others (1992).

B. Active volcanoes on the earth.

Figure 7-34 Distribution of volcanoes on Venus and the earth.

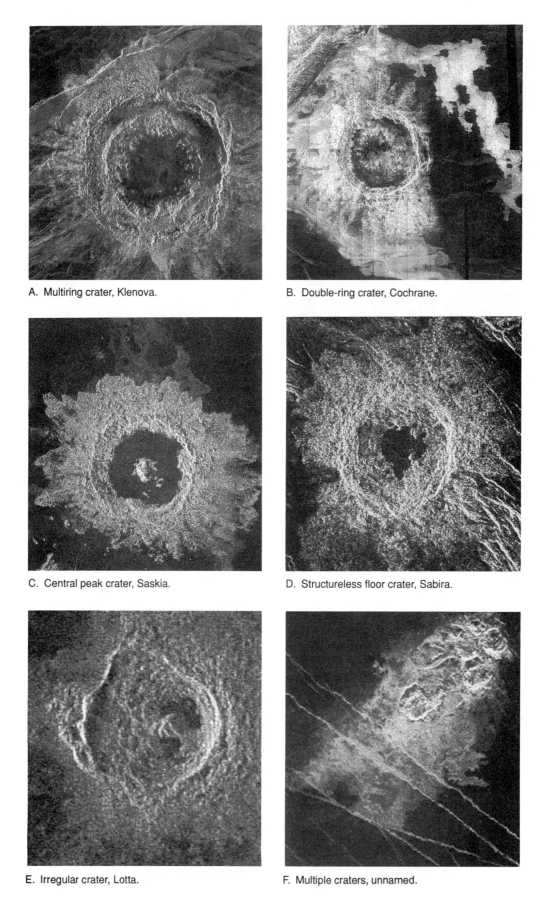

A. Multiring crater, Klenova.

B. Double-ring crater, Cochrane.

C. Central peak crater, Saskia.

D. Structureless floor crater, Sabira.

E. Irregular crater, Lotta.

F. Multiple craters, unnamed.

Figure 7-35 Magellan images of impact craters. From Schaber and others (1992).
Courtesy G. G. Schaber, U.S. Geological Survey.

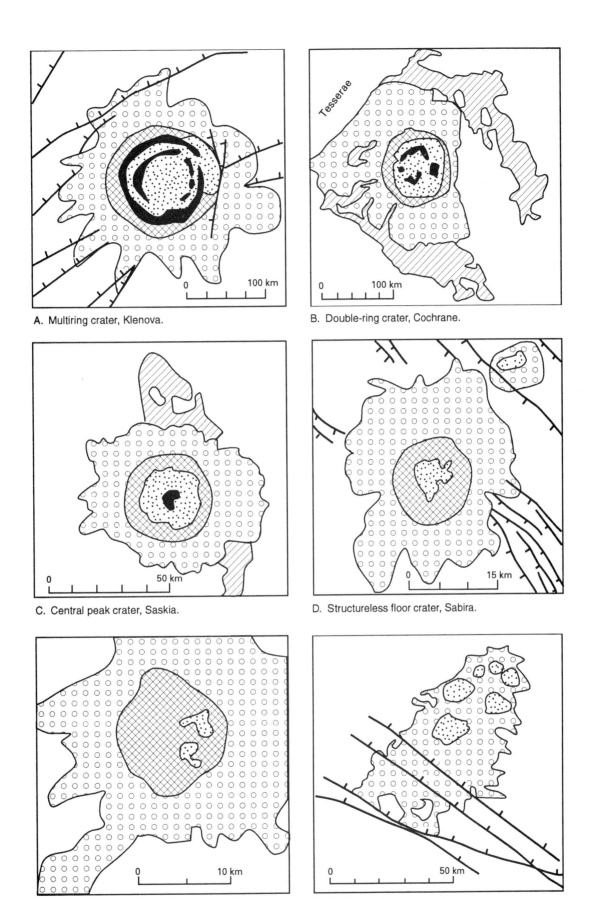

A. Multiring crater, Klenova.

B. Double-ring crater, Cochrane.

Tesserae

C. Central peak crater, Saskia.

D. Structureless floor crater, Sabira.

E. Irregular crater, Lotta.

F. Multiple craters, unnamed.

Figure 7-36 Interpretation maps of impact craters shown in Figure 7-35.
Patterns are identified in Figure 7-37. From Schaber and others (1992).
Courtesy G. G. Schaber, U.S. Geological Survey.

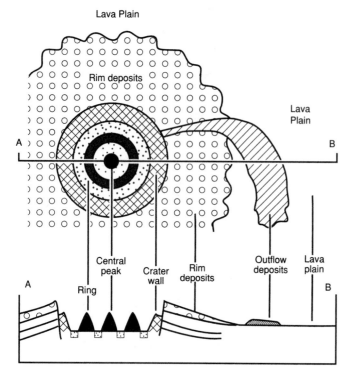

Figure 7-37 Generalized map and profile of impact crater showing materials and morphology.

Crater rings are concentric ridges and aligned hills within the crater (Figure 7-34A,B). **Central peaks** are isolated hills at the center of a crater (Figure 7-35C). Rings and peaks consist of shocked and crushed rock that was uplifted from beneath the floor in the aftermath of the impact explosion. Rings and peaks are embayed at their bases by crater floor materials. The radar signatures of rings and peaks are dominated by topographically controlled highlights and shadows.

The **wall** is the steep circular scarp that surrounds the crater and consists of slumped and downfaulted blocks and talus with bright signatures.

The **rim** is the area beyond the crater wall that slopes gently away from the crater (Figure 7-37). Rims are mantled by material ejected from the crater that may be deposited as far as three crater radii from the center. In the images of Figure 7-35 the bright signatures of the rough rim materials contrast with the dark lava plains upon which the ejecta were deposited.

Outflows originated as impact melt, which is a hot, fluidized mixture of gases, shock-melted rock, and crushed rock. The crater Cochrane (Figures 7-35B, 7-36B) has two prominent outflows that originated at the crater and flowed outward through the rim deposits onto the lava plain. The outflows have lobate outlines similar to lava flows. The

bright signatures indicate a surface roughness greater than the Magellan rough criterion of 3.2 cm (at the nominal depression angle of 65°).

Classes of Impact Craters Schaber and others (1992) assigned the impact craters to six classes, based on their morphology. Figures 7-35 and 7-36 show images and maps of typical craters for the classes, which are described below.

Multiring craters (Figures 7-35A, 7-36A) are characterized by several partial or complete rings within the central impact basin. There are only six of these craters, but they are the largest impact features on Venus and include all four craters greater than 140 km in diameter. At Klenova (Figure 7-35A) the outer ring is 95 km in diameter and is more complete and distinct than the inner ring. Dark material covers the central crater and the area between the wall and outer ring in the northeast sector. This smooth material may be lava or impact melt. The remainder of the outer floor is shocked material with bright signatures. Klenova clearly postdates the northeast-trending fractures.

Double-ring craters (Figures 7-35B, 7-36B) consist of an outer crater rim, or ring, and a single inner ring formed by a circular array of peaks and hills rising above a relatively flat floor. The origin of double-ring and multiring craters is controversial; Schaber and others (1992) review the various theories. Cochrane (Figure 7-35B) has well-developed outflow deposits. The outflows are connected to the crater by narrow channels cut through the rim deposits. This relationship shows the outflows are younger than the rim deposits.

Central peak craters (Figures 7-35C, 7-36C) have central peaks or mounds that rise above a level crater floor. Saskia (Figure 7-35C) shows the additional characteristics of terraced walls and dark crater floor.

Featureless floor craters (Figures 7-35D, 7-36D) have flat floors that lack peaks, rings, or other structures. Sabira has a relatively small dark floor surrounded by a broad zone of terraced wall deposits. Sabira postdates the northwest-trending fractures.

Irregular rim craters (Figures 7-35E, 7-36E) are characterized by scalloped and irregular rims. Many of these craters include several basins within their complex floors. These are among the smallest craters, but they comprise 29 percent of the population.

Multiple craters (Figures 7-35F, 7-36F) consist of two or more individual craters within a small area. Their rims may be separate or overlap.

The first four crater classes have distinct single, circular impact basins and were formed by high-velocity bolides of sufficient size and competence to survive passage through the atmosphere. In contrast, the "irregular" and "multiple crater" classes resulted from simultaneous impacts of tight clusters of asteroids that were disaggregated and slowed in transit through the atmosphere.

Splotches Splotches (not illustrated) are diffuse circular areas with no central crater and whose radar signatures are brighter or darker than the surrounding terrain. There are over 400 splotches on Venus that range from a few tens of kilometers to 200 km in diameter. As small incoming meteorites are destroyed in the atmosphere, the resulting shock waves interact with surface material in two ways:

1. They fracture material into a rough surface with a bright signature.
2. They melt and/or pulverize material finely enough to produce a smooth surface with a dark signature; this dark category comprises 90 percent of the splotches.

Splotches represent some of the "missing" class of small craters, whose parent meteors were destroyed within the Venusian atmosphere.

Resurfacing Event Strom, Schaber, and Dawson (1994) report that the impact crater record on Venus is unique among the terrestrial planets. Fully 84 percent of the craters are in pristine condition. Only 12 percent are fractured and 4 percent are embayed by lava, which is remarkable because intense volcanism and tectonism have affected the entire planet. Comparing the statistical distribution of Venusian craters with other planets strongly indicates that Venus had a global resurfacing event 300 million years ago that wiped out all older craters. The resurfacing event ended abruptly (on a geologic time scale) and was followed by a dramatic reduction in volcanism and tectonism. The present crater population appears to have accumulated since then and remains largely intact.

Availability of Data

Researchers interested in accessing Magellan data may contact

National Space Science Data Center
Goddard Space Flight Center
Mail Code 933.4
Greenbelt, MD 20771
Telephone: 301-286-6695
Fax: 301-286-4952

Summary

The Magellan mission was a major technologic and scientific accomplishment. The repeated high-quality images clearly show previously unknown features of the planet that most closely corresponds to the earth. Scientists can now make objective comparisons of Venus and the earth. Some major conclusions from Magellan images include the following:

1. The surface of Venus consists of lava flows and ejecta from impacts.
2. The random distribution of volcanic features shows that plate tectonism does not occur on Venus, at least at the rates that occur on earth.
3. Tectonism does occur on Venus, and many of the structures (fold belts and rift systems) resemble their counterparts on earth.
4. The essentially unaltered nature of the impact craters shows that Venus had a major resurfacing event 300 million years ago that has been followed by reduced volcanism and tectonism.

Additional interpretations of the Magellan images will modify and expand these initial observations.

COMMENTS

The short-lived Seasat mission in 1978 demonstrated the feasibility and utility of acquiring radar images from satellites. The sequence of manned SIR missions (A, B, and C) acquired images of increasing versatility and complexity. SIR-A images of Indonesia were interpreted to produce geologic maps of inaccessible regions with inclement weather that hampers acquisition of images in the visible and reflected IR. The Indonesian project also demonstrated that geologic mapping is feasible in tropical rain forests, although radar energy does not penetrate the tree canopy to any significant extent. SIR-A images in North African deserts also demonstrated that radar energy can penetrate up to 1.5 m of hyperarid sand and record images of the underlying bedrock surface. SIR-C acquired multiple images at different wavelengths and polarizations to investigate details of the complex interactions between microwave energy and the terrain. ERS-1 (European Space Agency) and JERS-1 (Japan) acquire images at steep depression angles that are optimal for oceanographic observations such as sea state, current patterns, and internal waves. The Magellan mission provided the first detailed images of Venus, despite the dense cover of CO_2 clouds. Geologists have interpreted the images to show significant differences as well as similarities between Venus and the earth. Radarsat (Canada) acquires images at a wide range of scales, spatial resolutions, and depression angles. The images will be useful for a variety of applications.

QUESTIONS

1. It was noted that Venus has much more complete radar coverage than the earth. Provide justification for a mission to acquire complete radar coverage of the earth.
2. Assume funds are available and that you are responsible for designing the earth-imaging radar system, which should consist of separate imagers for land and for water. For each imager give the following design characteristics: wavelength,

depression angle, polarization, and ground resolution cell. Explain your choices.

REFERENCES

Berlin, G. L., M. A. Tarabzouni, A. H. Al-Naser, K. M. Sheikho, and R. W. Larson, 1986, SIR-B subsurface imaging of a sand-buried landscape—Al Labbah Plateau, Saudi Arabia: IEEE Transactions on Geoscience and Remote Sensing, v. GE-24, p. 595–602.

Blom, R. G., R. E. Crippen, and C. Elachi, 1984, Detection of subsurface features in Seasat radar images of Means Valley, Mojave Desert, California: Geology, v. 12, p. 346–349.

Cimino, J. B. and C. Elachi, eds., 1982, Shuttle imaging radar–A (SIR-A) experiment: Jet Propulsion Laboratory Publication 82-77, Pasadena, CA.

Cimino, J. B., B. Holt, and A. H. Richardson, 1988, The Shuttle imaging radar–B (SIR-B) experiment report: Jet Propulsion Laboratory Report 88-10, Pasadena, CA.

European Space Agency, 1993, ERS user handbook: European Space Agency SP-1148, Revision 1, Noordwick, Netherlands.

Evans, D. L. and others, 1994, Earth from sky: Scientific American, v. 271, p. 70–75.

Ford, J. P., ed., 1988, Advances in Shuttle imaging radar–B research: Taylor and Francis Group, Bristol, PA.

Ford, J. P., R. G. Blom, M. L. Bryan, M. I. Daily, T. H. Dixon, C. Elachi, and E. C. Xenos, 1980, Seasat views North America, the Caribbean, and western Europe with imaging radar: Jet Propulsion Laboratory Publication 80-67, Pasadena, CA.

Ford, J. P., J. B. Cimino, and C. Elachi, 1983, Space Shuttle Columbia views the world with imaging radar—the SIR-A experiment: Jet Propulsion Laboratory Publication 82-95, Pasadena, CA.

Ford, J. P., J. B. Cimino, B. Holt, and M. R. Ruzek, 1986, Shuttle imaging radar views the earth from Challenger—the SIR-B experiment: Jet Propulsion Laboratory Publication 86-10, Pasadena, CA.

Ford, J. P. and F. F. Sabins, 1986, Satellite radars for geologic mapping in tropical regions: Environmental Research Institute of Michigan Fifth Thematic Conference, Remote Sensing for Geology, Proceedings, Ann Arbor, MI.

Ford, J. P. and others, 1989, Spaceborne radar observations—A guide for Magellan radar image analysis: Jet Propulsion Laboratory Publication 89-41, Pasadena, CA.

Ford, J. P. and others, 1993, Guide to Magellan image interpretation: Jet Propulsion Laboratory Publication 93-24, Pasadena, CA.

Head, J. W. and others, 1992, Venus volcanism—classification of volcanic features and structures, associations, and global distribution from Magellan data: Journal of Geophysical Research, v. 97, p. 13,153–13,197.

Lalonde, L., G. Posehn, and F. F. Sabins, 1993, Comparison and evaluation of ERS-1 and airborne radar images: Environmental Research Institute of Michigan, Ninth Thematic Conference on Geologic Remote Sensing, Proceedings, p. 21–33, Ann Arbor, MI.

McCauley, J. F. and others, 1982, Subsurface valleys and geoarcheology of the eastern Sahara revealed by Shuttle radar: Science, v. 318, p. 1004–1020.

Nazarenko, D., G. Staples, and C. Aspden, 1996, Radarsat—first images: Photogrammetric Engineering and Remote Sensing, v. 62, p. 143–146.

Radarsat International, 1995, Radarsat illuminated—your guide to products and services: Radarsat International, Richmond, B.C., Canada.

Sabins, F. F., 1983, Geologic interpretation of Space Shuttle radar images of Indonesia: American Association of Petroleum Geologists Bulletin, v. 67, p. 2076–2099.

Sabins, F. F., R. Blom, and C. Elachi, 1980, Seasat radar images of San Andreas fault, California: American Association of Petroleum Geologists Bulletin, v. 64, p. 619–628.

Schaber, G. G., J. F. McCauley, C. S. Breed, and G. R. Olhoeft, 1986, Shuttle imaging radar—Physical controls on signal penetration and subsurface scattering in the eastern Sahara: IEEE Transactions on Geoscience and Remote Sensing, v. GE-24, p. 603–623.

Schaber, G. G. and others, 1992, Geology and distribution of impact craters on Venus—what are they telling us?: Journal of Geophysical Research, v. 97, p. 13,257–13,301.

Solomon, S. C. and others, 1992, Venus tectonics—an overview of Magellan observations: Journal of Geophysical Research, v. 97, p. 13,199–13,255.

Strom, R. G., G. G. Schaber, and D. D. Dawson, 1994, The global resurfacing of Venus: Journal of Geophysical Research, v. 99, p. 10,899–10,926.

ADDITIONAL READING

Elachi, C., 1987, Spaceborne radar remote sensing—applications and techniques: IEEE Press, New York, NY.

Evans, D. and A. Freeman, 1995, Future directions for synthetic aperture radar: Proceedings, Land Satellite Information in the Next Decade, American Society for Photogrammetry and Remote Sensing, Falls Church, VA.

Ivanov, M. A. and J. W. Head, 1996, Tessera terrain on Venus—a survey of the global distribution, characteristics, and relation to surrounding units from Magellan data: Journal of Geophysical Research, v. 1012, p. 14,861–14,908.

Nishidai, T., 1993, Early results from Fuyo-1, Japan's Earth Resources Satellite (JERS-1): International Journal of Remote Sensing, v. 14, p. 1825–1833.

Raney, R. K. and others, 1991, Radarsat, IEEE Proceedings, v. 79, p. 839–849.

Sabins, F. F. and J. P. Ford, 1985, Space Shuttle radar images of Indonesia: Indonesian Petroleum Association, Proceedings, 14th annual convention, p. 470–476, Jakarta, Indonesia.

Saunders, R. S. and G. H. Pettengill, 1991, Magellan—mission summary: Science, v. 252, p. 247–249.

Singhroy, V. and others, 1993, Radarsat and radar geology in Canada: Canadian Journal of Remote Sensing, v. 19, p. 338–351.

DIGITAL IMAGE PROCESSING

Most remote sensing images are recorded in digital form and then processed by computers to produce images for interpreters to study. Less than a decade ago digital image processing was essentially confined to dedicated facilities with expensive mainframe computers operated by staffs of specialists. Today images can be processed with inexpensive desktop computers and software by users with average computer skills.

Digital processing did not originate with remote sensing and is not restricted to these data. Many image-processing techniques were developed in the medical field to process X-ray images and images from sophisticated body-scanning devices. For remote sensing, the initial impetus was the program of unmanned planetary satellites in the 1960s that *telemetered* (transmitted) images to ground receiving stations. The low quality of the images required the development of processing techniques to make the images useful. Another impetus was the Landsat program, which began in 1972 and provided the first sets of worldwide imagery in digital format. A third impetus was and remains the continued development of faster, more powerful, and less expensive computer systems for image processing.

This chapter describes and illustrates the major categories of image processing. Jensen (1986) and Schowengerdt (1983) describe in detail the methods and the mathematical transformations.

STRUCTURE OF DIGITAL IMAGES

One can think of any image as consisting of small, equal areas, or *picture elements*, arranged in regular rows and columns, called a *raster array*. The position of any picture element, or *pixel*, is determined on an *xy* coordinate system. Each pixel also has a numerical value, called a *digital number* (DN), that records the intensity of electromagnetic energy measured for the ground resolution cell represented by that pixel. Digital numbers range from zero to some higher number on a gray scale. This system records an image in strictly numerical terms on a three-coordinate system in which the *x*- and *y*-values locate each pixel and *z* gives the DN, which is displayed as a gray-scale intensity value. Scanner systems record images directly in a digital format where each ground resolution cell is represented by a pixel. Analog images, such as photographs and maps, may be converted into digital format by a process known as *digitization*. In this chapter digital image processing is largely illustrated using Landsat examples because the data are readily available and the images are familiar. The digital processes, however, are equally applicable to all forms of digital image data.

Landsat TM Images

Landsat TM data are available as *computer-compatible tapes* (CCTs) or as CD-ROMs, which can be read and processed by computers. Spectral bands and other characteristics of TM images are described in Chapter 3. Figure 8-1 illustrates the format of the digital data. The cross-track scanner has a ground resolution cell of 30 by 30 m. An oscillating scan mirror sweeps the cell alternately east and west across the terrain to produce scan lines oriented at right angles to the satellite orbit path. Scanning is continuous along the orbit; the data are subdivided into scenes consisting of 5667 scan lines that are 185 km long in the scan direction and 30 m wide in the orbit direction. The analog signal is sampled at intervals of 30 m to produce 6167 pixels (measuring 30 by 30 m) per scan line. Each TM image band consists of 34.9×10^6 pixels (Figure 8-1). The

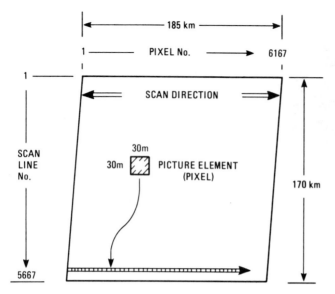

5667 scan lines × 6167 pixels = 34.9 × 10⁶ pixels per band
34.9 × 10⁶ pixels × 7 bands = 244.3 × 10⁶ pixels per scene

Figure 8-1 Arrangement of scan lines and pixels in a Landsat TM image.

seven bands have a total of 244.3 × 10⁶ pixels. Each of the six TM visible and reflected IR bands employs an array of 16 detectors; each sweep of the scan mirror records 16 lines of data. The thermal IR band 6 employs four detectors with 120-by-120-m ground resolution cells; these are resampled to 30-by-30-m pixels. (*Note:* in CCTs produced by EOSAT through 1989, pixel size was 28.5 by 28.5 m.) TM data are recorded on an 8-bit scale. Bernstein and others (1984) have analyzed the performance of the TM system; NASA (1983) gives details of the CCT format for TM data.

The structure of a TM digital image is illustrated in Figure 8-2, which is a greatly enlarged portion of band 4 from the Thermopolis, Wyoming, subscene that was described in Chapter 3. Figure 8-2A shows an array of 66 by 66 pixels. Bright pixels have high DNs, and dark pixels have lower DNs. Figure 8-2B is a map that shows the location for the printout of DNs in Figure 8-2C. The correlation of the image gray scale to the 8-bit DN scale is shown in Figure 8-2D, where a DN of 0 is displayed as black and 255 as white. A *histogram* (Figure 8-2D) is a useful way to display the statistics of an image. The vertical scale records the number of pixels associated with each DN value, shown on horizontal scale.

Referring again to the printout (Figure 8-2C), note the low values associated with the Wind River and the high values of the adjacent vegetation. These values are consistent with the strong absorption by water and strong reflection by vegetation of the IR energy recorded by TM band 4. In the river, note that the lowest digital numbers (DN = 8 to 17) correspond to the center of the stream; pixels along the stream margin have higher values (DN = 20 to 80). The 30-by-30-m ground resolution cells in the center of the river are wholly occupied by water, whereas marginal cells are occupied partly by water and partly by vegetation, resulting in a DN intermediate between water and vegetation. The mixed pixels form narrow gray bands along the river in Figure 8-2A. The TM format is typical of all multispectral digital data sets.

CCTs of Landsat images for the world are available from EOSAT at a cost of $4400 for a TM scene. CCTs are available as exabyte tape cartridges or as CD-ROMs. CCTs of TM images acquired from 1982 to 1986 are available from the EROS Data Center (EDC) for $650; these images are recorded as four quadrants that must be digitally mosaicked in order to process the entire scene. CCTs for MSS images are available from the EDC. TM images are indexed with the path-and-row system described in Chapter 3.

Digitization Procedure

Digitization is the process of converting an analog image into a raster array of pixels. Maps and other information may also be digitized. The digitized information is stored on a magnetic medium for subsequent computer processing. Digitizing systems belong to several categories: drum, flatbed, and linear array.

A typical *drum digitizing system* (Figure 8-3) consists of a rotating cylinder with a fixed shaft and a movable carriage that holds a light source and a detector similar to a photographic light meter. The positive or negative transparency is placed over the opening in the drum, and the carriage is positioned at a corner of the film. As the drum rotates, the detector measures intensity variations of the transmitted light caused by variations in film density. Each revolution of the drum records one scan line of data across the film. At the end of each revolution, the encoder signals the drive mechanism, which advances the carriage by the width of one scan line to begin the next revolution.

The width of the scan line is determined by the optical aperture of the detector, which typically is a square opening measuring 50 μm on a side. The analog electrical signal of the detector, which varies with film density changes, is sampled at 50-μm intervals along the scan line, digitized, and recorded as a DN value on magnetic tape. Each DN is recorded as a series of *bits*, which form an ordered sequence of ones and zeroes. Each bit represents an exponent of the base 2. The widely used 8-bit series represents 256 values on the gray scale ($2^8 = 256$ levels) with 0 for black and 255 for white. A group of 8 bits is called a *byte*. The film image is thus converted into a raster array of pixels that are referenced by scan-line number (y-coordinate) and pixel count along each line (x-coordinate). A typical black-and-white aerial photograph measuring 23 by 23 cm (9 by 9-in.) that is digitized with a 50-μm sampling interval is converted into 21.2 million pixels.

Pixel Number

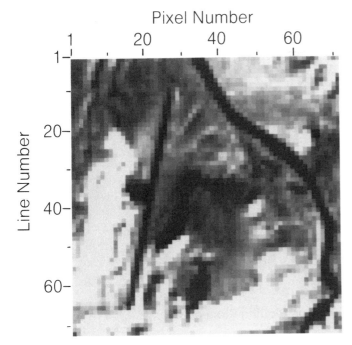

A. Gray-scale image.

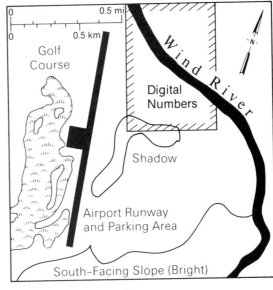

B. Location map.

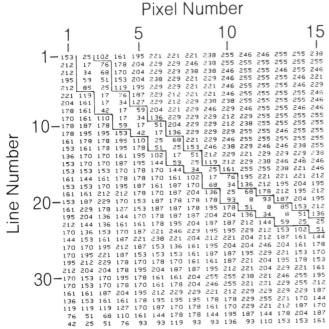

C. Array of DNs for the area shown in the location map.

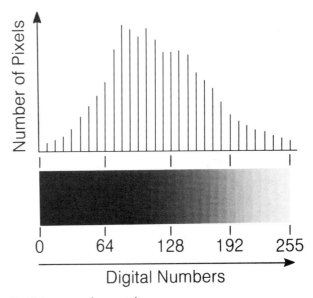

D. Histogram and gray scale.

Figure 8-2 Digital structure of a Landsat TM band 4 image. The area is a portion of the Thermopolis, Wyoming, subscene shown in Chapter 3.

The intensity of the digitizer light source is calibrated at the beginning of each scan line to compensate for any fluctuations of the light source. These drum digitizers are fast, efficient, and relatively inexpensive. Color images may be digitized into the values of the component primary colors on three passes through the system using blue, green, and red filters over the detector. A digital record is produced for each primary color, or wavelength band.

In *flatbed digitizers*, the film is mounted on a flat holder that moves in the *x* and *y* directions between a fixed light source

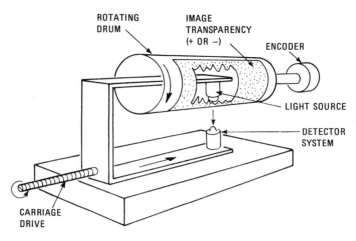

Figure 8-3 System for digitizing an analog image,
such as a photograph. From Bryant (1974, Figure 2).

and a detector. These devices are usually slower and more expensive than drum digitizers, but they are also more precise. Some systems employ reflected light in order to digitize opaque paper prints. The resulting voltage signal for each line is digitized and recorded.

Linear-array digitizers are similar to the along-track scanners described in Chapter 1. An optical system projects the image of the original photograph onto the focal plane of the system. A mechanical transport system moves a linear array of detectors across the focal plane to record the original photograph in digital form.

The digitized image is recorded magnetically and can be read into a computer for various processing operations. The processed data are then displayed as an image on a viewing device or plotted onto film as described in the following section. Inexpensive small-format digitizers are available as accessories for personal computers.

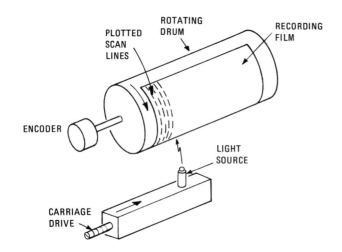

Figure 8-4 System for plotting an image from digital data. From Bryant (1974, Figure 3).

Image Plotting Procedure

Digital image data are plotted as images by *film writers* (Figure 8-4). Recording film is mounted on a rotating drum. With each rotation a scan line is exposed on the film by a light or laser, the intensity of which is modulated by the DNs of the pixels. On completion of each scan line, the carriage advances the light source to commence the next line. The exposed film is developed to produce a transparency from which one can make contact prints and enlargements. Modern plotters record data directly onto color film; three passes are made, each with a red, green, or blue filter over the light source. Advanced plotters mix the red, green, and blue components for each pixel and plot the color image in a single pass.

Color printers are available as accessories for personal computers and produce acceptable page-size prints. In order to produce high-quality prints of entire TM images at intermediate scales (1:250,000 or larger), film transparencies must be plotted and photographically enlarged.

IMAGE-PROCESSING OVERVIEW

Image-processing methods are grouped into three functional categories; these are defined next, together with lists of typical processing routines.

1. *Image restoration* compensates for data errors, noise, and geometric distortions introduced during the scanning, recording, and playback operations.
   a. Restoring line dropouts
   b. Restoring periodic line striping
   c. Restoring line offsets
   d. Filtering random noise
   e. Correcting for atmospheric scattering
   f. Correcting geometric distortions

2. *Image enhancement* alters the visual impact that the image has on the interpreter in a fashion that improves the information content.
   a. Contrast enhancement
   b. Density slicing
   c. Edge enhancement
   d. Making digital mosaics
   e. Intensity, hue, and saturation transformations
   f. Merging data sets
   g. Synthetic stereo images
3. *Information extraction* utilizes the decision-making capability of the computer to recognize and classify pixels on the basis of their digital signatures.
   a. Principal-component images
   b. Ratio images
   c. Multispectral classification
   d. Change-detection images

These routines are described and illustrated in the following sections. The routines are illustrated with Landsat examples, but the techniques are equally applicable to other digital image data sets. A number of additional routines are described in the reference publications.

IMAGE RESTORATION

Restoration processes correct the errors, noise, and geometric distortion introduced into the data during the scanning, transmission, and recording of images. The objective is to make the image resemble the original scene. In the early days of Landsat (mid-1970s) defects were not uncommon. Today, however, defects are relatively rare. For most TM centerpoints there are enough repeated images available that users can avoid defective data sets. Image restoration is relatively simple because the pixels from each band are processed separately.

Restoring Line Dropouts

The image in Figure 8-5A has a line dropout, which is a black scan line caused by a missing line of data; the resulting string of zero DNs produces the black line. The defect is restored by replacing each value of DN 0 with a DN calculated as the average of the DN values for the adjacent pixel in the nondefective scan lines that precede and succeed the missing line. Figure 8-5B shows the restored image.

Restoring Banding

Each eastbound and westbound sweep of the TM scan mirror acquires 16 scan lines for the detector arrays for each band. On some images the average brightness for the eastbound lines differs from the westbound lines, which causes alternating brighter and darker bands that are 16 scan lines wide. Figure 8-6A shows this defect, which is called *banding*. The banding is most apparent in uniform stretches of terrain such as water or desert. Restoration programs calculate the average histograms for the brighter and darker bands. The histograms are then adjusted to a uniform average brightness level for both sets of scan lines. Figure 8-6B shows the restored image.

Restoring Line Offsets

Figure 8-7A shows an uncommon but deleterious defect called *line offsets,* in which individual scan lines are offset east or west from their correct position. In Figure 8-7A lines are offset to the west, causing bright lines of land signatures to appear in the dark ocean. At first glance this image looked hopeless. Careful analysis of the digital data file showed that the defects could be restored. Each scan line should begin at pixel reference 1 on the west margin of the image, as shown in Figure 8-2C. The defective scan lines, however, originated at higher pixel numbers, which caused the offset on the image.

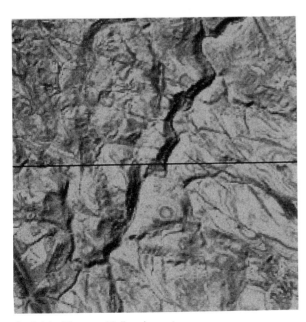

A. Original image with line dropout.

B. Restored image.

Figure 8-5 Restoring line dropouts on a Landsat TM image of central Yemen.

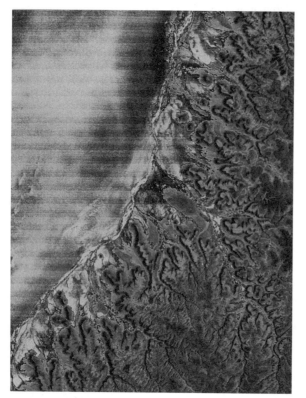

A. Original image with banding.

B. Restored image.

Figure 8-6 Restoring banding on a Landsat TM image of the Red Sea coast of Yemen.

A. Original image with line offsets.

B. Restored image.

Figure 8-7 Restoring offset scan lines. Landsat TM band 4 image of Oxnard, California.

The higher initial pixel numbers were corrected to 1, and the restored image in Figure 8-7B was generated.

Filtering of Random Noise

Nonrandom defects, such as banding, are readily recognized and restored by simple means. Random defects, on the other hand, require more sophisticated restoration methods. One random defect occurs as individual pixels with DNs that are much higher or lower than the surrounding pixels. These pixels produce scattered bright and dark specks called *random noise,* which mar the image. These specks also interfere with information-extraction procedures such as classification. Random-noise pixels are removed by digital filters.

Figure 8-8A shows an array of Landsat pixels, most of which have DN values ranging from 40 to 60. There are, however, two pixels with DN values of 0 and 90 that produce dark and bright noise in the image. This noise may be eliminated by applying a *moving average filter* in the following steps:

1. Design a *filter kernel*, which is a two-dimensional array of pixels with an odd number of pixels in both the *x* and *y* dimensions. The odd-number requirement means that the kernel has a central pixel, which will be modified by the filter operation. In Figure 8-8A, a 3-by-3 kernel is shown by the outlined box.
2. Calculate the average of the 9 pixels in the kernel; for the initial location, the average is 43.
3. Determine the difference between the average value and the central pixel (43 – 10 = 10).
4. If the difference exceeds (+ or –) a threshold value, in this example 20, replace the central pixel with the average value. In Figure 8-8B the original central pixel with a value of 10 has been replaced by 43.
5. Move the kernel to the right by one column of pixels, and repeat the operation. The new average is 50. The central pixel (40) is within the threshold limit of the new average and remains unchanged. In the third position of the kernel, the central pixel (DN = 90) is replaced by the average (53).
6. When the right margin of the kernel reaches the right margin of the pixel array, the kernel returns to the left margin, drops down one row of pixels, and the operation continues until the entire image has been subjected to the moving average filter.

Moving average filters are widely used in other image-processing applications.

Correcting for Atmospheric Scattering

Chapter 2 described how the atmosphere selectively scatters the shorter wavelengths of light, which causes haze and reduces the contrast ratio of images. For TM images, band 1 (blue) has the highest component of scattered light and band 7 (reflected IR) has the least. Figure 8-9 shows two techniques for determining the correction factor for different TM bands. Both techniques are based on the fact that band 7 is essentially free of atmospheric scattering, which can be verified by examining the DNs for shadows; these have values of either 0 or 1 on band 7. The first technique (Figure 8-9A) employs an area within the image that has shadows caused by irregular topography. For each pixel the DN in band 7 is plotted against the DN in band 1, and a straight line is fitted through the plot, using a least-squares technique. If there were no haze in band 1, the line would pass through the origin. Because there is haze, the

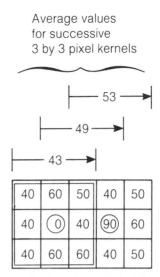

A. Original data with filter kernel (3 by 3 pixels).

B. Filtered data with noise pixels replaced by average values, shown in circles.

Figure 8-8 Filtering to remove random-noise pixels.

intercept is offset along the band 1 axis, as shown in Figure 8-9A. Haze has an additive effect on scene brightness. To correct the haze effect on band 1, the value of the intercept offset is subtracted from the DN of each band 1 pixel for the entire image. The procedure is repeated for the other TM bands.

Figure 8-9B shows a second restoration technique which also requires that the image have some shadows or other areas with DNs of 0 on band 7. The histogram of band 7 (Figure 8-9B) has pixels with DNs of 0. The histogram of band 1 lacks pixels in the range from 0 to approximately 20 because of light scattered into the detector by the atmosphere. The band 1 histogram also shows the characteristic, abrupt increase in pixels at a DN of approximately 20. This value of 20 is subtracted from all the DNs in band 1 to restore the effects of atmospheric scattering. The histograms of bands 2 and 3 are also restored in this manner. Band 2 (green) normally requires a subtraction of less than 10 DNs. Only a few DNs are typically subtracted from band 3 (red).

The amount of atmospheric correction depends upon wavelength of the bands and the atmospheric conditions. As mentioned earlier, scattering is more severe at shorter wavelengths. Humid, smoggy, and dusty atmospheres cause more scattering than clear, dry atmospheres.

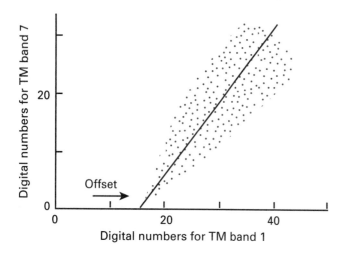

A. Plot of TM band 7 versus band 1 for an area with shadows. Offset of the line of least-squares fit along band 1 axis is caused by atmospheric scattering in that band.

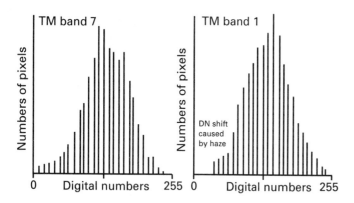

B. Histograms for TM bands 7 and 1. The lack of pixels with low DNs in band 1 is caused by illumination from light selectively scattered by the atmosphere.

Figure 8-9 Methods for determining atmospheric corrections on individual TM bands. From Chavez (1975, Figures 2 and 3).

Geometric Restorations

The scanning of TM images introduces a number of geometric irregularities that are classified as systematic and nonsystematic distortions. After these distortions are restored, the image may be rectified to match a specific map format.

Nonsystematic Distortions *Nonsystematic distortions* (Figure 8-10A) are caused by variations in the spacecraft attitude, velocity, and altitude and therefore are not predictable. These distortions must be evaluated from Landsat tracking data or ground-control information. Variations in spacecraft velocity cause distortion in the along-track direction only and are known functions of velocity that can be obtained from

tracking data. The amount of earth rotation during the 26 sec required to scan a TM image results in distortion in the scan direction that is a function of spacecraft latitude and orbit. In the correction process, successive groups of 16 scan lines are offset toward the west to compensate for earth rotation, which causes the parallelogram outline of the restored image.

Variations in attitude (roll, pitch, and yaw) and altitude of the spacecraft cause nonsystematic distortions that must be determined for each image in order to be corrected. The correction process employs geographic features on the image, called *ground control points* (GCPs), whose positions are known. Intersections of major streams, highways, and airport runways are typical GCPs. Latitude and longitude of GCPs may be determined from accurate base maps. Where maps are lacking we use portable devices called *gobal positioning systems* (GPSs) that determine latitude and longitude from *navigation satellites* (Leick, 1995). The user locates a GCP in the field and determines its position using the GPS. Differences between actual GCP locations and their positions in the image are used to determine the geometric transformations required to restore the image. The original pixels are resampled to match the correct geometric coordinates. The various resampling methods are described by Rifman (1973), Goetz and others (1975), and Bernstein and Ferneyhough (1975).

Systematic Distortions Geometric distortions whose effects are constant and can be predicted in advance are called *systematic distortions*. Scan skew, variations in scanner mirror velocity, and cross-track distortion belong to this category (Figure 8-10B). *Scan skew* is caused by the forward motion of the spacecraft during the time required for each mirror sweep. The ground swath scanned is not normal to the ground track but is slightly skewed, producing distortion across the scan line. The known velocity of the satellite is used to restore the correct geometric relationship.

Tests before Landsat was launched determined that the velocity of the scan mirror was not constant from start to finish of each scan line, resulting in minor systematic distortion along each scan line. The known mirror velocity variations are used to correct for this effect.

Cross-track distortion results from sampling pixels along a scan line at constant time intervals. The width of a pixel (in the scan direction) is proportional to the tangent of the scan angle and therefore is wider at either margin of the scan line. The data are recorded and displayed at a constant rate, however, which causes the pixels at margins of the image to be compressed relative to those at the center. Cross-track distortion occurs in all unrestored images acquired by cross-track scanners, whether from aircraft or satellites. Figure 8-11A is an original aircraft scanner image with cross-track distortion, which causes straight roads to be curved at the compressed margins. Figure 8-11C shows the constant width of the distorted pixels. Figure 8-11B is the image that was restored using trigonometric functions. Figure 8-11D shows the marginal

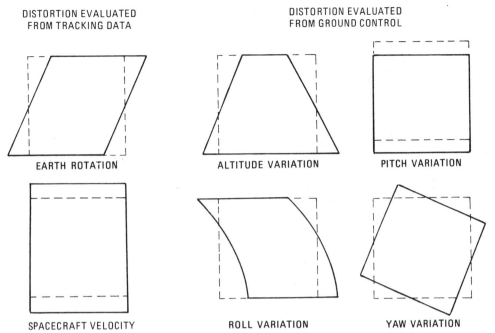

DISTORTION EVALUATED
FROM TRACKING DATA

DISTORTION EVALUATED
FROM GROUND CONTROL

EARTH ROTATION

ALTITUDE VARIATION

PITCH VARIATION

SPACECRAFT VELOCITY

ROLL VARIATION

YAW VARIATION

A. Nonsystematic distortions. Dashed lines indicate shape of distorted image.
Solid lines indicate shape of restored image.

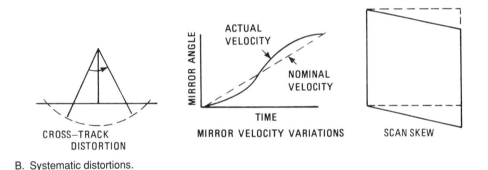

CROSS-TRACK
DISTORTION

MIRROR ANGLE

ACTUAL
VELOCITY

NOMINAL
VELOCITY

TIME

MIRROR VELOCITY VARIATIONS

SCAN SKEW

B. Systematic distortions.

Figure 8-10 Geometric distortions of Landsat images. From Bernstein
and Fernyhough (1975, Figure 3).

pixels plotted at greater widths than central pixels, which restores the geometry. Distortion is pronounced in this aircraft image, which was recorded with a 90° scan angle. Distortion is less pronounced in TM images that are recorded with a 14.9° scan angle. Nevertheless, cross-track distortion is present and must be restored to have an undistorted TM image.

For Landsat and other satellite images the systematic distortions are corrected before the data are distributed.

Map and Image Projections

Images and maps contain inherent geometric distortions because they record the curved surface of the earth on a flat display. Areas, distances, and angular relationships are distorted to varying degrees. A *map projection* is the systematic represen-

tation of a curved surface on a plane. Figure 8-12 shows the three basic projections, which are planar, conic, and cylindrical. The conic and cylindrical projections are "unrolled" into flat surfaces. Many different versions of the basic projections have been devised for different purposes. The history, characteristics, and mathematics of map projections are given by Snyder (1987), which forms the basis for the following discussion. The cylindrical projection is developed by unrolling a cylinder wrapped around the globe, touching at a great circle, with meridians projected from the center of the globe.

Mercator Projections A well-known type of cylindrical projection is the *Mercator projection,* which is also important because it has similar characteristics to images acquired by Landsat. A Mercator projection has the following characteristics:

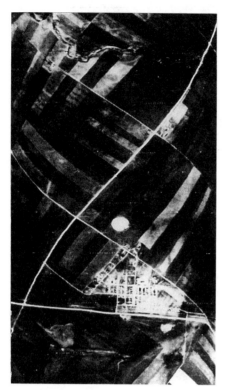

A. Original distorted image.

B. Restored image.

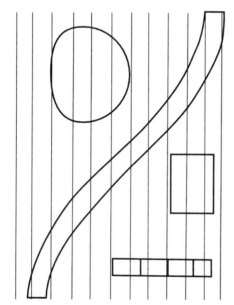

C. Geometry of distorted image.

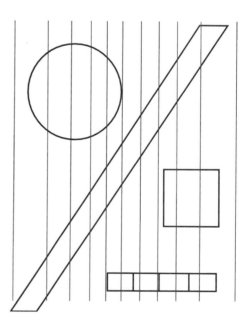

D. Geometry of restored image.

Figure 8-11 Cross-track distortion and restoration on images.

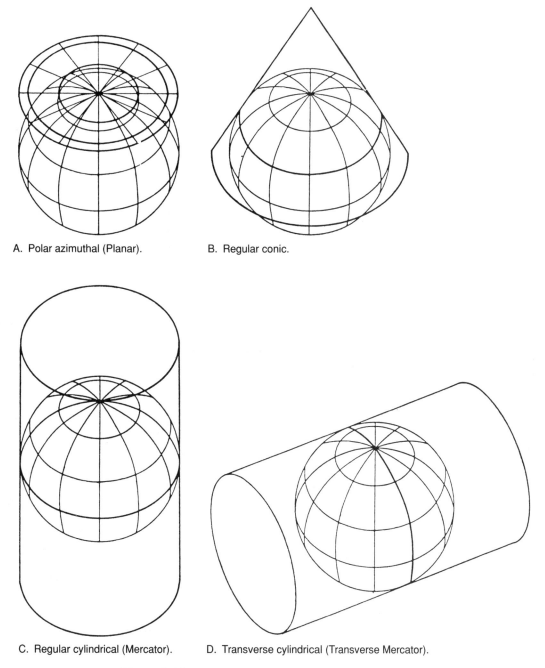

A. Polar azimuthal (Planar).   B. Regular conic.

C. Regular cylindrical (Mercator).   D. Transverse cylindrical (Transverse Mercator).

Figure 8-12 Basic map projections. From Snyder (1987, Figure 1).

1. The great circle of contact with the cylinder is the equator, as shown in Figure 8-12C.
2. Maps are *conformal;* that is, the relative local angles about every point are shown correctly. Although the shape of a large area is distorted, its small features are shaped essentially correctly.
3. Meridians are equally spaced parallel lines; parallels are unequally spaced parallel lines.
4. There is little error close to the equator. (The scale 10° north or south is only 1.5 percent larger than at the equator.)

Because of these useful attributes, other versions of the Mercator projection have been developed. In the *Transverse Mercator projection* the cylinder is oriented at a right angle to the equator. The great circle, of contact, is a longitude line called the central meridian. Map scale is essentially true within a zone 10° east or west of the central meridian. The most widely used maps are based on the *Universal Transverse Mercator* (UTM) projection in which the earth, between latitudes 84°N and 80°S, is divided into 60 zones, each generally 6° wide in latitude. Each zone is divided into 20 quadrangles generally 8° high in latitude. Each quadrangle is divided into grid squares 10,000 m on a side. Locations are readily identified by referring to the grid coordinates. Snyder (1987) provides additional information on the UTM projection.

Space Oblique Projection Landsat and similar polar-orbiting satellites acquire images in the *Space Oblique Mercator* (SOM) projection, which is a version of the transverse mercator projection. Figure 8-13 shows the SOM in which the cylinder is tangential to a great circle formed by the satellite ground track. The 185-km width of the image swath is determined by the angular field of view of the scanner and the altitude of the satellite. The scanning is a continuous process; the image swath is divided into individual TM scenes at intervals of 170 km along the orbit path (Figure 8-13). TM images are conformal and essentially true to scale throughout. The scale at the east and west margins averages 0.015 percent greater than along the central orbit track (Snyder, 1987).

Some users need images displayed in different projections. *Rectification* is the process of converting an image into a specified projection. Rectification employs GCPs (described in the section "Nonsystematic Distortions"), which are identified on the image and assigned latitude and longitude coordinates from existing maps or from GPS information. The computer program then resamples the original image pixels into the specified map projection.

For most uses images must be annotated with latitude and longitude coordinates and/or a UTM grid. The annotations may be added during the rectification process. If images are not rectified, the coordinates may be added to the original SOM projection by using GCPs. My associates routinely use annotated TM images (1:50,000 and 1:100,000 scale) for mineral exploration in Mexico and South America. Portable GPS receivers determine actual ground positions, which are compared with positions from the UTM grid on the images. The positions typically match within 100 m, which is suitable for much fieldwork.

IMAGE ENHANCEMENT

Enhancement is the modification of an image to alter its impact on the viewer. Generally, enhancement changes the original digital values; therefore, enhancement is not done until the restoration processes are completed.

Contrast Enhancement

Contrast enhancement modifies the gray scale to produce a more interpretable image. Virtually all bands of all images acquired by Landsat (and similar satellites) require contrast enhancement. Figure 8-14A is a TM band 4 image of the Thermopolis subscene that was plotted directly from the CCT. These original unenhanced images are sometimes called *raw*

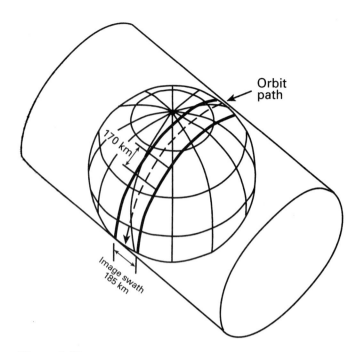

Figure 8-13 Space Oblique Mercator projection of Landsat images.

images. The image is very dark and lacks contrast. The histogram (Figure 8-14B) shows the statistical distribution of the raw data. The median value is 30, which explains the dark appearance of the image. The data occupy a limited range from 0 to 65. This range is only 25 percent of the TM system range (0 to 255) and explains the low contrast of the image.

The image in Figure 8-14A is representative of virtually all raw TM images. Contrary to appearances, the characteristics of raw images do not indicate a defect in the TM system. The sensitivity range of TM detectors was designed to record a wide range of terrain brightness from black basalt plateaus to white sea ice under a wide range of lighting conditions. No individual scene has a brightness range that covers the full sensitivity range of the TM detectors. Therefore, each original TM band requires contrast enhancement to produce useful images. Contrast enhancement is a subjective operation that can be strongly influenced by personal preferences of the operator and user. Histograms of the enhanced data give an objective view of the process. Many routines have been developed for enhancing contrast; three useful methods are described in the following sections.

Linear Contrast Stretch All contrast stretches are done with *lookup tables* (LUTs), which consist of an array of original (*input*) pixel values and a corresponding array of enhanced (*output*) values that are used to produce the stretched image. Figure 8-15 is a graphic display of an LUT that was used to enhance the original data shown in the histogram of Figure 8-14B. The vertical axis represents the original pixels, and the horizontal axis represents the enhanced pixels. The scale of the input axis is twice that of the output axis. The heavy solid and dashed lines show the enhancement transforms. In a simple (unsaturated) linear contrast stretch, the lowest original DN is assigned a new value of 0, the highest original DN is assigned a new value of 255, and the remaining original DNs are linearly reassigned new values ranging from 1 to 254. A disadvantage of this stretch is that the few percent of original pixels at the head and tail of the original histogram occupy an excessive portion of the new dynamic range.

This disadvantage is eliminated with the *saturated contrast stretch* shown by the solid line in the LUT (Figure 8-15). The darkest 2 percent (DN = 0 to 15) and brightest 2 percent (DN = 50 to 65) of the original input pixels are assigned output values of 0 and 255; in other words, they are *saturated* to pure black and white. The remaining 96 percent of input pixels (DN = 16 to 49) are linearly reassigned to the output range from 1 to 254. The resulting output image and histogram are shown in Figure 8-14C,D. The dramatic impact of contrast enhancement is emphasized by comparing the output image (Figure 8-14C) with the input image (Figure 8-14A). For this example, a saturation of 2 percent was selected as optimum after experimenting with other saturation levels. Other images may require higher or lower saturation levels.

The following sections illustrate other commonly used contrast enhancements.

Gaussian Contrast Stretch

Gaussian Contrast Stretch Variations in nature are commonly distributed in a normal (Gaussian) pattern, which is the familiar bell-shaped curve. This distribution is emulated for images by the *Gaussian contrast stretch,* in which the original pixels are reassigned to fit a Gaussian distribution curve. The image and histogram in Figure 8-14E,F show the effect of applying a Gaussian stretch to the original data.

Uniform Distribution Contrast Stretch

Uniform Distribution Contrast Stretch A *bin* of pixels refers to all pixels having the same DN. The linear and Gaussian stretches assign each bin of pixels to a uniform new DN range regardless of the number of pixels in the bin. For example, in the linear and Gaussian histograms the raw pixels with DN 20 and DN 30 are both given a uniform dynamic range, although the 30 bin contains 100 times more pixels. This disparity is compensated for in the *uniform distribution stretch* (or *histogram equalization stretch*), in which the input pixels are redistributed to produce a uniform population density of pixels along the output axis. The resulting output histogram (Figure 8-14H) has a wide spacing of bins in the center of the distribution curve and a close spacing of the less-populated bins at the head and tail of the histogram. In the resulting image (Figure 8-14G) the greatest contrast enhancement is applied to the most populated central range of DNs in the original image. Gray-level variations at the bright and dark extremes are compressed and become less distinct.

Other Contrast Stretches In addition to the standard methods described above, most software packages for image processing provide additional methods, such as exponential and logarithmic stretches. The user can also define custom contrast stretches. For example, TM images along the Red Sea coast include water (dark DNs), bedrock (intermediate DNs), and sand (bright DNs). An investigator who is interested in both water and sand could employ a custom stretch that assigns the dark half of the output DN range to water and the bright half to sand.

An important step in the process of contrast enhancement is for the user to inspect the original histogram and determine the elements of the scene that are of greatest interest. The user then chooses the optimum stretch for his or her needs. Experienced operators of image-processing systems bypass the histogram examination stage and interactively adjust the brightness and contrast of images that are displayed on a CRT. For some scenes a variety of stretched images are required to display fully the original data. It also bears repeating that contrast enhancement should not be done until other processing is completed, because the stretching modifies the original values of the pixels.

Density Slicing

Density slicing converts the continuous gray tone of an image into a series of density intervals, or slices, each corresponding to a specified range of DNs. Each digital slice is displayed as a separate color or outlined by contour lines. Qualitative analog displays are thus converted into quantitative digital displays, assuming that calibration data are available This technique also emphasizes subtle gray-scale differences that may be imperceptible to the viewer. Plate 9A is a density-sliced version of a black-and-white thermal IR image. The original continuous gray tones are shown in colors that represent discrete temperature ranges.

Edge Enhancement

Many interpreters are concerned with recognizing linear features in images. Geologists map faults, joints, and lineaments. Geographers map man-made linear features such as highways and canals. Some linear features occur as narrow lines against a background of contrasting brightness; others are the linear contact between adjacent areas of different brightness. In all cases, linear features are formed by edges. Some edges are marked by pronounced differences in brightness and are readily recognized. Typically, however, edges are marked by subtle brightness differences that may be difficult to recognize. *Edge enhancement* is the process of emphasizing the signatures of edges on images. Edges are enhanced in two ways:

1. Expanding the width of the linear feature
2. Increasing the DN difference across the feature

Two categories of digital filters are used for edge enhancement: directional filters and nondirectional filters.

A. Image from original data.

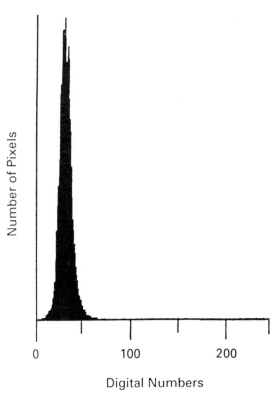

B. Histogram of original data.

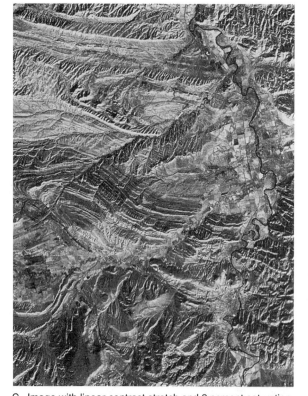

C. Image with linear contrast stretch and 2 percent saturation.

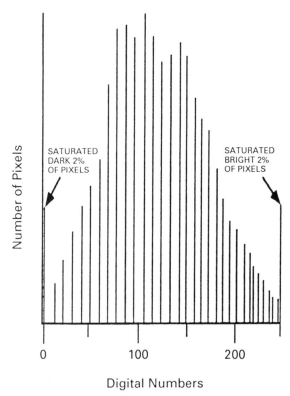

D. Histogram of linear contrast stretch and 2 percent saturation.

Figure 8-14 Contrast-enhancement methods. Landsat TM band 4 of the Thermopolis, Wyoming, subscene.

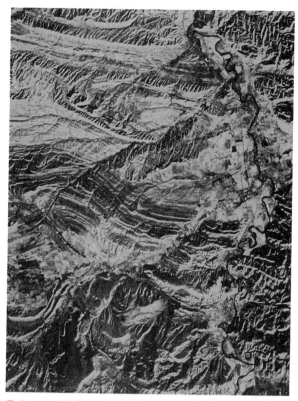

E. Image with Gaussian contrast stretch.

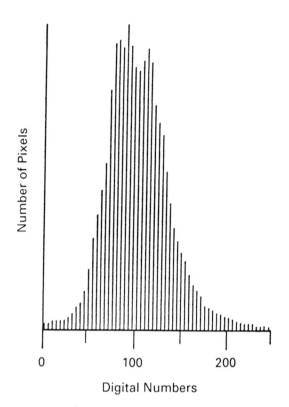

F. Histogram of Gaussian contrast stretch.

G. Image with uniform distribution stretch.

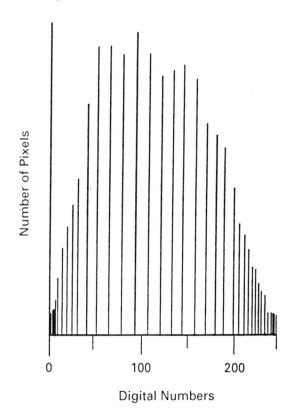

H. Histogram of uniform distribution stretch.

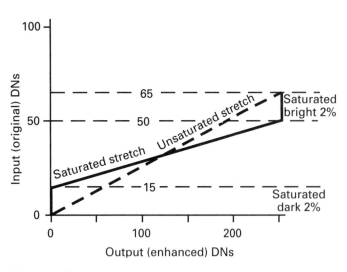

Input (original) DNs

Figure 8-15 Graphic plot of the lookup table for a linear contrast stretch with 2 percent saturation.

Nondirectional Filters *Nondirectional filters* (also called *Laplacian filters*) are named because they have no directional bias in enhancing linear features; almost all directions are enhanced. The only exception applies to linear features oriented parallel with the direction of filter movement; these features are not enhanced. Figure 8-16A shows a Laplacian filter that is a kernel of three lines and three pixels. The kernel is a template with 4 as the central value, 0 at each corner, and −1 at the center of each edge. The Laplacian kernel is placed over a 3-by-3 array of original pixels (Figure 8-16A), and each pixel is multiplied by the corresponding value in the kernel. The nine resulting values (four of which are 0 and four are negative numbers) are summed. The resulting value for the filter kernel is combined with the original central pixel of the 3-by-3 data array, and this new number replaces the original DN of the central pixel. For example, consider the Laplacian kernel placed over the array of 3-by-3 original pixels indicated by the box in Figure 8-16A. When the multiplication and summation are performed, the value for the filter kernel is 5. The original central pixel in the array (DN = 40) is combined with the filter

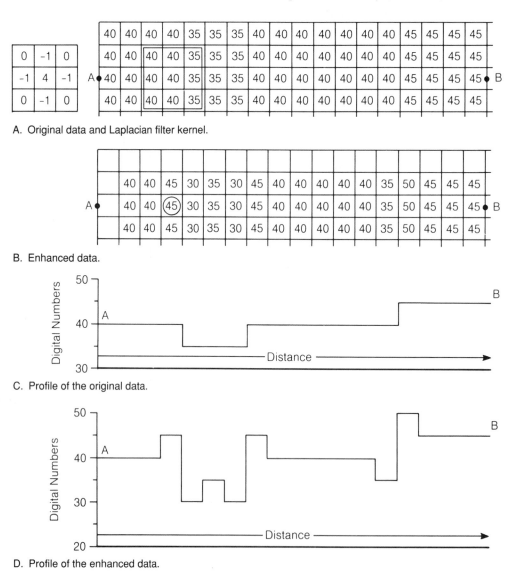

A. Original data and Laplacian filter kernel.

B. Enhanced data.

C. Profile of the original data.

D. Profile of the enhanced data.

Figure 8-16 Nondirectional edge enhancement using a Laplacian filter.

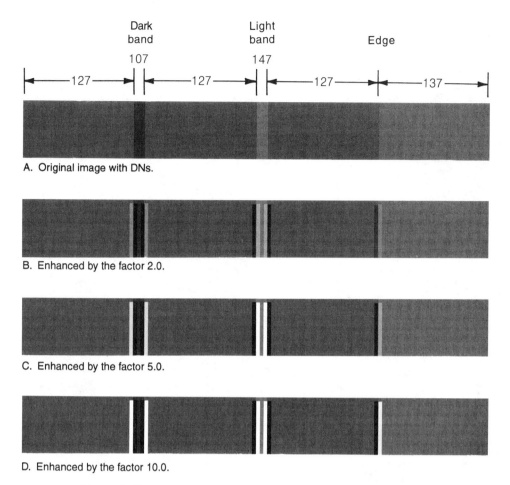

A. Original image with DNs.

B. Enhanced by the factor 2.0.

C. Enhanced by the factor 5.0.

D. Enhanced by the factor 10.0.

Figure 8-17 Computer-generated images illustrating nondirectional edge enhancement with a Laplacian filter and different weighting factors.

value to produce the new value of 45, which is used in the filtered data set (Figure 8-16B). The kernel moves one column of pixels to the right, and the process repeats until the kernel reaches the right margin of the pixel array. The kernel then shifts back to the left margin, drops down one line of pixels, and continues the same process. Figure 8-16B shows the enhanced pixels. The outermost column and line of pixels are blank because they cannot form central pixels in an array.

The effect of the edge-enhancement operation can be evaluated by comparing profiles A–B of the original and the filtered data. Figure 8-16C is a profile of the original data. The regional background (DN = 40) is intersected by a darker lineament (DN = 35) that is three pixels wide and has DN values of 35. The contrast ratio between the lineament and background, as calculated from Equation (1-3), is 40/35, or 1.14. In the profile of the enhanced data (Figure 8-16D), the contrast ratio is 45/30, or 1.50, which is an enhancement of 32 percent. The original lineament, which was three pixels wide, is five pixels wide in the filtered version.

The eastern (right-hand) portion of the original profile (Figure 8-16C) has a second lineament marked by a change in values from 40 to 45 along an edge that has no width. The original contrast ratio (45/40 = 1.13) is increased by 27 percent in the enhanced image (50/35 = 1.43). The original edge is expanded to a width of two pixels.

Figure 8-17A is a computer-generated synthetic image with a uniform background (DN = 127). The left portion of the image is crossed by a dark band (DN = 107). The central portion is crossed by a bright band (D = 147). In the right portion of the image, an edge is formed at the contact of background with a brighter surface (DN = 137). These three linear features have subtle expressions in the original image. The linear features of the synthetic image are similar to those in the digital arrays of Figure 8-16A. The Laplacian filter of Figure 8-16A was applied to the synthetic image. After the value of the filter kernel has been calculated and prior to combining it with the original central data pixel, the calculated value may be multiplied by a weighting factor. The factor weight may be less than 1 or

A. Original image.

B. Nondirectional enhancement.

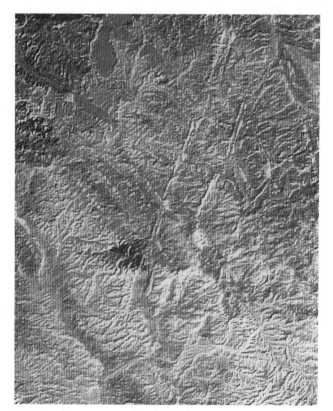

C. Directional enhancement of northeast-trending lineaments.

D. Directional enhancement of northwest-trending lineaments.

Figure 8-18 Edge enhancements of TM images of the Jabal an Naslah area, northwest Saudi Arabia.

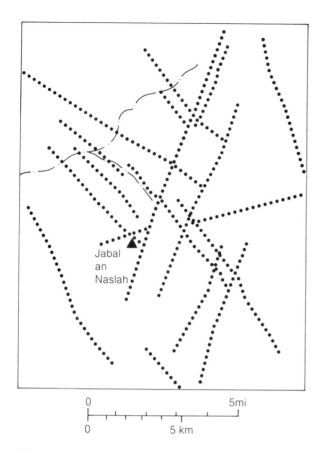

Figure 8-19 Lineaments interpreted from edge-enhanced images of the Jabal an Naslah area (Figure 8-18).

greater than 1 in order to diminish or to accentuate the effect of the filter. The weighted filter value is then combined with the central original pixel to produce the enhanced image.

In Figure 8-17B the calculated kernel value was weighted by a factor of 2.0 to produce the enhanced image. In Figure 8-17C a factor of 5.0 was used. In Figure 8-17D a factor of 10.0 was used, which saturates the brightest values to a DN of 255 and the darkest to a DN of 0. The filtering has significantly enhanced the expression of the original lineaments in Figure 8-17A.

Figure 8-18 shows nondirectional edge enhancement applied to a TM band 4 subscene in northwest Saudi Arabia. The area is a plateau of horizontal strata that is cut by fractures that trend northwest and northeast. Figure 8-19 is a map that shows the major fractures. Figure 8-18B shows the image after processing with the Laplacian filter kernel (Figure 8-16A). Both the northwest and northeast fracture directions are emphasized in the enhanced image.

Many image-processing facilities routinely apply a nondirectional filter to most images, which imparts a "crisper" appearance to the final product.

Directional Filters *Directional filters* are used to enhance linear features that trend in a specific direction, such as N 45°W. Figure 8-20A shows four directional filters that are designed to enhance the four cardinal directions. Figure 8-20B shows an array of original pixels for terrain with a background DN of 25 that is cut by two lineaments that trend northeast (NE-SW) and northwest (NW-SE). The lineaments have DNs of 30 and are brighter than the background. Figure 8-20C is a profile along line A-B that crosses the lineaments in an east-west direction. The profile shows that each lineament is one pixel wide with a brightness difference from the background of 5 DN.

The filter to enhance NE-SW edges is selected from Figure 8-20A and applied to the original data set. The filter kernel is placed over an array of nine original pixels (three lines by three pixels). Each original pixel is multiplied by the corresponding kernel cell and the results are summed. The resulting kernel value is then combined with the value of the original pixel in the center of the array. Figure 8-20D shows the enhanced pixels that have been processed in this manner. The DNs of the northeast-trending lineament are increased from 30 to 50. In addition, the filter has generated two bands of dark pixels (DN = 15), with a width of one pixel, that trend parallel with the original lineament. Enhanced profile A-B (Figure 8-20E) graphically shows the effects of the directional edge enhancement. Compare this profile with the profile of the original data (Figure 8-20C). The DNs of the original lineament differ from the background by 5 DN for a brightness difference of 20 percent (5/25 = 20%). The three parallel bands of the enhanced lineament differ from the background by a total of 35 DN (25 + 10 = 35) for a brightness difference of 140 percent (35/25 = 140%). The brightness difference is enhanced by 700 percent (140%/20% = 700%). The directional filter also enhances the geometric expression of the lineament. The original lineament is one pixel wide. The enhanced lineament is three pixels wide. Figure 8-20D shows the enhanced lineament (DN = 50) and the two parallel bands (DN = 15). Cross section A-B (Figure 8-20E) crosses the lineament obliquely and therefore gives a misleading impression of the geometric relationships. A cross section (not shown) drawn normal to the lineament (Figure 8-20D) shows that the enhanced lineament is three pixels wide. The geometric expression is enhanced by 300 percent (3 pixel width/1 pixel width = 300%). The key point is that the northeast-trending lineament is strongly enhanced, but the northwest-trending lineament is completely unchanged by this process.

In Figures 8-20F and 8-20G the filter kernel to enhance northwest-trending edges is applied to the original array of pixels. The brightness difference and the geometric expression of the northwest-trending lineament are enhanced in the same manner described above. The northeast-trending lineament is completely unchanged. In summary, the two directional filters have selectively enhanced the northeast- and northwest-trending lineaments by 300 percent in width and 700 percent in brightness difference.

The filters in Figure 8-20A can be modified to enhance linear features trending at directions other than the four cardinal directions, as described by Haralick (1984). Directional edge-enhancement filters were applied to the original Landsat

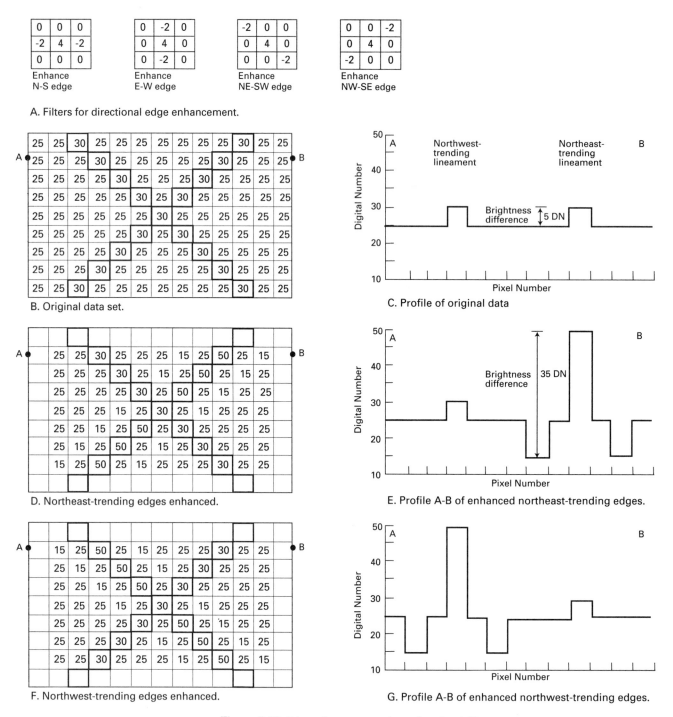

A. Filters for directional edge enhancement.

B. Original data set.

C. Profile of original data

D. Northeast-trending edges enhanced.

E. Profile A-B of enhanced northeast-trending edges.

F. Northwest-trending edges enhanced.

G. Profile A-B of enhanced northwest-trending edges.

Figure 8-20 Edge enhancement using a directional filter.

subscene in Saudi Arabia (Figure 8-18A). In Figure 8-18C, the northeast-trending lineaments are preferentially enhanced. In Figure 8-18D, the northwest-trending lineaments are preferentially enhanced. Comparing Figures 8-18C and 8-18D with the original image (Figure 8-18A) shows that both the brightness difference and the geometric width of the lineaments are selectively enhanced.

Figure 8-18 also compares the effects of directional filters with the effects of nondirectional filters. The nondirectional filter enhances both the northeast-trending lineaments and the northwest-trending lineaments (Figure 8-20B). The directional filters preferentially enhance either the northeast-trending lineaments (Figure 8-20C) or the northwest-trending lineaments (Figure 8-20D). Directional edge enhancement is a valuable

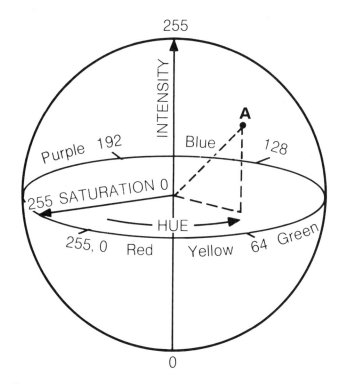

Figure 8-21 Coordinate system for the intensity, hue, and saturation (IHS) transformation. The color at point A has the following values: $I = 205$, $H = 75$, $S = 130$ (undersaturated).

technique for selectively improving both the brightness difference and the geometric width of linear features.

Intensity, Hue, and Saturation Transformations

The additive and subtractive systems of primary colors were described in Chapter 2. An alternate approach to color is the *intensity, hue, and saturation (IHS) system,* which is useful because it presents colors more nearly as they are perceived by humans. The IHS system is based on the color sphere (Figure 8-21) in which the vertical axis represents intensity, the radius represents saturation, and the circumference represents hue. The *intensity (I)* axis represents brightness variations and ranges from black (0) to white (255); no color is associated with this axis. *Hue (H)* represents the dominant wavelength of color. Hue values commence with 0 at the midpoint of red tones and increase counterclockwise around the circumference of the sphere to conclude with 255 adjacent to 0. *Saturation (S)* represents the purity of color and is represented by the radius that ranges from 0 at the center of the color sphere to 255 at the circumference. A saturation of 0 represents a completely impure color in which all wavelengths are equally represented as a shade of gray. Intermediate values of saturation are pastel shades, whereas high values are purer and more intense colors. Buchanan (1979) describes the IHS system in detail.

Thermopolis, Wyoming, Example When any three spectral bands of TM (or other multispectral data) are combined in the BGR system, the resulting color images typically lack saturation, even though the bands have been contrast-stretched. Plate 11A is a normal color image prepared from TM bands 1, 2, and 3 of the Thermopolis, Wyoming, subscene. The individual bands were contrast-enhanced, but the color image has the pastel appearance that is typical of many Landsat images. The undersaturation is due to the high degree of correlation between spectral bands. High reflectance values in the green band, for example, are accompanied by high values in the blue and red bands, so pure colors are not produced. To correct this problem, a method of enhancing saturation was developed that consists of the following steps:

1. Transform the three bands of data from the BGR system into the IHS system in which the three component images represent intensity, hue, and saturation. This IHS transformation was applied to TM bands 1, 2, and 3 of the Thermopolis subscene to produce the intensity, hue, and saturation images illustrated in Figure 8-22. The intensity image (Figure 8-22A) is dominated by albedo and topography. Sunlit slopes have high intensity values (bright tones), and shadowed areas have low values (dark tones). The airport runway and water in the Wind River have low values (see Figure 3-7H for locations). Vegetation has intermediate intensity values, as do most of the rocks. In the hue image (Figure 8-22B) red beds of the Chugwater Formation have conspicuous dark tones caused by low values assigned to red hues. Vegetation has intermediate to light-gray values assigned to the green hue. The original saturation image (Figure 8-22C) is very dark because of the lack of saturation in the original TM data. Only the shadows and river are bright, indicating high saturation for these features.

2. Apply a linear contrast stretch to enhance the original saturation image (Figure 8-22D). Note the overall increased brightness of the enhanced image. Also note the improved discrimination between terrain types.

3. Transform the intensity, hue, and enhanced saturation image from the IHS system back into the three images of the BGR system. These enhanced BGR images were used to prepare the new color composite image of Plate 11B, which is a significant improvement over the original version in Plate 11A. In the IHS version, note the wide range of colors and improved discrimination between colors. Some bright green fields are distinguishable in vegetated areas that were obscure in the original. The wider range of color tones helps separate rock units.

Figure 8-23 shows graphically the relationship between the BGR and IHS systems. Numerical values may be extracted from this diagram for expressing either system in terms of the other. In Figure 8-23 the circle represents a horizontal section through the equatorial plane of the IHS sphere, with the intensity axis passing vertically through the plane of the diagram.

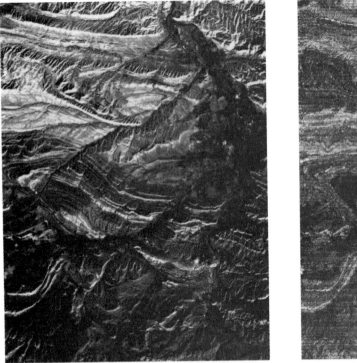

A. Intensity image.

B. Hue image.

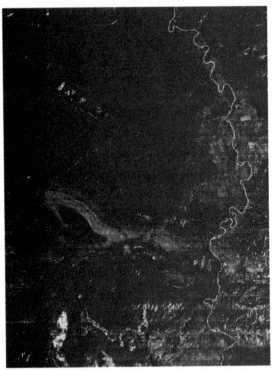

C. Saturation image, original.

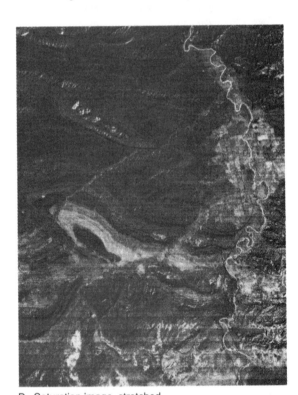

D. Saturation image, stretched.

Figure 8-22 Images created by the IHS transformation of TM bands 1, 2, and 3 of the Thermopolis, Wyoming, subscene.

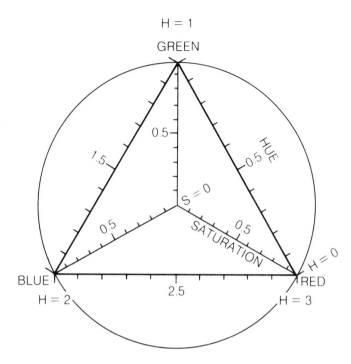

Figure 8-23 Diagram showing the relationship between the BGR and IHS systems.

The corners of the equilateral triangle are located at the positions of the red, green, and blue hues. Hue changes in a counterclockwise direction around the triangle, from red (0), to green (1), to blue (2), and again to red (3). Values of saturation are 0 at the center of the triangle and increase to a maximum of 1 at the corners. Any perceived color is described by a unique set of IHS values; in the BGR system, however, different combinations of additive primaries can produce the same color. The IHS values can be derived from BGR values through the transformation equations

$$I = R + G + B \qquad (8\text{-}1)$$

$$H = I - 3B \qquad (8\text{-}2)$$

$$S = I \qquad (8\text{-}3)$$

for the interval $0 < H < 1$, extended to $1 < H < 3$. After enhancing the saturation image, the IHS values are converted back into BGR images by inverse equations.

Western Bolivia Example Saturation enhancement is also effective in other regions and other combinations of spectral bands. Plate 11C is a TM 2-4-7 image in western Bolivia. The eastern portion consists of volcanic features of the Andes and the western portion is alluvial deposits. The bands were contrast-enhanced but are dominated by monotonous undersaturated hues of brown, tan, and gray. Plate 11D is the same image after saturation enhancement. Subtle color differences are

emphasized, and much more geologic information was interpreted from the enhanced image.

The IHS transform has other applications in addition to enhancing the saturation of images. The traditional method of enhancing edges on multispectral images, such as TM, is to enhance each of the three bands separately. It is more efficient to transform the three TM bands into IHS components, and then to apply edge enhancement to the intensity component. When the image is transformed back into BGR components, the edge enhancement is applied to each of the bands. This operation may be combined with the saturation enhancement.

The IHS transformation is also used to produce combination images, as described in the following section.

Combination Images

Different images that have been digitally merged are called *combination images*. For example, Landsat TM color images display three spectral bands but have a relatively coarse spatial resolution (30 m). SPOT pan images have good resolution (10 m) but display only a single spectral band. A combination image can display the spectral characteristics of the TM with the spatial resolution of SPOT. Plate 12A is a TM IR color image (bands 2, 3, and 4 shown as blue, green, and red) of the northern portion of Los Angeles. Figure 8-24 is a SPOT pan image and a map of the same area. Comparing the two images shows the advantages of each version and the reason for combining them. Spatial details of buildings and street patterns are clearly visible in the SPOT image but are not recognizable in the TM image. The circular streets in the large parking lot that surrounds Dodger Stadium are clearly seen in the SPOT but not in the TM. The sparse vegetation, however, has a distinctive red signature in the TM image but is not recognizable in the SPOT image. Various shades of blue in the TM indicate spectral signatures of different categories of land use. These spectral characteristics are absent in SPOT, which does, however, show details of size, shape, and spacing (density) of buildings and streets. The two images clearly complement each other; a combination of the two would be valuable for the interpreter.

The IHS transformation is used to combine such images in the following manner:

1. For the TM image, resample each original pixel (30 by 30 m) into nine pixels (10 by 10 m) to match the SPOT size.
2. Use ground control points to register the TM image to the SPOT image.
3. Transform the resampled TM color image into its IHS components.
4. Discard the TM intensity component and replace it with the SPOT data.
5. Transform the new set of IHS components back into BGR components.

Plate 12B is the combined TM/SPOT image with 10-m ground resolution cells and three spectral bands.

A. SPOT pan image (10-m resolution).

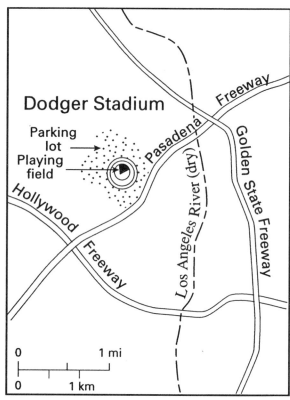

B. Location map.

Figure 8-24 SPOT panchromatic image and map, Los Angeles, California.

This technique is used to combine other images, such as radar and TM images. The combination image merges the shadows and highlights plus high spatial resolution of radar with the spectral information from TM. It is more efficient to interpret the combination image than to interpret TM and radar images separately and combine the results. Chavez, Sides, and Anderson (1991) compare three methods of merging images, including the IHS technique.

Digital Mosaics

Mosaics of Landsat images may be prepared by manually matching and splicing together individual images, as shown for the Central Arabian Arch in Plate 4. Differences in contrast and tone between adjacent images may cause the checkerboard pattern that is common on many mosaics. This problem is minimized by preparing mosaics directly from the digital CCTs.

Plate 13 is a digital mosaic of six TM images of southern California prepared by R. E. Crippen of the Jet Propulsion Laboratory (JPL). Earthquake data and traces of faults are added to the mosaic and are discussed in Chapter 13. The six images were acquired at different dates and seasons, which causes major color differences between adjacent scenes. For the area of sidelap between adjacent images, the computer op-

erator compared histograms for corresponding bands of each image and adjusted the histograms to match. This procedure was successful, as shown by the excellent match of colors throughout the mosaic. Adjacent images were then geometrically registered to each other by identifying common ground control points (GCPs) in the regions of sidelap. Pixels were then geometrically adjusted to match the desired map projection. The next step was to eliminate from the digital file the duplicate pixels within the areas of overlap. Optimum contrast stretching was then applied to all the pixels, producing a uniform appearance throughout the mosaic.

Synthetic Stereo Pairs

The stereo pairs of aerial photographs (see Chapter 2) and SPOT images (see Chapter 4) demonstrate the value of stereo images with vertical exaggeration. Landsat images, however, have no stereo capability, aside from negligible sidelap. *Synthetic stereo pairs* are stereo models produced by merging a single image with digital topographic data. Digital topographic data are produced by digitizing topographic contour maps, such as Figure 8-25A, into raster arrays of pixels. Each pixel represents a ground resolution cell and records the average elevation of the cell. Digital topographic data are also available

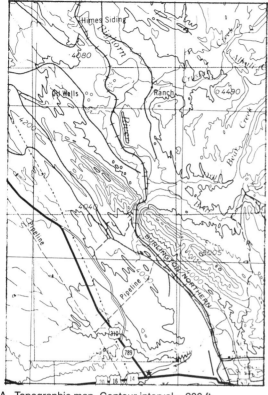

A. Topographic map. Contour interval = 200 ft.

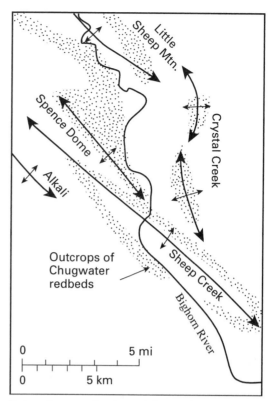

B. Location map showing anticlines and outcrops of the Chugwater Formation.

Figure 8-25 Maps of Sheep Mountain, Wyoming.

from the Topographic Division of the U.S. Geological Survey. Synthetic stereo pairs are produced for a TM image by the following procedure:

1. The digital topographic data are resampled into pixels that match the pixel size (30 by 30 m) of the TM image.
2. Ground control points are used to register Landsat pixels with topographic pixels. Each TM pixel now has an elevation value in addition to its spectral band values.
3. The operator selects a center point on the image (that will be analogous to the principal point of an aerial photograph) and designates the amount of vertical exaggeration for the stereo pair.
4. The computer program then radially displaces the pixels from the principal point to produce a new image with parallax, such as that shown for the aerial photographs in Chapter 2. The computer then shifts to a conjugate principal point in the image and generates the second image for the synthetic stereo pair.

Plate 12C is a synthetic stereo pair that was produced in this manner from a single TM 2-3-4 IR color image. View the images with a stereoscope to appreciate the three-dimensional effect and vertical exaggeration. Figure 8-25A is a topographic map of the area, which is located in the northeast portion of the

Bighorn Basin, Wyoming. Sedimentary strata are folded into an array of anticlines that are identified in the location map (Figure 8-25B). The major topographic and geologic feature is the northwest-trending Sheep Mountain anticline with a ridge of light-colored late Paleozoic strata exposed in the core. The core is surrounded by red beds of the Chugwater Formation (Triassic), which have a distinctive yellow signature in the image, and is indicated by the stippled pattern in Figure 8-25B. Sandstones and shales of Mesozoic age form ridges and valleys with gray and blue signatures. The fold axes and attitudes (dip and strike) of the strata are readily interpreted from the stereo pair.

Digital Perspective Images

Digital perspective images are three-dimensional views produced from digital topographic data that have been merged with other data sets. Plate 14A is a perspective image of the north flank of the San Bernardino Mountains and the Antelope Valley in southern California. R. E. Crippen of the JPL produced the image in the following manner:

1. A SPOT pan image was selected as the "master" image. Digital topographic data and a TM 2-4-7 image were resampled to 10-by-10-m pixels and registered with the SPOT data, using GCPs.

2. The merged data were geometrically transformed into a three-dimensional view of the terrain. The operator specified the look direction (toward the west) and vertical exaggeration.

The combination of topography and spectral information on the perspective image emphasizes differences between the rugged, vegetated mountains and the low-lying barren desert of the Antelope Valley. The two terrains are separated by a series of faults that strike parallel with the mountain front. The topographic scarps that mark the faults are accentuated by the perspective view and vertical exaggeration. Of special interest is the Blackhawk landslide, which forms a dark lobe in the lower right portion of the image. This massive prehistoric slide originated on the north flank of the mountains and moved far out into the desert. Details of the landslide are given in Chapter 13.

INFORMATION EXTRACTION

Image restoration and enhancement processes utilize computers to provide corrected and improved images for study by human interpreters. The computer makes no decisions in these procedures. However, the processes that comprise information extraction do utilize the computer's decision-making capability to identify and extract specific pieces of information. A human operator must instruct the computer and must evaluate the significance of the extracted information.

Principal-Component Images

If we compare individual TM bands such as the Thermopolis subscene (Chapter 3), we note a strong similarity. Areas that are bright or dark in one band tend to be bright or dark in the other bands. This relationship is shown diagrammatically in Figure 8-26, in which DNs for TM band 1 are plotted versus TM band 2. Data points are distributed in an elongate band, which shows that as DNs increase for one band, they increase for the other band. If for any pixel we know the DN for band 1, we can predict the approximate value in band 2. The data are said to be strongly correlated. This correlation means that there is much redundancy of information in a multispectral data set. If this redundancy were reduced, the amount of data required to describe a multispectral image could be compressed.

The *principal-components transformation*, originally known as the Karhunen-Loéve transformation (Loéve, 1955), is used to compress multispectral data sets by calculating a new coordinate system. For the two bands of data in Figure 8-26, the transformation defines a new axis (y_1) oriented in the long dimension of the distribution and a second axis (y_2) perpendicular to y_1. The mathematical operation makes a linear combination of pixel values in the original coordinate system that results in pixel values in the new coordinate system:

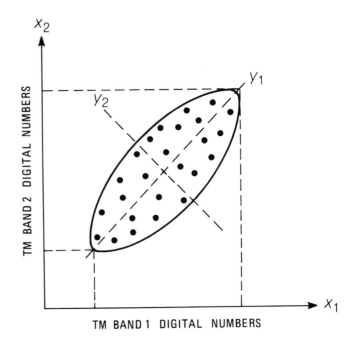

Figure 8-26 Plot of DNs for TM band 1 (axis x_1) and band 2 (axis x_2) showing correlation between these bands. The principal-components transformation was used to generate a new coordinate system (y_1, y_2). After Swain and Davis (1978, Figure 7.9).

$$y_1 = a_{11} x_1 + a_{12} x_2 \qquad (8\text{-}4)$$

$$y_2 = a_{21} x_1 + a_{22} x_2 \qquad (8\text{-}5)$$

where

(x_1, x_2) = pixel coordinates in the original system

(y_1, y_2) = coordinates in the new system

a_{11}, a_{12}, a_{22} = constants

In Figure 8-26 note that the range of pixel values for y_1 is greater than the ranges for either of the original coordinates, x_1 or x_2, and that the range of values for y_2 is relatively small.

The same principal-components transformation may be carried out for multispectral data sets consisting of any number of bands. Additional coordinate directions are defined sequentially. Each new coordinate is oriented perpendicular to all the previously defined directions and in the direction of the remaining maximum density of pixel data points. For each pixel, new DNs are determined relative to each of the new coordinate axes. A set of DN values is determined for each pixel relative to the first principal component. These DNs are then used to generate an image of the first principal component. The same procedure is used to produce images for the remaining principal components. The preceding description of the principal-components transformation is summarized from Swain and Davis (1978); additional information is given in Moik (1980).

A principal-components transformation was performed on the three visible and three reflected IR bands of TM data for the Thermopolis, Wyoming, subscene. Each pixel was assigned six new DNs corresponding to the first through the sixth principal-component coordinate axes. Figure 8-27 illustrates the six principal-component (PC) images, which have been contrast-enhanced. As noted earlier, each successive principal component accounts for a progressively smaller proportion of the variation of the original multispectral data set. These percentages of variation are indicated for each PC image in Figure 8-27 and are plotted graphically in Figure 8-28. The first three PC images contain 97 percent of the variation of the original six TM bands, which is a significant compression of data. PC image 1 (Figure 8-27A) is dominated by topography, expressed as highlights and shadows, that is highly correlated in all six of the original TM bands. PC image 2 (Figure 8-27B) is dominated by differences in albedo that also correlate from band to band because pixels that are bright in one TM band tend to be bright in adjacent bands. The least-correlated data are noise such as line striping and dropouts, which occur in different detectors and different bands. In Figure 8-27, noise dominates PC images 4, 5, and 6, which together account for only 2.6 percent of the original variation. Note, however, that PC image 6 (Figure 8-27F) displays parallel arcuate dark bands in the outcrop belt of the Chugwater Formation that may represent lithologic variations in this unit.

Any three PC images can be combined in red, green, and blue to create a color image. Plate 14B was produced by combining PC images from Figure 8-27 in the following fashion: PC 2 = red, PC 3 = green, PC 4 = blue. PC image 1 was not used in order to minimize topographic effects. As a result, the color PC image displays a great deal of spectral variation in the vegetation and rocks, although the three images constitute only 10.7 percent of the variation of the original data set.

Examination of the original TM bands for the Thermopolis subscene in Chapter 3 shows that the visible bands 1, 2, and 3 are highly correlated with each other, as are the three reflected IR bands 4, 5, and 7. Rather than transform all six bands into six PC images, it is useful to transform bands 1, 2, and 3 into three PC images and bands 4, 5, and 7 into three PC images.

In summary, the principal-components transformation has several advantages:

1. Most of the variance in a multispectral data set is compressed into the first two PC images.
2. Noise is generally relegated to the less-correlated PC images.
3. Spectral differences between materials may be more apparent in PC images than in individual bands.

Ratio Images

Ratio images are prepared by dividing the DN value in one band by the corresponding DN value in another band for each pixel. The resulting values are plotted as a ratio image. Figure 8-29 illustrates some ratio images prepared from TM bands of the Thermopolis subscene. In a ratio image the black and white extremes of the gray scale represent pixels having the greatest difference in reflectivity between the two spectral bands. The darkest signatures are areas where the denominator of the ratio is greater than the numerator. Conversely, the numerator is greater than the denominator for the brightest signatures. Where denominator and numerator are the same, there is no difference between the two bands.

For example, the spectral reflectance curve for vegetation (Figure 3-1) shows a maximum reflectance in TM band 4 (reflected IR) and a lower reflectance in band 2 (green). Figure 8-29C is the ratio image 4/2, which is produced by dividing DNs for band 4 by the DNs for band 2. The brightest signatures in this image correlate with the cultivated fields along the Wind River and Owl Creek (Figure 3-7H). Figure 8-29A is the ratio image 3/1 (red/blue), in which red beds of the Chugwater outcrops have very bright signatures.

Any three ratio images may be combined in red, green, and blue to produce a color image. In Plate 14C the ratio images 3/1, 5/7, and 3/5 are combined as red, green, and blue, respectively. Compare this image with the various Thermopolis color images (Plate 2) and the interpretation map (Figure 3-8). The signatures of the ratio color image express more geologic information and have greater contrast between units than do color images of individual TM bands. An advantage of ratio images is that they extract and emphasize differences in spectral reflectance of materials. A disadvantage of ratio images is that they suppress differences in albedo; materials that have different albedos but similar spectral properties may be indistinguishable in ratio images. Another disadvantage is that any noise is emphasized in ratio images.

Ratio images also minimize differences in illumination conditions, thus suppressing the expression of topography. In Figure 8-30 a red siltstone bed crops out on both the sunlit and shadowed sides of a ridge. In the individual Landsat TM bands 1 and 3, the DNs of the siltstone are lower in the shadowed area than in the sunlit outcrop, which makes it difficult to follow the siltstone bed around the hill. Values of the ratio image 3/1, however, are identical in the shadowed and sunlit areas, as shown by the chart in Figure 8-30; thus the siltstone has similar signatures throughout the ratio image. Highlights and shadows are notably lacking in the ratio images of Figure 8-29.

In addition to ratios of individual bands, a number of other ratios may be computed. An individual band may be divided by the average for all the bands, resulting in a normalized ratio image. Another ratio combination is produced by dividing the difference between two bands by their sum; for example, (band 4 − band 5)/(band 4 + band 5). Ratios of this type are used to process AVHRR data, as described in Chapter 9.

Multispectral Classification

For each pixel in a Landsat TM image, the spectral brightness is recorded for six spectral bands in the visible and reflected IR

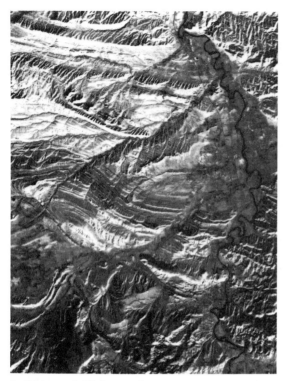

A. PC image 1 (88.4 percent).

B. PC image 2 (6.6 percent).

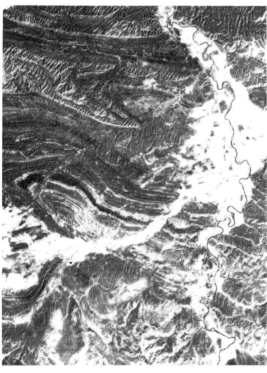

C. PC image 3 (2.4 percent).

D. PC image 4 (1.8 percent).

Figure 8-27 Principal-component (PC) images of the Thermopolis, Wyoming, subscene. PC images were generated from the six visible and reflected IR bands of TM images. The percentage of variance represented by each PC image is shown.

E. PC image 5 (0.5 percent).

F. PC image 6 (0.3 percent).

Figure 8-27 (*continued*)

regions. A pixel may be characterized by its *spectral signature*, which is determined by the relative reflectance in the different wavelength bands. *Multispectral classification* is an information-extraction process that analyzes these spectral signatures and then assigns pixels to classes based on similar signatures.

Procedure Multispectral classification is illustrated diagrammatically with a Landsat image (Figure 8-31A) of southern California that covers the Salton Sea, the Imperial Valley, and adjacent mountains and deserts. Figure 8-32A shows reflectance spectra for water, agriculture, desert, and mountains that are derived from TM bands. The data points are plotted at the center of each TM bandpass. In Figure 8-32B the reflectance ranges of bands 2, 3, and 4 form the axes of a three-dimensional coordinate system. The solid dots are the loci of the four terrain categories in Figure 8-32A. Plotting additional pixels of the different terrain types produces three-dimensional clusters or ellipsoids. The surface of an ellipsoid forms a *decision boundary*, which encloses all pixels for that terrain category. The volume inside the decision boundary is called the *decision space*. Classification programs differ in their criteria for defining decision boundaries. In many programs the analyst is able to modify the boundaries to achieve optimum results. For the sake of simplicity the cluster diagram (Figure 8-32B) is shown with only three axes. In actual practice the computer employs a separate axis for the six TM bands.

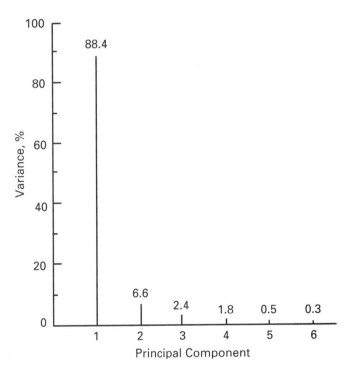

Figure 8-28 Plot showing the percentage of variance represented by PC images of the Thermopolis, Wyoming, subscene (Figure 8-27).

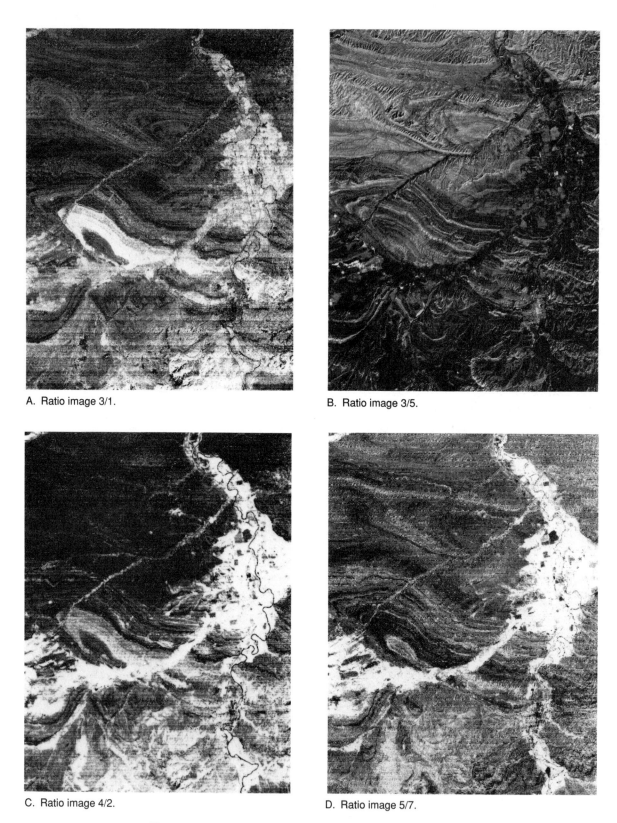

A. Ratio image 3/1.

B. Ratio image 3/5.

C. Ratio image 4/2.

D. Ratio image 5/7.

Figure 8-29 Ratio images of Landsat TM bands of the Thermopolis, Wyoming, subscene. The images have been contrast-enhanced.

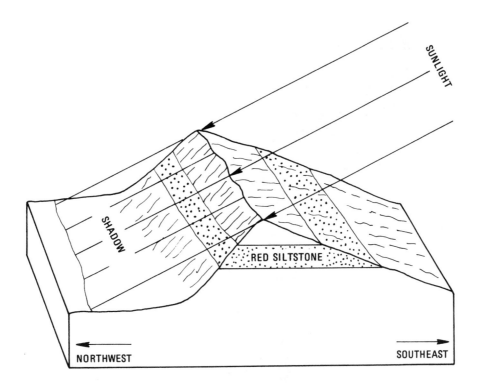

SILTSTONE REFLECTANCE			
ILLUMINATION	TM BAND 3	TM BAND 1	RATIO 3/1
Sunlight	94	42	2.24
Shadow	76	34	2.23

Figure 8-30 Suppression of illumination differences on a ratio image.

Once the boundaries for each cluster, or *spectral class*, are defined, the computer retrieves the spectral value for each pixel and determines its position in the classification space. Should the pixel fall within one of the clusters, it is classified accordingly. Pixels that do not fall within a cluster are considered unclassified. In practice the computer calculates the mathematical probability that a pixel belongs to a class; if the probability exceeds a designated threshold (represented spatially by the decision boundary), the pixel is assigned to that class. Applying this method to the digital data of the Salton Sea and the Imperial Valley scene produces the classification map of Figure 8-31B. The blank areas (unclassified category) occur at the boundaries between classes where the pixels include more than one terrain type.

There are two major approaches to multispectral classification:

1. **Supervised classification** The analyst defines on the image a small area, called a *training site*, which is representative of each terrain category, or class. Spectral values for each pixel in a training site are used to define the decision space for that class. After the clusters for each training site are defined, the computer then classifies all the remaining pixels in the scene.
2. **Unsupervised classification** The computer separates the pixels into classes with no direction from the analyst.

The two classification methods are compared by applying them to the TM data for the Thermopolis, Wyoming, subscene.

Supervised Classification The first step in the supervised classification is to select training sites for each of the terrain categories. In Figure 8-33 the training sites are indicated by black rectangles. Some categories are represented by more than one training site in order to cover the full range of reflectance characteristics. Figure 8-34 shows TM reflectance spectra for the terrain categories. Note the wider range of

A. Landsat image.

B. Classification map. A = agriculture, D = desert, M = mountains, W = water, blank = unclassified.

Figure 8-31 Multispectral classification of Landsat data for the Salton Sea and the Imperial Valley, California.

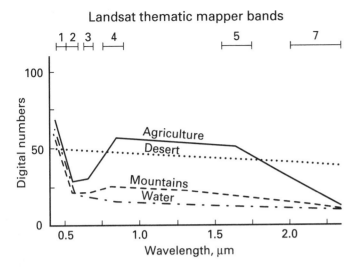

A. Spectral reflectance curves derived from TM bands. The high values in band 1 are caused by selective atmospheric scattering.

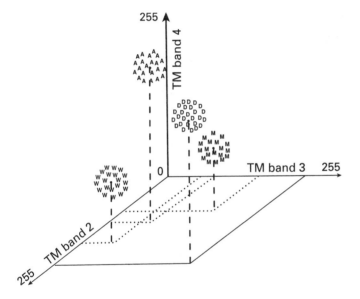

B. Three-dimensional cluster diagram for classification.

Figure 8-32 Landsat data used to classify image of the Salton Sea and the Imperial Valley, California.

spectral variation in the reflected IR bands (4, 5, and 7) than in the visible bands (1, 2, and 3). In the first attempt at supervised classification, training sites were selected for each of the geologic formations, but the results showed that individual formations of the same rock type had similar or overlapping spectral characteristics and were assigned to the same class. For example, many of the formations are sandstones, which proved to be spectrally similar despite their different geologic ages. The fi-

nal classification identified rocks as sandstone, shale, and red beds. The resulting supervised classification image (Plate 15A) uses colors to represent the six major terrain classes; black indicates unclassified pixels. The legend identifies each terrain class and the percentage of the map occupied by each class. The classification map may be evaluated by comparing it with Figure 8-33, which is based on geologic maps of the Thermopolis area. Plate 15A shows the major terrain categories accurately. Some narrow belts of the shale class are shown interbedded with the sandstone class; this is correct because several sandstone units, such as the Frontier Formation, include shale beds. The proportion of shale in the classification map is probably understated for the following reason. Shale weathers to very fine particles that are easily removed by wind and water. Sandstone, however, weathers to relatively coarse fragments of sand and gravel that may be transported downslope to partially conceal underlying shale beds. This relationship is shown where the course of dry streams that cross the outcrop of the Cody Shale is classified as sandstone because of the transported sand and sandstone gravel in the streambeds.

Unsupervised Classification Plate 15B shows an unsupervised-classification image, which was produced by a program that automatically defined 16 classes. After the 16 unsupervised classes were calculated, I reviewed them and combined them into the eight classes shown in Plate 15B. Patterns of the sandstone and shale classes are similar to those in the supervised map. The computer recognized two spectral subdivisions of the Chugwater Formation: a lower member shown in red and an upper member shown in orange (Plate 15B, Legend). The lower member is confined to the Chugwater outcrop belt, but the upper member is recognized beyond the Chugwater outcrops, where it also represents patches of reddish soil. The unsupervised classification recognized native vegetation and three distinct classes of agriculture.

Neither classification map recognized the outcrops of carbonate rocks shown in the map of Figure 8-33. The carbonate outcrops are classified as sandstone in both maps. Figure 3-1 shows reflectance spectra for sandstone and limestone, plus the spectral ranges of the TM bands. This figure demonstrates that the broad bandwidths recorded by TM bands lack the spectral resolution to distinguish sandstone from limestone. The supervised and unsupervised classifications failed to separate these rocks because the TM data were inadequate, not because the digital classification methods were inadequate. This example demonstrates the need for multispectral images with finer spectral resolution (narrower bandwidths), such as the images acquired by hyperspectral scanners.

Change-Detection Images

Change-detection images provide information about seasonal or other changes. The information is extracted by comparing two or more images of an area that were acquired at different

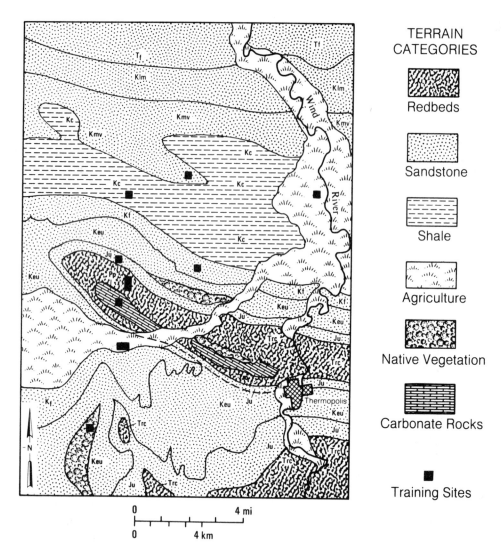

Figure 8-33 Terrain categories and training sites for supervised classification of the Thermopolis, Wyoming, subscene. Abbreviations for formations are given in Figure 3-8.

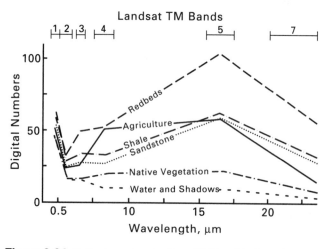

Figure 8-34 Reflectance spectra (from TM bands) of terrain categories shown in Figure 8-33

times. The first step is to register the images using corresponding GCPs. Following registration, the digital numbers of one image are subtracted from those of an image acquired earlier or later. The resulting values for each pixel are positive, negative, or zero; the latter indicates no change. The next step is to plot these values as an image in which a neutral gray tone represents zero. Black and white tones represent the maximum negative and positive differences, respectively. Contrast stretching is employed to emphasize the differences.

The change-detection process is illustrated with Landsat MSS band 5 images of the Goose Lake area of Saskatchewan, Canada (Figure 8-35). The DN of each pixel in the June 27, 1973, image (Figure 8-35B) is subtracted from the DN of the corresponding registered pixel in the September 7, 1973, image (Figure 8-35A). The resulting values are linearly stretched and displayed as the change-detection, or difference, image (Figure 8-35C). The location map aids in understanding

A. September 7, 1973, image.

B. June 27, 1973, image.

BLACK   GRAY   WHITE
$\overline{}$ ◄—————— 0 ——————► $+$
$(A < B)$ $(A > B)$

C. Difference image (image A minus image B).

D. Terrain map.

Figure 8-35 Change-detection image computed from seasonal Landsat MSS images, Saskatchewan, Canada. From Rifman and others (1975, Figures 2-14, 2-15, 2-17).

signatures in the difference image. Neutral gray tones representing areas of little change are concentrated in the northwest and southeast and correspond to forest terrain. Forest terrain has a similar dark signature on both original images. Some patches within the ephemeral Goose Dry Lake have similar light signatures on images A and B, resulting in a neutral gray tone on the difference image. The clouds and shadows that are present only on image B produce dark and light tones, respectively, on the difference image. The agricultural practice of seasonally alternating between cultivated and fallow fields is clearly shown by the light and dark tones on the difference image. On the original images, the fields with light tones have crops or stubble and the fields with dark tones are bare earth.

Change-detection processing is also used to produce difference images for other remote sensing data, such as between nighttime and daytime thermal IR images (Chapter 5).

HARDWARE AND SOFTWARE FOR IMAGE PROCESSING

The image-processing routines are implemented on computer systems that consist of hardware and software, both of which are evolving at a rapid rate. For example, when the second edition of this book was written, in 1986, a typical state-of-the-art image-processing system cost several hundred thousand dollars and was supported by a supercomputer and peripheral hardware costing tens of million dollars. In 1996 a stand-alone desktop computer system costing $15,000 replaces the earlier system.

Hardware

Personal computers can be classified according to their operating system: Mac OS (Macintosh Operating System), Windows 95, or Unix. This book is not the forum to debate the relative merits of these systems. Most systems are acceptable; the choice is largely based on personal preference. Figure 8-36 shows the basic components of a typical image processing system, such as the one I use.

The data input/output components are used to read original data, such as TM, into the system. The common current formats are 8-mm tape and CD-ROM. The second function of the input/output components is to record the digitally processed data for later playback into image prints (hard copy). The output function is also used to make backup records that safeguard against loss of data.

The original TM data are loaded onto the hard drive (not shown in Figure 8-36) which stores the original data. Several different data sets (TM, SPOT, digital terrain data) may be stored and used concurrently to create a combination image. During a processing session, a number of intermediate data sets are generated and stored on the hard drive. Therefore the hard drive should have several gigabytes (10^9 bytes) of memory, which is relatively inexpensive. The data are processed by

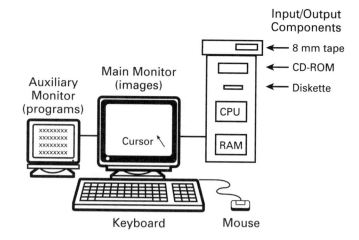

Figure 8-36 Components of a typical interactive image-processing system.

the central processing unit (CPU) which incorporates the random access memory (RAM), shown in Figure 8-36. The amount of RAM influences the speed at which the data are processed. RAM is relatively expensive, but several tens of megabytes are a minimum requirement. The system in Figure 8-36 has 96 megabytes of RAM and performs satisfactorily. New systems operate in a parallel-processing mode that accelerates processing speeds. Data are divided into subsets, typically four, which are processed simultaneously and recombined into output data.

Images and program information are shown on monitors, which are digitally controlled color display screens. Large-area, high-resolution monitors are desirable for most operations. The system in Figure 8-36 employs a second optional smaller monitor to display program information, which eliminates clutter on the large monitor. The keyboard and mouse enable the operator to interact with the system.

Software

Software is a set of instructions that commands the hardware system. A basic set of software (DOS, Apple, or Unix) is installed to operate the system. Additional application software is required for specific tasks, such as image processing. A variety of image-processing software is available for each of the three operating systems. Table 8-1 is a representative list of commercial vendors of image-processing software. All of the software packages include routines for the basic operations described in this chapter (restoration, enhancement, and information extraction), although the organization and nomenclature may be different. This book is not the forum to review the merits of different software packages, except to note that there are several ways to evaluate software:

1. Contact established users with applications similar to yours and get their opinions.
2. The Internet has bulletin boards that can be helpful.

3. Ask vendors for demonstration versions of their software that can be installed on your system for evaluation.
4. Many vendors have demonstration booths at the remote sensing conferences listed in Chapter 1, where you can try out the software.

In addition to commercial software, there are a variety of public-domain packages that are accessible via Internet. (*Note:* I am not suggesting "pirating," or copying copyrighted software. Not only is this practice illegal; it also denies compensation to the developers of the software. Without compensation, there is no incentive to develop new software that we will need in the future.)

Interactive Image-Processing Session

A typical Landsat image-processing session proceeds as follows:

1. After the CCT is loaded, the analyst selects three TM bands and assigns them to the blue, green, and red channels of the monitor. A typical combination is bands 2-4-7 shown in BGR (blue, green, and red). The TM image consists of 5667 lines, each with 6167 pixels, but the monitor in Figure 8-36 displays only 1152 lines by 870 pixels. The image is resampled to fit on the monitor; every eighth line and every eighth pixel are displayed. The mouse and on-screen cursor are used to select a representative subscene, which is displayed at full resolution with each line and pixel of original data shown on the monitor.

2. The display is examined for defects, such as banding and line dropouts, which are then restored.

3. The next step is to enhance the image. Each of the three spectral bands, together with its histogram, is viewed separately, and the contrast is enhanced. The three bands are then viewed as a color display, which typically is undersaturated. The image is transformed into its IHS components, and saturation is enhanced. At this stage it is generally useful to apply a nondirectional edge-enhancement filter to the intensity component. The image is transformed back into an enhanced BGR display. Coordinate grids (latitude and longitude; UTM) are added and the image is ready for plotting as hard copy.

4. Information extraction begins with the original data, not with the enhanced data from step 3. Images of principal components, ratios, and classifications are generated and interactively modified to suit the application.

Table 8-1 Image-processing software and vendors

Dimple
Cherwell Scientific Publishing, Inc.
744 San Antonio Road, Suite 27A
Palo Alto, CA 94303
Telephone: 415-852-0720
Fax: 415-852-0723

EASI/PACE
PCI Enterprises
50 West Wilmot Street
Richmond Hill, Ontario
Canada L4B 1M5
Telephone: 905-764-0614
Fax: 905-764-9604

ENVI
Research Systems, Inc.
2995 Wilderness Place
Boulder, CO 80301
Telephone: 303-786-9900
Fax: 303-786-9909

ER Mapper
Earth Resource Mapping
4370 La Jolla Drive, Suite 900
San Diego, CA 92122
Telephone: 619-558-4709
Fax: 619-558-2657

ImageStation
Intergraph Corp.
Huntsville, AL 35894
Telephone: 205-730-2000
Fax: 205-730-1263

IMAGINE
ERDAS
2801 Buford Highway NE, Suite 300
Atlanta, GA 30329
Telephone: 404-248-9000
Fax: 404-248-9400

TNTmips
MicroImages, Inc.
201 North 8th Street
Lincoln, NB 68508
Telephone: 402-477-9554
Fax: 402-477-9559

VI^{2}STA
International Imaging Systems
1500 Buckeye Drive
Milpitas, CA 95035
Telephone: 408-432-3400
Fax: 408-433-0965

COMMENTS

Digital image processing has been demonstrated in this chapter using examples of Landsat images that are available in digital form. It is emphasized, however, that any image can be converted into a digital format and processed in similar fashion. There are three major functional categories of image processing:

1. *Image restoration,* to compensate for data errors, noise, and geometric distortions introduced during the scanning, recording, and playback operations
2. *Image enhancement,* to alter the visual impact that the image has on the interpreter, in a fashion that improves the information content
3. *Information extraction,* to utilize the decision-making capability of the computer to recognize and classify pixels on the basis of their digital signatures.

In all of these operations the user should be aware of the trade-offs involved, as demonstrated in the discussion of contrast stretching.

A common query is whether the benefits of image processing are commensurate with the cost. This is a difficult question that can only be answered by the context of the user's needs. If digital filtering, for example, reveals a previously unrecognized fracture pattern that in turn leads to the discovery of major ore deposits, the cost benefits are obvious. On the other hand, it is difficult to state the cost benefits of improving the accuracy of geologic and other maps through digital processing of remote sensing data. However, it should also be noted that technical advances in software and hardware are steadily increasing the volume and complexity of the processing that can be performed, often at a reduced unit cost.

QUESTIONS

1. Many users advocate higher spatial resolution (smaller ground resolution cells) for imaging systems without considering the operational consequences. Assume that the Landsat TM ground resolution cell is reduced to 10 by 10 m. Refer to Figure 8-1, and calculate the following:
   a. Number of pixels per band
   b. Number of pixels per scene (seven bands)
   c. Ratio of new pixels divided by original (30 by 30 m) pixels
2. Chapter 1 described the AVIRIS hyperspectral aircraft scanner, which acquires images that cover 10.5 by 10.5 km with ground resolution cells that cover 20 by 20 m. Calculate the number of pixels in a single band of AVIRIS data. An AVIRIS scene consists of 224 spectral bands of data. Calculate the megabytes of data in an AVIRIS scene. Assume you have acquired complete AVIRIS coverage for the area of a TM scene. Calculate the megabytes of AVIRIS data. A CD-ROM stores 650 megabytes of data. Calculate the number of CD-ROMs required to store these AVIRIS data.
3. For Figure 8-8A replace 0 with 10 and 90 with 80. Calculate the filtered values to replace noise pixels.
4. For your particular application (forestry, geography, geology, oceanography, and so forth) of Landsat TM images, which of the contrast-enhancement methods in Figure 8-14 is optimum? Explain the reasoning for your selections and any trade-offs that might occur.
5. For Figure 8-16A replace all the 35s with 47s and all the 45s with 38s. Produce the following enhanced results:
   a. Filtered data set
   b. Profile of the revised original data
   c. Profile of your filtered data
6. For Figure 8-20B replace the 30s with 31; replace the 25s with 24. Use the appropriate filter (Figure 8-20A) to enhance the northeast-trending lineament. Record the enhanced pixels in a format similar to Figure 8-20D. Plot profile A–B for the original lineament and the enhanced lineament. Repeat the process for the northwest-trending original lineament. Calculate: a. the enhanced geometric expression; b. the enhanced brightness difference in percent.
7. For your particular application, assign priorities (most useful to least useful) for the information-extraction methods. Describe how you would apply and interpret each of the most useful procedures.

REFERENCES

Bernstein, R. and D. G. Ferneyhough, 1975, Digital image processing: Photogrammetric Engineering and Remote Sensing, v. 41, p. 1465–1476.

Bernstein, R., J. B. Lottspiech, H. J. Myers, H. G. Kolsky, and R. D. Lees, 1984, Analysis and processing of Landsat 4 sensor data using advanced image processing techniques and technologies: IEEE Transactions on Geoscience and Remote Sensing, v. GE-22, p. 192–221.

Bryant, M., 1974, Digital image processing: Optronics International, Publication 146, Chelmsford, MA.

Buchanan, M. D., 1979, Effective utilization of color in multidimensional data presentation: Proceedings of the Society of Photo-Optical Engineers, v. 199, p. 9–19.

Chavez, P. S., 1975, Atmospheric, solar, and MTF corrections for ERTS digital imagery: American Society for Photogrammetry and Remote Sensing, Proceedings of Annual Meeting, Phoenix, AZ.

Chavez, P. S., S. C. Sides, and J. A. Anderson, 1991, Comparison of three different methods to merge multiresolution and multispectral data—Landsat TM and SPOT panchromatic: Photogrammetric Engineering and Remote Sensing, v. 57, p. 295–303.

Goetz, A. F. H. and others, 1975, Application of ERTS images and image processing to regional geologic problems and geologic mapping in northern Arizona: Jet Propulsion Laboratory Technical Report 32-1597, Pasadena, CA.

Haralick, R. M., 1984, Digital step edges from zero crossing of second directional filters: IEEE Transactions on Pattern Analysis and Machine Intelligence, v. PAMI-6, p. 58–68.

Jensen, J. R., 1986, Introductory digital image processing—a remote sensing perspective: Prentice-Hall, Englewood Cliffs, NJ.

Leick, A., 1995, GPS satellite surveying, second edition: John Wiley & Sons, New York, NY.

Loéve, M., 1955, Probability theory: D. van Nostrand Co., Princeton, NJ.

Moik, H., 1980, Digital processing of remotely sensed images: NASA SP 431, Washington, DC.

NASA, 1983, Thematic mapper computer compatible tape: NASA Goddard Space Flight Center Document LSD-ICD-105, Greenbelt, MD.

Rifman, S. S., 1973, Digital rectification of ERTS multispectral imagery: Symposium on Significant Results Obtained from ERTS-1, NASA SP-327, p. 1131–1142.

Rifman, S. S. and others, 1975, Experimental study of application of digital image processing techniques to Landsat data: TRW Systems Group Report 26232-6004-TU-00 for NASA Goddard Space Flight Center Contract NAS 5-20085, Greenbelt, MD.

Schowengerdt, R. A., 1983, Techniques for image processing and classification in remote sensing: Academic Press, New York, NY.

Snyder, J. P., 1987, Map projections—a working manual: U.S. Geological Survey Professional Paper 1395.

Swain, P. H. and S. M. Davis, 1978, Remote sensing—the quantitative approach: McGraw-Hill Book Co., New York, NY.

ADDITIONAL READING

Arlinghaus, S. L., ed., 1994, Practical handbook of digital mapping: CRC Press, Boca Raton, FL.

Campbell, N. A., 1966, The decorrelation stretch transformation: International Journal of Remote Sensing, v. 17, p. 1939–1949.

Harris, J. R. and others, 1994, Computer-enhancement techniques for the integration of remotely sensed, geophysical, and thematic data for the geosciences: Canadian Journal of Remote Sensing, v. 20, p. 210–221.

Mayers, M. and L. Wood, 1988, Selected annotated bibliographies for adaptive filtering of digital image data: U.S. Geological Survey Open File Report 88-104.

Popp, T., 1995, Correcting atmospheric masking to retrieve the spectral albedo of land surfaces from satellite measurements: International Journal of Remote Sensing, v. 16, p. 3483–3508.

Russ, J. C., 1995, The image processing handbook: CRC Press, Boca Raton, FL.

Verbyla, D., 1995, Satellite remote sensing of natural resources: Lewis Publishers, Boca Raton, FL.

METEOROLOGIC, OCEANOGRAPHIC, AND ENVIRONMENTAL APPLICATIONS

Chapter 4 included descriptions of the major satellites and imaging systems that are designed for environmental applications, including meteorology. Remote sensing is a valuable source of environmental information about the atmosphere, continents, and oceans. All wavelength regions of the electromagnetic spectrum from ultraviolet (UV) through microwave are useful. In addition to the general-purpose imaging systems such as aerial photography and Landsat, a number of satellites have been deployed for specific environmental applications.

In 1960 the United States launched the first unmanned satellites for environmental and meteorologic monitoring. The early satellites carried miniature television cameras that produced low-resolution, visible-band images of cloud patterns for meteorologic use. There has been both a steady improvement in spatial resolution and an expansion of the spectral range into both the reflected IR and thermal IR bands. Because of these improvements, the satellite data are now used for oceanography, hydrology, and vegetation analysis in addition to meteorology. Other nations have launched improved satellites. Greenstone and Bandeen (1993) prepared a catalog of current and planned environmental satellites and their remote sensing payloads.

UV RADIATION AND OZONE CONCENTRATION

Ozone (O_3), a molecule made up of three atoms of oxygen, forms a layer in the atmosphere from 25 to 60 km above sea level that absorbs incoming UV radiation from the sun (Figure 9-1). The warm layer in the atmospheric temperature profile is caused by solar energy absorbed by the ozone layer. Without the ozone layer, incoming UV radiation from the sun would effectively sterilize the earth of all forms of life. The Halley Bay

Station on the Antarctic coast makes ground-based total ozone measurements during the daylight months. Farman, Gardiner, and Shanklin (1985) first reported the unexpected depletion of ozone over the Antarctic, commonly called the "ozone hole." These measurements were subsequently confirmed and monitored over time by satellite remote sensing. Probably no other global environmental change has caused as much worldwide concern as ozone depletion, which results from complex chemical reactions. Causes and processes of the depletion reactions are under debate (Biggs and Joyner, 1994). It is known that man-made chlorofluorocarbon gases (CFCs), used in refrigeration and other industries, are capable of depleting atmospheric ozone. As a precaution CFCs are being replaced by noninteractive substitutes in refrigeration and air-conditioning equipment.

Total Ozone Mapping Spectrometer (TOMS)

Global measurements of ozone are made by the *total ozone mapping spectrometer* (TOMS), which was first launched in 1978 on the polar-orbiting Nimbus-7. The original TOMS has operated for more than a decade past its design lifetime and made over 4100 daily global maps of total ozone concentration. In 1991 the former Soviet Union launched a Meteor-3 satellite carrying a TOMS instrument provided by NASA. Additional TOMS instruments will be carried on future satellites. The following summary is taken from Schoeberl (1993) and McPeters (1994).

TOMS operates only in daylight to measure solar UV energy backscattered from clouds or the ground. The multispectral instrument sequentially measures six narrow UV bands in the region of 312.5 to 380.0 nm. These bands form three pairs that transmit and absorb UV energy. Values for a pair of bands are compared to measure ozone concentration. If ozone were

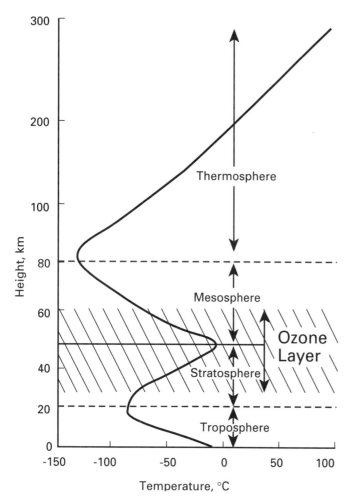

Figure 9-1 Vertical structure of the atmosphere.

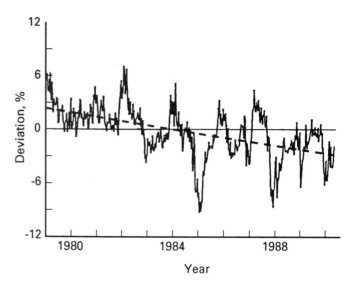

Figure 9-2 TOMS measurements of deviation in ozone concentration in latitude belt 30° to 40°N. Data are corrected for seasonal variations. The dashed trend line shows 4.5 percent ozone depletion per decade. From McPeters (1994, Figure 2).

totally absent, a UV absorption band would have the same reading as the adjacent transmission band. The difference between the pair of readings is a measure of ozone concentration. Recording three pairs of readings provides three separate measurements of ozone concentration. The data are calibrated relative to direct measurements of incoming solar UV radiation. The calibration procedure was revised to make TOMS data agree with data from a ground-based network.

The second European Remote-sensing Satellite (ERS-2), which was launched in 1995, carries a *global ozone monitoring experiment* (GOME) that covers a broader spectral range (240 to 790 nm) than TOMS 1.

Ozone Depletion Trends and the Antarctic Ozone Hole

Plate 16A shows a sequence of TOMS images (1979 to 1992) for the October monthly mean ozone concentration. The polar projection is centered near the South Pole. Ozone concentra-

tion is shown in *dobson units* (DU), which represent the physical thickness the ozone layer would have if it were brought to the earth's surface (100 DU = 1 mm thickness). The color scale in Plate 16A is calibrated in dobson units. The lowest ozone concentrations, shown in the colors black through blue, form the circular pattern centered at the South Pole that is the ozone hole. The images show the progressive development of the ozone hole, which has continued past 1992. In fact, the lowest recorded ozone concentration occurred on October 8, 1993 (not shown).

Although ozone depletion is most graphically shown at the Antarctic, it is a global trend. Figure 9-2 shows relative variation in TOMS data for the latitude belt 30° to 40°N, which includes the southern half of the United States. The data cover the period 1979 to 1990 and have been corrected for seasonal and other variations that obscure overall trends. The chart shows pronounced year-to-year fluctuations, but the dashed line shows a linear trend of –4.5 percent per decade. A similar analysis for nonpolar regions of the globe (65°N to 65°S) shows a trend of –2.6 percent. In 1992 and 1993 (not shown in Figure 9-2) global ozone dropped well below the average for 1979 to 1990. The cause of the anomalously low ozone is not certain, but the decrease may be related to atmospheric aerosols from the June 1991 eruption of Mt. Pinatubo. TOMS data are available on compact disc format from

Rich McPeters
Code 916
NASA Goddard Space Flight Center
Greenbelt, MD 20771

CLIMATE AND WEATHER

Daily and seasonal changes in the heating of earth by the sun cause atmospheric motions that are called *weather*. The sum of the seasonal variations is called *climate*. Clouds, precipitation, wind, and temperature are elements of weather and climate that are described in this section.

Cloud Mapping

Clouds are aerosols of tiny water droplets that have condensed from water vapor that has been evaporated from the surface (mostly oceans); under favorable conditions, clouds precipitate as rain or snow. Changes in cloud patterns signal weather changes, especially the advent of precipitation. Properties of clouds largely determine the amount of sunlight that reaches the earth and the amount of radiation that escapes; therefore, in addition to supplying water, clouds affect the surface temperature. Clouds are a key link in both the water and energy cycles that determine climate. Over very long time scales, climate is altered by natural causes, such as changes in the amount of sunlight, biologic changes of atmospheric composition, and geologic changes of continents and oceans. Today, however, there is evidence that human activities are affecting the natural balance of some atmospheric constituents (such as carbon dioxide and ozone), which may result in relatively fast climate changes. Predicting climate changes requires, among other factors, understanding how clouds will change and how their changes will affect the radiation balance of the earth.

Satellite remote sensing of clouds began with TIROS 1 in April 1960. Polar-orbiting NOAA satellites have routinely imaged the entire globe twice daily since the mid-1970s. A truly global array of satellites to observe the more rapid and detailed variations of clouds was formed in 1979 for the global weather experiment of the Global Atmospheric Research Program (GARP). In July 1983 GARP was succeeded by the International Satellite Cloud Climatology Project (ISCCP). Rossow's (1993) description of ISCCP is the basis for the following discussion.

Plate 16B is an ISCCP global image of thermal IR temperatures that was compiled from data acquired by an array of satellites for July 4, 1983. In order to reduce the huge volume of information transmitted every 30 min, the data is sampled at 3-hour intervals with a pixel spacing of 25 to 30 km. These data are then analyzed in three steps:

1. **Cloud detection** examines the visible and IR data to determine whether some clouds are present in each pixel and classifies the pixels as "clear" and "cloudy."
2. **Radiation analysis** compares the radiation measurements to those calculated with a radiative transfer model of the clouds, atmosphere, and surface. Atmospheric properties are specified from other data, so the clear pixels can be used to measure properties of the surface, namely temperatures

of water, ice or snow, and land. Once the atmosphere and surface are identified, cloud properties are extracted from the cloudy pixels.
3. **Statistical analysis** summarizes the information about clouds, atmosphere, and surface and displays it as a map.

Rossow (1993) describes the steps and their limitations. Plate 16B shows the results of this analysis for a period of 3 hours on July 4, 1983. Cloudy pixels are shown in a gray scale with bright signatures for cooler temperatures and dark for warmer temperatures. Clear pixels are classified as water, ice and snow, or land and are shown in their respective temperature scales.

ISCCP maps are prepared in various formats. The basic 3-hour maps are aggregated into longer time intervals showing average cloud cover on daily, weekly, monthly, and annual bases. Other maps show seasonal deviations of clouds from their annual means. Additional maps show cloud optical thickness, cloud top pressure, and temperature and reflectance of the noncloud surface.

All ISCCP data sets are available from

NOAA Satellite Data Services Division
Princeton Executive Square, Room 100
Washington, DC 20233
Telephone: 301-763-8400
Fax: 301-763-2635

Rainfall Mapping

Precipitation is the nearly universal ultimate source of the water that supports life on the earth. A better understanding of precipitation patterns and how they are affected by global changes is essential for the health and survival of humankind. The process of precipitation can change atmospheric circulation, especially in the tropics, because condensation of water vapor releases the latent heat of condensation. Thus the processes of atmospheric circulation, cloud formation, and precipitation are mutually interactive components of the global climate system.

Measuring precipitation is one of the most challenging tasks for meteorologists. Rain gauges are the earliest, and still most common, measuring device. The most serious drawback to rain gauges is that they record point measurements, whereas precipitation is highly variable in space. Even in developed regions, such as Europe and the United States, the number and distribution of gauges provides only an approximation of moisture patterns. Essentially no rain-gauge measurements are made over the oceans. For these reasons various remote sensing techniques have been developed. The first was ground-based radar systems operating at wavelengths of millimeters, which are backscattered by rain and ice particles but not by cloud aerosols. As a rotating antenna scans the atmosphere, any backscattered return signal is interpreted in terms of precipitation. Estimating rainfall rates is difficult, however, because the relationship between backscatter and rain rate is highly nonlinear. Also, a radar beam does not curve with the surface of the earth, which limits the maximum range to a few hundred kilometers and nearer to 100 km for quantitative data.

Meteorologists began to attempt estimations of precipitation from the very earliest satellite images. Today rainfall is determined from two spectral regions (visible and reflected IR, and passive microwave), which are described in the next two sections.

Rainfall Estimates from Visible and Reflected IR Images

Rainfall is related to the amount and types of cloud cover. From our own experience we can look at clouds and estimate the probability of rain. Similar estimates can be made globally and quantitatively using satellite visible and thermal IR images that record the amount and radiant temperatures of clouds. Arkin and Ardanuy (1989) reviewed the numerous methods for estimating precipitation based on cloudiness that have been developed over the past 20 years. A simple, and therefore easily and widely applied, technique is based on comparing satellite images with rain-gauge and radar measurements. In the eastern tropical Atlantic Ocean, Richards and Arkin (1981) found that coverage of large areas (280 by 280 km and larger) by clouds with radiant temperatures less than 235°K was highly correlated with accumulated rainfall at hourly to daily intervals. Richards and Arkin (1981) developed an algorithm, called the GOES precipitation index (GPI), that estimates rainfall from cloud cover on satellite images. The GPI algorithm is applied to images, such as the ISCCP example in Plate 16B, which are resampled to pixels that represent 2.5° latitude by 2.5° longitude (280 by 280 km). The images are collected every 3 hours and accumulated over 5-day periods. The fractional coverage of cold clouds for each area is determined as the ratio of the number of pixels representing temperatures less than 235°K to the total number of pixels. Rainfall is then estimated for each pixel using the GPI algorithm.

Arkin and Janowiak (1993) used this technique to prepare Plate 17A, which shows mean annual estimated rainfall (millimeters per day) for 1986 to 1994. Areas of high rainfall occur near the equator in central Africa, northern South America, and Southeast Asia. The rainiest region in the tropics (red pixels) is located in the eastern Indian Ocean west of Sumatra, where annual rainfall exceeds 14 mm · day^{-1}. Major dry areas (white pixels) are the subtropical high-pressure zones in the southern portions of the Atlantic, Pacific, and Indian Oceans and in the Sahara and Arabian Deserts (< 0.5 mm · day^{-1}).

Arkin and Janowiak (1993) present additional GPI maps that graphically show monthly, seasonal, and annual variations in rainfall. They also describe the following limitations of the method. Data are available from 40°N to 40°S latitude, but the maps are truncated at 25°N. The GPI method was calibrated for the tropical Atlantic Ocean and is useful primarily for convective rainfall. The method is less useful for the cold, extra-tropical landmasses of the Northern Hemisphere; therefore, the maps end at 25°N. Another limitation is that the maps do not truly show precipitation but rather spatial distribution of cold clouds with a color code that is highly correlated with rainfall, at least for certain areas and seasons. The spatial precision of the maps is imprecise because the expanse of cloud is often much greater than the associated precipitation. Also, the calibration was derived from a relatively small area and may not be reliable elsewhere.

The GPI maps of Arkin and Janowiak (1993) are available from

Lola Olsen
NASA Goddard Space Flight Center
Greenbelt, MD 20771

Roy Jenne
National Center for Atmospheric Research
P.O. Box 3000
Boulder, CO 80307

Phil Arkin or John Janowiak
National Center for Environmental Prediction, NOAA/NWS
Washington, DC 20233

Rainfall Estimates from Passive Microwave Images

Passive microwave systems are cross-track scanners that use radiometers to detect energy radiated from the surface at microwave wavelengths. At these longer wavelengths the intensity of radiant energy is very low; therefore, ground resolution cells are large, on the order of many square kilometers.

There are presently two distinct methods for estimating precipitation from passive microwave data (J. E. Janowiak, NOAA/ NWS, personal communication):

1. **Emission methods** use observations of the thermal emissions from precipitation to estimate rainfall rates over ocean surfaces, where surface emissivities are uniformly low. This method cannot estimate rainfall over land because of large variations in emissivity of different land surfaces. The method is capable of detecting both convective and "warm process" (stratiform, orographically induced) rainfall.

2. **Scattering methods** infer precipitation from the scattering of upwelling radiation due to precipitation-size ice particles above the rain layer in convective systems. This method is useful for quantifying precipitation over both land and water, but it can only detect rainfall associated primarily with convective systems that penetrate the freezing level. Although this method is incapable of detecting shallow, "warm" rainfall, such as that associated with orographic lifting, it will typically account for the stratiform component of convective rainfall in a statistical sense.

The special sensor microwave/imager (SSM/I) is carried on Defense Meteorologic Satellite Program (DMSP) polar-orbiting satellites. SSM/I records four wavelengths of microwave energy (0.35, 0.81, 1.35, and 1.55 cm) with an image swath width of 1400 km and a nominal ground resolution cell of 55 by 55 km (Table 4-1). Both vertically (V) and horizontally (H) polarized energy is recorded, except for the 1.35-cm band, which records vertically polarized energy only. No cross-polarized energy is recorded because no energy is transmitted in this passive system. Grody (1991) developed a decision tree that identified precipitation from all other atmospheric and surface features by using the following four SSM/I bands: 1.55 cm (V and H), 1.35 cm (V), 0.35 cm (V). Weng, Ferraro, and Grody (1994) developed an SSM/I scattering index that was calibrated against known radar rates of rainfall (millimeters per hour). The scattering index is calculated for each pixel identified as precipitation to determine its rainfall rate. Weng, Ferraro, and Grody (1994) show global rainfall maps for August 1993. These maps cover virtually the entire globe, unlike the cloud-derived maps (Plate 17A), which are limited to an equatorial belt. Another advantage of the microwave method is that it measures the actual response of rain rather than inferring rainfall from cloud temperatures. SSM/I data are available from

National Snow and Ice Data Center
University of Colorado, Campus Box 449
Boulder, CO 80309
Telephone: 303-492-2378
Fax: 303-492-2468

Wind Patterns over the Oceans

Accurate information on surface wind patterns over the oceans is essential because:

1. Winds drive the ocean currents.
2. Fluxes of heat, moisture, and momentum across the air-sea boundary are important factors in forming, moving, and modifying water masses. These interactions also intensify storms near coasts and over the open ocean.

Prior to satellite remote sensing observations, ocean winds were measured by anemometers on ships and buoys, which lack coverage, accuracy, and timeliness. Unlike anemometers, remote sensing systems do not record winds directly. However, wind blowing across the ocean produces waves ranging in wavelength and height from capillary waves to ocean swells. The size and direction of waves is determined by the velocity and direction of the surface wind. Active and passive satellite systems operating in the microwave region record surface characteristics of the ocean. These data are processed and interpreted to produce timely maps of wind patterns. Radar scatterometers (described later in the section "Sea Ice) record the

roughness of the sea, which is used to estimate surface wind vectors, with some ambiguity in direction. Scatterometer data from satellites are very limited at present but will be available from future systems.

Passive microwave systems record energy radiated from the ocean at microwave wavelengths. The intensity of microwave energy is the product of emissivity times kinetic temperature of the water. Emissivity varies with the roughness of the water and is used to estimate wind velocity. Table 4-3 lists characteristics of the SSM/I carried on the polar-orbiting DMSP satellite. SSM/I records passive microwave data for measuring surface wind speeds, but not directions, over the oceans. SSM/I data are combined with conventional surface wind observations and analyses from the European Center for Medium-Range Weather Forecasting (ECMWF). Atlas, Hoffman, and Bloom (1993) describe how these data are combined in a filtering procedure. The resulting analysis is used to assign directions to the wind velocities, which are shown in maps as vectors called *streamlines*. Plate 18A is a map of the annual average of surface wind streamlines (shown by arrows) and vector magnitude (shown in meters per second by color code). The map covers the period from July 1987 through June 1988. The *intertropical convergence zone* (ITCZ) is a major feature in Plate 18A. In the eastern Pacific Ocean, northerly winds from the coast of South America converge with southerly winds from the coast of North America. Off Central America the merged streamlines abruptly turn westward as a zone of lower velocity, that is the ITCZ (Plate 18A). The ITCZ has a similar pattern in the Atlantic Ocean, where northerly and southerly streamlines along the west coast of Africa converge at the equator and turn westward at lower velocities. Plate 18A shows that the average wind speed varies significantly along the ITCZ; note the increase in velocity westward across the Pacific Ocean.

Weather Prediction

Images from GOES East and West satellites are acquired every 30 min, 24 hours daily, but a faster schedule may be used during periods of severe weather. Data are also transmitted to National Weather Service satellite service units, which distribute information to subscribers. Epstein and others (1984) summarized the following applications of GOES data:

1. Cloud-motion wind vectors are automatically computed for low-level winds and are manually processed for mid- and high-level winds. Over 1200 vectors are generated each day.
2. Precipitation resulting from convective storm systems and hurricanes is estimated.
3. Freeze warnings are provided by defining surface-temperature patterns associated with rapid radiative cooling in citrus and vegetable growing areas. The nearly real-time, continuous monitoring allows farmers to delay the start of

A. Visible-band image of Hurricane Alicia approaching Galveston, Texas, August 17, 1983.

B. Thermal IR image of a severe weather outbreak in western Kansas and southeast Oklahoma, April 2, 1982.

Figure 9-3 Enlarged GOES images of storms. From Epstein and others (1984, Figures 3 and 5). Courtesy W. M. Callicott, NOAA Satellite Services Division.

preventive measures until a freeze is imminent, thus saving thousands of dollars.

4. Snow-cover analyses are used to predict the snowmelt runoff potential in river basins across the nation. Analyses are derived from both the meteorologic satellites and from Landsat images.

5. Hurricane classification is performed using a semiobjective method of classifying the stage of development and wind intensity from cloud patterns. Figure 9-3A shows Hurricane Alicia in August 1983 when its center was about to make landfall at the Texas-Louisiana border.

6. Messages describing weather systems and features interpreted from satellite images are prepared and distributed over the National Weather Service communications links at regular intervals daily.

7. The 30-min image sequences are combined to form animated cloud images. The animation reveals the dynamics associated with the cloud development, providing the analyst with an added dimension for interpreting the data.

8. Severe convective storm development is monitored by combining cloud-top temperatures with cloud motions at different levels to describe and classify the events in terms of severity and to anticipate where severe weather will eventually develop. Figure 9-3B is an enhanced GOES infrared image showing a severe weather outbreak over the midwestern states from which tornadoes developed in April 1982. The enhancement scheme reverses the gray scale at key temperatures to accentuate important features and re-

sults in the coldest clouds with the deepest convection having black signatures.

9. Detection of fog and the rate of dissipation are determined to support shipping, fishing, and aviation interests.

OCEAN PRODUCTIVITY

The food chain in the world's oceans begins with simple one-celled microscopic plants called phytoplankton that are concentrated in the upper few tens of meters of sunlit water. The distribution patterns of phytoplankton are determined by concentrations of nutrients in seawater that in turn are determined by upwelling currents that bring cold, nutrient-rich water to the surface. Phytoplankton have the characteristic spectral signature of chlorophyll in the visible region: blue and red wavelengths are absorbed; green wavelengths are reflected. The coastal zone color scanner (CZCS, see Chapter 4) records images at these wavelengths that are processed to map phytoplankton concentration.

Chlorophyll in the Ocean

Near coastlines and the mouths of rivers, color of the ocean may determined by the amount of suspended clay and silt. For most of the ocean, however, suspended organic constituents influence ocean color. The most important organic constituent is phytoplankton, which contain chlorophyll. Figure 9-4 shows

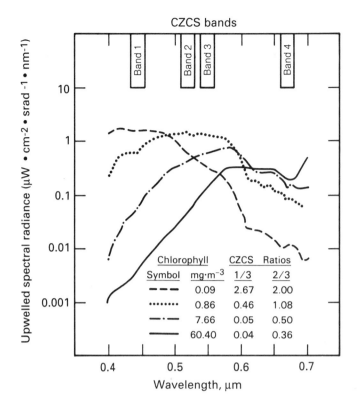

Figure 9-4 Spectra of seawater with various concentrations of chlorophyll. Note positions of CZCS visible bands. From Hovis and others (1980, Figure 1).

radiance spectra in the visible region for four types of seawater, in which chlorophyll content ranges from 0.09 to 60.40 mg · m⁻³. Chlorophyll content increases by approximately an order of magnitude for each of the four water samples. In the blue band, water with the lowest chlorophyll content has a radiance 1000 times that of water with the highest chlorophyll content. In the green band, however, the radiance difference from minimum to maximum chlorophyll is only 10 times. These spectral differences are seen in the color of seawater, which is deep blue in the open ocean where concentrations of phytoplankton are low. Water in coastal zones and zones of upwelling, however, is rich in nutrients that support phytoplankton. The resulting absorption of blue light by chlorophyll causes green hues in the water. The spectral radiance curves in Figure 9-4 indicate that CZCS bands 1, 2, and 3 are most suitable for interpreting chlorophyll concentrations.

Gordon and others (1980, 1983) described the following procedure to derive chlorophyll concentrations from CZCS data:

1. Pixels are classified as land, clouds, or water.
2. The water pixels are corrected for atmospheric scattering.
3. Ratio images partially compensate for the influence of non-chlorophyllous material, such as mud and silt, in the water.

The ratio of CZCS bands 1/3 is sensitive to chlorophyll variations at low concentrations. At the two highest chlorophyll concentrations in Figure 9-4, the differences in this ratio are too small (0.01) to be reliable. Ratio 2/3 is more sensitive to chlorophyll variations at high concentrations. Therefore, both ratios are calculated: ratio 1/3 is used for chlorophyll concentrations of 1.5 mg · m³ or less; ratio 2/3 is used for higher concentrations.

4. The ratio values are converted into chlorophyll concentrations, using observations of chlorophyll and spectral radiance in coastal waters of the United States. For chlorophyll concentrations of 0.08 to 1.5 mg · m⁻¹, errors range from 30 to 40 percent.
5. Chlorophyll concentrations are displayed as maps such as Plate 17B.

Although the algorithms are based on a limited number of stations in U.S. coastal waters, later studies have shown this method to be valid for much of the world's oceans. McClain, Feldman, and Esaias (1993) describe limitations of the method and show global maps of chlorophyll distribution for various seasons. Hovis (1984) edited a collection of CZCS images for coastal regions.

Productivity and Upwelling along the California Coast

Plate 17B is a CZCS image of the California coast that Abbott (1984) prepared by the method described above. The color scale shows high chlorophyll concentrations in red, orange, and yellow hues; low concentrations are shown in blue and magenta. Nutrient-rich water is concentrated in a narrow zone along the coast that contrasts with nutrient-poor water farther offshore. White patches are clouds. Plate 17C shows sea-surface temperature from an AVHRR thermal IR image that was acquired a few hours after the CZCS image. The narrow belt of cool temperatures (blue and green hues) along the coast is caused by upwelling of cool nutrient-rich water. There is a very close correlation between the cool water and the high concentrations of chlorophyll shown in Plate 17B. Even the narrow offshore filaments of cool water are accompanied by high chlorophyll concentrations. Along the southern California coast (not shown in Plate 17B,C) periods of strong upwelling in the summer are accompanied by "red tides" caused by blooms of dinoflagellates in the nutrient-rich colder water.

The correlation between cool water and high apparent productivity is not perfect. One mismatch occurs at the mouth of San Francisco Bay. A plume of warm water is accompanied by high values on the CZCS image. At this locality the CZCS image actually shows colors caused by suspended sediment rather than chlorophyll.

Global Productivity of the Oceans

Plate 18B is a composite image of the globe. Average annual chlorophyll concentration of the oceans, determined from the CZCS, is shown with a color scale similar to that in Plate 17B. Vast expanses of the ocean have very low productivity, shown by magenta and dark-blue hues. Areas of higher productivity occur at higher latitudes and in narrow coastal belts, such as the west coasts of Africa and North and South America. Color patterns on the continents show density of vegetation determined from AVHRR data, as described in Chapter 12.

In many parts of the world, upwelling and the associated productivity are seasonal events. Brock and McClain (1992) analyzed a series of CZCS images of the northwestern Arabian Sea along the coast of Yemen and Oman. In August and September southwesterly monsoon winds cause strong upwelling and dense phytoplankton blooms that extend far offshore. For much of the year, however, productivity is low.

Along the coast of Peru, offshore winds normally cause strong upwelling in the Pacific Ocean that results in high productivity and abundant fish. Periodically, however, the winds weaken, upwelling ceases, and coastal waters become warm and unproductive. This event is called "El Niño" because it begins around Christmas time, when the infant Jesus (El Niño) was born. The effects of El Niño are felt as far north as the California coast, where unusually warm water brings subtropical fish species (yellowfin tuna, dorado, and wahoo) well north of their normal ranges. El Niño events are now predicted from images showing anomalously warm ocean temperatures of Peru in the fall. Fiedler (1984) used seasonal images acquired by CZCS and AVHRR to monitor the 1982 to 1983 El Niño along the U.S. Pacific coast. Subsequent events have been predicted and monitored in a similar fashion.

OCEAN CURRENTS

Traditional methods for mapping oceanic circulation patterns employ current meters, drift floats, and direct temperature measurements. In addition to being expensive, these methods are hampered by the difficulty of obtaining simultaneous data over broad expanses of water. These problems are largely overcome by satellite remote sensing systems that provide essentially instantaneous images of circulation patterns over very large areas. Current systems are mapped by recognizing some property of a water current that differs from that of the surrounding water. Remote sensing systems record the following properties of water:

1. **Color due to suspended material such as sediment and plankton** Images in the visible region acquired by Landsat, CZCS, and other satellites are employed.
2. **Radiant temperature** The thermal IR bands of GOES, AVHRR, Landsat TM, and others are used. Chapter 5 shows HCMM day and night thermal IR images of the Pacific Ocean off central California.
3. **Surface roughness** The faster moving water of a current system commonly has a pattern of small waves that differs from the pattern of adjacent water. These roughness differences are readily detected by their brightness patterns in radar images such as Seasat, JERS-1, and ERS-1,2. Chapter 7 shows an ERS-1 image of currents in the Saint Lawrence River.

Visible-Band Images

The Landsat MSS band 1 (green) images in Figure 9-5 show seasonal changes in current patterns of the Cape Mendocino region, California. On both images sediment plumes originate at the mouths of the Mad, Eel, and Mattole Rivers. The California Current flows southward at overall speeds of generally less than 0.25 m · sec^{-1} and controls circulation patterns for much of the year, as shown by the southward movement of sediment plumes on the April image (Figure 9-5B). The Davidson Current is a deep countercurrent that flows northward along the California coast. For most of the year it travels at a depth below 200 m. In the late fall and winter, however, north winds are weak or absent and the countercurrent appears at the surface, inshore from the main California Current. The January image (Figure 9-5A) shows the northward movement of sediment plumes propelled by the Davidson Current. Figure 9-5C shows the seasonal current patterns that have been interpreted from the images.

The CZCS image of Plate 17B shows the California Current on July 7, 1981, based on the distribution of chlorophyll. This image shows the southward flow of the California Current that prevails for much of the year. The CZCS image covers a more extensive area than the Landsat images; the Cape Mendocino region is a small area near the center of the CZCS image. CZCS pixels are 10 times larger than those of Landsat MSS, but this coarser resolution is not a problem at the regional scale of current patterns.

Thermal IR Images

Most oceanic current systems are significantly warmer or cooler than the adjacent water and are therefore easily recognized in thermal IR images. Plate 17C is an AVHRR image of the California coast that was acquired shortly after the CZCS image (Plate 17B). The band of cool water along the coast is the California Current, which was originally thought to be a broad, slow, southward-moving current. It is apparent from satellite images that this conception is far too simple. The California Current actually consists of a series of eddies and jets with a net southward drift. The jets themselves are interesting features because they are associated with the larger promontories along the coast. Many of the jets extend several hundred kilometers offshore. At the time the AVHRR and CZCS images were acquired, an unusually strong high-pressure center in the Pacific Northwest was coupled with a low-pressure center off the coast of southern California. The result was strong southerly winds along the coast with speeds of 15 m · sec^{-1} at Cape Mendocino (Abbott, 1984).

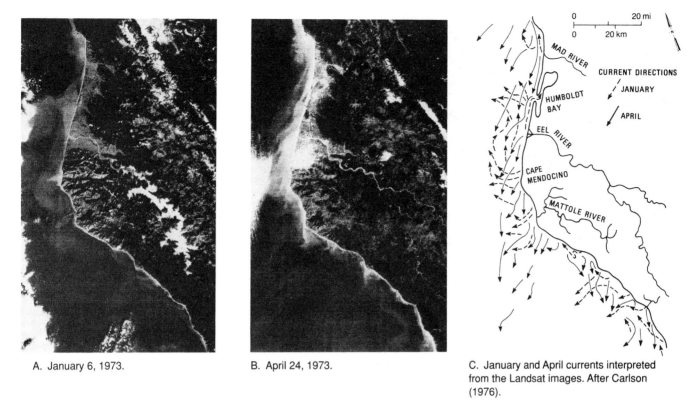

A. January 6, 1973.

B. April 24, 1973.

C. January and April currents interpreted from the Landsat images. After Carlson (1976).

Figure 9-5 Landsat MSS band 1 images showing seasonal changes in current and sedimentation patterns off Cape Mendocino, California.

Satellite thermal IR images are routinely employed to monitor oceanic phenomena, including upwelling patterns and El Niño currents in the eastern Pacific Ocean, which have major impacts on commercial fisheries. The daily acquisition of AVHRR images enables oceanographers to monitor detailed changes in current patterns.

Radar Images of Internal Waves

Waves form at the interface between fluids of different density; the well-known example is surface waves, or wind waves, which form at the interface between water and air. Within water bodies, the *thermocline* is the interface between the surface layer of warmer, less-dense water and the underlying layer of colder, denser water. In shallow seas, tidal currents that encounter seafloor irregularities, such as submarine canyons and breaks in slope, may cause waves at the thermocline (Figure 9-6). These waves are commonly called *internal waves* but are also known as *solitons* or *internal gravity waves*. The circulation pattern of an internal wave causes a low linear bulge at the surface that is accompanied by distinctive patterns of small-scale surface waves. The circulation cell rises toward the sea surface at the trailing edge of an internal wave, which decreases the roughness of the small-scale surface waves and results in a linear band of smooth water, called a *slick*, parallel

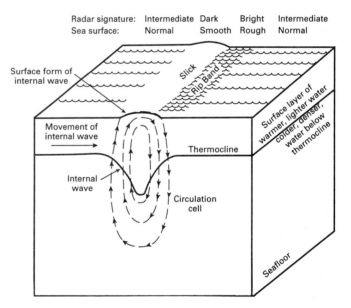

Figure 9-6 Model of an internal wave showing relationships between circulation pattern, surface roughness, and radar signatures. Modified from Osborne and Burch (1980, Figure 3).

A. Oblique aerial photograph showing research vessel *Endeavor*.

B. Aircraft L-band image (23.5 cm). Very bright spot is the *Endeavor*. Sea conditions are similar but not identical to those in the photograph.

Figure 9-7 Internal waves in the Gulf of Georgia, British Columbia, Canada, July 1978. Courtesy J. F. R. Gower, Canada Institute of Ocean Sciences.

with the crest of the internal wave (Figure 9-6). The circulation cell descends at the leading edge, causing a band of rougher water, called a *rip band*.

Figure 9-7A is an oblique aerial photograph of alternating slicks and rip bands accompanying internal waves in the Gulf of Georgia on the Pacific coast of Canada. The rip bands have bright signatures because of increased sunglint from the rougher water, and the slicks are dark (Hughes and Gower, 1983). Figure 9-7B is an aircraft radar image that was acquired 8 min after the photograph. The rip bands have bright radar signatures, and the slicks are dark. The Canadian research vessel *Endeavor* is present in both images; the metal hull and superstructure produce the very bright radar signature.

Figure 9-8A is a Seasat image of the Mid-Atlantic Bight that spans the continental slope on the east (right) and the shallow shelf on the west (left). The shelf break trends northeast across the image. Internal waves are generated where tidal currents encounter the shelf break. The waves occur in *packets* of parallel waves, each packet having up to 30 waves. The arcuate wave fronts are convex toward the direction of travel, which is generally shoreward (westward), and they travel at an estimated speed of 0.3 m • sec^{-1}. Length along the crests is about 50 km, and maximum wavelength within packets is 1.3 km. A few east-trending wave packets may have been generated by currents that interacted with east-trending submarine canyons.

Figure 9-8B is a SIR-A image of a packet of large internal waves in the Andaman Sea with crest lengths of 100 km and wavelengths of 2.5 to 6.5 km. The pattern of rip bands (bright) and slicks (dark) is clearly seen. Osborne and Burch (1980)

studied internal waves during offshore oil-drilling operations in the Andaman Sea. Under normal sea conditions the small surface waves were 0.6 m high; the rip bands had wave heights of 1.8 m and traveled at speeds of 2.2 m • sec^{-1}; the slicks had ripples approximately 0.1 m high. Additional satellite radar images of internal waves are illustrated in Fu and Holt (1982) and in Ford, Cimino, and Elachi (1983).

Internal waves are also expressed in Landsat images by bright and dark bands that correspond to rip bands and slicks, respectively. Figure 9-9 is a Landsat image with a number of internal wave packets present. The image is of the Atlantic shelf off Cape Cod, Massachusetts, acquired several years before and 500 km to the northeast of the Seasat image in Figure 9-8A. Notice that the patterns and dimensions of the waves in both images are similar. Mariners and oceanographers knew about internal waves for many decades before satellites were launched; satellite images, however, have shown the waves to be far more common and widely distributed than previously suspected.

SEA ICE

Increased shipping activity and petroleum exploration in Arctic waters have increased the need for information on sea-ice conditions. Global weather predictions require information about the thermal conditions of the polar seas, which in turn are related to ice abundance. Thus information on ice cover can aid meteorologists. Remote sensing, especially from satel-

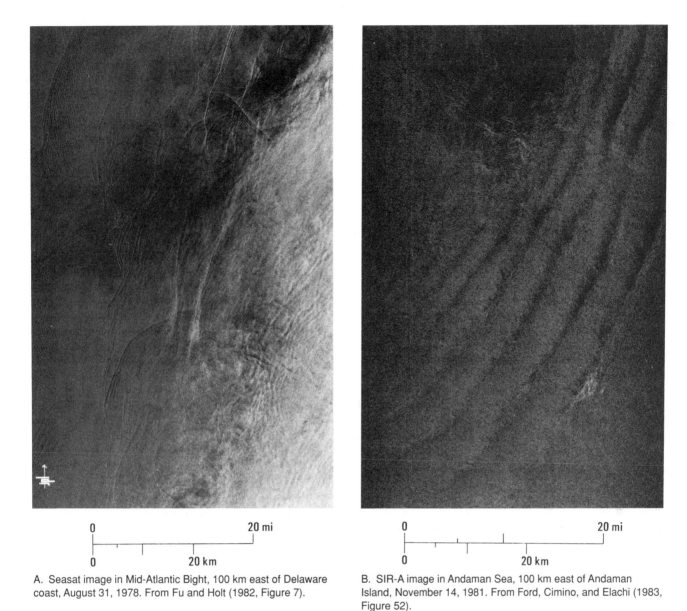

A. Seasat image in Mid-Atlantic Bight, 100 km east of Delaware coast, August 31, 1978. From Fu and Holt (1982, Figure 7).

B. SIR-A image in Andaman Sea, 100 km east of Andaman Island, November 14, 1981. From Ford, Cimino, and Elachi (1983, Figure 52).

Figure 9-8 Satellite radar images of internal waves. Courtesy J. P. Ford, JPL (Retired).

lites, is the only practical way to map sea ice on a regional, repetitive basis. The most important sea-ice features are defined in Table 9-1, which is summarized from the more extensive nomenclature of the World Meteorological Organization. Most of these features are illustrated in the images of the following sections.

Landsat Images

Figure 9-10 is a Landsat MSS band 7 image of Dove Bay, on the east coast of Greenland, which illustrates many of the features in Table 9-1. The prominent flaw lead separates the pack ice to the east from the fast ice to the west that is attached to the shore. The flaw lead and many other leads in the pack ice are refrozen, as indicated by their dark gray tone; open leads have dark signatures. Fragments of broken floes occur along the southern part of the flaw lead and between large floes of the pack ice. Individual floes have a wide range of sizes and shapes. This pack has a concentration of over 90 percent ice floes. The pack is classified as *consolidated* because the leads are refrozen and there is no open water. *Open pack ice* has approximately equal proportions of ice and water, and the floes are not in contact. A swarm of icebergs in the bottom-left corner of Figure 9-10 calved from coastal glaciers in the summer of 1972 but were locked in the fjord by winter fast ice before they could enter Dove Bay.

0   20 mi

0   20 km

Figure 9-9 Landsat MSS band 5 image of internal waves on the Atlantic shelf off Cape Cod, Massachusetts. Courtesy J. R. Apel, The Johns Hopkins University.

Landsat orbits provide images as far north and south as 81° latitude, although above 70° latitude illumination is insufficient to acquire images from late October to late March. Convergence of orbits at these high latitudes provides up to

three or four consecutive days of coverage of the same area during each 16- or 18-day cycle. These repeated images may be used to measure movement of sea ice. Figure 9-11 shows images acquired on March 20 and 21, 1973, in the Davis Strait between Greenland and Baffin Island. The images were registered to each other by the latitude-and-longitude grid. The eastern one-quarter of both images is covered with brash ice and very small floes. Most of the area is covered by large floes up to 40 km long. On the March 20 image (Figure 9-11A) there are few open leads, indicated by black signatures, and many of the leads are refrozen, shown by the gray signatures. The open leads are wider and more abundant on the March 21 image (Figure 9-11B), and some of the larger floes have broken up. The arrows in Figure 9-11B are vectors that show the direction and amount of ice movement during the 24-hour period. Ice movement was consistently southeastward at an average rate of 0.4 km $\cdot$ h^{-1}. This rate assumes that the ice traveled a straight path; if it had followed an irregular course, the actual rate of movement would be higher.

Thermal IR Images

Thermal IR systems can acquire images during periods of polar darkness but not when heavy clouds and fog are present. One can estimate relative ice thickness from thermal IR signatures. Figure 9-12 shows thermal IR images acquired in November 1983 with an aircraft cross-track scanner of the Arctic Ocean north of Alaska. The bright horizontal line records the aircraft flight path. The irregular bright line is a topographic profile of the surface recorded with a laser altimeter along the flight path; peaks directed toward top of page are topographic highs.

Open water has the highest radiant temperature and brightest signature in the images. Sea ice insulates the relatively

Table 9-1 Sea-ice terminology

Feature	Description
Fast ice	Ice that forms adjacent to and remains attached to the shore. May extend seaward for a few meters to several hundred kilometers from the coast.
Floe	Any relatively flat piece of sea ice 20 m or more across. Floes are classified according to size.
Ice concentration	Percentage of total sea-surface area that is covered by ice.
Pack ice	General term for any area of sea ice, other than fast ice, regardless of form or forms present. Pack ice is classified by concentration of the floes.
Lead	Any fracture or passageway through sea ice that is navigable by surface vessels. Leads may be open or refrozen. A *flaw lead* separates fast ice from pack ice.
First-year ice	Sea ice of not more than one winter's growth. Thickness ranges from 0.3 to 2.0 m.
Multiyear ice	Old ice that has survived more than one summer's melt. Compared to first-year ice, floes of multiyear ice are thicker and rougher and have rounder outlines.
Pressure ridge	Wall of broken ice forced up by pressure.
Brash ice	Accumulations of floating ice made up of fragments not more than 2 m across. The wreckage of other forms of ice.

Figure 9-10 Landsat MSS band 4 image of sea ice in Dove Bay, east coast of Greenland, March 25, 1973.

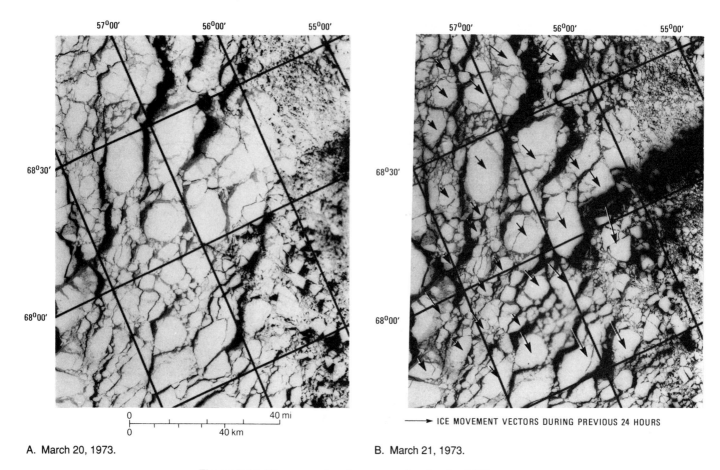

57°00'　　　56°00'　　　55°00'　　　　　57°00'　　　56°00'　　　55°00'

68°30'

68°00'

0　　　　　　　　　40 mi
0　　　　　　　　　40 km

→ ICE MOVEMENT VECTORS DURING PREVIOUS 24 HOURS

A. March 20, 1973.　　　　　　　　　　　B. March 21, 1973.

Figure 9-11　Movement of sea ice measured on Landsat MSS
images of Davis Strait, west of Greenland.

warm water beneath it. Larger amounts of radiant energy are transmitted to the surface of thin ice and smaller amounts to the surface of thicker ice. As a result, thin ice appears warmer than thick ice. Figure 9-12A illustrates these relationships between ice thickness and radiant temperatures. The center of the image has an extensive area of open water with a bright signature (warm temperature). The right portion of the image is covered with a thin sheet of first-year ice with a gray signature (intermediate temperature). The left portion is a large rounded floe of thicker, multiyear ice with a dark signature (cool temperature). The open water contains both multiyear floes and first-year floes. The multiyear floes are larger and cooler and have rounded outlines. The first-year floes are smaller and warmer with angular outlines. The general relationship between radiant temperature and ice thickness can be altered by other factors, such as variations in emissivity and the presence of water on the ice surface during the summer melting period.

The laser profile shows a relationship between ice type and surface roughness. Patches of open water in the center of Figure 9-12A have smooth profiles; the first-year ice in the right portion has minor irregularities; the multiyear floes have rough surfaces caused by their history of fracturing, thawing, refreezing, and ablation by Arctic storms.

Figure 9-12B shows multiyear floes separated by leads that are largely refrozen and have intermediate temperatures. A few narrow open leads are recognizable by their relatively warm temperature. The faint diagonal pattern from lower left to upper right results from wind streaks (Chapter 5). The laser profile drifts toward the top of the image because of changing aircraft altitude. Figure 9-12C shows extensive open and refrozen leads. The multiyear floe in the lower right corner of the image is cut by several pressure ridges, one of which is crossed by the profile that records several meters of elevation. The thicker ice of the ridge has an irregular narrow thermal IR signature that is cooler than the surrounding floe. Some ridges are accompanied by parallel fractures with warmer signatures.

Thermal IR images acquired by satellite systems such as AVHRR are used to compile regional maps of ice concentration in polar regions and are especially useful during periods of winter darkness. The thermal IR band 6 of the Landsat TM, with its 120-m spatial resolution, is also useful for mapping ice.

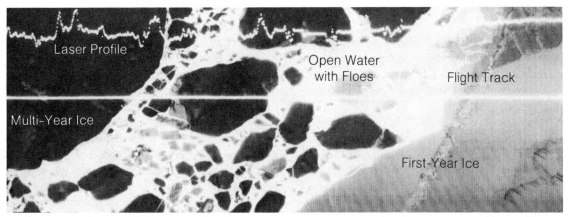

A. First-year and multiyear ice with open water.

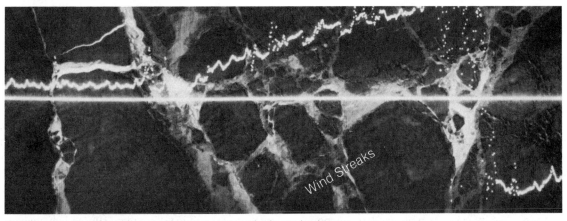

B. Floes of multiyear ice separated by open water and refrozen leads.

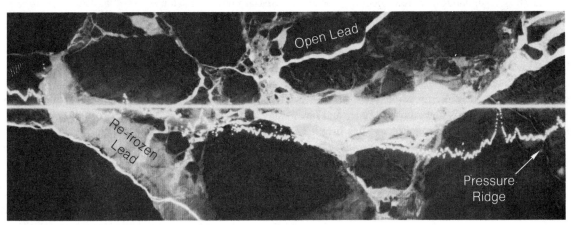

C. Multiyear ice with leads and pressure ridges.

Figure 9-12 Aircraft thermal IR images (8 to 12 μm) of sea ice in the Arctic Ocean north of Alaska, November 1983. Bright signatures are warm; dark signatures are cool. Cross-track width is 480 m. Courtesy Chevron Corporation, Marathon Oil Company, Shell Development Company, and Sohio Petroleum Company.

Radar Images

Radar is capable of acquiring images during both darkness and bad weather, which is essential for investigating sea ice. Radar images record surface roughness as a function of radar wavelength and depression angle as defined by the roughness criteria in Chapter 6.

Aircraft Radar Images Radar signatures of sea-ice features are shown in aircraft X-band images of the Beaufort Sea off northern Canada. Figure 9-13A shows an extensive sheet of first-year ice that encloses some multiyear floes and is cut by open and refrozen leads and by pressure ridges. The rough surface of multiyear ice causes strong backscatter and bright radar signatures. Calm water in open leads causes little or no backscatter and has very dark signatures. First-year ice has minor roughness and a dark gray signature. The very narrow, very bright lines crossing the first-year ice are highlights from pressure ridges. Broader bands with light-gray signatures are leads filled with brash ice.

Figure 9-13B is located near the terminus of glaciers in the Arctic islands that produce icebergs similar to those in the Landsat image (Figure 9-10). The left two-thirds of Figure 9-13B show a sheet of first-year fast ice attached to a small island of bedrock in the northwest. The fast ice encloses a number of icebergs calved from nearby glaciers. Tabular icebergs in the upper left area have flat but rough surfaces; the other icebergs are irregular. The right portion of the image has floes of first-year ice separated by calm open water (dark signature) and by brash ice (bright signature). A few icebergs are included with the floes; three large tabular icebergs occur along the lower right margin of the image, and several irregular icebergs occur in the upper right area. This example demonstrates the value of radar for detecting and tracking bergs as they enter shipping lanes. Kirby and Lowry (1981) published radar images of bergs and discussed their interpretation.

As stated earlier, younger ice is generally smoother and has darker radar signatures; older ice is rougher and has brighter signatures. During the summer melting season, however, thin sheets of water may cover the ice and partially mask the roughness characteristics. Even under these circumstances, experienced interpreters can recognize ice types based on morphology and distribution patterns.

Satellite Radar Images In 1978, Seasat acquired repetitive images of Arctic sea ice. The L-band Seasat system had a longer wavelength (23.5 cm) and steeper depression angle (70°) than do typical aircraft X-band systems. As a result the Seasat smooth and rough criteria (1 cm and 6 cm, respectively) were higher than for the aircraft images in Figure 9-13. Figure 9-14 illustrates two Seasat images acquired at a 3-day interval in October 1978. Fu and Holt (1982) describe these and other images. The northwest corner of Banks Island in the Beaufort Sea occurs in the southeast corner of each image. A sheet of smooth fast ice is attached to the west coast of Banks Island. Smooth ice is distinguished from calm water by the presence of bright pressure ridges. The pack ice consists of floes of multiyear ice, many of which are aggregates of smaller floes that are separated by brash ice. The very rough brash ice has a brighter signature than the floes. A conspicuous floe is Fletcher's ice island, a tabular iceberg 7 by 12 km in size. The bright signature of much of the iceberg is attributed to low corrugated ice ridges and scattered rock debris inherited from its glacial origin on Ellsmere Island. This iceberg, which is also called T-3, was discovered in 1946 and has been tracked since then, remaining in the clockwise circulation pattern of the Beaufort Sea.

Considerable ice movement occurred during the 3-day interval in the acquisition of the two Seasat images. The vectors in Figure 9-14B were plotted by connecting positions of individual floes, using the technique applied to repetitive Landsat images. The pack directly north of Banks Island was stable, but on the west and northwest, floes moved southward approximately 20 km at an average rate of $0.3 \text{ km} \cdot \text{h}^{-1}$. In the earlier Seasat image (Figure 9-14A) the leads were narrow, but 3 days later they were extensive in the moving portion of the pack. Many of the leads in Figure 9-14B have dark signatures indicating smooth, calm water. The gray patches within the leads represent the formation of new ice as the leads began to freeze.

In other Seasat images of the Arctic seas, areas of open water commonly have bright signatures due to small-scale waves generated by wind; these areas should not be mistaken for patches of rough ice. Ketchum (1984) illustrated and described Seasat images of sea ice.

Radar Scatterometer Data A *radar scatterometer* is a nonimaging active system for quantitatively measuring the radar backscatter of terrain as a function of the incidence angle. Scatterometer data are useful for characterizing the surface roughness of materials and are particularly useful for identifying types of sea ice. A scatterometer transmits a continuous radar signal directly along the flight path. The 3°-wide beam illuminates terrain both ahead and behind the aircraft, but for simplicity Figure 9-15 shows only the forward portion. The 0° incidence angle is directly beneath the aircraft, and the maximum incidence angle is 60°. At an altitude of 900 m, the ground resolution cell at a 30° incidence angle is a 54-by-54-m square. The return signal is a composite of the backscattering properties of all the terrain features within the cell. The wavelength of this scatterometer system is 2.25 cm (X band), and both the transmitted and received energy are vertically polarized. As the aircraft advances along the flight path, a ground resolution cell is illuminated initially at a 60° incidence angle and then at successively decreasing angles. The recorded amplitude and frequency of the successive returns are processed with Doppler techniques to obtain the scattering coefficient at each of several incidence angles. Details of scatterometer theory and operation are given in Moore (1983).

A. First-year ice (dark) and multiyear ice (bright) with open and refrozen leads. Beaufort Sea north of Liverpool Bay, Canada, January 1984.

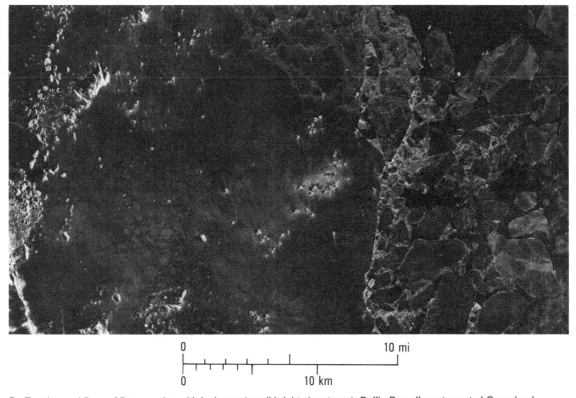

B. Fast ice and floes of first-year ice with icebergs (small bright signatures). Baffin Bay off west coast of Greenland, June 9, 1984.

Figure 9-13 Aircraft X-band radar images (3-cm wavelength) of sea ice. Courtesy M. A. Wride, Intera Technologies, Ltd.

A. October 3, 1978.

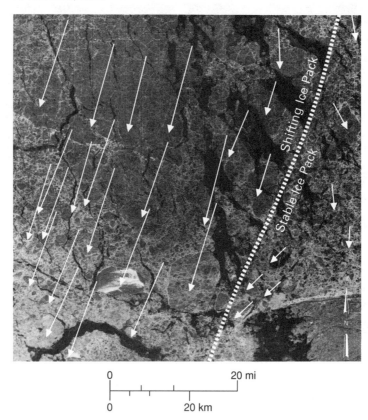

B. October 6, 1978. Arrows show movement during previous three days.

Figure 9-14 Movement of sea ice interpreted from Seasat images (23.5-cm wavelength). Beaufort Sea northwest of Baffin Island, Canada. Courtesy J. P. Ford, JPL (Retired).

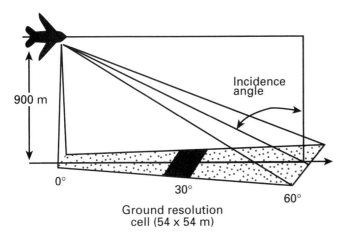

Figure 9-15 Geometry of the radar scatterometer system.

52.0°. Scattering coefficient curves (Figure 9-17) display the returns at different incidence angles for an area on the ground.

Scatterometer profiles and aerial photographs of sea ice off Point Barrow, Alaska, were acquired in 1967 and interpreted by Rouse (1968). The photomosaic (Figure 9-16A) shows the following features:

A and F	Open water
A to C	Smooth first-year ice
B	Narrow open leads
C to F	Slightly ridged first-year ice
C	Major pressure ridge separating the smooth ice and ridged first-year ice
D	Floe of smooth ice within the ridged ice
B, E, and F	Areas of smooth ice, ridged ice, and open water respectively; scattering coefficient curves shown in Figure 9-18

Scatterometer data may be displayed either as profiles or as scattering coefficient curves. Profiles (Figure 9-16B) display the scattering coefficient of the terrain along the flight line for a particular incidence angle. In this example the incidence angles range from 2.5° (almost directly beneath the aircraft) to

The scatterometer profiles (Figure 9-16B) are plotted at the same horizontal scale as the photomosaic. At low incidence angles (2.5° and 6.7°), open water has a strong return. Similarly the smooth ice has stronger returns than ridged ice at these low incidence angles. At the higher incidence angles (25.0° and

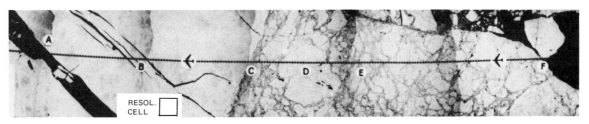

A. Photomosaic showing scatterometer flight line.

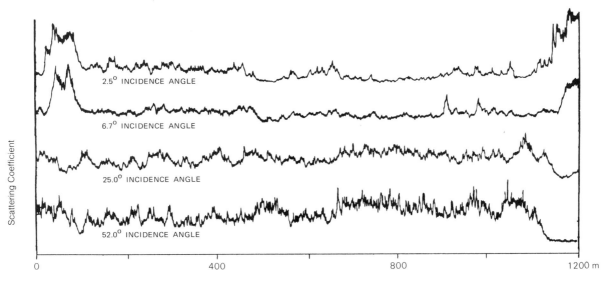

B. Scatterometer profiles.

Figure 9-16 Scatterometer flight line and profiles of sea ice off Point Barrow, Alaska. From Rouse (1968, Figure 10).

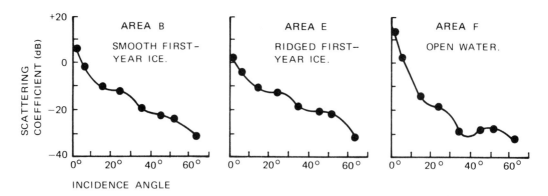

Figure 9-17 Scattering coefficient curves of various types of sea ice off Point Barrow, Alaska. Areas B, E, and F are localities shown in Figure 9-17A. From Rouse (1968, Figure 10).

52.0°), most of the microwave energy encountering the water and smooth ice is specularly reflected away from the antenna and produces little return. Energy backscattered from the rough ridged ice, however, produces a relatively strong and characteristically spiked profile at these incidence angles. Note that the floe of smooth ice at D can be recognized on the 52.0° profile by its reduced return within the ridged ice.

Passive Microwave Images Passive microwave images acquired by SSM/I and other satellite systems are used to prepare global maps of sea-ice coverage. In the microwave region, radiant temperature, or brightness temperature, is the product of kinetic temperature times emissivity at the wavelength band of the detector. At microwave wavelengths, brightness temperature is determined by a version of Equation 5-8,

$$T_{rad} = \varepsilon \cdot T_{kin} \quad (9\text{-}1)$$

where

T_{rad} = radiant temperature

ε = emissivity (at microwave wavelengths)

T_{kin} = kinetic temperature

Emissivity has a stronger influence on radiant temperature in the microwave region than in the thermal IR region. In the microwave region, ice has a higher emissivity than water; therefore, ice has a higher microwave brightness temperature than water. Different types of ice have different emissivity values. In the microwave region of 1 to 3 cm, using horizontal polarization, typical brightness temperatures are as follows: open ocean, 100° to 140°K; multiyear ice, 180° to 210°K; and first-year ice, 220°K.

Parkinson and Gloersen (1993) show seasonal images of ice cover in the Arctic and Antarctic regions that were acquired by several passive microwave systems.

BATHYMETRY

Oceans cover 70 percent of the earth, but much of the seafloor is mapped only in a general fashion, and some charts are inaccurate. Accurate, updated charts are needed for shallow shelf areas where deposition, erosion, and growth of coral reefs can change bottom topography within a few years after a *bathymetric survey* (which measures ocean depths) has been completed. Knowledge of the geomorphology of oceanic spreading centers, transform faults, and submarine volcanoes is needed to improve our understanding of plate tectonics.

Landsat Images

Remote sensing of the seafloor from satellites or aircraft is restricted by the fact that water absorbs or reflects most wavelengths of electromagnetic energy. Only visible wavelengths penetrate water, and the depth of penetration is influenced by the turbidity of the water. Figure 9-18 shows penetration of different wavelengths through 10 m of different types of water. This figure demonstrates that penetration is essentially restricted to the visible region (0.4 to 0.7 µm). Energy at reflected IR wavelengths (greater than 0.7 µm) is almost totally absorbed by water, as illustrated by the black-and-white IR photographs in Chapter 2. Ten meters of clear ocean water transmit almost 50 percent of the incident blue and green wavelengths (0.4 to 0.6 µm) but less than 10 percent of the red light (0.6 to 0.7 µm). Figure 9-18 shows that optical density, or turbidity, becomes progressively higher as one moves from ocean to coastal to bay water. The increased turbidity causes a decrease in light transmittance and also shifts the wavelength of maximum transmittance toward longer wavelengths (0.5 to 0.6 µm).

Figure 9-19 shows TM band images for a subscene of Joulters Key at the north end of Andros Island, which is the largest island in the Bahamas. Plate 7C is a TM normal color image with bands 1, 2, and 3 shown in blue, green, and red, re-

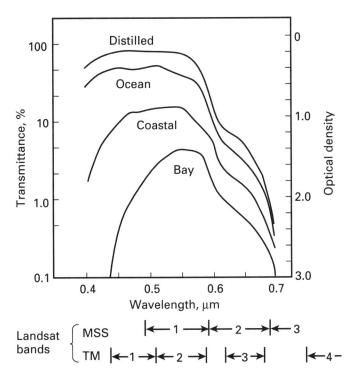

Figure 9-18 Spectral transmittance through 10 m of various types of water. From Specht, Needler, and Fritz (1973, Figure 1).

spectively. The map and profile (Figure 9-19G,H) show that the small islands of Joulters Key are located at the east edge of a shallow platform that drops abruptly eastward into the great depths of Tongue of the Ocean. To the west are tidal flats of brilliant white coral sand and mud that are partially emergent during minimum low tides. West of the tidal flats are stretches of very shallow seafloor that are always submerged. In the color image (Plate 6C) Tongue of the Ocean is very dark blue, the steep slope at the east margin of the platform is medium blue, the tidal flats are white, and the shallow areas are light blue. Vegetation on Andros Island, Joulters Key, and scattered small islands is green. Shorelines are difficult or impossible to define on the visible bands because those wavelengths penetrate the shallow water and are reflected from the bottom to produce signatures that are similar to signatures of the adjacent land. Shorelines are sharply defined, however, on reflected IR bands 4, 5, and 7 (Figure 9-19D, E, F), because these wavelengths are absorbed by water and are reflected by the land. The profile (Figure 9-19H) shows relative water penetration by TM bands, which increases with decreasing wavelength (from IR to blue), consistent with the transmittance curves for ocean water (Figure 9-18).

The relationship between wavelength and water penetration is also seen by comparing the individual TM bands in Figure 9-19. In band 1 the seafloor of the platform throughout the western portion of the image is shown in various tones of gray.

In band 3, only the shallowest seafloor has gray tones. The longest-wavelength IR images, bands 5 and 7, are identical; the islands appear slightly larger in band 4, which indicates a few centimeters of water penetration.

Sabins (1987, Figure 9-9) interpreted bathymetry from Landsat MSS images of the Sibutu Island group in the Celebes Sea off the northeast coast of Borneo.

Bathymetric signatures on Landsat images are also influenced by factors other than water depth. Water clarity, sunglint from the sea surface, atmospheric conditions, and reflectance of sediment and vegetation on the seafloor are important factors. Therefore, the depth ranges associated with TM bands at Joulters Key will not necessarily be the same in other areas. Lyzenga (1981), Nordman and others (1990), and Harris and Kowalik (1994) describe methods for extracting depth information from Landsat images. All methods should be calibrated with sounding information for the area under investigation.

Landsat images have revealed errors on hydrographic charts. In the Chagos Archipelago of the Indian Ocean, Landsat images revealed the presence of a previously uncharted reef and also showed that a known bank was charted 18 km east of its true position (Hammack, 1977). The U.S. Defense Mapping Agency incorporated these changes on new editions of their published hydrographic charts. Landsat images of the Georgia coast disclosed an island that was not present on existing maps or on aerial photographs acquired several years earlier. The island had formed recently by the accumulation of sand from drift along the shore. In the Red Sea, Chevron's marine seismic surveys in the late 1970s were hampered by uncharted submerged coral heads that snagged cables. Digitally processed Landsat images were used to locate and avoid these hazards.

Visible light images of the deeper seafloor are acquired by camera and television systems with their own light sources, which are lowered on cables or operated from submersible research vessels. These systems acquire high-resolution images, but their coverage is restricted by the limited penetration of light in water.

Side-Scanning Sonar Images

Sound waves are readily transmitted through water and have long been used for detection of submarines and for *fathometers* that measure depth. The general term for this form of active remote sensing is *sound navigation ranging* (sonar). Figure 9-20 shows a *side-scanning sonar,* which is an active system for imaging the seafloor with pulses of sonic energy. The system is contained in a torpedo-shaped housing, called a *sonar fish,* that is towed near the seafloor from a cable that also provides power and transmits data from the sonar fish to a recording system on the ship. The fish contains *transducers,* which are devices that convert electrical energy into sound energy; they also convert received sound into electrical energy. Transducers on each side of the fish transmit a narrow pulse of sound at right angles (look direction) to the ship track (azimuth direction). The pulse

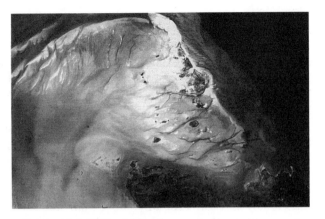

A. Band 1, blue (0.45 to 0.52 μm).

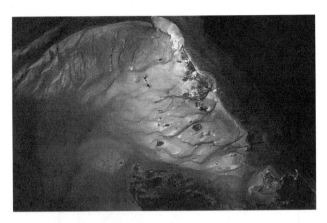

B. Band 2, green (0.52 to 0.60 μm).

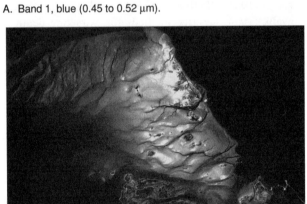

C. Band 3, red (0.63 to 0.69 μm).

D. Band 4, reflected IR (0.76 to 0.90 μm).

E. Band 5, reflected IR (1.55 to 1.75 μm).

F. Band 7, reflected IR (2.08 to 2.35 μm).

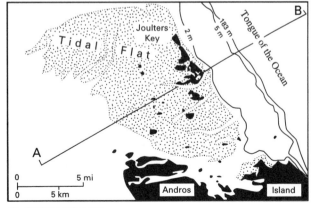

G. Bathymetric map.

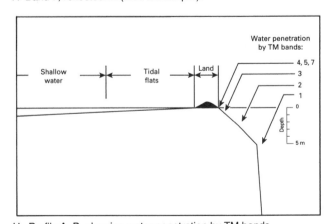

H. Profile A–B, showing water penetration by TM bands.

Figure 9-19 Landsat TM bands of Joulters Key, Bahamas.

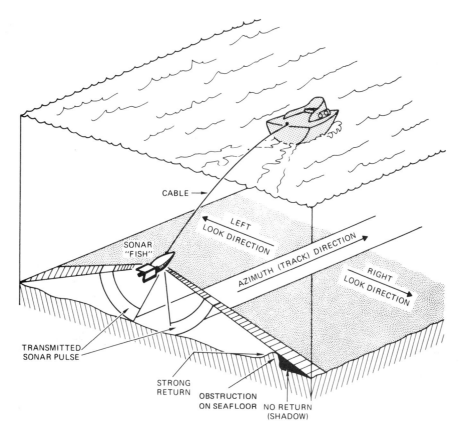

Figure 9-20 Side-scanning sonar system.

encounters the seafloor and is reflected back to the transducer, where the received sound generates electrical signals that vary in amplitude proportional to the strength of the received sound. As the ship moves forward, the process continues, generating two strips of imagery separated by a narrow blank strip directly beneath the sonar fish. As in radar terminology, the direction of travel is the azimuth direction and the direction of transmitted pulses is the look direction. The data may be recorded on magnetic tape and played back to produce images. The analogy between sonar and radar systems extends to the geometry of the images; sonar images are also subject to slant-range distortion, which may be corrected during playback. Belderson and others (1972) published a standard reference to side-scanning sonar. Kleinrock (1992) published a survey of operational systems together with examples of images, interpretation techniques, and a bibliography.

In sonar images, strong sonic returns are recorded as dark signatures and weak returns as bright signatures. Topographic scarps facing the sonic look direction produce strong returns, while surfaces facing or sloping away from the look direction produce weak returns. Smooth surfaces of mud or sand reflect the sonic pulse specularly; hence these surfaces have weak returns and are recorded with bright signatures. Rough surfaces such as boulder fields and lava scatter much of the incident energy back to the sonar fish and are recorded with dark signatures.

Figure 9-21 is a sonar image of the Make It Burn (MIB) submarine volcano, or seamount, on the East Pacific Rise at depths of 2000 m and greater. As shown in the interpretation map (Figure 9-22A), the ship track was toward the west; the upper half of the image was recorded with a look direction toward the right of the ship, and the lower half was recorded with a look direction toward the left. The narrow blank strip is the area directly beneath the fish that was not imaged. The profile (Figure 9-22B) shows that the seamount has sloping flanks, a relatively horizontal summit plateau, and a large caldera surrounded by a scarp that is formed by ring fractures. There are two pit craters and a small cone on the flat floor of the caldera. The floors of the caldera and pit craters are relatively smooth and thus have uniform, bright signatures in the image (Figure 9-21). The summit plateau has a rough, hummocky surface, which is recorded as irregular bright and dark spots. The scarp enclosing the caldera is a steep slope that faces the left and right look directions and is recorded with arcuate dark signatures, as are the walls of the pit craters. The northern and southern flanks of the seamount are in an acoustic

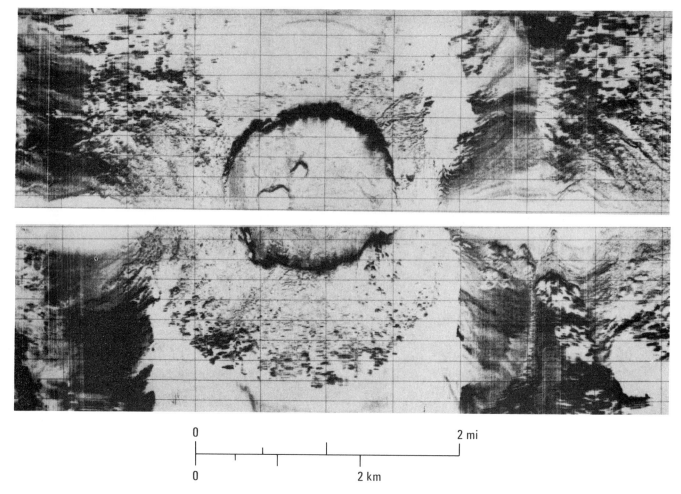

| 0 | | | | | | | | | | 2 mi |
| 0 | | | | | 2 km | | | | | |

Figure 9-21 Side-scanning sonar image of the MIB seamount, East Pacific Rise. Dark signatures are strong sonic returns; bright signatures are weak sonic returns. From Fornari, Ryan, and Fox (1984, Figure 3A). Courtesy D. J. Fornari, Lamont-Doherty Geological Observatory, Columbia University.

shadow zone because the summit plateau blocks the transmitted sonar energy. For this reason the flanks have bright signatures. Smaller features such as gullies, slump blocks, landslides, and volcanic cones are recognizable in the image.

The image in Figure 9-21 was acquired with the Sea MARC I system, which operates at sonic frequencies of 27 and 30 kHz. The system is towed 200 to 400 m above the bottom and images a swath with maximum total width of 5 km at a spatial resolution of several meters. Another sonar system, Gloria, operates at 6.5 kHz and is towed near the sea surface to acquire images with a total swath width of 60 km and a spatial resolution of 45 m. Gloria was used to acquire sonar mosaics of the continental shelves of the United States for a distance of 200 nautical miles (370 km) offshore. These mosaics are published by the U.S. Geological Survey as a series entitled *Atlas of the U.S. Exclusive Economic Zone* and listed in Table 9-2. In addition to sonar mosaics, each atlas includes bathymetric contour maps and profiles of gravity, magnetic, and seismic data. Sea MARC and Gloria are large and complex systems. Smaller sonar systems are commercially available and are used for surveys in shallow water. Additional sonar images are shown in Chapter 13.

Table 9-2 *Atlas of the U.S. Exclusive Economic Zone* published as miscellaneous investigation maps by the U.S. Geological Survey

USGS Miscellaneous Map Number	*Locality*	*Date*
I-1792	Western conterminous USA	1986
I-1864-A,B	Gulf of Mexico, eastern Carribean	1987
I-2053	Bering Sea	1991
I-2054	Atlantic continental margin	1991

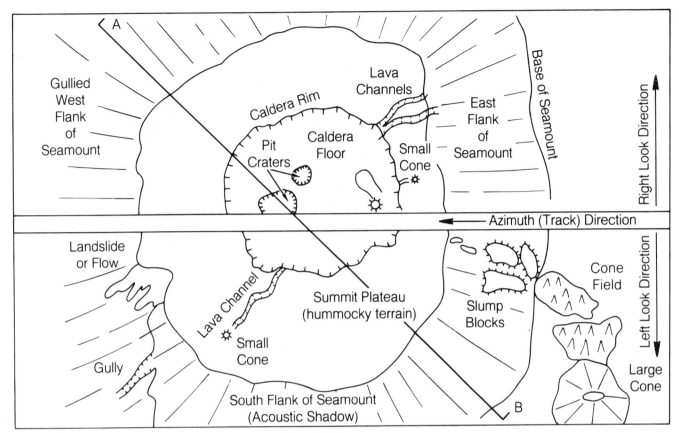

A. Interpretation map. After Fornari, Ryan, and Fox (1984, Figure 3b).

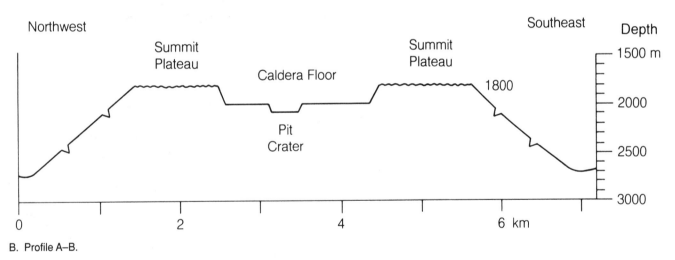

B. Profile A–B.

Figure 9-22 Interpretation map and profile from sonar image of MIB seamount, East Pacific Rise (Figure 9-21).

Shallow Bathymetric Features on Seasat Images

Radar energy does not penetrate water but is reflected and scattered by the water surface. Seasat images of shallow seas, however, commonly have bright and dark signatures that record the presence of underwater features such as channels and sandbars at depths of several tens of meters. Figure 9-23 is a Seasat image of the English Channel in which the linear bright and dark patterns record the submerged sandbars and depressions at depths down to 30 m. The coincidence between radar patterns and bathymetry is seen by comparing the Seasat

Figure 9-23 Seasat image of the English Channel. From Fu and Holt (1982, Figure 32A). Courtesy J. P. Ford, JPL (Retired).

image with the corresponding bathymetric chart (Figure 9-24). This relationship between radar signatures and shallow bathymetry also occurs in Seasat images of the Bahama Banks, Bermuda, the Nantucket Shoals, and Cook Inlet of Alaska. The following explanation of this relationship is summarized from Kasischke and others (1983).

Under normal conditions, the ocean surface is covered by small wind-generated waves called *small-scale waves*. Their wavelengths range from less than 1 cm to 1.5 m. The resulting surface roughness interacts with the depression angle and wavelength of Seasat to produce a typical intermediate gray signature. Local variations in velocity of surface ocean currents, however, can change the size and spacing of the small-scale waves with attendant changes in surface roughness and radar signature. An increase in current velocity stretches the small-scale waves, which decreases their amplitude and reduces surface roughness. Decreased current velocity results in increased roughness. Figure 9-25 is a model cross section of the ocean surface and a shallow sandbar, such as found in the English Channel. A tidal current flows from right to left with

relative velocity indicated by the length of the arrows. Over the seafloor at the right side of Figure 9-25, the tidal velocity and small-scale waves are in equilibrium, and the resulting ocean surface roughness has an intermediate gray signature in the radar image. As the current flows toward the left and approaches the shallower water over the sandbar, a constant volume of water must flow through a smaller cross-sectional area, which results in a velocity increase. The higher velocity expands the spacing between the small-scale waves, which results in smoother water. This strip of smoother water results in a dark radar signature parallel with the crest of the sandbar. On the down-current flank of the sandbar (left side of Figure 9-25), the velocity decreases and the roughness increases, which results in a bright radar signature. Farther down-current from the sandbar, the normal pattern of small-scale waves resumes. In summary, the radar signature of a shallow sandbar is a bright band on the down-current flank and a dark band on the up-current flank.

In the bathymetric chart (Figure 9-24), arrows show direction and velocity of tidal currents at the time the Seasat image

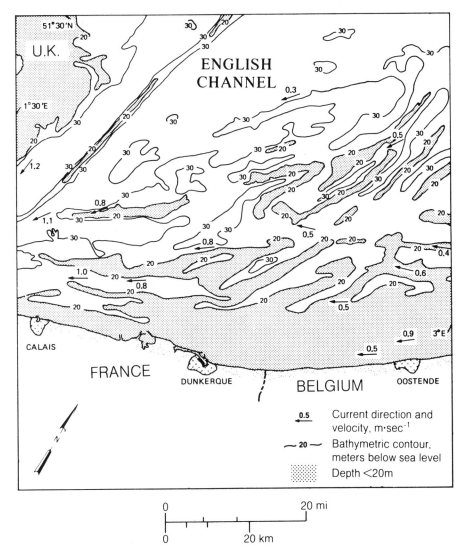

Figure 9-24 Bathymetric chart of the English Channel (Figure 9-23). From Fu and Holt (1982, Figure 31B).

was acquired. The tidal current was flowing from east to west, or right to left, which matches conditions of the model. In the Seasat image the dark bands are on the up-current (east) flank of the sandbars and the bright bands are on the down-current (west) flank. This model for radar expression applies to shallow water (less than 50 m deep) with a tidal current of at least $0.4 \text{ km} \cdot \text{sec}^{-1}$ and a wind velocity of 1.0 to 7.5 m · sec⁻¹ (Schuchman, 1982).

Satellite Radar Altimetry

Radar altimeters are nonimaging systems carried on aircraft and satellites to measure altitude with great precision. A pulse of microwave energy is transmitted vertically downward. The

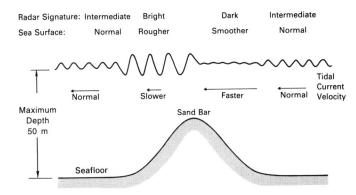

Figure 9-25 Cross-section model showing relationships among sandbar, tidal current, sea-surface roughness, and radar signatures. Modified from Kasischke and others (1983, Figure 4).

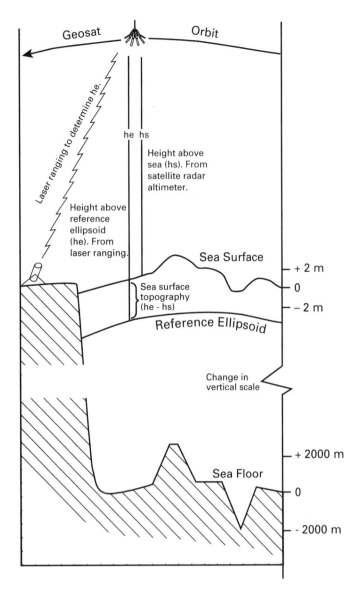

Figure 9-26 Mapping bathymetry from satellite radar altimeter data. Based on Gahagan and others (1988, Figure 2).

Radar altimeters on the Geosat and ERS satellites have provided a wealth of data that are corrected and calibrated to produce bathymetric maps of the world's oceans. Smith and Sandwell (1994) describe the digital processing of the data and illustrate the resulting bathymetric maps. The accuracy of the maps is 5 to 10 km in location and 100 to 250 m in depth. The left side of Figure 9-26 shows how the maps are generated. The height above the sea surface (*hs*) is measured by the altimeter, which is carried on the Geosat satellite in this example. The height of Geosat above a datum (*he*), called the reference ellipsoid, is determined from precise laser measurements of the satellite orbit (Figure 9-26). Subtracting *hs* from *he* provides elevations for mapping the sea surface.

Figure 9-27A is a bathymetric model of a portion of the Southern Ocean that was generated from Geosat radar altimeter data. The model is illuminated toward the lower margin to produce highlights and shadows that enhance bathymetric features. Figure 9-27B is an interpretation map that shows bathymetric scarps that are the expression of faults. The Foundation seamounts are a chain of undersea volcanoes that were discovered from this map. The seamounts were unknown because large portions of the world's seafloor are inadequately charted. Bathymetric maps are traditionally compiled from sonar depth-sounding profiles recorded by oceanographic survey vessels. In the Southern Ocean, for example, there are many gaps of hundreds of kilometers between tracks of bathymetric surveys, which explains how major features, such as the Foundation seamount chain, were not charted. In this region, however, satellite orbits are 2 to 4 km apart, and very accurate bathymetric maps are now feasible.

ENVIRONMENTAL POLLUTION

Remote sensing is an effective means of monitoring environmental pollution, as described in the following sections.

Hazardous Waste Sites

The term *hazardous waste* refers to by-products of human activity that range from mine dumps to nuclear waste repositories. At most sites the major problem is chemicals leaching from the dump by rain and entering the surface and subsurface water supplies. Major applications of remote sensing are to

1. locate unreported waste sites;
2. map the extent and severity of damage to adjacent areas;
3. monitor cleanup operations and recovery of the environment.

Many sites are covered by soil and vegetation, as are the adjacent areas. The actual chemicals are rarely detectable by remote sensing, but they can cause stress to vegetation that is initially detectable by decreased reflectance in the reflected IR

two-way travel time of the return pulse is recorded and converted into distance above the surface, or height. In 1978 the Seasat satellite (Chapter 7) carried a radar altimeter that recorded altitude profiles along orbits over the oceans. Dixon and others (1983) demonstrated that variations in sea-surface elevation on these Seasat profiles correlate with bathymetric features on the seafloor. The right side of Figure 9-26 shows that bathymetric relief on the order of a few kilometers causes relief on the sea surface on the order of a few meters. This strong correlation between bathymetry and the shape of the sea surface is caused by differences in the gravity field that are associated with bathymetric features (Dixon and others, 1983).

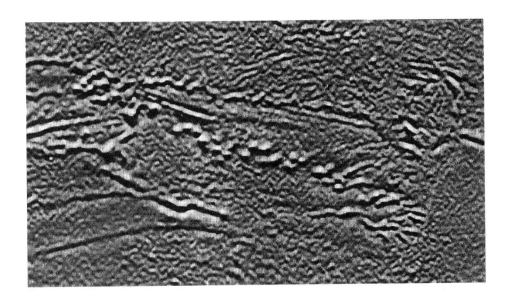

A. Digital bathymetric model compiled from Geosat radar altimeter data. The model is illuminated from the upper margin toward the lower margin of the image. Courtesy E. Hurwitz, NOAA.

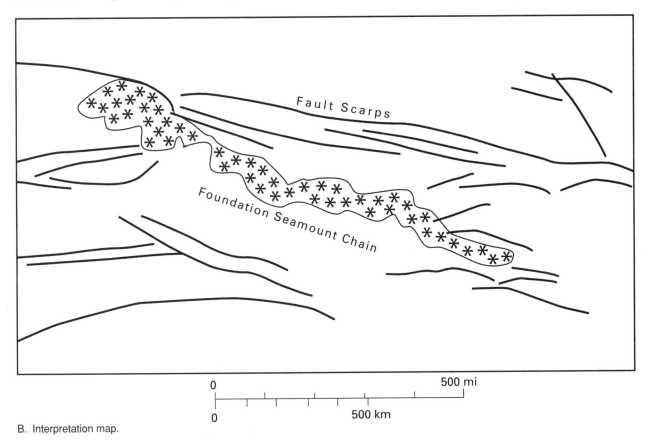

B. Interpretation map.

Figure 9-27 Bathymetric model and map of the previously uncharted Foundation seamounts in the Southern Ocean.

band. Vegetation stress can be recognized on IR color photographs and multispectral images.

IR color photographs were used to investigate a Superfund site in Michigan. *Superfund sites* have been designated as especially hazardous by the U.S. Environmental Protection Agency (EPA), which provides federal funds for cleanup. Industrial wastewater containing toxic organic chemicals was dumped into unlined ponds and contaminated surface water and groundwater. Herman and others (1994) analyzed a series of IR color photographs acquired from 1969 to 1986. The photographs were scanned and converted into green, red, and reflected IR bands of digital data. Ratio images were produced of the reflected IR band divided by the red band (RIR/R). On the ratio images, stressed vegetation has a darker signature than healthy vegetation. The stressed vegetation correlated with contaminated surface water and groundwater as shown by chemical tests. Dumping was stopped and the site was remediated by pumping the contaminated groundwater. Subsequent photographs showed a marked decrease in stressed vegetation.

Marsh and others (1991) investigated a waste site near Phoenix, Arizona. IR color photographs and airborne video images were acquired 2.5 years apart and documented both cleanup efforts and additional dumping.

Another EPA Superfund site is the old Virginia City gold-mining district in north-central Nevada. The ore was ground and treated with mercury to extract the gold. The waste, called mill tailings, is contaminated with hazardous levels of mercury. During the past 130 years, runoff has spread the tailings up to 80 km from the mill sites. Fenstermaker and Miller (1994) used aircraft hyperspectral scanner images to map the redistributed tailings, which have a distinctive spectral signature.

Thermal Plumes

Many industrial plants withdraw water from lakes, rivers, and the ocean to cool their processes and then return it to the same water bodies at higher temperatures. The heated water discharges, called *thermal plumes*, may be monitored by airborne thermal IR scanners in the same manner as natural water currents of different temperatures. Nuclear and fossil-fuel electrical power plants, refineries, and chemical and steel plants use large volumes of water. Aside from any chemicals or suspended matter, the heated water affects the environment in two ways:

1. Excessively high temperatures may kill organisms or inhibit their growth and reproduction. In some areas, however, the heated discharge water is used for commercial cultivation of lobsters and oysters.
2. The heated water has a lower content of the dissolved oxygen that is essential for aquatic animals and for oxidation of organic wastes.

Environmental legislation has been enacted to regulate thermal discharges. In California coastal waters, for example, the maximum temperature of thermal discharges must not exceed the natural temperature of receiving waters by more than 11°C. At a distance of 300 m from the point of discharge, the surface temperature of the ocean must not increase by more than 2.2°C. The temperature of the discharge water may be lowered by passing it through cooling towers or by mixing it with cooler water before it is discharged. Thermal IR surveys are an ideal way to monitor the temperature and pattern of the discharge outfalls.

Figure 9-28A is a calibrated aircraft thermal IR image of the heated plume discharged into Montsweag Bay, Maine, from the Maine Yankee nuclear power plant. Figure 9-28B is a contour map of surface water temperatures that was derived from the image data. The branching or merging contour lines occur where the horizontal thermal gradient is so steep that contour lines coalesce. Both image and map show the location of the plume and the temperature distribution within it. Note, for example, that some parts of the bay are not affected by the plume during low tide. The upstream and downstream extent and temperature level of the plume are precisely shown. The images and map provide no information about the thickness of the plume because the IR energy is radiated from the surface of the water.

To appreciate the practical value of monitoring thermal plumes using IR surveys, imagine undertaking the following exercise. Design a monitoring system using conventional surface thermometers that will produce a thermal map, with the precision and detail of Figure 9-28B, throughout a tidal cycle. You must use several hundred surface thermometers, precisely positioned and located; you must calibrate them and record from them to an accuracy of 0.5°C; and they must all be read at the same times to provide simultaneous data for contouring. Finally, you must deploy and retrieve the thermometers during strong tidal currents. This survey would be impractical to conduct but is readily accomplished with several IR scanning flights. Davies and Mofor (1993) used airborne density-sliced thermal IR images to monitor thermal plumes from power plants on the coast of Scotland.

Oil Spills

Major industrialized nations and regions of the world (such as Japan, the United States, and Europe) obtain much or all of their energy from petroleum imported via tanker ships from overseas sources. The National Research Council (1985) has estimated the amount of oil that enters world oceans annually from various sources (Table 9-3). Transportation of oil plus municipal and industrial wastes and runoff account for 81.6 percent of the oil entering the oceans each year. Natural sources (seeps and sediment erosion) contribute five times more oil than do offshore production operations.

A. Thermal IR image (8 to 14 μm).

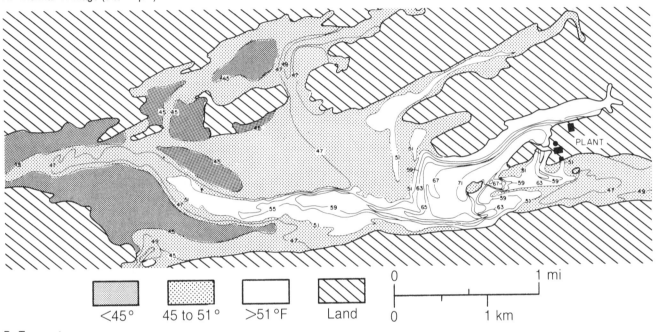

| <45° | 45 to 51° | >51 °F | Land |

0   1 mi

0   1 km

B. Temperature map.

Figure 9-28 Thermal plume from the Maine Yankee nuclear power plant, Montsweag Bay, Maine. Image was acquired at low tide. Courtesy Maine Yankee Power Company and Daedalus Enterprises, Inc.

Table 9-3 Sources of oil in the oceans

Source	Metric tons per year, millions	Annual input, percent
Transportation	1.47	45.3
Municipal and industrial wastes and runoff	1.18	36.3
Atmospheric fallout	0.3	9.2
Natural sources	0.24	7.7
Offshore production	0.05	1.5
Totals	3.24	100.0

Source: National Research Council (1985, Table 2-22).

Characteristics and Interaction Mechanisms of Oil Spills The EPA uses the following nonquantitative classification of oil spills, given in the order of decreasing thickness:

Mousse Brown emulsion of oil, water, and air that forms thick streaks and resembles the dessert called chocolate mousse.

Slick Relatively thick layer with a definite brown or black color.

Sheen Thin, silvery layer on the water surface with no black or brown color.

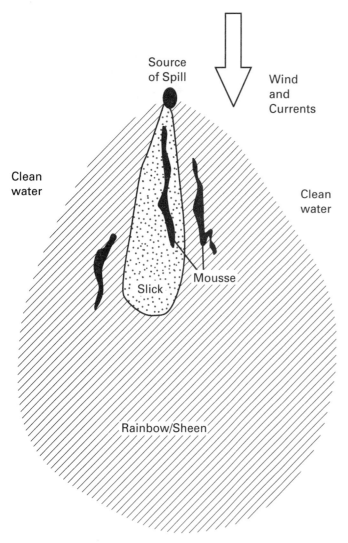

Figure 9-29 Diagram showing the characteristics of a typical oil spill.

Rainbow Very thin iridescent multicolored bands visible on the water surface.

Rainbow/sheen Two categories commonly lumped together because they are volumetrically insignificant and are difficult to distinguish.

Figure 9-29 shows the distribution and relative extent of these components in a typical oil spill, where 90 percent of the volume is concentrated in 10 percent of the area, largely in the form of slicks and mousse. Table 9-4 lists the interactions between oil and electromagnetic energy at wavelength regions ranging from UV through radar. Figure 9-30 shows these interactions diagramatically for oil and water to explain their different signatures. The following sections describe the detection of oil slicks on images acquired in the various spectral regions.

UV Images Figure 9-30A shows that incoming UV radiation from the sun stimulates fluorescence in oil. Figure 9-31 shows spectral radiance curves for water (solid) and a thin layer of oil (dashed). The higher radiance of oil at wavelengths of 0.30 to 0.45 μm (long-wavelength UV to visible blue) is due to fluorescence. Individual crude oils and refined products have peak fluorescence at different wavelengths that depend upon composition and weathering of the hydrocarbons. Figure 9-32 is a daytime image acquired by a cross-track aircraft scanner equipped with a UV detector (using wavelengths of 0.32 to 0.38 μm). The area is the Santa Barbara Channel off the coast of southern California, where numerous natural oil seeps occur on the seafloor and form widespread slicks on the surface. Fluorescence from the oil slicks at UV wavelengths causes the bright signatures in the image. The platforms shown in the image are oil production facilities.

UV images are the most sensitive remote sensing method for monitoring oil on water and can detect films as thin as 0.15 μm (Maurer and Edgerton, 1976). Daylight and very clear atmosphere are necessary to acquire UV images. UV energy is

Table 9-4 Remote sensing of oil spills

Spectral region	Oil signature	Oil property detected	Imaging requirements	False signatures
UV, passive (0.3 to 0.4 μm)	Bright	Fluorescence stimulated by sun	Day; good weather	Foam
UV, active (0.3 to 0.4 μm)	Bright	Fluorescence stimulated by laser	Day and night; good weather	Foam
Visible and reflected IR (0.4 to 3.0 μm)	Bright—mousse Dark—slick Bright—sheen	Reflection and absorption of sunlight	Day; good weather	Wind slicks, discolored water
Thermal IR (8 to 14 μm)	Bright—mousse Dark—slick	Radiant temperature controlled by emissivity	Day and night; good weather	Warm and cool currents
Radar (3 to 30 cm)	Dark	Dampening of capillary waves	Day and night; all weather	Wind slicks, current patterns

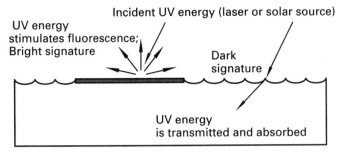

A. UV region.

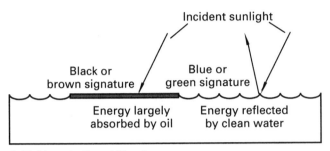

B. Visible and reflected IR regions.

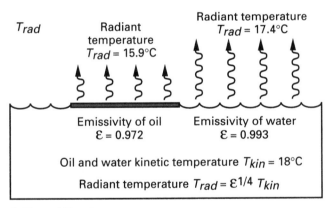

C. Thermal IR region.

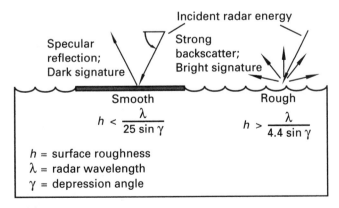

D. Radar region.

Figure 9-30 Interaction mechanisms among oil, water, and electromagnetic energy at different wavelength regions.

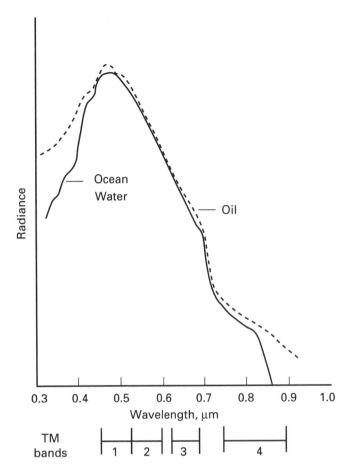

Figure 9-31 Spectral radiance for ocean water and a thin film of oil. From Vizy (1974, Figure 5).

strongly scattered by the atmosphere but usable images can be acquired at altitudes below 1000 m. Floating patches of foam and seaweed have bright UV signatures that may be confused with oil (Table 9-4). Foam and seaweed can be recognized on images in the visible band acquired simultaneously with the UV images.

Figure 9-32 is a passive image that records energy stimulated by the sun. Active UV systems have been developed for aircraft use. A laser irradiates the water with UV energy that stimulates any oil to fluoresce. The fluorescence is recorded as a spectrum, rather than an image. The spectrum can be compared with a reference library of spectra to identify the unknown hydrocarbon. These active systems can acquire images both day and night. O'Neil, Buja-Bijunas, and Rayner (1980) describe a system the Canadian government used to monitor oil pollution. Quinn and others (1994) describe an active system developed for the Kuwait government.

The British Petroleum Company used an active UV system, called *airborne laser fluorosensor* (ALF), for oil exploration in offshore basins around the world. Many oil accumulations leak

PLATFORM B

PLATFORM A

PLATFORM HILLHOUSE   WAKE

SEEPAGE

```
0                1 mi
├────────────────┤
0        1 km
├────────┤
```

Figure 9-32 UV scanner image (0.32 to 0.38 μm) of natural oil seeps in the Santa Barbara Channel, California, July 31, 1974. Bright signatures are fluorescence stimulated by sunlight interacting with the oil. From Maurer and Edgerton (1976, Figure 9).

hydrocarbons that form slicks in offshore basins, such as the Santa Barbara Basin. The ability to detect slicks by remote sensing in unexplored basins can contribute to an exploration project. British Petroleum and Pertamina (the Indonesian national oil company) used ALF to survey seven offshore basins in Indonesia (Thompson and others, 1991). Six of the basins have slicks; three of those basins were judged to have the highest exploration potential.

Visible and Reflected IR Images The interaction between oil and electromagnetic energy in the visible and reflected IR regions is determined by the absorption and reflection of sunlight (Figure 9-30B). A complicating factor is sunglint from calm water caused by the oil. During their invasion of Kuwait in January 1991, Iraqi forces deliberately released 4 to 6 million barrels of crude oil into the Arabian Gulf, where it covered approximately 1200 km² of water and 500 km along the Saudi Arabian coastline. This spill is probably the largest in history; for comparison the 1989 *Exxon Valdez* spill in Alaska (<300,000 barrels) was an order of magnitude smaller. The Iraqis also set fire to onshore oil wells, which created giant smoke plumes and spilled additional oil onto the desert. The smoke plumes were tracked on images from AVHRR, Meteosat, and Landsat (Limaye and others, 1992; Cahalan, 1992).

Figure 9-33 shows Landsat TM-band images for a subscene of the Arabian Gulf and Saudi Arabian coast that was acquired February 16, 1991. These images show signatures of the spill in the visible, reflected IR, and thermal IR bands. Figure 9-34 is an interpretation map that I prepared from the images and from personal experience with other spills; no field information was available. Figure 9-35 shows reflectance spectra of water, mousse, and oil sheen that were measured from the TM digital data. Mousse has a higher reflectance in band 5 than in band 4 or 7. Stringer and others (1992, Figure 14) showed similar spectra for the *Exxon Valdez* spill in Alaska.

The individual TM bands of the Arabian Gulf spill were digitally processed to extract additional information. Three different sets of principal-component (PC) images were created: six PC images from the six visible and reflected IR bands, three PC images from the three visible bands, and three PC images from the three reflected IR bands. The PC images of the reflected IR bands 4, 5, and 7, shown in Figure 9-36A, B, C, extract the most information. This relationship is explained by the spectra in Figure 9-35, which have maximum differences in the reflected IR region. PC image 1 (Figure 9-36A) emphasizes the heaviest concentrations of mousse with bright signatures. PC image 2 emphasizes fine details of the mousse plus slick, as shown by the narrow black tendrils adjacent to the coast in the top portion of Figure 9-36B. Although PC image 3 (Figure 9-36C) is dominated by noise, the northwest-trending brighter patch is atmospheric haze or smoke. A number of band-ratio images were created, but in general these were not as effective as the PC images. In ratio 4/3 (Figure 9-36D) the mousse is extracted with a bright signature that is comparable to PC image 1 (Figure 9-36A).

Plate 19A, B, C shows the combinations of individual TM bands that are most effective in discriminating the oil. In the 1-2-3 normal color image (Plate 19A) the dark signature of mousse is difficult to discriminate from the dark blue of water. Water penetration of these visible bands causes bright signatures from the shallow shoals (Figure 9-33H), which complicates interpretation. In the 4-5-7 image of reflected IR bands (Plate 19B), mousse has a distinctive bluish green signature. Mousse has a strong reflectance in band 5 (Figure 9-35), which is shown in green.

The National Geographic Society (Williams, Heckman, and Schneeberger, 1991) published an extensive collection of images and photographs of the environmental destruction caused in Kuwait by the Iraq invasion.

Thermal IR Images Figure 9-30C shows the interaction of thermal IR energy with oil and water. Both liquids have the same kinetic temperature because they are in direct contact. Chapter 5 points out that radiant temperature is a function of both radiant temperature and emissivity. Table 5-1 shows that the emissivity of water is 0.993, but a thin film of petroleum reduces water's emissivity to 0.972. Radiant temperature is calculated from Equation 5-8 as

$$T_{rad} = \varepsilon^{1/4} T_{kin}$$

For water at a kinetic temperature of 291°K (18°C), the radiant temperature is

$$T_{rad} = 0.993^{1/4} \times 291°K$$
$$= 290.5°K, \qquad \text{or } 17.5°C$$

For an oil slick at the same kinetic temperature of 291°K, the radiant temperature is

$$T_{rad} = 0.972^{1/4} \times 291°K$$
$$= 288.9°K, \qquad \text{or } 15.9°C$$

This difference of 1.6°C in radiant temperature between the oil (15.9°C) and water (17.5°C) is readily measured by thermal IR detectors, which are typically sensitive to temperature differences of 0.1°C. Figure 9-37 shows two images acquired by an aircraft multispectral scanner of an oil slick in the Gulf of Mexico off Galveston, Texas. The oil tanker *Burmah Agate* collided with another vessel, spilling crude oil. Figure 9-37B is a thermal IR image in which the oil slick has a cooler signature (darker) than the surrounding water (brighter signature). The warm streaks are caused by mousse, which reradiates absorbed sunlight at thermal IR wavelengths. The oil slick and mousse are identified much better in this thermal IR image than in the matching visible image (Figure 9-37A).

Figure 9-33F is a TM thermal IR band 6 image of the Arabian Gulf spill. The bright (warm) sinuous streaks are mousse. Plate 19D is a color density-sliced version of the thermal IR image. The mousse has warm signatures (red and yellow). A broad, northwest-trending cool band (dark blue signature) correlates with bright tones in the normal color image (Plate 19A) and is attributed to thin, high clouds. On the color density-sliced image, sinuous cool streaks associated with the warm mousse have dark signatures on the normal color image and are attributed to oil slicks.

The aircraft and TM images record a single radiant temperature value for oil over a broad spectral range. Salisbury, D'Aria, and Sabins (1993) measured thermal IR reflectance spectra for five oils, plus seawater, foam, and mousse. At wavelengths of 8 to 14 μm, spectra of the oils and mousse are essentially flat and featureless. Water and foam, however, have a broad reflectance minimum from 10 to 12 μm; this feature could be useful in discriminating oil from water on multispectral IR images such as TIMS.

The daytime and nighttime capability of thermal IR systems is valuable for surveillance around the clock. Rain and fog, however, prevent image acquisition. Also, the interpreter must be careful to avoid confusing cool water currents with oil slicks. This problem can be minimized by interpreting simultaneously acquired UV and IR images.

Radar Images Figure 9-30D shows that an oil slick eliminates the roughness caused by small-scale waves, which results in an area of low backscatter (dark signature) surrounded by the stronger backscatter (bright signature) from rough, clean water. Figure 9-38 shows the characteristic dark signatures of oil slicks on two ERS-1 images of European waters. ERS-1 radar images are particularly sensitive to differences in water roughness because of their steep depression angle and VV polarization. The Seasat image of the Santa Barbara coast shown in Chapter 7 accurately records the oil slicks. On SIR-A images, slicks are less apparent but can be enhanced by image processing (Estes, Crippen, and Star, 1985). Aircraft images are also effective for recognizing oil slicks.

Radar images may be acquired day or night under any weather conditions, which is an advantage over other remote sensing systems for monitoring oil spills. As with other images, however, radar images must be interpreted carefully, because dark streaks may be signatures of smooth water that is not caused by oil. Internal waves and shallow bathymetric features are two other possible causes of dark signatures. This problem can be reduced by comparing radar signatures with signatures on simultaneously acquired images in other wavelength regions.

Images of Other Oil Spills On June 3, 1979, the Ixtoc 1 offshore well in the Bay of Campeche, off the Yucatan Peninsula of Mexico, blew out and was not capped until March 24, 1980.

A. Band 1, blue (0.45 to 0.52 µm).

B. Band 2, green (0.52 to 0.60 µm).

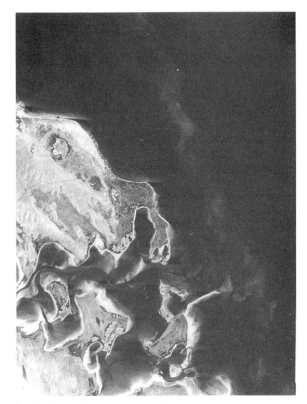

C. Band 3, red (0.63 to 0.69 µm).

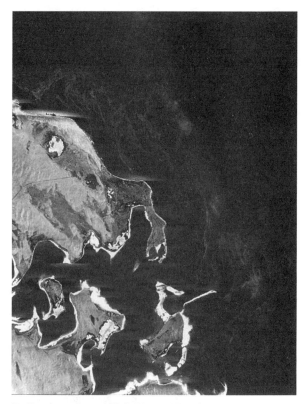

D. Band 4, reflected IR (0.76 to 0.90 µm).

Figure 9-33 Landsat TM-band images of an oil spill in the Arabian Gulf, February 16, 1991.

E. Band 5, reflected IR (1.55 to 1.75 μm).

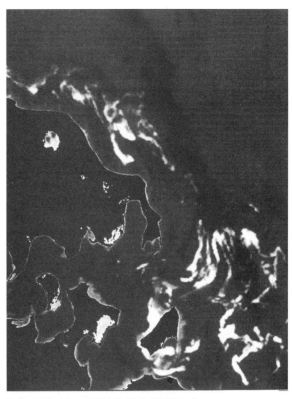

F. Band 6, thermal IR (10.40 to 12.50 μm).

G. Band 7, reflected IR (2.08 to 2.35 μm).

H. Location map.

Figure 9-34 Map of the Arabian Gulf oil spill interpreted from the TM images in Figure 9-33.

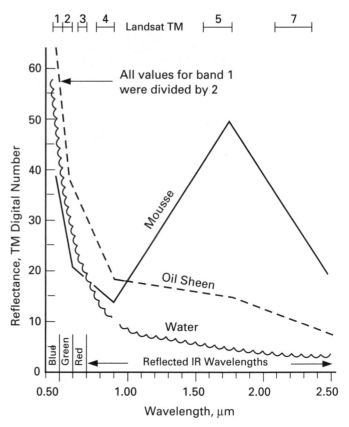

Figure 9-35 Reflectance spectra derived from TM bands of the oil spill in the Arabian Gulf (Figure 9-33).

It is estimated that a half million metric tons of oil were discharged. Between 30 and 50 percent burned at the wellhead; less than 10 percent was recovered. A number of Landsat MSS images of the oil spill were analyzed by Deutsch, Vollmers, and Deutsch (1980). In these visible and reflected IR bands, most of the oil film has a dark signature relative to the bright water, which is attributed to oil dampening of the small-scale waves and sunglint. Patches with brighter signatures than the water are thought to be mousse.

On March 24, 1989, the tanker *Exxon Valdez* ran aground in Prince William Sound, Alaska, spilling over 262,000 barrels of oil that eventually covered more than 1.9×10^4 km². Investigators from the University of Alaska (Stringer and others, 1992) processed and interpreted a number of different images (AVHRR, Landsat TM, and SPOT). Despite their coarse spatial resolution (1 km) AVHRR images were useful because of the large extent of the spill and the daily acquisition of im-

ages, clouds permitting. The most useful AVHRR images were thermal IR band 4 (10.5 to 11.5 µm). The oil has cool signatures because of its lower emissivity. Several TM images were acquired. Thermal IR band 6 of the TM has a finer spatial resolution (120 m) than AVHRR; most of the oil is cool, but "filaments" are warm and may represent mousse that has characteristic warm signatures in other spills. The Alaskan oil has a very strong reflectance in TM band 5, as does the Arabian Gulf oil (Figure 9-35E). SPOT and aircraft radar images were also employed.

Perspectives Oil spills are "media events" accompanied by dire predictions of ecologic disasters for future generations. The media are not interested in revisiting sites of former spills to report long-term impacts on the environment. The Congressional Research Service of the Library of Congress prepared a report for Congress, *Oil in the Ocean: The Short- and Long-Term Impacts of a Spill* (Mielke, 1990). Few people are aware of this report; therefore the Summary section is quoted in its entirety.

This report describes the short- and long-term impacts of an oil spill. The short-term impact is the incident as generally

A. PC 1 image of bands 4, 5, and 7.

B. PC 2 image of bands 4, 5, and 7.

C. PC 3 image of bands 4, 5, and 7.

D. Ratio 4/3 image.

Figure 9-36 TM images that were digitally processed to extract information for the oil spill in the Arabian Gulf.

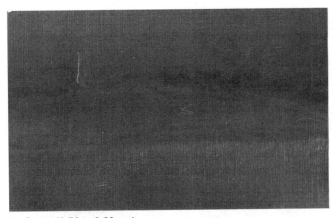

A. Green (0.52 to 0.60 μm).

B. Thermal IR (10.40 to 12.50 μm).

Figure 9-37 Aircraft multispectral scanner images of an oil spill in the Gulf of Mexico near Galveston, Texas. Courtesy NASA Johnson Space Center.

reported immediately following the spill, and the long-term impact is the life cycle of the spilled oil. Not surprisingly, the impacts are often different. The media presentation is commonly one of a catastrophic occurrence, and a major oil spill is indeed that. Media coverage tends to focus on the more emotional aspects of destruction to the local environment, to which irreparable harm is often claimed. Rarely does media coverage convey the fact that oil is a natural substance, and that natural processes, over time, will do much to remove it.

Oil that is spilled or seeps naturally into the ocean is eventually accommodated by natural physical, chemical, and biological processes, including spreading, evaporation, solution, emulsification, tar lump formation, photochemical oxidation, microbial degradation, uptake by organisms, and sedimentation and shoreline stranding. Factors particular to each spill influence the effectiveness of these processes, and

determine the severity of the ecological impact. Although human intervention can help to make a shoreline look clean, it has rarely been very effective in removing oil, and improper clean-up can be detrimental to the ecological restoration of the area. Historically, it has been unusual for more than 10 to 15 percent of the oil to be recovered from a large spill.

The life cycles of six major spills were chosen for examination because they occurred sufficiently long ago for long-term effects to have become apparent and for attention from the media to have subsided. Because they are still being studied, the *Exxon Valdez* and the *Mega Borg* incidents are not discussed. The six events chosen are the Santa Barbara and Ixtoc 1 blowouts and the *Argo Merchant, Burmah Agate, Alvenus,* and *Amoco Cadiz* tanker spills. Each event received extensive media coverage at the time and are still thought of by many as major environmental catastrophes. In fact, the environmental damage and socioeconomic consequences were relatively modest, and, as far as can be determined, of relatively short duration.

The longest residence time spilled oil appears to have in the marine and coastal environment is generally less than a decade—often much less. The major ecological impact comes at the time of the spill or within the first few months. Beyond a few months, most of the oil is reduced to tarry residues or is chemically detectable in sediments and resident organisms, which may be of scientific interest, but in terms of further ecological impact, likely to be fairly insignificant. Short-term impacts on marine animal life are dramatic but recovery of species populations in almost every case studied has been swift.

Because many physical and biological processes in the marine and coastal environment are poorly understood, it is difficult for scientists to measure the full impacts of an oil spill and sometimes the results appear contradictory.

This unedited summary from a U.S. government study is quoted to bring some objectivity to the topic of oil spills. Every reasonable effort must be made to prevent spills. When the inevitable accidents occur, however, it is useful to know their long-term consequences from scientific studies rather than media accounts.

COMMENTS

For many environmental applications, remote sensing is the best means of acquiring basic information, particularly on a regional scale and repetitive schedule. Environmental satellites

Figure 9-38 (facing page) ERS-1 radar images of oil spills. For these images acquired at a wavelength of 5.7 cm and a depression angle of 67°, the smooth criterion is 0.25 cm and the rough criterion is 1.41 cm. Courtesy European Space Agency.

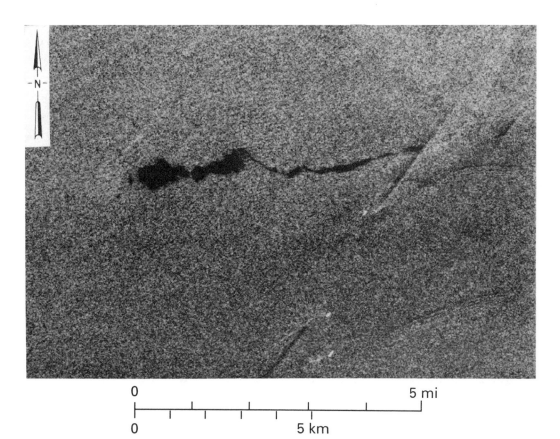

A. Offshore from Gothenberg, Sweden.

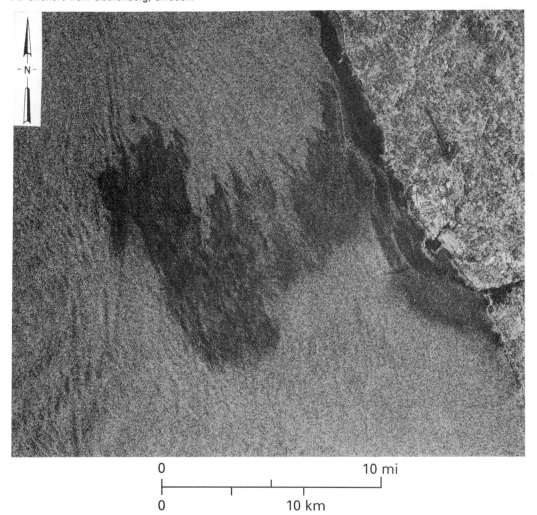

B. Oporto, Portugal, October 4, 1994.

have been launched specifically for these purposes. Digital-processing methods are becoming increasingly important for extracting and interpreting desired information from the extensive databases.

Many aspects of the oceans that were once poorly understood (such as circulation, sea state, productivity, sea-ice distribution, and bathymetry) are becoming better known through remote sensing techniques. Pollution in the form of thermal plumes and oil spills may be monitored.

QUESTIONS

1. Prepare a satellite comparison chart in the following manner. In a vertical column at the left margin, list the following systems:

> Landsat MSS
> Landsat TM
> GOES
> AVHRR
> CZCS

Across the top of the chart, list the following characteristics as headings:

> Spectral range, μm
> Number of spectral bands
> Ground resolution cell, m
> Image-swath width, km
> Repetition rate, days

Complete the chart by listing the characteristics for each satellite.

2. From this chart, select the optimum satellite system for your particular application(s). Give reasons for your selection.

3. Describe how you would employ CZCS images and AVHRR thermal IR images to increase the catches of commercial fisheries in the oceans. (*Hint*: Upwelling occurs where deeper water comes to the surface carrying dissolved nutrients. Under the right conditions, upwelling will support plankton growth, which is the first link in the oceanic food chain.)

4. Describe the various conditions under which you would *not* rely on the following images for bathymetric information in a shallow sea:

> Landsat images
> Seasat images

5. List the various properties of water currents that enable one to detect them on images. List the imaging system or systems that record each property.

6. Your agency is responsible for monitoring the environment of San Francisco Bay, which has daily tidal fluctuations and inflows from several natural streams. Along the shore are numerous fossil-fuel power plants, chemical plants, oil re-fineries, and sewage plants. Fog and rain are dense at times. Describe the remote sensing system you would recommend to monitor thermal and chemical pollution in this situation.

7. For monitoring types, distribution, and movement of sea ice in the Arctic Ocean, list the advantages and disadvantages of following images:

> Landsat TM band 4
> Landsat TM band 6
> Seasat

REFERENCES

Abbott, M. R., 1984, Northern California/Cape Mendocino *in* Hovis, W. A., ed., Nimbus-7 CZCS coastal zone color scanner imagery for selected coastal regions: NASA Goddard Space Flight Center, Greenbelt, MD.

Arkin, P. A. and P. E. Ardanuy, 1989, Estimating climate-scale precipitation from space—a review: Journal of Climate, v. 2, p. 1229–1238.

Arkin, P. A. and J. E. Janowiak, 1993, Tropical and subtropical precipitation *in* Gurney, R. J., J. L. Foster, and C. L. Parkinson, eds., Atlas of satellite observations related to global change: p. 165–180, Cambridge University Press, Cambridge, Great Britain.

Atlas, R., R. N. Hoffman, and S. C. Bloom, 1993, Surface wind velocity over the oceans *in* Gurney, R. J., J. L. Foster, and C. L. Parkinson, eds., Atlas of satellite observations related to global change: p. 129–139, Cambridge University Press, Cambridge, Great Britain.

Belderson, R. H., N. H. Kenyon, A. H. Stride, and A. R. Stubbs, 1972, Sonographs of the seafloor, a picture atlas: Elsevier Publishing Co., New York, NY.

Biggs, R. H. and M. E. B. Joyner, 1994, Stratospheric ozone depletion/UV-B radiation in the atmosphere: Springer-Verlag, Berlin, Germany.

Brock, J. C. and C. R. McClain, 1992, Interannual variability in phytoplankton blooms observed in the northwestern Arabian Sea during the southwest monsoon: Journal of Geophysical Research, v. 97, p. 733–750.

Cahalan, B., 1992, The Kuwait oil fires as seen by Landsat: Journal of Geophysical Research, v. 97, p. 14,565–14,571.

Carlson, P. R., 1976, Mapping surface current flow in turbid near-shore waters of the northeast Pacific *in* Williams, R. S. and W. D. Carter, eds., ERTS-1, a new window on our planet: U.S. Geological Survey Professional Paper 929, p. 328–329.

Davies, P. A. and L. A. Mofor, 1993, Remote sensing observations and analyses of cooling water discharges from a coastal power station: International Journal of Remote Sensing, v. 14, p. 253–273.

Deutsch, M., R. R. Vollmers, and J. P. Deutsch, 1980, Landsat tracking of oil slicks from the 1979 Gulf of Mexico oil well blowout: Proceedings of the 14th International Symposium on Remote Sensing of the Environment, p. 1197–1211, Environmental Research Institute of Michigan, Ann Arbor, MI.

Dixon, W. S., M. Naraghi, M. K. McNutt, and S. M. Smith, 1983, Bathymetric prediction from Seasat altimeter data: Journal of Geophysical Research, v. 88, p. 1563–1571.

Epstein, E. S., W. M. Callicott, D. J. Cotter, and H. W. Yates, 1984, NOAA satellite programs: IEEE Transactions on Aerospace and Electronic Systems, v. AES 20, p. 325–344.

Estes, J. E., R. E. Crippen, and J. L. Star, 1985, Natural oil seep detection in the Santa Barbara Channel, California, with Shuttle imaging radar: Geology, v. 13, p. 282–284.

Farman, J. C., B. C. Gardiner, and J. D. Shanklin, 1985, Large losses of total ozone in Antarctica reveal seasonal ClO_x/NO_x interaction: Nature, v. 315, p. 207–210.

Fenstermaker, L. K. and J. R. Miller, 1994, Identification of fluvially redistributed mill tailings using high resolution aircraft data: Photogrammetric Engineering and Remote Sensing, v. 60, p. 989–995.

Fiedler, P. C., 1984, Satellite observations of the 1982–1983 El Nino along the U.S. coast: Science, v. 224, p. 1251–1254.

Ford, J. P., J. B. Cimino, and C. Elachi, 1983, Space Shuttle *Columbia* views the world with imaging radar—the SIR-A experiment: Jet Propulsion Laboratory Publication 82-95, Pasadena, CA.

Fornari, D. J., W. B. F. Ryan, and P. J. Fox, 1984, The evolution of craters and calderas on young seamounts—insights from Sea MARC I and Sea Beam sonar surveys of a small seamount group near the axis of the East Pacific Rise at 10°N: Journal of Geophysical Research, v. 89, p. 11,069–11,083.

Fu, L. L. and B. Holt, 1982, Seasat views oceans and sea ice with synthetic aperture radar: Jet Propulsion Laboratory Publication 81-120, Pasadena, CA.

Gahagan, L. M. and others, 1988, Tectonic fabric of the ocean basins from satellite altimetry data: Tectonophysics, v. 155, p. 1–26.

Gordon, H. R., D. K. Clark, J. L. Mueller, and W. A. Hovis, 1980, Phytoplankton pigments from the Nimbus-7 coastal zone color scanner—comparison with surface measurements: Science, v. 210, p. 63–66.

Gordon, H. R. and others, 1983, Phytoplankton pigment concentrations in the middle Atlantic Bight—comparison of ship determinations and CZCS estimates: Applied Optics, v. 22, p. 20–36.

Greenstone, R. and B. Bandeen, 1993, Operational and research satellites for observing the earth in the 1990s and thereafter *in* Gurney, R. J., J. L. Foster, and C. L. Parkinson, eds., Atlas of satellite observations related to global change: p. 449–457, Cambridge University Press, Cambridge, Great Britain.

Grody, N. C., 1991, Classification of snow cover and precipitation using a special microwave imager: Journal of Geophysical Research, v. 96, p. 7423–7435.

Hammack, J. C., 1977, Landsat goes to sea: Photogrammetric Engineering and Remote Sensing, v. 43, p. 683–691.

Harris, P. M. and W. S. Kowalik, eds., 1994, Satellite images of carbonate depositional settings: American Association of Petroleum Geologists, Tulsa, OK.

Herman, J. D., J. E. Waites, R. M. Ponitz, and P. Etzler, 1994, A temporal and spatial resolution study of a Michigan Superfund site: Photogrammetric Engineering and Remote Sensing, v. 60, p. 1007–1017.

Hovis, W. A., ed., 1984, Nimbus-7 CZCS coastal zone color scanner imagery for selected coastal regions: NASA Goddard Space Flight Center, Greenbelt, MD.

Hovis, W. A. and others, 1980, Nimbus-7 coastal zone color scanner—system description and early imagery: Science, v. 210, p. 60–63.

Hughes, B. A. and J. F. R. Gower, 1983, SAR imagery and surface truth comparisons of internal waves in Georgia Strait, British Columbia, Canada: Journal of Geophysical Research, v. 88, p. 1809–1824.

Kasischke, E. S., R. A. Schuchman, D. R. Lyzenga, and G. A. Meadows, 1983, Detection of bottom features on Seasat synthetic aperture radar imagery: Photogrammetric Engineering and Remote Sensing, v. 49, p. 1341–1353.

Ketchum, R. D., 1984, Seasat SAR sea ice imagery—summer melt to autumn freeze-up: International Journal of Remote Sensing, v. 5, p. 533–544.

Kirby, M. E. and R. T. Lowry, 1981, Iceberg detectability problems using SAR and SLAR systems *in* Deutsch, M., D. R. Weisnet, and A. Rango, eds., Satellite hydrology: p. 200–212, American Water Resources Association, Minneapolis, MN.

Kleinrock, M. C., 1992, Capabilities of some systems used to survey the deep-sea floor *in* R. A. Geyer, ed., CRC handbook of geophysical exploration at sea, second edition: ch 2, p. 36–86, CRC Press, Boca Raton, FL.

Limaye, S. S. and others, 1992, Satellite monitoring of smoke from the Kuwait oil fires: Journal of Geophysical Research, v. 97, p. 14,551–14,563.

Lyzenga, D. R., 1981, Remote sensing of bottom reflectance and water attenuation parameters in shallow water using aircraft and Landsat data: International Journal of Remote Sensing, v. 2, p. 71–82.

Marsh, S. E., J. L. Walsh, C. T. Lee, and L. A. Graham, 1991, Multi-temporal analysis of hazardous waste sites through the use of a new bi-spectral video remote sensing system and standard color-IR photography: Photogrammetric Engineering and Remote Sensing, v. 57, p. 1221–1226.

Maurer, A. and A. T. Edgerton, 1976, Flight evaluation of U.S. Coast Guard airborne oil surveillance system: Marine Technology Society Journal, v. 10, p. 38–52.

McClain, C. R., G. Feldman, and W. Esaias, 1993, Oceanic biological productivity *in* Gurney, R. J., J. L. Foster, and C. L. Parkinson, eds., Atlas of satellite observations related to global change: p. 251–263, Cambridge University Press, Cambridge, Great Britain.

McPeters, R., 1994, Monitoring UV-B radiation from space *in* Biggs, R. H. and M. E. B. Joyner, eds., Stratospheric ozone depletion/UV-B radiation in the atmosphere: Springer-Verlag, Berlin, Germany.

Mielke, J. E., 1990, Oil in the ocean; the short- and long-term impacts of a spill: Congressional Research Service, Report to Congress 90-356 SPR, Library of Congress, Washington, DC.

Moore, R. K., 1983, Radar fundamentals and scatterometers *in* Colwell, R. N., ed., Manual of remote sensing, second edition: ch. 9, p. 369–427, American Society for Photogrammetry and Remote Sensing, Falls Church, VA.

National Research Council, 1985, Oil in the sea—inputs, fates, and effects: National Academy Press, Washington, DC.

Nordman, M. E., L. Wood, J. L. Michalek, and J. J. Chriaty, 1990, Water depth extraction from Landsat-5 imagery: Proceedings of 23rd International Symposium on Remote Sensing of the Environment: p. 1129–1139, Environmental Research Institute of Michigan, Ann Arbor, MI.

O'Neil, R. A., L. Buja-Bijunas, and D. M. Rayner, 1980, Field performance of a laser fluorosensor for the detection of oil spills: Applied Optics, v. 19, p. 863–870.

Osborne, A. R. and T. L. Burch, 1980, Internal solitons in the Andaman Sea: Science, v. 208, p. 451–460.

Parkinson, C. L. and P. Gloersen, 1993, Global sea ice coverage *in* Gurney, R. J., J. L. Foster, and C. L. Parkinson, eds., Atlas of satellite observations related to global change: p. 141–163, Cambridge University Press, Cambridge, Great Britain.

Quinn, M. F. and others, 1994, Measurement and analysis procedures for remote identification of oil spills using a laser fluorosensor: International Journal of Remote Sensing, v. 15, p. 2637–2658.

Richards, F. and P. A. Arkin, 1981, On the relationship between satellite-observed cloud cover and precipitation: Monthly Weather Review, v. 109, p. 1081–1093.

Rossow, W. B., 1993, Clouds, in Gurney, R. J., J. L. Foster, and C. L. Parkinson, eds., Atlas of satellite observations related to global change: p. 141–163, Cambridge University Press, Cambridge, Great Britain.

Rouse, J. W., 1968, Arctic ice type identification by radar: University of Kansas Center for Research Technical Report 121-1, Lawrence, KN.

Sabins, F. F., 1987, Remote sensing—principles and interpretation, second edition: W. H. Freeman and Co., New York, NY.

Salisbury, J. W., D. M. D'Aria, and F. F. Sabins, 1993, Thermal remote sensing of crude oil slicks: Remote Sensing of Environment, v. 45, p. 225–231.

Schoeberl, M. A., 1993, Stratospheric ozone depletion in Gurney, R. J., J. L. Foster, and C. L. Parkinson, eds., Atlas of satellite observations related to global change: p. 59–65, Cambridge University Press, Cambridge, Great Britain.

Schuchman, R. A., 1982, Quantification of SAR signatures of shallow water ocean topography: Ph.D. dissertation, University of Michigan, Ann Arbor, MI.

Specht, M. R., D. Needler, and N. L. Fritz, 1973, New color film for water penetration photography: Photogrammetric Engineering and Remote Sensing, v. 40, p. 359–369.

Stringer, W. J. and others, 1992, Detection of petroleum spilled from the MV *Exxon Valdez*: International Journal of Remote Sensing, v. 13, p. 799–824.

Thompson, M. C., C. Remington, J. Purnomo, and D. Macgregor, 1991, Detection of liquid hydrocarbon seepage in Indonesian offshore frontier basins using airborne laser fluorosensor (ALF)—the results of a Pertamina/BP joint study: Indonesian Petroleum Association, Proceedings 20th Annual Convention, p. 663–689, Jakarta, Indonesia.

Vizy, K. N., 1974, Detecting and monitoring oil slicks with aerial photos: Photogrammetric Engineering, v. 40, p. 697–708.

Weng, F., R. R. Ferraro, and N. C. Grody, 1994, Global precipitation estimations using Defense Meteorologic Satellite program F10 and F11 special sensor microwave imager data: Journal of Geophysical Research, v. 99, p. 14,493–14,502.

Williams, R. S., J. Heckman, and J. Schneeberger, 1991, Environmental consequences of the Persian Gulf war: National Geographic Society, Washington, DC.

ADDITIONAL READING

Barale, V. and P. M. Schlittenhardt, eds., 1993, Ocean colour—theory and application in a decade of CZCS experience: Kluwer Academic Publishers, Dordrecht, the Netherlands.

Bukata, B. P., K. Y. Kondratyev, D. V. Pozdyakov, and J. H. Jerome, 1995, Optical properties and remote sensing of inland and coastal waters: CRC Press, Boca Raton, FL.

Carleton, A. M., 1991, Satellite remote sensing in climatology: CRC Press, Boca Raton, FL.

Carsey, F., ed., 1992, Microwave remote sensing of sea ice: American Geophysical Union, Geophysical Monograph Series, v. 68.

Clark, C. D., 1993, Satellite remote sensing of marine pollution: International Journal of Remote Sensing, v. 14, p. 2985–3004.

Davis, H. H., P. D. Caldwell, P. B. Goodwin, and E. Karver, 1994, Use of SPOT satellite imagery to obtain GIS input for oil spill models: 10th Thematic Conference on Geologic Remote Sensing, Environmental Research Institute of Michigan, v. II, p. 55–64, Ann Arbor, MI.

Gloersen, P. and others, 1992, Arctic and Antarctic sea ice, 1978–1987 satellite passive microwave observations and analysis: NASA, Washington, DC.

Gurney, R. J., J. L. Foster, and C. L. Parkinson, eds., 1993, Atlas of satellite observations related to global change: Cambridge University Press, Cambridge, Great Britain.

Hall, D. K. and J. Martinac, 1985, Remote sensing of ice and snow: Chapman and Hall, New York, NY.

Hobbs, R. J. and H. A. Mooney, eds., 1990, Remote sensing of biosphere functioning: Springer Verlag, New York, NY.

Ikeda, M. and F. Dobson, eds., 1995, Oceanographic applications of remote sensing: CRC Press, Boca Raton, FL.

Jones, I. S. F., Y. Sugimori, and R. W. Stewart, eds., 1993, Satellite remote sensing of the oceanic environment: Seibutsu Kenkyusha, Tokyo, Japan.

Kidder, S. Q. and T. H. von der Haar, 1995, Satellite meteorology: Academic Press, Orlando, FL.

Lodge, A. E., ed., 1988, The remote sensing of oil slicks: John Wiley & Sons, New York, NY.

Mobley, C. D., 1994, Light and water—radiative transfer in natural waters: Academic Press, New York, NY.

Robinson, I. S., 1985, Satellite oceanography: John Wiley & Sons, New York, NY.

10

OIL EXPLORATION

Remote sensing has become an accepted technique for oil exploration. The following case histories describe several exploration projects that established major new oil provinces. Each project is an example not only of remote sensing, but also of a multidisciplinary team approach. Field and subsurface geology, geochemistry, and geophysics are essential disciplines. Oil explorationists must be prepared to perform in this technical environment.

EXPLORATION PROGRAMS

The search for oil in unexplored onshore areas normally begins with regional reconnaissance followed by progressively more detailed (and expensive) steps that culminate by drilling a wildcat well. *Wildcat wells* are exploratory tests in previously undrilled areas, whereas *development wells* are drilled to produce oil from a previously discovered field. A typical exploration program proceeds as follows:

1. **Regional remote sensing reconnaissance** Small-scale Landsat mosaics covering hundreds of thousands of square kilometers are especially useful in this phase. The objective is to locate *sedimentary basins*, which are areas underlain by thick sequences of sedimentary rocks. These basins are essential for the formation of oil fields.
2. **Reconnaissance geophysical surveys** *Aerial magnetic surveys* produce maps that record the intensity of the earth's magnetic field. Sedimentary basins have lower magnetic intensities than do areas underlain by basement rocks such as granite and metamorphic rocks. The aerial magnetic maps thus can confirm the presence of sedimentary basins. Surface *gravity surveys* record the intensity of the earth's gravity field. Sedimentary rocks have a lower specific gravity than basement rocks; hence sedimentary basins are shown by lower values on gravity maps. Gravity and magnetic maps may also show regional structural features.
3. **Detailed remote sensing interpretation** Individual digitally processed Landsat images are interpreted to identify

and map geologic structures, such as anticlines and faults, that may form oil traps. Promising structures may be mapped in detail using stereo pairs of SPOT images or aerial photographs. Radar images are used in regions of poor weather where it is difficult to acquire good photographs and Landsat images. At this stage, geologists go into the field to check the interpretation and collect samples of the exposed rocks.
4. **Seismic surveys** Explosives or mechanical devices are used to transmit waves of sonic energy into the subsurface, where they are reflected by geologic structures. The reflected waves are recorded at the surface and processed to produce *seismic maps and cross sections* that show details of subsurface geologic structure. Images and maps produced from images facilitate logistics and navigation in uncharted regions. For instance, topographic maps derived from stereo SPOT images were employed for a seismic survey in Yemen (Chapter 4).
5. **Drilling** Wildcat wells are drilled to test the *oil prospects,* or drilling targets, that are defined by the preceding steps. Because of the inevitable uncertainties of predicting geologic conditions thousands of kilometers below the surface, less than 20 percent of wildcat wells are successful.

Each of the following case histories illustrates a particular aspect of remote sensing for oil exploration. The Sudan project was the first major discovery using remote sensing and is a classic example of the systematic exploration process described above. The Papua New Guinea project was the first reported oil discovery using radar images. The Central Arabian Arch project resulted in new discoveries using Landsat TM images in a region with very subtle surface expression of oil traps.

SUDAN PROJECT

The Sudan project originated with a Chevron exploration program in northeastern Kenya that used Landsat MSS images to

Figure 10-1 Mosaic of Landsat MSS band 2 images of southern Sudan, showing the boundaries of the original Chevron exploration concession granted in 1974.

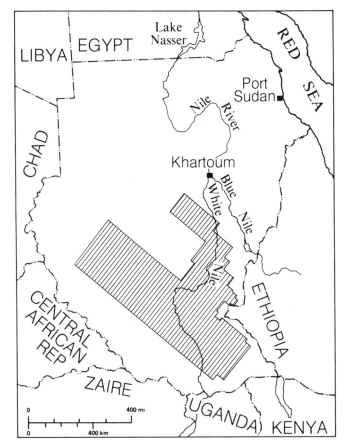

A. Original Chevron exploration concession, 1974.

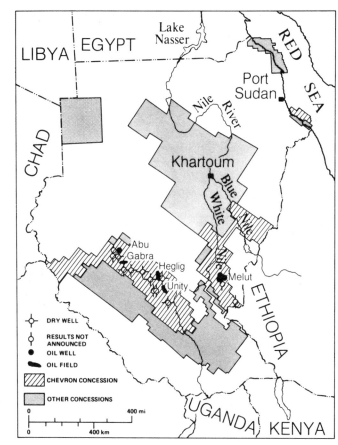

B. Status of exploration and concessions, 1982.

Figure 10-2 Maps showing the status of oil exploration in Sudan.

map regional geology, as described by Miller (1975). Geophysical surveys defined several prospects that were tested by wildcat wells. The wells were dry because subsurface conditions were not conducive to hydrocarbon generation. The Kenya project did, however, contribute the following valuable information:

1. Landsat images were shown to be reliable sources of geologic information.
2. The geophysical surveys and wells outlined a sedimentary basin in northeast Kenya that could extend northward into the adjacent part of Sudan.

These results encouraged Chevron to begin a major project in Sudan, which borders Kenya on the north.

Exploration Program

Sudan is the largest nation in Africa, with an area equal to the United States east of the Mississippi River. Sudan is incompletely mapped, both geographically and geologically. When the project began in 1974 there were no oil fields and little oil exploration had been done. The first step was to compile an

analog mosaic of Landsat MSS images at a scale of 1:1,000,000 for the southern part of Sudan. Figure 10-1 is a portion of the mosaic at a reduced scale. The White Nile River flows northward from the highlands of Uganda into the vast Sudd Swamp, which occupies much of the eastern portion of the mosaic. The river emerges from the swamp and flows northward toward Khartoum. The following description refers to the exploration steps outlined above.

1. J. B. Miller (1975) of Chevron interpreted the mosaics and inferred the presence of a previously unknown sedimentary basin in the vicinity of the present-day Sudd Swamp. Miller also analyzed regional drainage patterns on the mosaic and noted that, while there were numerous local bends and meanders, the major streams were relatively straight at the scale of the Landsat mosaic. These stream lineaments were interpreted as the faults that bounded the subsurface basins. Based on this regional Landsat interpretation, Chevron obtained an exploration concession from the Sudanese government for the area outlined in the mosaic (Figure 10-1). The concession boundaries were drawn to include the inferred sedimentary basins. Figure 10-2A shows the location of the original concession, which covered an area of

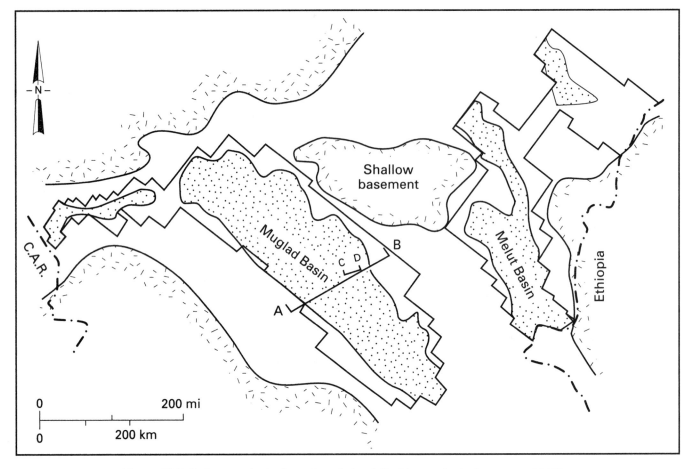

Figure 10-3 Sedimentary basins in southern Sudan defined by geophysical surveys and drilling. From Schull (1988, Figures 3, 9).

516,000 km². For comparison, note that the state of California covers an area of less than 410,000 km².

The individual Landsat MSS images covering the concession area were then digitally processed into IR color images and registered to ground control points. A base map for the concession was produced from the images and used throughout the project (Miller and Vandenakker, 1977).

2. An aerial magnetic survey confirmed the existence of the sedimentary basins. A gravity survey provided details on the configuration of the basins. A map showing the basins (Figure 10-3) was compiled from these surveys, supplemented with information from later seismic surveys and exploration wells. The elongate Melut and Muglad Basins are separated and surrounded by shallow basement rocks. Figure 10-3 also shows the boundaries of the concession in 1979 after it was reduced to an area of 259,000 km². Regional cross section A–B (Figure 10-4) shows the thick sedimentary section in the Muglad Basin. The cross section also shows the major high-angle normal faults that bound the basin. These faults are expressed as subtle lineaments on

the Landsat mosaic. Other faults with less displacement off-set sedimentary rocks within the basin and form oil traps.

3. The individual Landsat images were interpreted in detail to identify surface expression of faults. This information, plus the gravity and magnetic maps, guided the following seismic surveys.

4. Seismic surveys were difficult and expensive to conduct in this area of few roads and towns. The images were also employed to assess terrain conditions for the seismic surveys. Southern Sudan has wet and dry seasons with drastically different terrain conditions. Plate 20A is a Landsat subscene in the Sudd Swamp acquired during the dry season. The sinuous red band is a stream with vegetation and associated small lakes that have dark signatures. The dark terrain on either side of the stream is grassland that was burned by the local people to produce better forage during the ensuing wet season. Plate 20B is a supervised-classification map that was prepared from a wet-season image in the following manner. Field crews in the area noted localities of major terrain categories and communicated this information to the

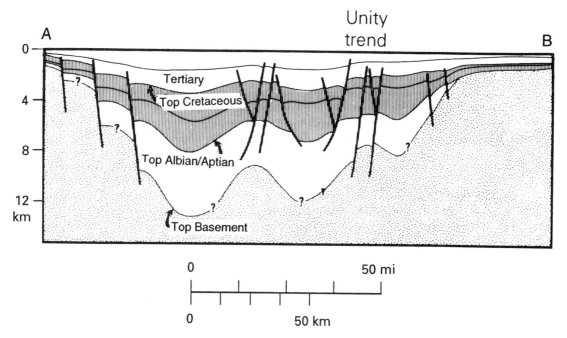

Figure 10-4 Regional cross section (A–B) across the Muglad Basin. From Schull (1988, Figure 11).

Chevron remote sensing group in San Francisco. The group used the field localities as training sites to produce the supervised-classification map of Plate 20B. Classification colors and corresponding terrain categories are listed below.

Color	Terrain category
Dark blue	Open water
Light blue	Shallow water with vegetation
Red	Papyrus and water hyacinth
Orange	Wet grass with standing water
Green	Bullrushes
Black	Bullrushes with standing water
Dark yellow	Dry grass
Light yellow	Upland areas, driest areas

These terrain classification maps were valuable to the seismic crews for planning operations. For example, swamp buggies were used in marsh areas, but where papyrus plants occurred (red signature), the stalks jammed the drive mechanisms. Seismic lines were laid out to avoid papyrus areas by referring to the classification image. The resulting cost savings more than paid for the entire remote sensing program.

Oil Discoveries

Step 5 was to drill wildcat wells to test the prospects that were defined by the preceding steps of the project. Drilling began in October 1977 and the first oil was recovered in May 1978 from the second well, which was noncommercial. The first significant oil flow occurred in the fifth well in August 1979. The first important discovery was made in early 1980 at Unity 2. Figure 10-5 is a seismic section across the Unity field. The irregular subhorizontal lines are the sonic responses of alternating strata of sandstone and shale of Cretaceous and Tertiary ages. The strata are offset by high-angle normal faults, shown by heavy lines. Vertical lines are oil wells; black circles are sandstone beds that produce oil. The Unity oil field is a gentle anticline bounded by two faults that dip toward each other (Figure 10-5). Subsequent drilling discovered the Heglig field, shown in Figure 10-2B, where the oil is trapped by normal faults. A total of 86 wells were drilled, mostly in the Muglad Basin. Recoverable reserves are estimated at 250 to 300 million barrels of oil (Schull, 1988). Figure 10-2B shows the status of oil exploration in 1982. Black patterns are oil discoveries. The Chevron/Shell concession is shown by the diagonal pattern. The stippled pattern shows concessions acquired by competing oil companies.

Construction of a pipeline to Port Sudan on the Red Sea was ready to begin in the early 1980s when a major civil war began between the north and south regions of Sudan. The war, which continues to this day, has halted all production and development of Sudan's only oil reserves. The technical accomplishments of the Sudan project should not be overshadowed by this political unrest.

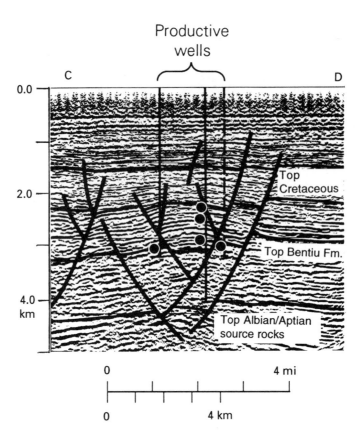

Figure 10-5 Seismic section C–D across the Unity oil field, showing wells. See Figure 10-3 for location of section. Adapted from Schull (1988, Figure 13).

PAPUA NEW GUINEA PROJECT

When the Chevron Corporation acquired Gulf Oil Company in 1984, it also acquired Gulf's interest in two petroleum prospecting licenses (PPLs) located in the eastern portion of the island of New Guinea, which is part of the nation of Papua New Guinea (PNG). Chevron operates the project on behalf of the other partners who own major interests.

Background

Figure 10-6 includes an index map of New Guinea and a location map of the PPLs, which extend for more than 350 km southeast from the border with Indonesia. The PPLs cover an area of approximately 14,400 km² that spans the boundary between the Fly-Strickland Lowlands in the southwest and the Papuan Fold and Thrust Belt in the northeast. Oil explorationists have been interested in the region for many years because

1. numerous oil and gas seeps occur in the fold belt;
2. some wildcat wells had indications of hydrocarbons, which are called "shows."

Although a number of wells had been drilled, no commercial oil fields had been discovered in PNG when this project began.

Oil exploration in the PPLs is difficult for several reasons. The area is remote and inaccessible with few roads; transportation is largely by foot or by aircraft that use the scattered airstrips. The area is covered by tropical rain forest. The climate is hot and humid with persistent cloud cover and heavy rainfall. The terrain is dominated by rugged linear ridges and steep stream channels. Much of the region is underlain by limestone that weathers to karst topography. The pinnacles and steep-walled pits cannot be crossed by vehicles and are dangerous to traverse on foot. Seismic surveys, which are the routine method for defining drilling prospects, cannot be recorded in most of the PPLs for the following reasons:

1. The inaccessibility and rugged topography result in seismic survey costs that exceed $100,000 per kilometer.
2. Subsurface caverns in the karst terrain strongly scatter the seismic energy, which results in very poor data.

For these reasons, it was decided at the outset of the project to rely on remote sensing and limited fieldwork.

Radar Images

The persistent cloud cover precluded timely acquisition of aerial photographs or Landsat or SPOT images. Earlier Chevron experience with SIR-A images in Indonesia (Chapter 7) had demonstrated that radar images are optimum for geologic interpretation in equatorial regions. In 1985 a contractor conducted an airborne radar survey of the PPLs. The X-band (3-cm wavelength), synthetic-aperture images have a spatial resolution of 12 m. Data were recorded digitally and processed into image strips that were combined to produce mosaics using the methods described in Chapter 6. Figure 10-7 is a greatly reduced version of the three mosaics that were produced at an original scale of 1:250,000.

In Chapter 6 the Venezuela images demonstrated the importance of selecting the optimum radar look direction to enhance the expression of geologic structure. The Papuan Fold and Thrust Belt, which underlies most of the project area, consists of northwest-trending ridges that are the expression of anticlines and thrust faults. Most of the thrust faults are exposed along the southwest flanks of anticlines, whereas the northeast flanks are predominantly dipslopes. For these reasons, the radar look direction was oriented toward the north to illuminate the faults, while the less-important dipslopes were partially shadowed. The north look direction is oblique to the northwest regional trend. Chevron interpreters considered using a northeast look direction, which is normal to the regional trend. However, this orientation would have resulted in extremely bright highlights from the southwest-facing slopes, which would have obscured important information. Radar shadows are oriented toward the upper margin of the mosaic (Figure 10-7), which may cause topographic inversion for some viewers. Rotate the page 180° to eliminate this problem. Adjacent radar swaths were acquired with 60 percent sidelap, which produced stereo images. The radar mosaic and stereo image strips were interpreted using the criteria developed earlier in Indonesia (Chapter 7).

Exploration Program

The exploration program focused on the Papuan Fold and Thrust Belt because of the numerous surface anticlines and abundant oil and gas seeps (Ellis and Pruett, 1986). The initial wildcat well was located on the crest of the Mananda anticline, shown in Figure 10-6, which is a prominent feature on the mosaic. The well was dry, but there were encouraging shows of oil in the Toro Sandstone (Cretaceous).

Exploration then shifted southeast from Mananda to the area south of Lake Kutubu where several promising anticlines are expressed on the radar mosaic. The location map (Figure 10-6) shows the position of the Lake Kutubu area. Figures 10-8 and 10-9 are the image and interpretation map for the Lake Kutubu area, shown at the 1:250,000 scale of the original mosaic. Most the terrain in the image (Figure 10-8) is underlain by the Darai Limestone (middle Tertiary), which weathers to karst topography with a distinctive pitted signature on the image. Smooth terrain in the valleys is formed by alluvial deposits. A few outcrops of clastic strata of the Orubadi Formation (late Tertiary) occur along southwest-facing slopes. Topography is generally a reliable expression of the underlying geologic structure because the region was deformed fairly recently and erosion has not obliterated structural landforms. Northwest-trending ridges are anticlines, and the intervening valleys are synclines.

The Darai Plateau in the southwest portion of the area is within the relatively undeformed Fly-Strickland Lowlands which forms a generally uniform surface. Northeast-trending lineaments in the plateau are fractures or faults that have been enlarged by solution of the limestone bedrock. The Hegegio thrust fault (Figure 10-9) is the structural boundary between the lowlands and the Papuan Fold Belt to the northeast. The numerous anticlines of the fold belt are expressed on the image as slightly curving linear topographic swells that are outlined by highlights and shadows. The anticlines are capped by the Darai Limestone with its distinctive karst topography. The synclines are floored by alluvial deposits with characteristic smooth texture on the image. Near the Hegegio fault thrust, the anticlines and synclines are of comparable width, but to the northeast, toward Lake Kutubu, the synclines become narrower because they are overridden by thrust plates that have moved relatively southwestward (Figure 10-9).

Figure 10-10 is a SPOT multispectral image (20-m spatial resolution) that was acquired during a rare morning with relatively few clouds. Compare the SPOT image (with a sun elevation of 62°) with the radar image (with a depression angle of 17°). Topographic and textural features of geologic significance are greatly enhanced on the radar image because of the low depression angle, as demonstrated earlier in Irian Jaya (Chapter 6). The IR color version of the SPOT image did provide useful information on vegetation cover and trafficability. The SPOT image was digitally merged with the radar image. The resulting radar/SPOT image (not shown) has the shadow enhancement of radar with the spectral information of SPOT and was useful in field operations.

Oil Discoveries

The first commercial oil well in PNG was completed in 1986 on the crest of the Iagifu anticline, which is now the Iagifu field (Figure 10-9). Additional wells were drilled to delineate the Iagifu field. Next, the Hedinia anticline was successfully drilled and is now the Hedinia field. Subsequently, the Agogo

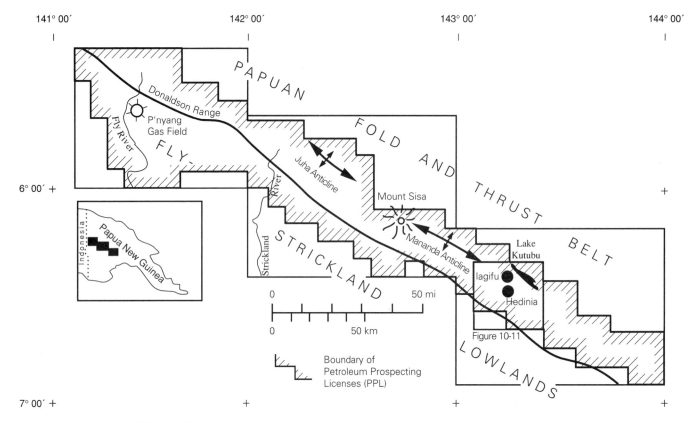

Figure 10-6 Location map for radar mosaics of the Papuan Fold Belt, PNG (Figure 10-8).

field and Southeast Hedinia field were discovered. Figure 10-9 shows these fields with stippled patterns. Figure 10-11 is a subsurface structure map of the Iagifu and Hedinia fields with contours on top of the Toro Sandstone, which is the oil-producing formation. Figure 10-9 shows the location of the subsurface map. Figure 10-12 is a geologic section across the Iagifu and Hedinia fields, which are trapped in the crests of anticlines located above the Hedinia thrust fault. On the cross section note that the subsurface anticlines are expressed as ridges on the topographic profile, which helps explain the accuracy of the radar interpretation. A number of additional structures within the PPLs remain to be drilled. In 1992, Chevron and its partners completed a major pipeline to transport the oil to a tanker terminal that was built on the coast of the Arafura Sea.

In 1990 Chevron used radar images to discover the P'nyang gas field in the northwest portion of the PPL (Figure 10-6). Valenti (1996) showed the image and interpretation map of the P'nyang anticline.

CENTRAL ARABIAN ARCH PROJECT

Saudi Arabia has the world's largest oil reserves and is a major exporter of oil. Therefore, it is surprising to realize that most of

the kingdom was relatively unexplored for oil by the late 1980s, for reasons given in the following section. The Central Arabian Arch project helped to correct this situation and aided in the discovery of significant new oil fields.

Background

In 1933 the Chevron Corporation, then called Standard Oil Company of California, was a small West Coast company in desperate need of new oil reserves. At that time Chevron obtained from King Ibn Saud the exploration rights to a huge area in eastern and central Saudi Arabia. In 1936 Texaco joined the venture to create the partnership that later became the Arabian American Oil Company (Aramco). Chevron geologists mapped the Dammam dome at Dhahran, shown as "D" in Figure 10-13. In 1939, after several dry holes, the first oil was discovered in the kingdom at Dammam. Oil exploration and development halted during World War II but resumed shortly after the war. Aramco discovered the giant Ghawar field in 1948. Chevron and Texaco realized that large capital investments were required to develop this resource. Therefore, the Aramco partnership was expanded to include Exxon and Mobil. Aramco proceeded to discover a number of major fields both onshore and offshore in

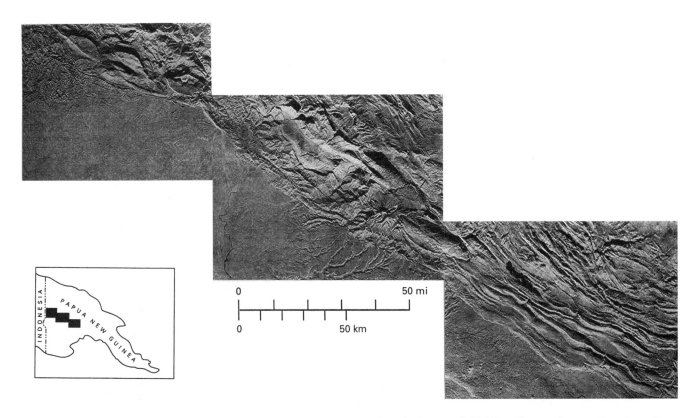

Figure 10-7 Mosaic of X-band (3-cm) aircraft radar images of the Papuan Fold Belt. Courtesy J. M. Ellis, Chevron Overseas Petroleum, Inc.

the Arabian Gulf. In 1972 the government of Saudi Arabia began to acquire the assets of Aramco. By 1980 the oil reserves and equipment were wholly owned by the kingdom. The company name was changed to Saudi Aramco in 1989. The original partners continue to provide personnel and technology to Saudi Aramco on a contract basis.

For many years the Saudi government had restricted exploration to "Retained Areas" in the vicinity of the existing fields, which are shown by hachured outlines in Figure 10-13. This restriction accounts for the lack of exploration in much of the kingdom. For a number of years through 1986 no new fields were discovered. At that time the government instructed Aramco to begin exploring the vast region with oil potential that lies outside the Retained Areas. In early 1988 Aramco requested the Remote Sensing Research Group of Chevron to conduct a Landsat study to aid the new exploration program. The initial study site, which is outlined in Figure 10-13, was an area in the Central Arabian Arch.

Major objectives of the Central Arabian Arch project were

1. to map the regional geology at a scale of 1:250,000;
2. to interpret local anomalies that may be expressions of subsurface structures that are potential oil fields.

The project was conducted by L. E. Wender (Aramco) and F. F. Sabins (Chevron). The TM images were digitally processed using techniques described by Sabins (1991).

Regional Geologic Mapping

Seven TM 2-4-7 images were digitally processed by the Chevron Remote Sensing Research Group (Sabins, 1991). The images were interpreted at a scale of 1:250,000 using the procedures described by Wender and Sabins (1991). The individual geologic maps were compiled into the generalized small-scale map in Figure 10-14, which shows locations of the individual images. The project area is located on the east flank of the Arabian shield, which is a regional uplift of Precambrian basement rocks. The basement rocks extend far to the east in the subsurface and provided a subsiding platform upon which were deposited an aggregate thickness of nearly 5500 m (18,000 ft) of strata ranging from Paleozoic through Tertiary ages.

The structure in the Paleozoic and Mesozoic strata is dominated by a gentle regional dip of approximately 1° away from the Arabian shield. In the southern part of the project area, dips are toward the east; in the north, dips are toward the northeast. The area where the dip direction changes is called the Central Arabian Arch (Figure 10-14).

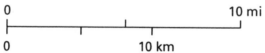

Figure 10-8 Radar image of the Lake Kutubu region, which includes oil fields discovered by the PNG project. The look direction is toward the north; the average depression angle is 17°. See Figure 10-6 for location. Courtesy J. M. Ellis, Chevron Overseas Petroleum, Inc.

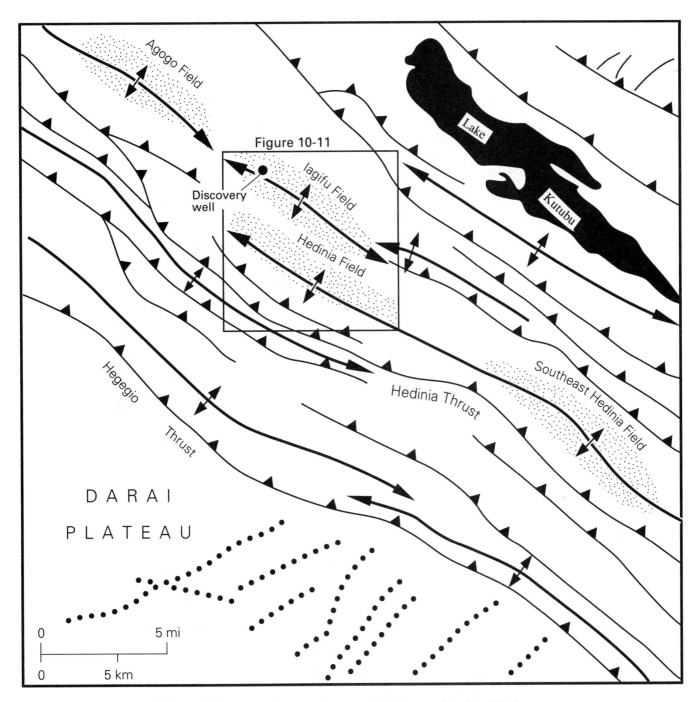

Figure 10-9 Structural interpretation map of radar image of the Lake Kutubu region (Figure 10-9), showing locations of oil fields discovered by the PNG project.

Landsat Anomalies

The second objective of the project was to recognize image features on the images that may be oil prospects. Prior to this project, my image interpretation experience was largely in regions of moderate to high structural relief such as the Saharan Atlas Mountains, Papuan Fold Belt, and the western United States. In these areas, surface dips of 5° or more are common and subsurface structures are expressed at the surface by opposing dip directions, arcuate outcrop patterns, and offset beds, as seen in the Thermopolis image (Chapter 3).

In the Central Arabian Arch, however, surface dips are 1° or less and the regional structural pattern is dominated by broad uniform dipslopes that extend for several hundred kilometers

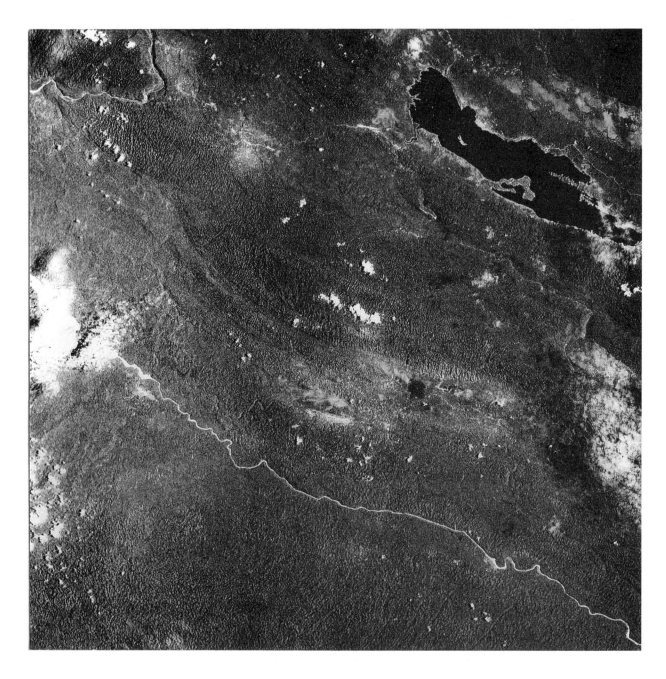

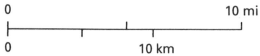

Figure 10-10 SPOT band 3 multispectral image (20-m spatial resolution) of the Lake Kutubu region acquired February 27, 1987. The sun azimuth is 96°, and the sun elevation is 62°. Compare with the radar image (Figure 10-8) to evaluate the impact of different illumination geometry.

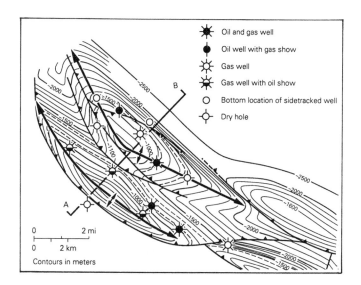

Figure 10-11 Subsurface structure map of the Iagifu and Hedinia oil fields. Contours are drawn on top of the productive Toro Sandstone. From Lamerson (1990, Figure 12).

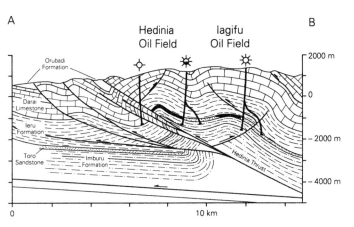

Figure 10-12 Cross section A–B across the Iagifu and Hedinia oil fields. From Lamerson (1990, Figure 10-13)

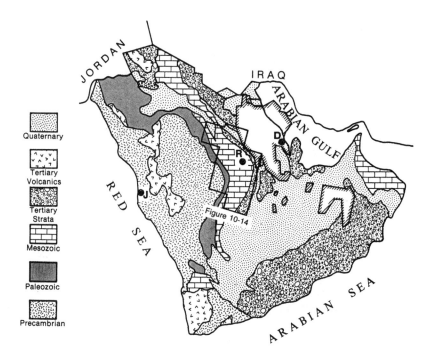

Figure 10-13 Location map of the Arabian Peninsula showing the location of the Central Arabian Arch project. Retained Areas are indicated by hachured borders. D = Dhahran, J = Jiddah, R = Riyadh.

along the strike and several tens of kilometers in the dip direction. In order to recognize structural anomalies in this region of low structural relief, we needed a three-dimensional model to show the relationship between subsurface structures and their expression on Landsat images. Within the Retained Areas, oil fields are simple drape anticlines in Mesozoic strata formed over high-angle faults that offset basement rocks and Paleozoic strata. The anticlines grade upward into flattening of the regional dip to form subtle structural terraces at the surface. Using this information, we developed the model shown in Figure 10-15. Erosion of the structural terrace produces the anomalous surface features of the model that are recognizable on Landsat images.

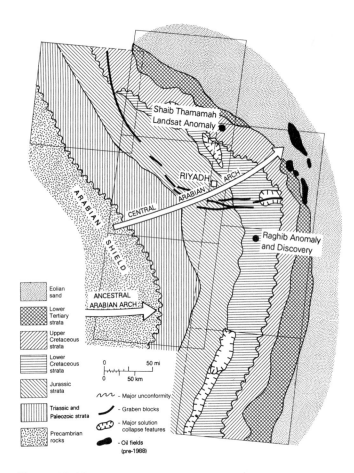

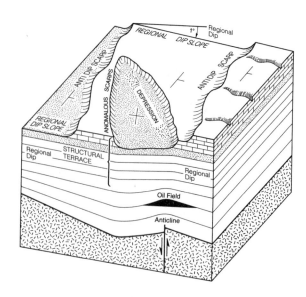

Figure 10-15 Geologic model for oil exploration in terrain with subhorizontal strata. The topographic and geologic anomaly at the surface is caused by the same subsurface structure that forms the oil field.

Figure 10-14 Geologic map of the Central Arabian Arch interpreted from TM images. Images boundaries are shown. From Wender and Sabins (1991, Figure 2).

The Shaib Thamamah anomaly, located 70 km north of Riyadh (Figure 10-14), demonstrated the validity of the structure model. Plate 20C and Figure 10-16 are the image and interpretation map of the anomaly, which occurs within the outcrop belt of the Aruma Formation (upper Cretaceous); Figure 10-17 gives the symbols for Figure 10-16. Prior to this project, the Aruma Formation was mapped as a single unit. On the Landsat 2-4-7 image, however, it is readily divided into two members with the following characteristics:

Upper member Resistant limestone that forms extensive dipslopes and steep antidip scarps with a dark greenish brown signature. In the field the color is medium to light tan.

Lower member Poorly resistant shaley limestone that weathers to slopes with a very light blue signature. In the field the color is medium to light tan.

The ability to separate the Aruma Formation into two members on the image is a major factor in recognizing the Shaib Thamamah anomaly. The anomaly is defined on the image (Plate 20C) by geomorphology and outcrop patterns.

1. **Geomorphology** The regional dipslope of the upper member terminates as a steep, southwest-facing antidip scarp that is shown in the field photograph of Figure 10-18A. The anomaly is a flat-floored topographic depression, bounded by erosional scarps of the upper member. These scarps are anomalous because they face east, which opposes the westward orientation of the regional antidip scarps. Figure 10-18B is a field photograph showing the depression in the foreground bounded by anomalous scarps in the background. The map (Figure 10-16) shows the scarps that enclose the depression which forms the anomaly.

2. **Outcrop pattern** The normal outcrop pattern of the Aruma Formation is a broad uniform dipslope formed by the upper member. At Shaib Thamamah, however, this pattern is interrupted by an inlier of the lower member that is 20 km long and 10 km wide. In Plate 20C the light-blue signature of the lower member distinguishes it from the upper member. In the field we observed two small exposures of the Wasia Sandstone, shown by small stippled patterns and the symbol Kw in Figure 10-16, that underlies the lower member. The presence of these older rocks within the inlier em-

phasizes the structural significance of the Shaib Thamamah anomaly.

The anomaly is so large that it is difficult to recognize in the field, but it is easily identified in the image.

In the model (Figure 10-15) the surface anomaly is underlain by an anticline and a fault. In order to evaluate subsurface structure at Shaib Thamamah, Saudi Aramco recorded seismic line 5624 at the location shown in Figure 10-16. The seismic section (Figure 10-19) shows that the anomaly is underlain by closely spaced vertical faults that offset the Khuff Formation (Permian) and adjacent beds. The offset dies out upward into a gentle arching of the Jilh Formation (Triassic). Higher in the section the arch becomes a structural terrace, or flattening of the northeast regional dip, at the Arab Formation (Jurassic) and overlying units. The surface profile of the seismic line (Figure 10-19) shows the depression that coincides with the geomorphic anomaly at Shaib Thamamah.

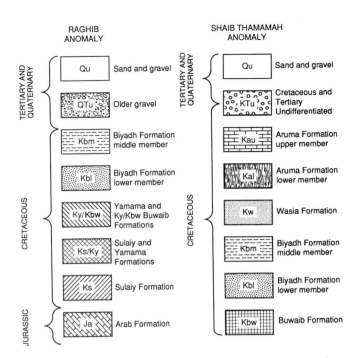

Figure 10-17 Symbols for geologic maps of the Shaib Thamamah and Raghib anomalies (Figures 10-16 and 10-20).

Elsewhere in the project area, we compared Landsat anomalies with seismic lines and found relationships similar to those at Shaib Thamamah.

Oil Discoveries

We used the model concept to interpret a number of anomalies which we then checked in the field. A few anomalies were eliminated as potential structures, but most were confirmed as targets for additional evaluation. Saudi Aramco followed up with seismic surveys of the most promising anomalies.

Plate 20D and Figure 10-20 are the image and interpretation map of the Raghib anomaly, which is located 110 km southeast of Riyadh (Figure 10-14). Limestones of the Arab and Sulaiy Formations crop out in the west portion of the image and are eroded to an irregular surface. The blue signature is locally obscured by patches of yellow windblown sand. The green circles in the northwest corner are wheat fields with centerpoint irrigation systems. Because of solution and collapse, it is locally difficult to separate the upper Sulaiy and lower Yamama Formations. The Yamama Formation (Cretaceous) consists of limestone with a medium-brown signature. The Buwaib Formation (Cretaceous) is a thin sequence of limestone with alternating bands of light blue and medium brown. The lower member of the Biyadh Formation is a resistant sandstone with a heavy coating of desert varnish and a dark-brown signature. The middle member is nonresistant, light-to-medium-gray sandstone with irregular dark-brown patches. Windblown sand covers much of the middle member.

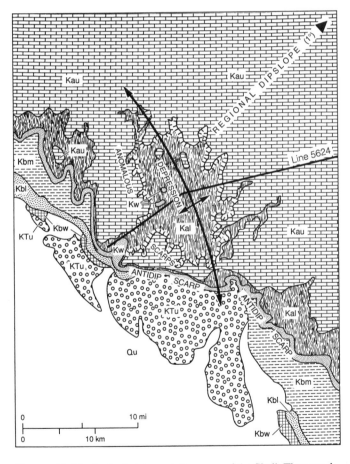

Figure 10-16 Geologic interpretation map of the Shaib Thamamah Landsat anomaly (Plate 20C). The location of the anomaly is shown in Figure 10-14. Symbols are explained in Figure 10-17. From Wender and Sabins (1991, Figure 4).

A. View east along antidip scarp formed by the upper member of the Aruma Formation. Note the gentle dipslope.

B. View west across center of anomaly. Depression in the foreground is underlain by the lower member of the Aruma Formation. The anomalous scarps in the background are underlain by the upper member of the Aruma Formation.

Figure 10-18 Field photographs of the Shaib Thamamah anomaly.

In this region south of the Central Arabian Arch, strata dip east at 1°; erosion produces linear north-trending antidip scarps. At Raghib this regional pattern is interrupted in the Sulaiy outcrops by arcuate scarps that are concave to the west (Plate 20D). The depression on the west side of the arcuate scarps is mantled by windblown sand. On the map (Figure 10-20) the scarps are shown by bold hachured lines with the hachures pointing toward the depression. These features are the erosional remnants of an anomaly that originally resembled Shaib Thamamah. Erosion has removed the updip western margin of the anomaly, leaving the central depression and the arcuate scarps on the downdip eastern margin.

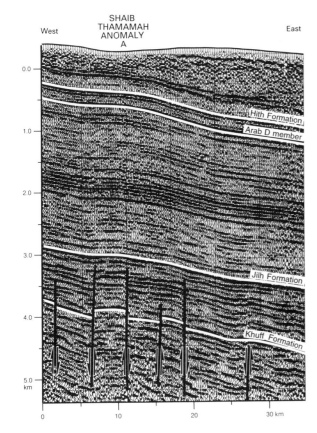

Figure 10-19 Seismic section 5624 across the Shaib Thamamah anomaly (Plate 20D). The location of the section is shown in Figure 10-16.

In late 1988 we checked the Landsat interpretation in the field. Early in 1989 Saudi Aramco completed a seismic survey of the Raghib area. Figure 10-21 is a seismic structure map at the same scale as the image (Plate 20D) and the interpretation map (Figure 10-20). Seismic structure contours are analogous to topographic contours but show the elevations of a rock formation rather than topography. Contours in Figure 10-21 are drawn on top of the Khuff Formation. Contour values are shown as two-way travel times in seconds (relative to sea-level datum). Shorter travel times indicate higher elevations of the Khuff Formation. Contours with travel times less than 1.020 sec define a northwest-trending, doubly-plunging anticline that is 35 km long and 10 km wide. The axis of the subsurface anticline is annotated on the interpretation map (Figure 10-20), which shows that the east flank of the anticline underlies the anomalous arcuate scarps on the image. The depression west of the scarps coincides with the structurally highest part of the anticline. Figure 10-22 is seismic section 4134, which crosses the Raghib surface anomaly at the location shown in Figure 10-21. The seismic section shows that the anticline at the Khuff Formation overlies a vertical fault that offsets older

strata. The anticline dies out upward into a structural terrace at the surface. Erosion of the terrace caused the Raghib image anomaly. These relationships match those of the structure model (Figure 10-15).

In late 1989 Saudi Aramco drilled Raghib 1 wildcat well located at the base of the surface scarp that coincides with the crest of the subsurface anticline. The location is shown in the geologic interpretation map (Figure 10-20) and the seismic map (Figure 10-21). Commercial hydrocarbons were discovered in sandstones of the Unayzah Formation (Pennsylvanian and Permian) that directly underlies the Khuff Formation (Figure 10-22). Sandstone of the Unayzah reservoir has an average porosity greater than 20 percent, and permeabilities of several darcies are common (McGillivray and Husseini, 1992). Production tests flowed 2984 barrels of oil, 1180 barrels of condensate, and 26.2 million cubic feet of gas per day. Similar results were obtained for the Raghib 2 confirmation well drilled 10 km northwest of Raghib 1 (Figure 10-21). Organic-rich shale of the Qusaiba Formation (Figure 10-22) of Silurian age is the source of the oil in the Raghib field (Abu-Ali and others, 1991).

OIL EXPLORATION RESEARCH

The Sudan, PNG, and Central Arabian Arch projects represent a broad category of oil fields that are expressed on images by geologic structures and anomalies. Another category of fields is expressed by seeps of oil and gas that have leaked to the surface. Exploration based on recognizing oil and gas seeps is called *direct detection*, and much research has been done on this method.

Oil and gas seeps may interact with surface rocks, soil, and vegetation to produce anomalous conditions that are clues to the hydrocarbon deposit. Classic examples of rock alteration occur at the Cement and Velma oil fields in southern Oklahoma. Sandstone outcrops in the area are typically red, but over the fields they are tan and gray. Gypsum ($CaSO_4 \cdot nH_2O$) is locally replaced by calcite ($CaCO_3$) over the fields. The color change from red to gray is caused by escaping hydrocarbons that chemically reduced the red iron oxide in the sandstone to a nonred iron compound. Donovan (1974) recognized an additional surface alteration effect; namely, the secondary calcite and dolomite ($Ca_{1/2}Mg_{1/2}CO_3$) in the surface rocks have unusual carbon isotope values. These values indicate that hydrocarbons leaking from the reservoir were oxidized and the carbon was incorporated into the secondary carbonate minerals of the surface rocks. Similar carbon isotopic values occur over the Davenport oil field in central Oklahoma (Donovan, Friedman, and Gleason, 1974).

Several investigators, including myself, have digitally processed Landsat data to identify any spectral signatures of the color and mineralogic alteration patterns at the Cement oil field. Such signatures could then be used to explore for other

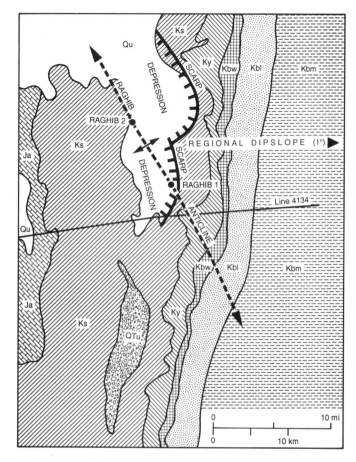

Figure 10-20 Geologic interpretation map of the Raghib Landsat anomaly (Plate 20D). Raghib 1 and 2 are the discovery well and confirmation well for the Raghib oil field. Symbols are explained in Figure 10-17. Location of Raghib anomaly is shown in Figure 10-14. From Wender and Sabins (1991, Figure 7).

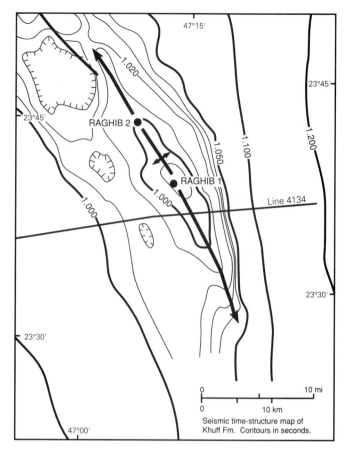

Figure 10-21 Seismic structure map of the Raghib oil field. The area coincides with the Landsat image (Plate 20D) and the geologic map (Figure 10-20). Contours are drawn on top of Khuff Limestone (Permian). From Wender and Sabins (1991, Figure 8)

fields. No successes have been reported. In the field we found that the color changes at the Cement and Velma fields are visible only at limited exposures in road cuts and streambeds. Most of the area is covered with soil and agriculture that obscures the alteration effects.

In the early 1980s a NASA/Geosat project interpreted airborne multispectral data and Landsat images over three oil and gas fields to develop exploration methods, including direct detection. At the Coyanosa field in west Texas, Lang, Nicolais, and Hopkins (1985) found no evidence of surface alteration caused by hydrocarbon seepage; however, some significant structural features were expressed in the images.

The Lost River gas field in West Virginia was studied because it is located in forested terrain and provided the opportunity to investigate possible vegetation anomalies associated with a gas field. Lang, Curtis, and Kovacs (1985) prepared a supervised-classification map that shows the distribution of tree species. Maple trees occur over the gas field in environ-

ments normally occupied by oak and hickory trees. A soil gas survey identified unusually high concentrations of methane and ethane that coincide with the anomalous maple trees. The gas in the soil may cause anaerobic soil conditions that inhibit growth of oak and hickory trees but not maples.

At the Patrick Draw field in southwest Wyoming, sagebrush is the predominant vegetation. Lang, Alderman, and Sabins (1985) described a ratio color image that indicated an anomalous area of sagebrush at the west margin of the field. Field investigation found a few square kilometers of blighted sagebrush with stunted growth and small leaves. A soil gas survey showed high concentrations of gas that could be responsible for this blighted sagebrush. The blighted vegetation, however, does not extend over a significant proportion of the field, and soil gas concentrations may occur with no associated vegetation anomaly.

In summary, these and other investigations have shown that remote sensing for direct detection is an unproven exploration method.

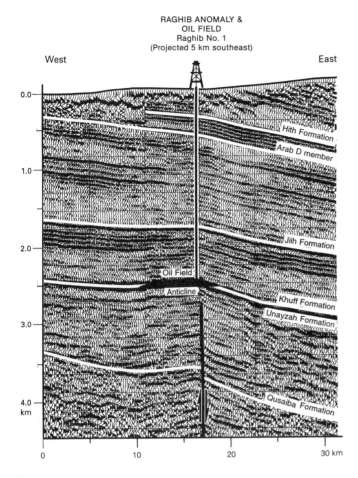

RAGHIB ANOMALY &
OIL FIELD
Raghib No. 1
(Projected 5 km southeast)

West   East

Hith Formation
Arab D member
Jilh Formation
Oil Field
Anticline
Khuff Formation
Unayzah Formation
Qusaiba Formation

Figure 10-22 Seismic section 4134 across the Raghib oil field. The location of the line is shown in Figures 10-20 and 10-21. Sandstone of the Unayzah Formation (Pennsylvanian and Permian) is the reservoir for the field. Organic-rich shale of the Qusaiba Formation (Silurian) is the source of the oil.

OTHER ENERGY RESOURCES

Remote sensing has been used to explore for other energy resources, as summarized in the following sections.

Geothermal Energy

Geothermal energy is obtained from subsurface reservoirs of steam or hot water that are shallow enough to be drilled and produced economically. Most geothermal energy is used to generate electricity; in Iceland, however, some energy is used for heating. Some geothermal reservoirs have no visible or thermal expression at the surface and are not detectable by remote sensing methods. Other reservoirs, however, have surface thermal expressions ranging in intensity from a minor increase in ground temperature to the presence of hot springs and geysers. Hot springs and geysers commonly occur along faults and fractures that allow hot water to reach the surface.

Thermal springs and hydrothermally altered rocks associated with a geothermal area in Mexico have been interpreted from thermal IR images by Valle and others (1970). At the Geysers geothermal area in northern California, hot springs and fumaroles were also detected on thermal IR images, and there was evidence of higher ground temperatures (Moxham, 1969). However, there was little evidence of a regional surface-temperature anomaly on images of the Geysers area. Reykjavik, Iceland utilizes energy from high-temperature geothermal areas along zones of active rifting and volcanism. Friedman and others (1969) recognized these features on thermal IR images. Geothermal vents and hot springs have been detected on thermal IR images in such diverse localities as Japan, Italy, Ethiopia, and the United States. Remote sensing is especially useful in remote areas where the surface expression of geothermal activity has not been located by conventional means.

Some geothermal areas lack fumaroles, steaming ground, or hot springs; instead the surface temperature is only slightly warmer than surrounding areas. These low-intensity temperature anomalies are difficult to detect on IR images. Watson (1975) noted that these subtle anomalies can be masked by temperature variations caused by geology, vegetation, and topography. To minimize this masking effect, IR images should be acquired at least three times during a diurnal cycle (Watson, 1975). This technique located a weak thermal anomaly in the Raft River area of Idaho that was confirmed by ground measurements.

Coal, Oil Shale, and Tar Sands

The location and distribution of large reserves of coal, oil shale, and tar sands are already known in the United States and Canada. Therefore, exploration is relatively inactive and there is little application of remote sensing. Landsat images are potentially useful during the mining of coal. The status of strip mining and land reclamation can be periodically monitored on repeated Landsat images. Rock falls are hazards in underground coal mines. In Indiana, Wier and others (1973) demonstrated that areas of dense fractures on Landsat images coincide with areas of rock falls in the mines and could be used to predict the hazards.

COMMENTS

Remote sensing has played a key role in major oil discoveries. In the Sudan a previously unknown sedimentary basin was recognized on a mosaic of Landsat MSS images. The images also provided structural and terrain information that contributed to the first oil discoveries in Sudan.

In Papua New Guinea the persistent cloud cover inhibits acquisition of visible and reflected IR images, but aircraft radar images provided structural interpretations that resulted in the first oil discoveries. A major gas field was also discovered from radar image interpretations.

The interior of Saudi Arabia was sparsely explored in the late 1980s. Local anomalies interpreted on Landsat TM images proved to be the expression of subsurface structures. Two of the anomalies were drilled and are now commercial oil fields, part of a new production trend.

QUESTIONS

1. You are employed by an international oil exploration company that plans to evaluate the petroleum potential of the western portion of the People's Republic of China in preparation for negotiating an exploration concession. Because of limited accessibility, deadlines, and competitor pressure, concession areas must be selected solely on the basis of remote sensing evaluations. Prepare such a plan for your management, giving reasons for each step.
2. Assume that western China is completely covered by Landsat TM images, SIR images, and aircraft black-and-white stereo photographs. List the advantages and disadvantages of each of these kinds of satellite images for your project.

REFERENCES

Abu-Ali, M. A., V. A. Franz, J. Shen, F. Monnier, M. D. Mahmoud, and T. M. Chambers, 1991, Hydrocarbon generation and migration in the Paleozoic system in Saudi Arabia: Proceedings Middle East Oil Show, Bahrain, SPE 21376, p. 345–356, Society of Petroleum Engineers, Dallas, TX.

Donovan, T. J., 1974, Petroleum microseepage at Cement field, Oklahoma evidence and mechanism: Bulletin of the American Association of Petroleum Geologists, v. 58, p. 429–446.

Donovan, T. J., I. Friedman, and J. D. Gleason, 1974, Recognition of petroleum-bearing traps by unusual isotopic compositions of carbonate-cemented surface rocks: Geology, v. 2, p. 351–354.

Ellis, J. M. and F. D. Pruett, 1986, Application of synthetic aperture radar (SAR) to southern Papua Fold Belt exploration: Proceedings Fifth Thematic Conference on Geologic Remote Sensing, p. 15–34, Environmental Research Institute of Michigan, Ann Arbor, MI.

Friedman, J. D., R. S. Williams, G. P. Palmason, and C. D. Miller, 1969, Infrared surveys in Iceland: U.S. Geological Survey Professional Paper 650-C, p. C89–C105.

Lamerson, P. R., 1990, Evolution of structural interpretations in Iagifu/Hedinia field, Papua New Guinea in Carman, G. J. and Z. Carman, eds., Petroleum exploration in Papua New Guinea: Proceedings First PNG Petroleum Convention, p. 283–300, PNG Chamber of Mines and Petroleum, Port Moresby, Papua New Guinea.

Lang, H. R., W. H. Alderman, and F. F. Sabins, 1985, Patrick Draw, Wyoming, petroleum test case report: The NASA/Geosat test case project, section 11, American Association of Petroleum Geologists, Tulsa, OK.

Lang, H. R., J. B. Curtis, and J. S. Kovacs, 1985, Lost River, West Virginia, petroleum test site: The NASA/Geosat test case project, section 12, American Association of Petroleum Geologists, Tulsa, OK.

Lang, H. R., S. M. Nicolais, and H. R. Hopkins, 1985, Coyanosa, Texas, petroleum test site: The NASA/Geosat test case project, section 13, American Association of Petroleum Geologists, Tulsa, OK.

McGillivray, J. G. and M. I. Husseini, 1992, The Paleozoic petroleum geology of central Arabia: American Association of Petroleum Geologists Bulletin, v. 76, p. 1473–1490.

Miller, J. B., 1975, Landsat images as applied to petroleum exploration in Kenya: NASA Earth Resources Survey Symposium, NASA TM X-58168, v. 1-B, p. 605–624.

Miller, J. B., and J. Vandenakker, 1977, Sudan interior exploration project—planimetry and geology: Third Pecora Conference, American Association of Petroleum Geologists, Sioux Falls, SD.

Moxham, R. M., 1969, Aerial infrared surveys at the Geysers geothermal steam field, California: U.S. Geological Survey Professional Paper 630-C, p. C106–C122.

Sabins, F. F., 1991, Digital processing of satellite images of Saudi Arabia: Proceedings Middle East Oil Conference, Bahrain, SPE 21357, p. 207–212, Society of Petroleum Engineers, Dallas, TX.

Schull, T. J., 1988, Rift basins of interior Sudan—petroleum exploration and discovery: American Association of Petroleum Geologists Bulletin, v. 72, p. 1128–1141.

Valenti, G. L., I. C. Phelps, and L. I. Eisenberg, 1996, Geological remote sensing for hydrocarbon exploration in Papua New Guinea: Proceedings Eleventh Thematic Conference on Applied Remote Sensing, p. I–97 to I–108, Environmental Institute of Michigan, Ann Arbor, MI.

Valle, R. G., J. D. Friedman, S. J. Gawarecki, and C. J. Banwell, 1970, Photogeologic and thermal infrared reconnaissance surveys of the Los Negritos–Ixtlan de Los Hervores geothermal area, Michoacan, Mexico: Geothermics, special issue 2, p. 381–398.

Watson, K., 1975, Geologic applications of thermal infrared images: Proceedings of the IEEE, n. 501, p.128–137.

Wender, L. E. and F. F. Sabins, 1991, Geologic interpretation of satellite images of Saudi Arabia: Proceedings Middle East Oil Conference, Bahrain, SPE 21358, p. 213–218, Society of Petroleum Engineers, Dallas, TX.

Wier, C. E., F. J. Wobber, O. R. Russell, R. V. Amoto, and T. V. Leshendok, 1973, Relationship of roof falls in underground coal mines to fractures mapped on ERTS-1 imagery: Third ERTS-1 Symposium, NASA SP-351, p. 325–843.

ADDITIONAL READING

Berger, Z., 1994, Satellite hydrocarbon exploration: Springer Verlag, Berlin, Germany.

Ellis, J. M. and L. L. Dekker, 1988, Petroleum exploration with airborne radar (SAR) and geologic field work, Sinu Basin of northwest Colombia: Proceedings Sixth Thematic Conference on Geologic Remote Sensing, p. 79–89, Environmental Research Institute of Michigan, Ann Arbor, MI.

Ellis, J. M., W. Narr, P. B. Goodwin, and G. Perez, 1994, Remote sensing technology for geologic mapping and field operations, Colombia: Proceedings Tenth Thematic Conference on Geologic

Remote Sensing, v. 1, p. 3–13, Environmental Research Institute of Michigan, Ann Arbor, MI.

Krishnamurthy, J., N. V. Kumar, V. Jayaraman, and M. Manivel, 1996, An approach to evaluate ground water potential through remote sensing and a geographical information system: International Journal of Remote Sensing, v. 17, p. 1867–1884.

Matzke, R. H., J. G. Smith, and W. K. Foo, 1992, Iagifu/Hedinia field—first oil from the Papuan Fold and Thrust Belt *in* M. T. Halbouty, ed., Giant oil and gas fields of the decade 1978–1988, American Association of Petroleum Geologists, Memoir 54, p. 471–482.

Prost, G. L.,1993, Remote sensing for geologists: Gordon and Breach Science Publishers, Lausanne, Switzerland.

Yergin, D., 1991, The prize—the epic quest for oil, money, and power: Simon & Schuster, New York, NY.

CHAPTER

MINERAL EXPLORATION

Remote sensing is valuable for mineral exploration in at least four ways:

1. Mapping regional lineaments and structural trends along which groups of mining districts may occur
2. Mapping local fracture patterns that may control individual ore deposits
3. Detecting hydrothermally altered rocks associated with ore deposits
4. Providing basic geologic data

These applications are illustrated by case histories in the following sections.

REGIONAL LINEAMENTS AND ORE DEPOSITS OF NEVADA

Prospectors and mining geologists have long realized that, in many mineral provinces, mining districts occur along linear trends that range from tens to hundreds of kilometers in length. These trends are referred to as *mineralized belts* or *zones*, and many deposits have been found by exploring along the projections of such trends. The state of Nevada has many historic and active mining districts of great wealth. In the late 1800s, rich gold and silver deposits (such as Virginia City and Goldfield) were discovered in the western part of the state. Later, porphyry copper deposits such as the Ruth, Ely, Eureka, and Yerington mines were discovered. Exploration has continued, resulting in the more recent discovery of large deposits of gold at Carlin, Alligator Ridge, Cortez, and elsewhere.

It was long recognized that Nevada mining districts were not randomly distributed but tended to occur in linear zones or belts. Landsat images have enabled geologists to evaluate the relationship between mineral deposits and regional linear structural features. Rowan and Wetlaufer (1975) interpreted a mosaic of Landsat MSS images of Nevada (Figure 11-1); they recognized 367 lineaments, 80 percent of which correlated with previously mapped faults. Fifty-seven percent of the lin-

eaments are formed by the linear contact between the bedrock of mountain ranges and the alluvium of adjacent valleys. The lineaments are the expression of basin-and-range faults that dominate the structure of Nevada. Other lineaments are formed by linear ridges, aligned ridges, and tonal boundaries.

Figure 11-2A shows seven major lineaments that cross Nevada and are several hundred kilometers in length. The Walker Lane, Las Vegas, Midas Trench, and Oregon-Nevada lineaments have previously been documented as major crustal features. Rowan and Wetlaufer (1975) described all the lineaments.

The maps in Figure 11-2 illustrate the relationship between Landsat lineaments and ore deposits. Figure 11-2B shows Nevada mining districts, ranked by dollar value of ore produced, based on prices at time of production. During the late 1800s and early 1900s, when much of the ore was mined, the price of gold was $20 per ounce and silver was less than $1 per ounce. For comparison, today the price of gold is near $400; silver is near $6. One linear belt of mining districts coincides with the northeast-trending Midas Trench lineament. The districts in the southwest part of the state occur in a broad belt parallel with the northwest-trending Walker Lane lineament.

The contours in Figure 11-2C show the number of mining districts per unit area. Comparing this map with the lineament map (Figure 11-2A) shows the concentration of districts along the Midas Trench lineament. The high concentration is interrupted by an area of low mining density at the intersection with the Oregon-Nevada lineament. The lack of mining districts along the Oregon-Nevada lineament may be caused by the extensive cover of young volcanic rocks that mask any underlying deposits. The highest density of ore deposits along the Midas Trench occurs at the intersection with the Rye Patch lineament. In south-central Nevada the East-Northeast lineament system coincides with two of the three east-trending belts of high mining density of Figure 11-2C. Ore deposits along the Walker Lane lineament are generally concentrated at the intersections with east-trending lineaments. Figure 11-2D ranks the districts per unit area according to dollar value of production. These value trends closely resemble the trends of mining

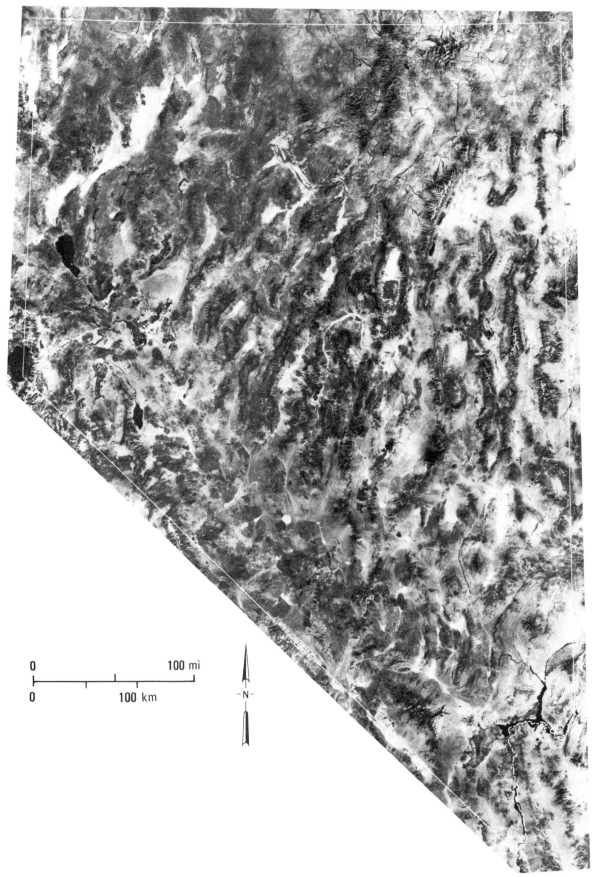

0   100 mi

0   100 km

- N -

Figure 11-1 Mosaic of Nevada compiled from Landsat MSS band 5 images. From Rowan and Wetlaufer (1975, Figure 4).

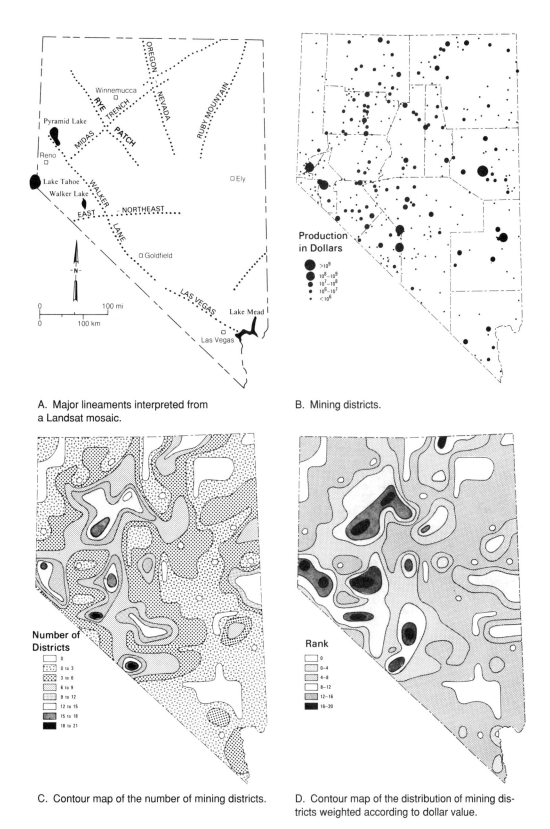

A. Major lineaments interpreted from a Landsat mosaic.

B. Mining districts.

Production in Dollars

- $>10^9$
- 10^8-10^9
- 10^7-10^8
- 10^6-10^7
- $<10^6$

Number of Districts

- 0
- 0 to 3
- 3 to 6
- 6 to 9
- 9 to 12
- 12 to 15
- 15 to 18
- 18 to 21

Rank

- 0
- 0–4
- 4–8
- 8–12
- 12–16
- 16–20

C. Contour map of the number of mining districts.

D. Contour map of the distribution of mining districts weighted according to dollar value.

Figure 11-2 Landsat lineaments and mining districts of Nevada. Maps from Rowan and Wetlaufer (1975). Mining data from Horton (1964).

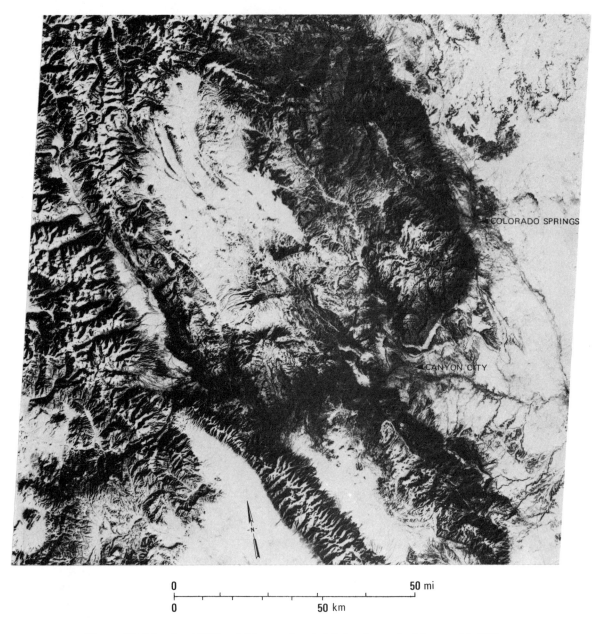

0 ———————————————————— 50 mi

0 ———————————————————— 50 km

Figure 11-3 Landsat MSS band 7 image of central Colorado acquired January 11, 1973, at a sun elevation of 23°.

density. The Midas Trench and Walker Lane lineaments are marked by aligned concentrations of mining districts.

REGIONAL LINEAMENTS AND ORE DEPOSITS OF SOUTH AFRICA AND AUSTRALIA

A previously unknown major lineament was discovered on a Landsat MSS image of southern Namibia and the Cape Province of South Africa by Viljoen and others (1975, Figure 3). The lineament was found to be a fault that was named the

Tantalite Valley fault zone. The fault zone appears to have right-lateral strike-slip displacement and has been mapped for 450 km along the strike. A number of large mafic intrusives have been emplaced along the Tantalite Valley fault zone and are recognized on Landsat images.

On a Landsat color mosaic of the northwest Cape Province of South Africa, Viljoen and others (1975, Figures 11 and 12) mapped a pronounced structural discontinuity, now called the Brakbos fault zone, that separates the Kaapvaal craton on the east from the Bushmanland Metamorphic Complex on the west. The contact between these structural provinces is obscure in the

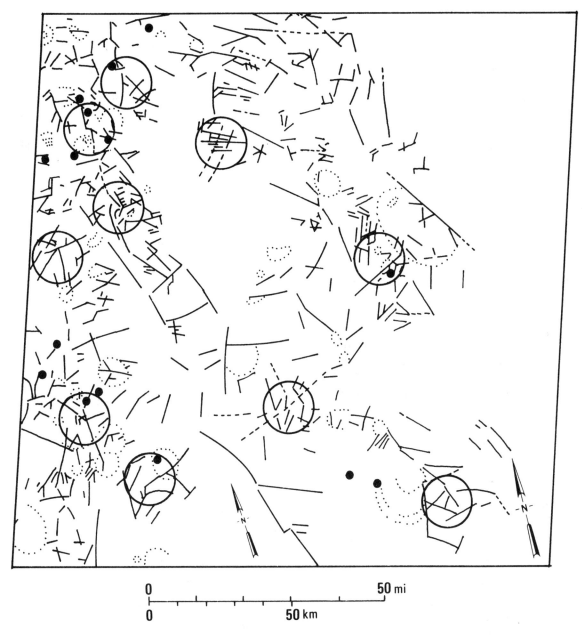

Figure 11-4 Interpretation map for the Landsat image of central Colorado (Figure 11-3). Solid lines are distinct lineaments; dashed lines are indistinct lineaments; dotted lines are curvilinear features. Large circles (165 km^2) are selected target areas for exploration; solid dots are major mining districts. From Nicolais (1974, Figure 3)

field and had previously been drawn approximately 30 km to the east of the Brakbos fault zone, which is also defined on gravity maps. The use of seasonal Landsat images for mapping rock types is illustrated for the Transvaal Province in Chapter 3.

LOCAL FRACTURES AND ORE DEPOSITS

Within a mineral province, areas with numerous fractures and fracture intersections are good prospecting targets because fractures are conduits for ore-forming solutions. Local fracture patterns are mappable on enlarged Landsat images, especially those acquired at low sun angles and those that have been digitally enhanced to emphasize fractures.

Nicolais (1974) evaluated the relationships between Landsat fracture patterns and ore deposits in central Colorado. Figure 11-3 is a winter image with snow cover and low sun elevation, which enhance the fractures. On the interpretation map (Figure 11-4), fractures and circular features are plotted together with the location of major mining districts. The Landsat interpretation reduced the original MSS image (34,225 km^2) to 10 target areas, each 165 km^2 in area. These target areas were selected because they show concentrations of fracture intersections or intersections of fractures and circular features. Five of these target areas coincide with, or are directly adjacent to, major mining districts. The other five target areas may be sites of undiscovered ore deposits.

Rowan and Bowers (1995) interpreted linear features from Landsat TM and aircraft radar images of the Reno, Nevada, quadrangle (1:250,000). The linear features were digitized and analyzed statistically to show patterns of orientation and density. These patterns were compared with the distribution of gold and silver mineralization. The comparison showed that in most areas the linear patterns are expressions of structural features that controlled the mineralization.

MAPPING HYDROTHERMALLY ALTERED ROCKS

Many ore bodies are deposited by hot aqueous fluids, called *hydrothermal solutions,* that invade the host rock, or *country rock.* During formation of the ore materials, these solutions also interact chemically with the country rock to alter the mineral composition for considerable distances beyond the site of ore deposition. The hydrothermally altered country rocks contain distinctive assemblages of secondary minerals, called *alteration minerals,* that replace the original minerals. Alteration minerals commonly occur in distinct sequences, or *zones of hydrothermal alteration,* relative to the ore body. These zones are caused by changes in temperature, pressure, and chemistry of hydrothermal solutions at progressively greater distances from the ore body. At the time of ore deposition, the zones of altered country rock may not extend to the surface of the ground. Later uplift and erosion expose successively deeper alteration zones and eventually the ore body itself. Not all alteration is associated with ore bodies, and not all ore bodies are marked by alteration zones, but these zones are valuable indicators of possible deposits. Fieldwork, laboratory analyses of rock samples, and interpretation of aerial photographs have long been used to explore for hydrothermal alteration zones.

In regions where bedrock is exposed, multispectral remote sensing is useful for recognizing altered rocks because their reflectance spectra differ from those of the country rock. The Goldfield mining district, Nevada, is a classic example of remote sensing of hydrothermal alteration zones.

GOLDFIELD MINING DISTRICT, NEVADA

The Goldfield district in southwest Nevada (Figure 11-5) was noted for the richness of its ore. Over 4 million troy ounces (130,000 kg) of gold with silver and copper were produced, largely in the boom period between 1903 and 1910. During peak production the town had a population of 15,000, but today Goldfield is almost a ghost town.

Geology, Ore Deposits, and Hydrothermal Alteration

The geology and hydrothermal alteration of the district have been thoroughly mapped and analyzed by the U.S. Geological Survey (Ashley, 1974, 1979), which makes Goldfield an excellent locality to develop and test methods for mineral exploration. Volcanism began in the Oligocene epoch with eruption of rhyolite and quartz latite flows and the formation of a small caldera and ring-fracture system. Hydrothermal alteration and ore deposition occurred during a second period of volcanism in the early Miocene epoch when the dacite and andesite flows that host the ore deposits were extruded. Heating associated with volcanic activity at depth caused convective circulation of hot, acidic, hydrothermal solutions through the rocks. Fluid movement was concentrated in the fractures and faults of the ring-fracture system. Following ore deposition, the area was covered by younger volcanic flows. Later doming and erosion exposed the older volcanic center with altered rocks and ore deposits.

In the generalized map (Figure 11-5), the hydrothermally altered rocks are cross-hatched and the unaltered country rocks are blank. The map also identifies alluvial deposits and postore volcanic rocks (those that formed after ore deposition). Approximately 40 km^2 of the area is underlain by altered rocks, but less than 2 km^2 of the altered area contains mineral deposits, which are shown in black. The oval band of altered rocks was controlled by the circular ring-fracture system, with a linear extension toward the east. The central patch of alteration shown in Figure 11-5 was controlled by closely spaced faults and fractures.

Figure 11-6 is a vertical section through a typical ore-bearing vein and the associated altered rocks. Alteration is concentrated at faults and fractures where hydrothermal solutions penetrated the country rocks and gold-bearing veins were deposited. Intensity of alteration decreases laterally away from the vein. Characteristics of the alteration zones are summarized as follows:

Silicic zone Predominantly quartz, which replaces the ground mass of host rock; subordinate amounts of alunite and kaolinite replace feldspar phenocrysts. Fresh rock of this zone, which is gray and resembles chert, is resistant to erosion and weathers to ridges with conspicuous dark coatings of desert varnish. Contact with adjacent argillic zones is sharp. All ore deposits occur in veins of the silicic zone, but not all veins contain ore.

Argillic zone Alteration minerals are predominantly clay. The argillic zone is divided into three subzones (Figure 11-6) based on the predominant clay species. Disseminated grains of pyrite (iron sulfide) are present that weather to iron oxides. The argillic rocks generally have a bleached appearance, but the secondary iron oxide minerals (limonite, jarosite, and goethite) impart local patches of red, yellow, and brown to the outcrops. At Goldfield, no ore deposits occur in the argillic

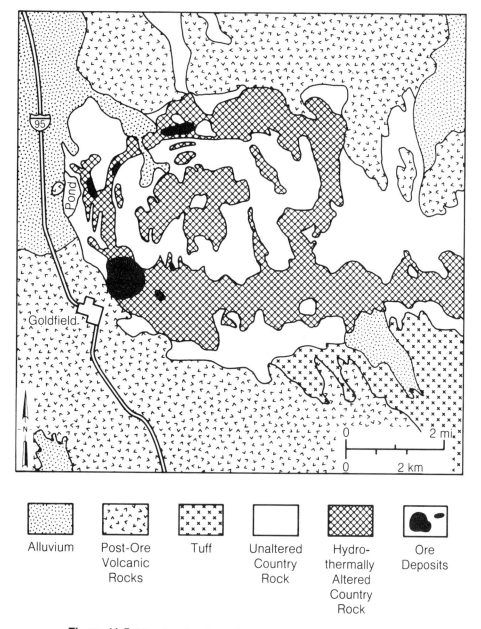

Figure 11-5 Map showing the geology and hydrothermal alteration of the Goldfield mining district, Nevada. From Ashley (1979, Figures 1 and 8).

zone, but the presence of these altered rocks may be a clue to the occurrence of veins.

Alunite-kaolinite subzone Relatively narrow and locally absent. In addition to alunite and kaolinite, some quartz is present.

Illite-kaolinite subzone Marked by the occurrence of illite.

Montmorillonite subzone Montmorillonite is the dominant clay mineral in this subzone, which has a pale yellow color due to jarosite, an iron sulfate mineral.

Propylitic zone These rocks represent regional alteration of lower intensity than the argillic and silicic zones. Chlorite, calcite, and antigorite are typical minerals in this zone and impart a green or purple color to the rocks. Propylitic alteration is ab-

sent at numerous localities in Goldfield, where the argillic zone grades directly into unaltered rocks.

Country rock Dacite and andesite. These unaltered gray volcanic rocks are hard and resistant to erosion. Country rocks, shown by the blank pattern in the map (Figure 11-5), surround the outcrops of altered rocks.

Secondary iron minerals Pyrite (iron sulfide) is deposited along with ore minerals in and adjacent to the silicic zone. Weathering oxidizes pyrite to form limonite and hematite (iron oxide), which have distinctive red and brown signatures. Jarosite (iron sulfate) is also common in hydrothermally altered rocks and has a yellow signature.

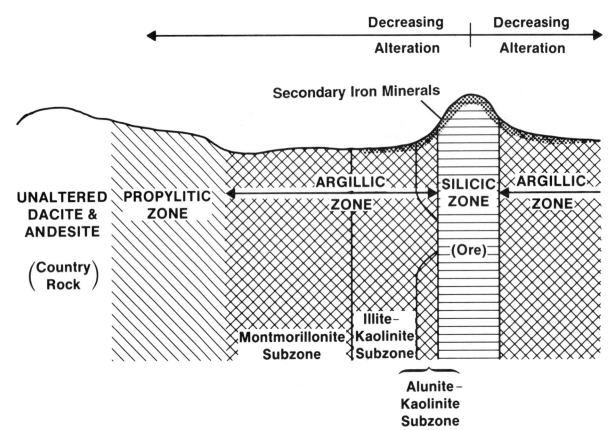

Figure 11-6 Cross section of typical hydrothermal alteration zones at the Goldfield mining district. Not to scale. The silicic zone has a maximum width of a few meters. The argillic zone extends for several tens of meters on each side of the silicic zone. After Ashley (1974) and Harvey and Vitaliano (1964).

The orderly sequence of subzones shown in Figure 11-6 rarely occurs in the field, because the veins are so closely spaced that subzones coalesce and overlap to form the altered outcrops shown in the map. Prospectors have long been aware of the association between hydrothermally altered rocks and ore deposits. Many mines were discovered by recognizing outcrops of altered rocks, followed by assays of rock samples. Prior to remote sensing, altered rocks were recognized by their appearance in the visible spectral bands. Today remote sensing and digital image processing enable the use of other spectral bands for mineral exploration. Remote sensing relies on recognizing two components of the hydrothermal system:

1. Alunite and clay minerals
2. Iron minerals

Landsat TM images of Goldfield are used to illustrate mineral exploration.

Recognizing Hydrothermal Alteration on Landsat Images

Figure 11-7 shows the visible and reflected IR band images for the Goldfield district, which is a small subscene of the full TM image. Plate 21A is an enhanced TM band 1-2-3 image (normal color). A yellow patch directly northeast of the town of Goldfield (Figure 11-5) is caused by the mine dumps and disturbed ground of the main mineralized area. A white patch 3 km north of Goldfield is the dry tailings pond of the abandoned Columbia Mill, where gold was separated from the altered host rock. The tailings pond is a useful reference standard since it contains a concentration of altered rock material. The dark signatures in the margins of the image are volcanic rocks that are younger than the ore deposits and altered rocks. Distinctive light-blue signatures in the southeast portion are outcrops of volcanic tuff. Neither the normal color TM image nor the IR color image (not illustrated) are diagnostic for recognizing the hydrothermally altered rocks. Therefore, additional digital processing is required in order to use TM data for mineral exploration.

Alunite and Clay Minerals on 5/7 Ratio Images Figure 11-8A shows reflectance spectra of alunite and the three common hydrothermal clay minerals. These minerals have distinctive absorption features (reflectance minima) at wavelengths within the bandpass of TM band 7, which is shown with a stippled pattern in Figure 11-8A. The minerals have higher reflectance values within TM band 5. Ratio images,

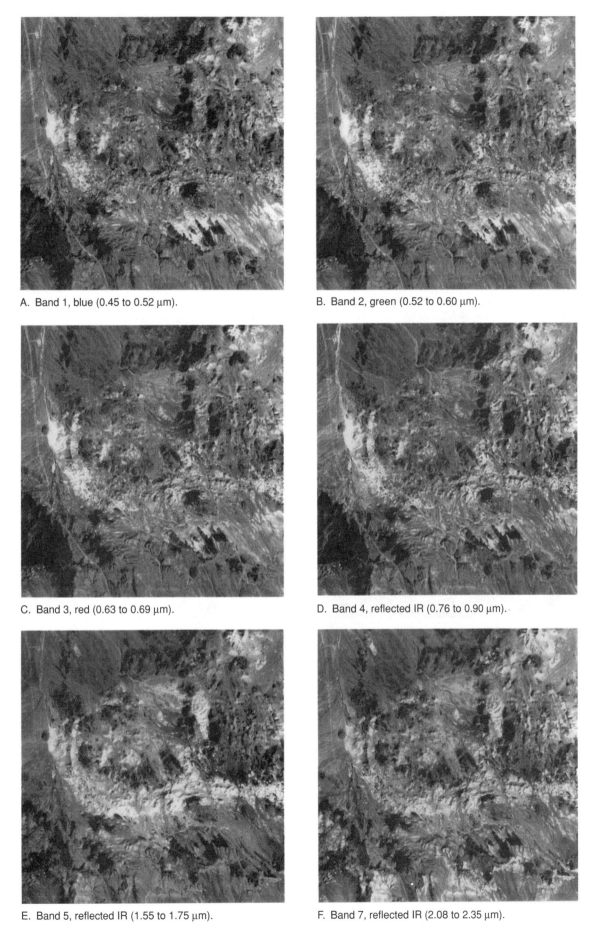

A. Band 1, blue (0.45 to 0.52 μm).

B. Band 2, green (0.52 to 0.60 μm).

C. Band 3, red (0.63 to 0.69 μm).

D. Band 4, reflected IR (0.76 to 0.90 μm).

E. Band 5, reflected IR (1.55 to 1.75 μm).

F. Band 7, reflected IR (2.08 to 2.35 μm).

Figure 11-7 Landsat TM band images of the Goldfield mining district.

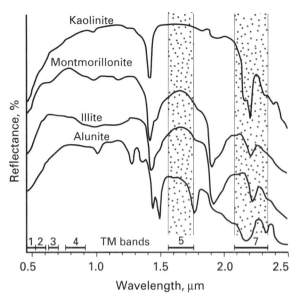

A. Laboratory reflectance spectra. TM bands 5 and 7 (stippled) are used to calculate the 5/7 ratio image.

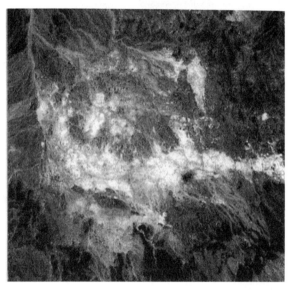

B. TM 5/7 ratio image.

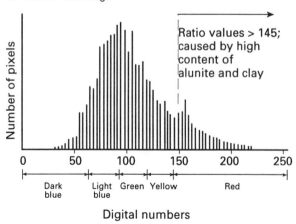

Ratio values > 145; caused by high content of alunite and clay

C. Histogram for 5/7 image. The colors are used in Plate 21C.

Figure 11-8 Recognition of hydrothermal clays and alunite, Goldfield mining district.

described in Chapter 8, can emphasize and quantify these spectral differences.

The TM ratio 5/7 is effective for recognizing clays and alunite. Table 11-1 explains how the 5/7 ratio distinguishes altered rocks from unaltered rocks. Both rocks have similar values in band 5. The reflectance of unaltered rocks in band 7 is similar to that in band 5. Therefore the 5/7 ratio for unaltered rocks is unity (1.00). Altered rocks, however, have lower reflectance in band 7 because of absorption caused by the minerals shown in Figure 11-8A. Therefore the 5/7 ratio for altered rocks is much greater than unity (1.45). The numbers in Table 11-1 are typical and will differ for other examples. The decimal ratio values are converted to 8-bit digital numbers (DNs) and displayed as images.

Figure 11-8B is a 5/7 ratio image of Goldfield with higher ratio values shown in brighter tones. Comparing the image with the map (Figure 11-5) shows that the high (bright) ratio values correlate with hydrothermally altered rocks. Figure 11-8C is a histogram of the 5/7 ratio image that shows the higher ratio values (DNs > 145) of the altered rocks. Low ratio values represent unaltered rocks.

Plate 21C is a color density-sliced version of the 5/7 image using the color assignments shown in Figure 11-8C. The highest ratio values (DN > 145) are shown in red, with the next highest values (DN = 125 to 145) shown in yellow. The red and yellow ratio color slices correlate with the altered rocks.

Iron Minerals on 3/1 Ratio Images Iron minerals are the second indicators of hydrothermally altered rocks. Figure 11-9A shows spectra of the iron minerals that have low blue reflectance (TM band 1) and high red reflectance (TM band 3). Iron-stained hydrothermally altered rocks therefore have high values in a 3/1 ratio image. Figure 11-9B is a 3/1 ratio image with high DN values shown in bright tones. Plate 21D is a density-sliced version of the 3/1 image, with color assignments shown in the histogram of Figure 11-9C. Highest ratio values (DN > 150) are shown in red, with the next highest values (DN = 135 to 150) shown in yellow. The red and yellow slices correlate with the hydrothermally altered rocks in the map.

Color Composite Ratio Images Color composite ratio images are produced by combining three ratio images in blue, green, and red as described in Chapter 8. Plate 21B shows the ratios 3/5, 3/1, and 5/7 in red, green, and blue, respectively. The orange and yellow tones delineate the outer and inner areas of altered rocks in a pattern similar to that of the density-sliced ratio images. An advantage of the color ratio image is that it combines the distribution patterns of both iron-staining and hydrothermal clays. A disadvantage is that the color patterns are not as distinct as in the individual density-sliced images.

Classification Images

An unsupervised multispectral classification was applied to the TM bands and resulted in 12 classes. These classes were

Table 11-1 Calculation of TM 5/7 ratio values

	Band 5 reflectance (typical)	Band 7 reflectance (typical)	Ratio 5/7 (typical)	DNs for ratio 5/7
Unaltered rocks (without clays and alunite)	160	160	1.00	100
Altered rocks (with clays and alunite)	160	110	1.45	145

aggregated into the six classes shown in the classification image of Plate 21E. Table 11-2 explains the colors of Plate 21E. Two types of altered rocks were classified. The class shown in red and labeled "Altered rocks, A" is confined to altered rocks but does not indicate the full extent of alteration. The class shown in orange and labeled "Altered rocks, B" includes all of the remaining altered rocks, as well as some rocks outside the alteration zone. Basalt (blue), volcanic tuff (purple), and unaltered rocks (green) are reasonably classified. Alluvium (yellow) is considerably more extensive in the classification image (Plate 21E) than in the geologic map (Figure 11-5). Field checking and comparison with the normal color image (Plate 21A) shows that much of the bedrock is thinly covered with detritus and is correctly classed as alluvium by the computer. The map, however, shows the lithology of the underlying bedrock that was interpreted by the field geologist.

Summary

The spectra of alteration minerals (Figures 11-8A and 11-9A) were recorded in the laboratory using pure minerals. Remote sensing images record data from weathered outcrops of mixtures of rocks and minerals together with soil and vegetation. Despite these problems, the digitally processed images give an accurate picture of the alteration pattern at Goldfield. In order to bridge the gap between laboratory and outcrop, Rowan, Goetz, and Ashley (1979) used a portable spectrometer in the field to record spectra of several hundred representative outcrops of altered and unaltered rocks at Goldfield. Figure 11-10 summarizes their results as average reflectance curves for altered and unaltered outcrops. Because of the averaging effect, these curves lack the fine spectral detail of the laboratory curves, but the differences between altered and unaltered rocks are clearly shown. The altered rocks have distinctly lower reflectance in band 7 than in band 5. Unaltered rocks have similar values in those bands. In the visible portion of the spectrum, altered rocks have a higher red reflectance because of the iron minerals.

The Landsat TM study of the well-known Goldfield district has provided techniques for mineral exploration in many regions. At least one major porphyry copper deposit has been discovered by these techniques.

PORPHYRY COPPER DEPOSITS

Most of the world's copper is mined from *porphyry deposits,* which occur in a different geologic environment from gold deposits of the Goldfield type. Porphyry deposits are named for the *porphyritic* texture of the granitic host rock, in which large feldspar crystals are surrounded by a fine-grained matrix of quartz and other minerals. Granitic porphyries occur as plugs (or *stocks*) up to several kilometers in diameter that intruded the older country rock and reached to within several kilometers of the surface. Intensive fracturing of the porphyry and country rock occurred during the emplacement and cooling of the stock. Heat from the magma body caused convective circulation of hydrothermal fluids through the fracture system to form ore deposits and alter the porphyry stock and adjacent country rock.

Figure 11-11 is a model of hydrothermal alteration of porphyry copper deposits that was developed by Lowell and Guilbert (1970). The most intense alteration occurs in the core of the porphyry body and diminishes radially outward in a series of zones described below.

Potassic zone This zone contains the most intensely altered rocks in the core of the stock. Characteristic minerals are quartz, sericite, biotite, and potassium feldspar.

Phyllic zone Quartz, sericite, and pyrite are common.

Ore zone This zone consists of disseminated grains of chalcopyrite, molybdenite, pyrite, and other metal sulfides. Much of the ore occurs in a cylindrical *ore shell* near the boundary between the potassic and phyllic zones. Copper typically constitutes 1 percent, or less, of the rock, but the large volume of ore is suitable for open pit mining. Where the ore zone is

Table 11-2 Explanation of colors in classification image of Goldfield mining district (Plate 21E)

Color	Classification	Percent of Image
Yellow	Alluvium	39.2
Blue	Basalt	14.0
Purple	Tuff	6.6
Red	Altered rocks, A	5.3
Orange	Altered rocks, B	18.3
Green	Unaltered rocks	16.6

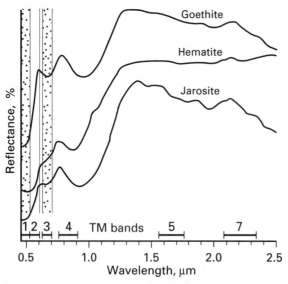

A. Laboratory reflectance spectra. TM bands 1 and 3 (stippled) are used to calculate 3/1 ratio image.

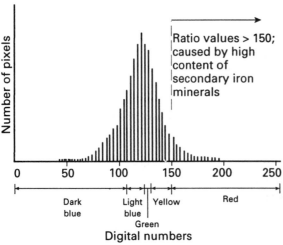

B. TM 3/1 ratio image.

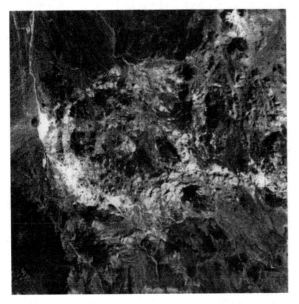

Ratio values > 150; caused by high content of secondary iron minerals

C. Histogram for 3/1 image. The colors are used in Plate 21D.

Figure 11-9 Recognition of hydrothermal iron minerals, Goldfield mining district.

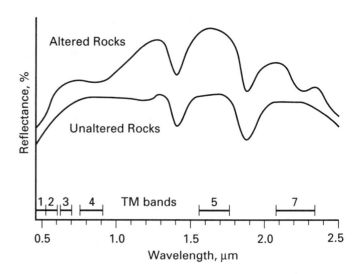

Figure 11-10 Field spectra (averaged) of altered and unaltered rocks at the Goldfield mining district. From Rowan, Goetz, and Ashley (1979, Figure 2A).

exposed by erosion, pyrite oxidizes to form a red to brown limonitic crust called a *gossan*. Gossans can be useful indicators of underlying mineral deposits, although not all gossans are associated with ore deposits.

Argillic zone Quartz, kaolinite, and montmorillonite are characteristic minerals of the argillic zone in porphyry deposits, just as they are associated with the argillic zone at Goldfield and elsewhere.

Propylitic zone Epidote, calcite, and chlorite occur in these weakly altered rocks. Propylitic alteration may be of broad extent and have little significance for ore exploration.

Few porphyry deposits have the symmetry and completeness of the model in Figure 11-11. Structural deformation, erosion, and deposition commonly conceal large portions of the system. Nevertheless, recognition of small patches of altered rock on remote sensing images can be a valuable exploration clue.

In the early 1980s, NASA and the Geosat Committee evaluated satellite and airborne multispectral images for porphyry copper deposits in southern Arizona. At the Silver Bell mining district, Abrams and Brown (1985) used color ratio images to separate the phyllic and potassic alteration zones from the argillic and propylitic zones. A supervised classification map defined the outcrops of altered rocks.

COLLAHUASI MINING DISTRICT, CHILE

For a number of years prior to 1992 a subsidiary of the Chevron Corporation explored for mineral deposits around the world. Chevron conducted a copper exploration project in the Collahuasi mining district, Chile, that used remote sensing to

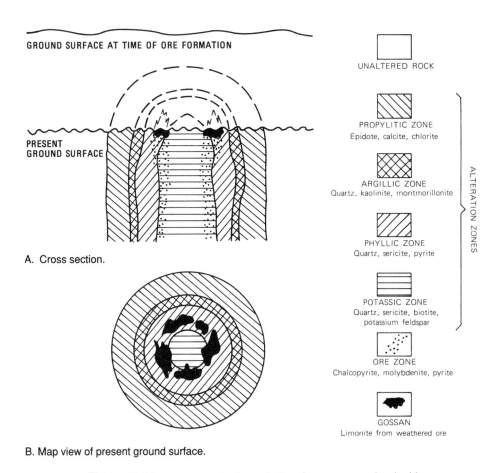

GROUND SURFACE AT TIME OF ORE FORMATION

PRESENT
GROUND SURFACE

A. Cross section.

B. Map view of present ground surface.

UNALTERED ROCK

PROPYLITIC ZONE
Epidote, calcite, chlorite

ARGILLIC ZONE
Quartz, kaolinite, montmorillonite

PHYLLIC ZONE
Quartz, sericite, pyrite

POTASSIC ZONE
Quartz, sericite, biotite,
potassium feldspar

ORE ZONE
Chalcopyrite, molybdenite, pyrite

GOSSAN
Limonite from weathered ore

ALTERATION ZONES

Figure 11-11 Model of hydrothermal alteration zones associated with porphyry copper deposits. From Lowell and Guilbert (1970, Figure 3).

discover major ore reserves. Much of the following account is summarized from a publication by Dick and others (1993).

Geologic and Exploration Background

The Collahuasi mining district is located in northern Chile, 180 km southeast of the city of Iquique. The district lies within a north-trending belt of porphyry copper deposits that includes the major mines at El Teniente, Disputada, El Salvador, Escondida, and Chuquicamata. The Collahuasi district is bounded on the west by a major regional fault system that also passes through the open pit at the Chuquicamata mine. Figure 11-12 is a geologic map showing the distribution of the Macata, Capella, and Collahuasi Formations, which are of Jurassic and Cretaceous age. These country rocks are intruded by granitic stocks of late Cretaceous to early Tertiary age that are hosts for the porphyry copper deposits.

Mineral production in the Collahuasi district began in the late 1800s when copper was mined from veins now known to be related to the porphyry system at Rosario (Figure 11-12). During the 1930s these veins were Chile's third largest producer of copper. Modern exploration began in 1976 when a joint venture

between Superior Oil Company and Falconbridge Limited acquired the Collahuasi properties. The joint venture discovered a porphyry deposit at Rosario. In 1985, ownership of the district changed to a three-way joint venture among Falconbridge Limited, Shell Oil Company, and Chevron Corporation. From 1985 to 1990 exploration concentrated on evaluating the Rosario deposit. Rosario, however, occupies only a small portion of the 28,000 hectares of the Collahuasi district. There were indications of other mineralized centers within the district, but geologic information was incomplete.

Exploration Program

In 1991 the partners began an exploration program to

1. process satellite images to show alteration anomalies;
2. conduct geophysical surveys to evaluate anomalies;
3. drill core holes to locate ore deposits.

Process Images In step 1 the Remote Sensing Research Group of Chevron processed satellite images of the Collahuasi district and adjacent areas. Northern Chile is ideally suited for

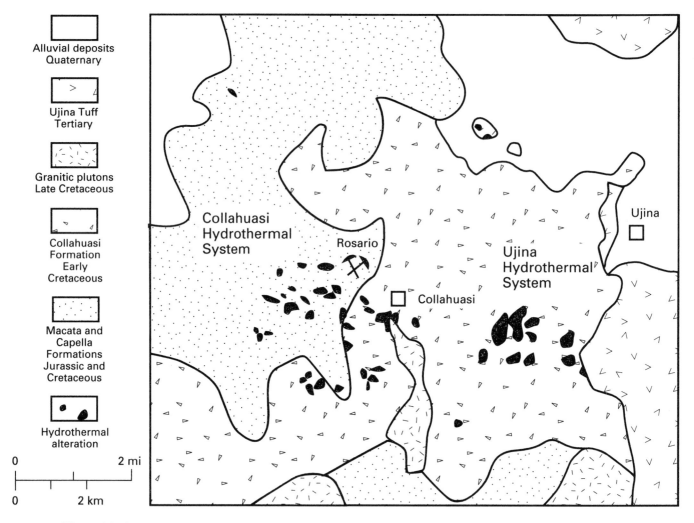

Figure 11-12 Geologic map of the Collahuasi mining district, Chile. Hydrothermal alteration anomalies are edited from Landsat TM ratio images. Geology generalized from Vergara (1978A,B).

such studies, because vegetation, soils, and clouds are virtually absent in this arid environment of the high Andes Mountains. Landsat TM data were processed into a 2-4-7 color image that is optimum for geologic interpretation in this arid terrain. A SPOT panchromatic image (10-m spatial resolution) was merged with the TM image, using the technique described in Chapter 8. Plate 22 shows the Collahuasi portion of this TM/SPOT image.

TM 3/1 and 5/7 ratio images were produced using the methods described for the Goldfield mining district. The 5/7 ratio image was density-sliced to display the highest ratio values that record maximum concentrations of alunite and clays. The 3/1 ratio image was density-sliced to display maximum concentrations of iron minerals. Areas with high values on both ratio images were identified and called *alteration anomalies*.

The TM anomalies were evaluated to eliminate false anomalies, which consist of:

1. Sedimentary rocks, such as shale, that are rich in clay.
2. Rocks with an original red color, such as iron-rich volcanic rocks and sedimentary red beds.
3. Detritus eroded from outcrops of altered rocks; these recent deposits in alluvial fans and channels may indicate the proximity of altered rocks.

The revised anomalies are shown in black on the geologic map (Figure 11-12). A circular cluster of anomalies, over 6 km in diameter, occurs south and west of Collahuasi and Rosario and is now called the Collahuasi hydrothermal system. The Rosario deposit, with a diameter of 1.5 km, occupies only a small portion of the north margin of the system. The remainder of the Collahuasi hydrothermal system was largely unexplored.

A second cluster of anomalies, 3 km wide, occurs southwest of Ujina (Figure 11-12) and is called the Ujina hydrothermal

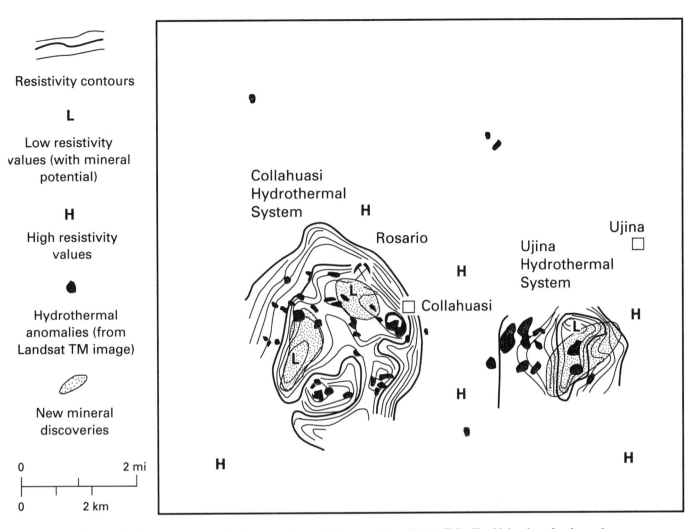

Figure 11-13 Contour map of resistivity values, Collahuasi mining district, Chile. H = high values, L = low values. Hydrothermal alteration anomalies are edited from Landsat TM ratio images. From Dick and others (1993).

system. Minor alteration was recognized earlier at Ujina, but the area has received very limited exploration attention in the past. The alteration shown on the ratio images is much more extensive than previously recognized at Ujina.

Geophysical Surveys In step 2, geophysical surveys were made to record subsurface properties of the rocks at the Collahuasi and Ujina hydrothermal systems. Dick and others (1993) provide details of the configuration and results of the geophysical surveys. The entire district was covered by a helicopter-borne aeromagnetic survey that mapped subsurface geologic structures and the distribution of magnetic minerals. The aeromagnetic map shows that the Collahuasi and Ujina hydrothermal systems are localized at intersections of major northeast- and northwest-trending faults. The Ujina System has a circular rim of high magnetic values that is interpreted as an ore

shell within the porphyry deposit similar to that shown in the porphyry model (Figure 11-11).

A ground-based survey measured resistivity of the rocks. Unmineralized rocks typically have high resistivity values. Metallic minerals, such as copper, have low values; therefore, mineralized rocks have low resistivity values. Known porphyry copper deposits are characterized by circular patterns of low resistivity. Figure 11-13 is a contour map of the resistivity survey at the same scale as the image (Plate 22) and map (Figure 11-12). High resistivity values are shown by H; the very important low values are shown by L.

Results of the resistivity survey are outstanding. Circular patterns of low resistivity contours occur at both the Collahuasi and Ujina hydrothermal systems (Figure 11-12). These patterns are analogous to those of classic porphyry copper deposits. At Collahuasi the resistivity pattern is 5 km in diameter. The lowest values form a marginal rim that may represent the ore shell

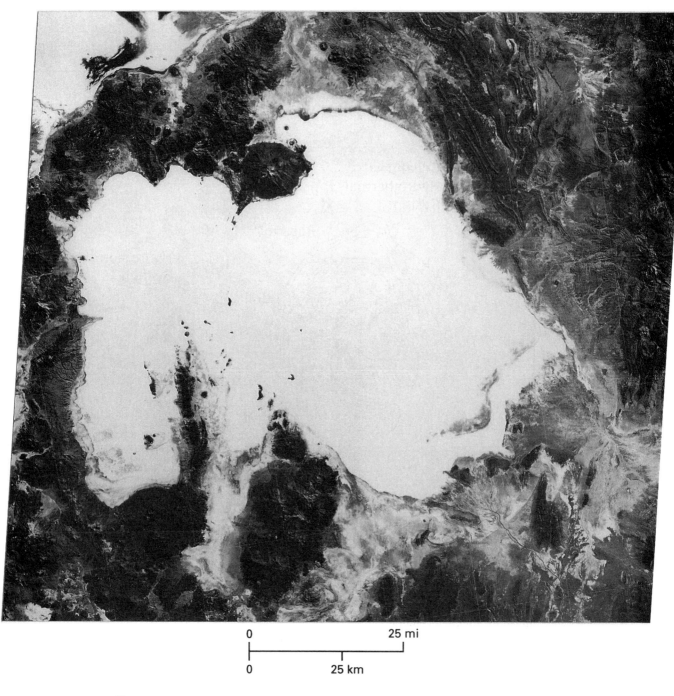

```
0                    25 mi
├──────┬──────┬──────┤
0            25 km
```

Figure 11-14 Landsat TM band 4 image of Salar de Uyuni and vicinity, southwest Bolivia. From Sabins and Miller (1994, Figure 2).

of the porphyry model. The very low overall resistivity of the Collahuasi system is interpreted as an extensive development of veinlet mineralization.

The Ujina hydrothermal system has a circular pattern of low resistivity contours 3 km in diameter. The geologic map (Figure 11-12) shows outcrops of the Ujina Tuff, which post-dates the hydrothermal activity. The Ujina Tuff covers the eastern portion of the Ujina resistivity feature. The Landsat anomalies coincide with the exposed western portion of the feature.

Core Drillings and Ore Discoveries In step 3, core holes were drilled to evaluate the hydrothermal systems outlined by

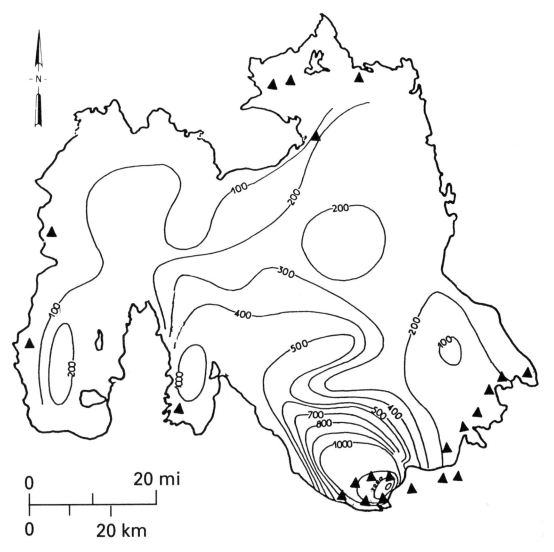

Figure 11-15 Map of Salar de Uyuni. Triangles show high values for TM ratio 4/7 that correlate with high concentrations of ulexite. Contours show boron concentration (mg · l^{-1}) near surface brine. From Risacher (1989, Figure 34).

the remote sensing and geophysical studies. The first holes tested the low resistivity values at Rosario, on the north rim of the Collahuasi system, where the drills found zones of structurally controlled copper mineralization. These results led to the discovery of two major ore bodies within the Collahuasi system, which are shown by stippled patterns in Figure 11-13.

At Ujina, drilling of the resistivity feature discovered a major new porphyry copper deposit shown by the stippled pattern in the eastern portion of Figure 11-13. The primary ore deposit is overlain by secondary enriched ore. By early 1993, drilling had outlined over 150 million tons of enriched ore with a grade of 1.8 percent copper.

In late 1992 Chevron decided to sell its mineral properties in order to concentrate on its energy business. Chevron sold its one-third interest in the undeveloped Collahuasi district to Minorco for $190 million cash. Chevron's total investment in the property is estimated at $23 million. The remote sensing

work that contributed so much to the increased value of the property cost less than $50,000. In 1995 Minorco and Falconbridge purchased Shell's one-third interest for $195 million. Minorco and Falconbridge will spend $1.3 billion to develop Collahuasi into a world-class copper mine. Production will start in late 1998 and last for 45 years. Total current minable reserves are 14 million tons of copper with a value of $36.4 billion at 1994 copper prices. Remote sensing played a key role in defining this valuable property.

BORATE MINERALS, SALAR DE UYUNI, BOLIVIA

Boron and its compounds occur as borate minerals in the crust and brine of certain ancient and modern dry salt lakes, called *salars* in Spanish. Figure 11-14 is a TM image of Salar de Uyuni in southwest Bolivia, which is the world's largest salar

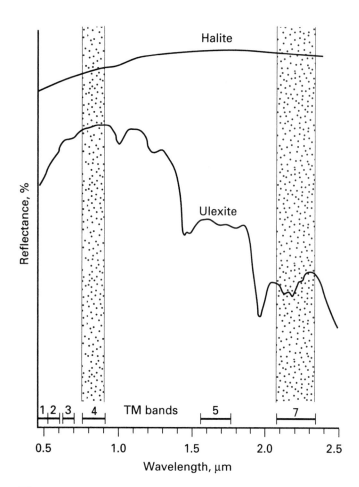

Figure 11-16 Reflectance spectra for halite (NaCl) and ulexite (NaCaB$_5$O · 8H$_2$O). TM bands 4 and 7 were used to calculate the 4/7 ratio image.

(10,000 km^2). The salar is known to contain borate minerals, but the ore reserves and economic potential were incompletely evaluated. Risacher (1989) analyzed brine samples from 68 shallow drill holes and prepared a map of boron concentration, shown in Figure 11-15. Had the holes been uniformly distributed over the salar, each hole would represent 147 km^2, which is very sparse sampling. The Landsat TM, however, covers the salar with more than 11×10^6 ground resolution cells that represent 0.0009 km^2 each. The Bolivian government contracted with Intercontinental Resources, Inc., to conduct a Landsat evaluation of the salar.

A major question in the evaluation was whether borate minerals in the crust of the salar have spectral features that can be recognized in TM data. Figure 11-16 shows the spectrum of ulexite (NaCaB$_5$O · 8H$_2$O), which is the principal borate mineral in the salar. Figure 11-16 also shows the spectrum of halite (NaCl), or rock salt, which constitutes more than 90 percent of the crust. A TM 4/7 ratio image should have high values for ulexite and low values for halite. A 4/7 ratio image was gener-

ated and density-sliced to highlight the highest ratio values, which are shown in the map (Figure 11-15) by triangles. The highest ratio values coincide with the contours of maximum boron concentration in an embayment at the south margin of the salar. Additional triangles elsewhere around the margin of the salar indicate potential borate reserves that were not detected by the sample program.

This ratio method should be useful for borate exploration in other dry lakes.

ADDITIONAL MINERAL EXPLORATION PROJECTS

A number of investigators have used remote sensing and image processing to explore for various minerals at many localities. Table 11-3 summarizes representative projects. Two of the more extensive projects were in South America.

Knepper and Simpson (1992), of the U.S. Geological Survey, processed 13 Landsat TM scenes of southwest Bolivia to produce a variety of images for mineral exploration. Ratio images 3/4, 3/1, and 5/7 were combined in blue, green, and red to produce color ratio images. The confusing effects of vegetation, clouds, water, and shadows were minimized by using a mask of pixels from TM band 4 that identified these features. On the color ratio image the masked features are replaced by gray signatures, which improves the recognition of hydrothermally altered rocks. Knepper and Simpson (1992, Plate 2) prepared a map showing areas of altered rocks interpreted from the images.

Eiswerth and Rowan (1993), also of the U.S. Geological Survey, processed portions of three TM images in west-central Bolivia, using methods similar to those of Knepper and Simpson (1992). Their objective was to identify volcanic and structural features and hydrothermally altered rocks. For their project area, Eiswerth and Rowan (1993) found the 5/1 ratio to be more effective than 3/1.

HYPERSPECTRAL SCANNERS FOR MINERAL EXPLORATION

The models in Figures 11-6 and 11-11 show that hydrothermal alteration systems consist of distinct zones characterized by one or two secondary minerals. An image showing the distribution of these individual minerals would be an improvement over the broad alteration maps produced from TM data. These detailed images can be produced from data acquired by hyperspectral scanners. Figure 11-17 shows laboratory spectra of common alteration minerals in the atmospheric window from 2.0 to 2.5 μm and the 50 spectral bands recorded by the AVIRIS hyperspectral scanner, which is described in Chapter 1. The bandpass of TM band 7 is also shown for comparison. Figures 1-24 and 1-25 show that AVIRIS has the spectral reso-

Table 11-3 Representative mineral exploration studies using remote sensing

Locality	Reference	Comments
Western North and South America	Spatz and Wilson (1994)	Summarizes published remote sensing studies of 12 major mining districts from British Columbia to Chile.
Altiplano, Bolivia	Knepper and Simpson (1992)	TM color ratio composite images used to recognize hydrothermally altered rocks.
Canada	Singhroy (1991)	Ten papers on mineral exploration using Landsat and radar.
Chile, Peru, and Bolivia	Eiswerth and Rowan (1993)	TM color ratio composite images used to recognize hydrothermally altered rocks. Field studies evaluated results.
Jordan	Abdelhamid and Rabba (1994)	A variety of digitally processed TM images identified a historic copper-manganese deposit and located prospects.
Sonora, Mexico	Bennett, Atkinson, and Kruse (1993)	TM data were integrated with field and laboratory data to discover several prospects.
Nevada	Watson, Kruse, and Hummer-Miller (1990)	TIMS data were processed to recognize silicified rocks associated with gold deposits.
Spain	Goosens and Kroonenberg (1994)	TM ratio images were used to identify altered rocks overlain by residual soil.
Sudan	Griffiths and others (1987)	Landsat MSS images and fieldwork showed gold occurrences are concentrated along regional shear zones in mafic metavolcanics.
Zaire, Zambia, and Angola	Unrug (1988)	Major lead-zinc vein deposits occur at intersections of Landsat lineaments with folds and thrust faults. Unexplored intersections are potential targets.

lution to record the spectra of alteration minerals. The following section describes how to derive images from AVIRIS data that show the abundance and distribution of individual minerals. There are, however, two major technical challenges to producing such images.

1. Figure 11-17 shows that many alteration minerals, especially the clays, have spectra with similar shapes. The major absorption feature near 2.2 μm occurs at slightly different wavelengths for the different clays and for alunite. There are minor absorption features that also help distinguish the spectra. Image-processing programs can identify the spectrum of an AVIRIS pixel by comparing it with a library of reference spectra for known minerals. This procedure is a form of supervised classification. The procedure is effective, however, only for the rare ground resolution cells in which only a single mineral occurs.

2. Each ground resolution cell of AVIRIS measures 20 by 20 m. In areas of complex geology, such as Goldfield, the 400 m² of a cell includes a range of different minerals. The resulting pixel is called a *mixed pixel* because its spectrum is a mixture of the spectra for the different minerals that occupy the ground resolution cell. These individual mineral species are called *spectral endmembers*. Digital *unmixing* programs are used to derive the spectra of the endmembers for each mixed pixel. For each endmember, an *endmember abundance* image is derived which shows the relative abundance of the endmember.

Figure 11-18 shows AVIRIS endmember abundance images of the Goldfield mining district that were digitally processed at Analytical Imaging and Geophysics LLC, using a spectral unmixing program of Boardman (1993). For each spectral endmember, the brightness signature for each pixel is proportional to the abundance of that endmember. Figure 11-18A is the endmember abundance image for kaolinite, which is the most abundant alteration mineral at Goldfield. The kaolinite image accurately shows the pattern of hydrothermal alteration that Ashley (1979) mapped. Figure 11-18D is a version of Ashley's map at the scale and coverage of the AVIRIS images. Figure 11-18B is the endmember abundance image for illite, which is less abundant than kaolinite. The highest concentration of illite occurs in the central area of altered rocks (Figure 11-18D). Figure 11-18C is the endmember abundance image for alunite, which is the least abundant mineral. Patches of alunite occur in the south and west portions of the alteration belt that surrounds the Goldfield district. A few patches of alunite also occur within the central area of alteration. Plate 21F is a color composite made by combining the endmember abundance image of illite in blue, alunite in green, and kaolinite in red. The black-and-white base is AVIRIS band 30 (visible red). The primary colors show areas with high concentrations of the assigned mineral. Other colors indicate co-occurrence of endmember minerals. Yellow, for example, indicates a mixture of kaolinite and alunite.

The AVIRIS color image covers the western two-thirds of the TM 5/7 ratio image shown in Plate 21C. It is instructive to

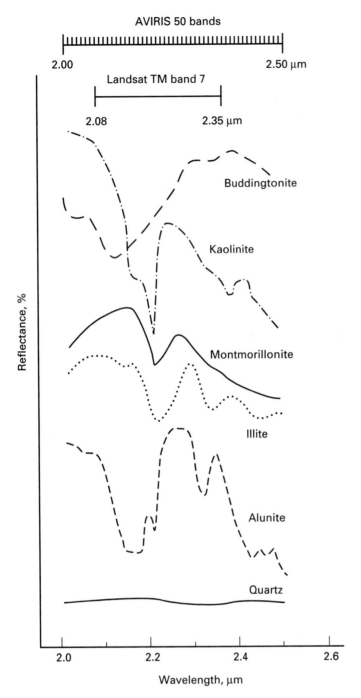

AVIRIS 50 bands

2.00   2.50 μm

Landsat TM band 7

2.08   2.35 μm

Buddingtonite

Kaolinite

Montmorillonite

Illite

Alunite

Quartz

Reflectance, %

2.0   2.2   2.4   2.6

Wavelength, μm

Figure 11-17 Laboratory spectra of alteration minerals in the atmospheric window from 2.0 to 2.5 μm. Spectra are offset vertically for clarity. Bandpasses of AVIRIS and TM band 7 are shown.

compare these images. The red and yellow signatures of the 5/7 image show the aggregate distribution of clays and alunite. The colors of the AVIRIS image show the distribution of individual alteration minerals. In summary, TM images show the broad pattern of hydrothermal alteration; AVIRIS images show the distribution of the individual alteration minerals.

STRATEGY FOR MINERAL EXPLORATION

A strategy is needed to employ remote sensing in actual mineral exploration and development. I was a co-owner of Intercontinental Resources Inc. (IRI—now a subsidiary of Queenstake Resources Ltd., of Canada), which was founded in late 1992 to explore for minerals worldwide using remote sensing and digital image processing. Our original exploration strategy had three stages.

1. Use TM ratio images to identify hydrothermally altered areas. Combine alteration information with structural interpretations from TM 2-4-7 images to identify targets that are potential mineral deposits.
2. Acquire, process, and interpret aircraft hyperspectral data for the targets to identify prospects with the greatest potential.
3. Do field evaluation and stake claims for the prospects identified in stage 2.

IRI's actual experience in Chile, Peru, and Mexico was as follows. In these active exploration areas many companies are competing for claims. Unless we staked claims immediately after stage 1, no attractive unclaimed areas would be available by the time stage 2 was completed. IRI changed its plan and did field evaluation and staked claims immediately after stage 1, which proved to be a successful strategy. Incidentally, our follow-up work has shown the TM ratio images to be very effective exploration tools. We now anticipate that hyperspectral surveys may be useful to map details of alteration zones and guide the core drilling and development of mineral properties.

Figure 11-17 points out a limitation of all multispectral and hyperspectral imaging systems that operate in the visible and reflected IR regions, regardless of their spectral resolution. Secondary silica in the form of quartz is an important component of hydrothermal alteration systems but has no diagnostic spectral features in the band from 2.0 to 2.5 μm. This inability to detect quartz is a handicap. A solution lies in the thermal IR region (8 to 14 μm), where silica content is indicated by the wavelength where the greatest energy absorption occurs, as shown in Figures 5-35 to 5-37. The thermal IR multispectral scanner (TIMS) acquires images that can be used to map silica content (Plate 9C).

MINERAL EXPLORATION IN COVERED TERRAIN

The Collahuasi and Goldfield districts are in arid terrain with extensive exposures of bedrock and little soil or vegetation.

Figure 11-18 *(facing page)* Endmember abundance images derived from 1995 AVIRIS data of Goldfield mining district. Brighter signatures represent higher concentrations of the mineral. Courtesy F. A. Kruse, Analytical Imaging and Geophysics LLC, Boulder, Colorado.

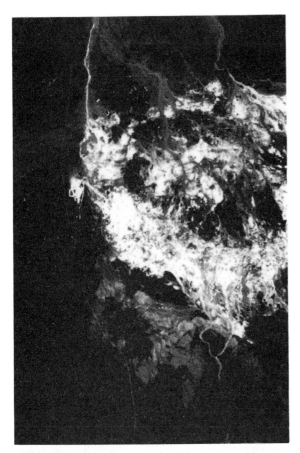

A. Kaolinite image.

B. Illite image.

C. Alunite image.

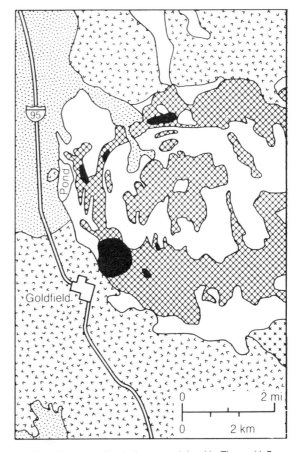

D. Alteration map. Symbols are explained in Figure 11-5.

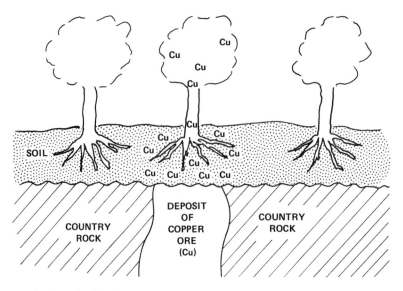

Figure 11-19 Copper enrichment of vegetation and soil overlying a concealed copper deposit.

Most of the world has temperate to humid climates, however, and mineral deposits are obscured or concealed by soil and vegetation. The composition of residual soil reflects the composition of the underlying bedrock from which the soil was derived by weathering processes.

Geochemical exploration techniques analyze the metal content of water and soil samples. Areas with high metal concentrations are then tested by core drilling. Figure 11-19 illustrates the copper enrichment of soil overlying a copper deposit in the bedrock. Vegetation growing in mineralized soil may have a higher metal content in its tissue than vegetation in normal, or background, soil, as the figure shows. This concentration of metals in vegetation is the basis for biogeochemical and geobotanical prospecting methods. However, these techniques for testing soil and analyzing vegetation are not applicable in areas where the soil has been transported rather than formed in place. Alluvial and glacial soils are examples of transported soils.

Biogeochemical and Geobotanical Exploration

Biogeochemical exploration consists of collecting vegetation samples that are analyzed chemically for high metal concentrations that may indicate a concealed ore deposit.

Geobotanical exploration searches for unusual vegetation conditions that may be caused by high metal concentrations in the soil. Sampling and chemical analyses of vegetation are not required, but skill and experience are needed to recognize the more subtle vegetation anomalies. Remote sensing techniques are being investigated as possible geobotanical exploration methods. The major geobotanical criteria for recognizing concealed ore deposits are as follows:

1. **Lack of vegetation** This may be caused by concentrations of metals in the soil that are toxic to plants. These areas are sometimes called *copper barrens* when they are caused by high concentrations of that metal. Areas that lack vegetation may be seen on aerial photographs. These barren areas may result from causes other than mineralization, however.
2. **Indicator plants** These are species that grow preferentially on outcrops and soils enriched in certain elements. Cannon (1971) prepared an extensive list of indicator plants. For example, in the Katanga region of southern Zaire, a small blue-flowered mint, *Acrocephalus robertii*, is restricted entirely to copper-bearing rock outcrops.
3. **Physiological changes** High metal concentrations in the soil may cause abnormal size, shape, and spectral reflectance characteristics of leaves, flowers, fruit, or entire plants. A relationship between spectral reflectance properties of plants and the metal content of their soils could form the basis for remote sensing of mineral deposits in vegetated terrain.

Remote Sensing for Minerals in Vegetated Terrain

Chlorosis, or yellowing of leaves, is an example of a spectral change visible to the eye. Chlorosis results from an upset of the iron metabolism of plants that may be caused by excess concentrations of copper, zinc, manganese, or other elements. Relatively high metal concentrations are required to produce chlorosis, which has been observed in plants growing near mineral deposits. Chlorosis is not a reliable prospecting criterion, however, for these reasons: (1) Many areas of known mineralized soil support apparently healthy plants with no visible toxic symptoms. (2) Chlorosis may result from conditions unrelated to mineral deposits, such as soil salinity.

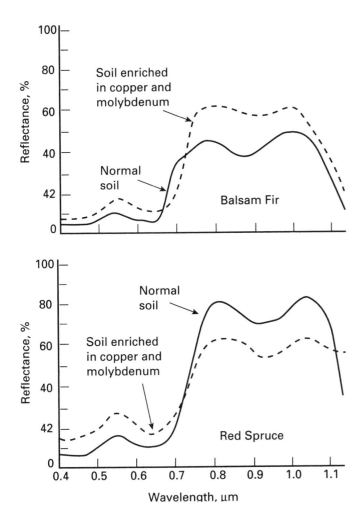

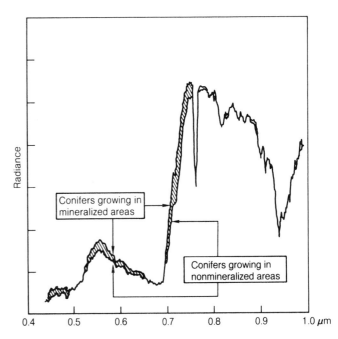

Figure 11-21 Airborne reflectance spectra for conifers in Cotter Basin, Montana. Note the "blue shift" for conifers growing in a mineralized area. From Collins and others (1983, Figure 4B).

Figure 11-20 Reflectance spectra of balsam fir and red spruce growing in normal soil and in soil enriched in copper and molybdenum. From Yost and Wenderoth (1971, Figures 5, 6).

The large, low-grade, copper-molybdenum deposit at Catheart Mountain, Maine, has been used as a geobotanical remote sensing test site (Yost and Wenderoth, 1971). Field spectrometers measured reflectance of trees growing in normal soil and in mineralized soil overlying the deposit (Figure 11-20). Red spruce and balsam fir growing in the mineralized soil both had higher metal concentrations than trees in unmineralized soil. In the reflected IR spectral region, the mineralized balsam firs have a higher reflectance than the normal trees, whereas mineralized red spruce have a lower reflectance than the normal trees (Figure 11-20). In the green spectral region, the mineralized trees of both species have a higher reflectance.

In the years following the 1970 study, a number of additional investigations have been made and were summarized by Labovitz and others (1983, Figure 1). With some exceptions, vegetation reflectance in the green and red bands generally in-

creased with increasing metal concentration in the soil. In the reflected IR region, however, there is less agreement; some studies show increased vegetation reflectance and others show decreased reflectance. Labovitz and others (1983) also noted that the geobotanical model of Figure 11-19 is not universally correct. In Virginia they found that the leaves of oak trees growing in metal-rich soil may have a lower metal content than leaves from trees in normal soil.

Geophysical Environmental Research used a nonimaging airborne system that acquires detailed reflectance spectra. The spectra in Figure 11-21 were acquired for conifers growing in a mineralized area and in an adjacent nonmineralized area. In the green band (0.5 to 0.6 μm) reflectance is higher for trees in the mineralized area, which is consistent with the results of other studies. Beginning at a wavelength of about 0.7 μm, vegetation spectra slope steeply upward to the high reflectance values in the IR region. In Figure 11-21, this steep slope is shifted slightly toward shorter wavelengths for the conifers growing in the mineralized area. This shift, called the *blue shift*, has been noted in vegetation over several mineralized areas (Collins and others, 1983) and may have exploration potential.

In summary, remote sensing of mineral deposits in vegetated terrain is in the research-and-development stage. Relationships among mineral deposits, soil chemistry, and reflectance properties of vegetation are complex but are being investigated at several research centers.

COMMENTS

Remote sensing has proven a valuable aid in exploring for mineral resources. Many ore deposits are localized along regional and local fracture patterns that provided conduits along which ore-forming solutions penetrated host rocks. Landsat and radar images are used to map these fracture patterns. Hydrothermally altered rocks associated with many ore deposits have distinctive spectral features that are recognizable on digitally processed TM images. Hyperspectral scanners may enable us to identify specific alteration minerals.

In vegetated areas hydrothermally altered rocks are not detectable on images; therefore, other remote sensing methods are needed. Reflectance spectra of foliage growing over mineralized areas may differ from spectra of foliage in adjacent nonmineralized areas. The spectral differences, however, are variable for different plant species. Additional research and development is needed for remote detection of mineral deposits in vegetated terrain.

QUESTIONS

1. You are employed by an international mineral exploration company that plans to explore for hydrothermal gold deposits in the southern third of the Andes Mountains of Chile. Your assignment is to plan the remote sensing phase of the exploration campaign ranging from regional reconnaissance to definition of individual prospects. You can utilize the image-acquisition systems and image-processing systems described in this book. Prepare the remote sensing exploration campaign, with reasons and justification for each step.

2. As part of your Chile project, you need to identify remotely the alteration minerals: kaolinite, montmorillonite, illite, and alunite. Your company has an airborne scanner that can digitally record five bands of data in the region from 2.0 to 2.5 μm. Each band has a spectral range of 0.05 μm, such as 2.10 to 2.15 μm, that you can designate. Use Figure 11-17 to designate the five optimum bands for identifying the alteration minerals. List your reasons for selecting each band. Describe how you would digitally process the resulting airborne data to produce maps showing distribution of the different alteration minerals.

3. What is the major disadvantage of the system in question 2 for mapping hydrothermally altered rocks?

4. Describe an improved imaging system that eliminates the disadvantage identified in question 3. How would you process and interpret data from the improved system?

REFERENCES

Abdelhamid, G. and I. Rabba, 1994, An investigation of mineralized zones revealed during geological mapping, Jabal Hamra Faddan—Wadi Araba, Jordan, using Landsat TM data: International Journal of Remote Sensing, v. 15, p. 1495–1506.

Abrams, M. J. and D. Brown, 1985, Silver Bell, Arizona, porphyry copper test site: Joint NASA/Geosat test case study, section 4, American Association of Petroleum Geologists, Tulsa, OK.

Ashley, R. P., 1974, Goldfield mining district: Nevada Bureau of Mines and Geology Report 19, p. 49–66.

Ashley, R. P., 1979, Relation between volcanism and ore deposition at Goldfield, Nevada: Nevada Bureau of Mines and Geology Report 33, p. 77–86.

Bennett, S. A., W. W. Atkinson, and F. A. Kruse, 1993, Use of thematic mapper imagery to identify mineralization in the Santa Teresa district, Sonora, Mexico: International Geology Review, v. 35, p. 1009–1029.

Boardman, J. W., 1993, Automated spectral unmixing of AVIRIS data using convex geometry concepts: Summaries of Fourth Annual JPL Airborne Geoscience Workshop, v. 1, p. 11–14, Pasadena. CA.

Cannon, H. L., 1971, The use of plant indicators in ground water surveys, geologic mapping, and mineral prospecting: Taxon, v. 20, p. 227–256.

Collins, W., S. H. Chang, G. Raines, F. Canney, and R. Ashley, 1983, Airborne biogeophysical mapping of hidden mineral deposits: Economic Geology, v. 78, p. 737–749.

Dick, L. A., G. Ossandon, R. G. Fitch, C. M. Swift, and A. Watts, 1993, Discovery of blind copper mineralization at Collahuasi, Chile in Romberger, S. B. and D. I. Fletcher, eds., Integrated methods in exploration and discovery: Society of Economic Geologists, Abstracts, p. AB 21–23, Denver, CO.

Eiswerth, B. A. and L. C. Rowan, 1993, Analyses of Landsat thematic mapper images of study areas located in western Bolivia, northern Chile, and southern Peru: Investigation de Metales Preciosos en El Complejo Volcanico Neogeno-Cuaternario de Los Andes Centrales, U.S. Geological Survey Project BID/TC-88-02-32-5, p. 19–44.

Goosens, M. A. and S. B. Kroonenberg, 1994, Spectral discrimination of contact metamorphic zones and its potential for mineral exploration, province of Salamanca, Spain: Remote Sensing of the Environment, v. 47, p. 331–344.

Griffiths, P. S., P. A. S. Curtis, S. E. A. Fadul, and P. D. Scholes, 1987, Reconnaissance geological mapping and mineral exploration in northern Sudan using satellite remote sensing: Geological Journal, v. 22, p. 225–249.

Harvey, R. D. and C. J. Vitaliano, 1964, Wall-rock alteration in the Goldfield district, Nevada: Journal of Geology, v. 72, p. 564–579.

Horton, R. C., 1964, An outline of the mining history of Nevada, 1924–1964: Nevada Bureau of Mines, Report 7, pt. 2, Reno, NV.

Knepper, D. H. and S. L. Simpson, 1992, Remote sensing in Geology and mineral resources of the Altiplano and Cordillera Occidental, Bolivia: U.S. Geological Survey Bulletin 1975, p. 47–55.

Labovitz, M. L., E. J. Masuoka, R. Bell, A. W. Segrist, and R. F. Nelson, 1983, The application of remote sensing to geobotanical exploration for metal sulfides—results from the 1980 field season at Mineral, Virginia: Economic Geology, v. 78, p. 750–760.

Lowell, J. D. and J. M. Guilbert, 1970, Lateral and vertical alteration-mineralization zoning in porphyry ore deposits: Economic Geology, v. 65, p. 373–408.

Nicolais, S. M., 1974, Mineral exploration with ERTS imagery: Third ERTS-1 Symposium, NASA SP-351, v. 1, p. 785–796.

Risacher, F., 1989, Economic study of the Salar de Uyuni: Institute Français de Recherche Scientifique pour le Developpement en Cooperation, Informe 32, translated by E. Jackson-Reardon.

Rowan, L. C. and T. L. Bowers, 1995, Analysis of linear features mapped in Landsat thematic mapper and side-looking airborne radar images of the Reno, Nevada 1° by 2° quadrangle, Nevada and California—implications for mineral resource studies: Photogrammetric Engineering and Remote Sensing, v. 61, p. 749–759.

Rowan, L. C., A. F. H. Goetz, and R. P. Ashley, 1979, Discrimination of hydrothermally altered and unaltered rocks in the visible and near infrared: Geophysics, v. 42, p. 533–535.

Rowan, L. C. and P. H. Wetlaufer, 1975, Iron-absorption band analysis for the discrimination of iron-rich zones: U.S. Geological Survey, Type III Final Report, Contract S-70243-AG.

Rowan, L. C., P. H. Wetlaufer, A. F. H. Goetz, F. C. Billingsley, and J. H. Stewart, 1974, Discrimination of rock types and detection of hydrothermally altered areas in south central Nevada by the use of computer-enhanced ERTS images: U.S. Geological Survey Professional Paper 883.

Sabins, F. F. and R. M. Miller, 1994, Resource assessment—Salar de Uyuni and vicinity: Proceedings Tenth Thematic Conference on Geologic Remote Sensing, p. I92–I103, Environmental Research Institute of Michigan, Ann Arbor, MI.

Singhroy, V. H., ed., 1991, Geological remote sensing in Canada: Canadian Journal of Remote Sensing, v. 17, p. 71–200.

Spatz, D. M. and R. T. Wilson, 1994, Exploration remote sensing for porphyry copper deposits, western America Cordillera: Proceedings Tenth Thematic Conference on Geologic Remote Sensing, p. 1227–1240, Environmental Research Institute of Michigan, Ann Arbor, MI.

Unrug, R., 1988, Mineralization controls and source metals in the Lufilian Fold Belt, Shaba (Zaire), Zambia, and Angola: Economic Geology, v. 83, p. 1247–1258.

Vergara, H., 1978A, Cuadrangulo Ujina: Carta Geologica de Chile, no. 33, Escala 1:50,000, Instituto de Investigaciones Geologicas, Santiago, Chile.

Vergara, H., 1978B, Cuadrangulo Quehuita y sector occidental del cuadrangulo Volcan Mino: Carta Geologica de Chile, no. 32, Escala 1:50,000, Instituto de Investigaciones Geologicas, Santiago, Chile.

Viljoen, R. P., M. J. Viljoen, J. Grootenboer, and T. G. Longshaw, 1975, ERTS-1 imagery an appraisal of applications in geology and mineral exploration: Minerals Science and Engineering, v. 7, p. 132–168.

Watson, K., F. A. Kruse, and S. Hummer-Miller, 1990, Thermal infrared exploration in the Carlin trend, northern Nevada: Geophysics, v. 55, p. 70–79.

Yost, E. and S. Wenderoth, 1971, The reflectance spectra of mineralized trees: Proceedings Seventh International Symposium on Remote Sensing of Environment, v. 1, p. 269–284, University of Michigan, Ann Arbor, MI.

ADDITIONAL READING

Abrams, M. J., D. Brown, L. Lepley, and R. Sadowski, 1983, Remote sensing for porphyry copper deposits in southern Arizona: Economic Geology, v. 78, p. 591–604.

Hunt, G. R. and R. P. Ashley, 1978, Spectra of altered rocks in the visible and near infrared: Economic Geology, v. 74, p. 1613–1629.

Kaufmann, H., 1988, Mineral exploration along the Aqaba-Levant structure by use of TM data; concepts, processing, and results: International Journal of Remote Sensing, v. 9, p. 1639–1658.

Rowan, L. C., C. A. Trautwein, and T. L. Purdy, 1991, Maps showing association of linear features and metallic mines and prospects in the Butte 1° by 2° quadrangle, Montana: U.S. Geological Survey Miscellaneous Investigations Series Map I-2050-B.

Segal, D. B. and L. C. Rowan, 1989, Map showing exposures of limonitic rocks and hydrothermally altered rocks in the Dillon 1° by 2° quadrangle, Idaho and Montana: U.S. Geological Survey Miscellaneous Investigations Series Map I-1803-A.

12

LAND USE AND LAND COVER: GEOGRAPHIC INFORMATION SYSTEMS

Remote sensing from aircraft and satellites provides our first opportunity to inventory the surface resources of the earth in a systematic repetitive manner. Categories of land use and land cover are mapped at scales and resolutions ranging from worldwide to local municipalities. Geographic information systems use computer technology to merge remote sensing images with other data sets and produce inventory maps. Many of these techniques are also used for archaeologic investigations.

NEED TO CLASSIFY LAND USE AND LAND COVER

Land use describes how a parcel of land is used (such as for agriculture, residences, or industry), whereas *land cover* describes the materials (such as vegetation, rocks, or buildings) that are present on the surface. For example, the land cover of an area may be evergreen forest, but the land use may be lumbering, recreation, oil extraction, or various combinations of activities. Accurate, current information on land use and cover is essential for many planning activities. Remote sensing methods are becoming increasingly important for mapping land use and land cover for the following reasons:

1. Large areas can be imaged quickly and repetitively.
2. Images can be acquired with a spatial resolution that matches the degree of detail required for the survey.
3. Remote sensing images eliminate the problems of surface access that often hamper ground surveys.
4. Images provide a perspective that is lacking for ground surveys.
5. Image interpretation is faster and less expensive than conducting ground surveys.
6. Images provide an objective, permanent data set that may be interpreted for a wide range of specific land uses and land covers, such as forestry, agriculture, and urban growth.

There are some disadvantages to remote sensing surveys:

1. Some types of land use may not be distinguishable on images.
2. Most images lack the horizontal perspective that is valuable for identifying many categories of land use.

Remote sensing interpretations should be supplemented by ground checks. The following section describes a system for classifying land use and land cover that is based on the interpretation of remote sensing images. A succeeding section uses the system to interpret images of the Los Angeles region.

MULTILEVEL CLASSIFICATION SYSTEM

A classification system based on remote sensing images should meet the following requirements (Anderson and others, 1976):

1. Both activities (land use) and resources (land cover) should be classified.
2. The minimum level of accuracy in identifying land-use and land-cover categories from remote sensing data should be at least 85 percent.
3. The accuracy of interpretation for all categories should be approximately equal.
4. Repeatable results should be obtainable from one interpreter to another and from one time of sensing to another.
5. The system should be applicable over extensive areas.
6. The system should be usable for remote sensing data obtained at different times of the year.
7. The system should allow use of subcategories that can be derived from ground surveys or from larger-scale remote sensing data.

Table 12-1 Images employed in multilevel classification

Level	System	Image scale
I	Landsat TM and MSS images; AVHRR images	Smaller than 1:250,000
II	High-altitude aerial photographs; TM and SPOT images AVHRR images (with ancilliary data)	1:80,000 to 1:250,000
III	Medium-altitude aerial photographs	1:20,000 to 1:80,000
IV	Low-altitude aerial photographs	Larger than 1:20,000

Source: Modified from Anderson and others (1976).

8. Aggregation of categories should be possible.
9. Comparison with future land-use data should be possible.
10. Multiple land uses should be recognizable.

Anderson and others (1976) of the U.S. Geological Survey developed a multilevel classification system that meets these requirements. The user can select the type and scale of image that suits the objectives of the project. Table 12-1 lists the image systems and image scales employed for each of the four classification levels. For example, level I is suitable for an entire state, whereas level III is suitable for a municipality.

The levels I and II categories of Table 12-2 were slightly modified from those defined by Anderson and others (1976). The level III categories were modified from those defined by the Florida Bureau of Comprehensive Planning (1976). The classification system appears straightforward and definitive on paper, but there are some problems and uncertainties in using the system. On the images it may be difficult to recognize definitive examples of the various categories. Distinguishing land from water may seem simple until one considers the problem of classifying seasonally wet areas, tidal flats, or marshes with various kinds of plant cover. Another problem is to define boundaries between different land-use categories that grade into each other. The boundary between the "light industrial" and "heavy industrial" categories (131 and 132 of Table 12-2) may be difficult to draw. In such cases, the boundary is drawn to separate areas where one land use predominates. The categories listed in Table 12-2 are clearly defined, but commonly two or more categories are intermixed and cannot be mapped separately at the scale of the images. For example, warehouses of the "commercial and services" category (120) are commonly located within predominantly industrial areas (category 130). Where the minor use occupies more than one-third of the area, the "mixed" category (180) is used at level II. An alternative treatment is a compound category (130/120), with the minor category given in the second position. Multiple use of a parcel of land presents another problem to the interpreter. For example, forest lands are commonly used for recreational purposes, such as hunting and camping; a compound category

(410/170) may be employed for this situation. The interpreter must explain the "ground rules" and any modifications of the classification system that were used in a project.

The classification may be modified to accommodate specific needs. For example, one can precede irrigated orchards with an "i" (i221) and nonirrigated orchards with an "n" (n211).

It is difficult to map and display any area on an image smaller than 5 mm on a side. As shown in the following examples, however, one can still designate narrow linear features such as freeways and waterways.

MULTILEVEL CLASSIFICATION SYSTEM OF THE LOS ANGELES REGION

The Los Angeles, California, region is used to demonstrate the multilevel system for classifying land use and land cover. A Landsat TM image at a scale of 1:1,000,000 (Figure 12-1) was interpreted to produce the level I land-use map (Figure 12-2). Figure 12-3 is a SPOT pan image of the western portion of the Los Angeles region at a scale of 1:120,000. Figure 12-4 is a high-altitude aerial photograph of the same area at the same scale. Both images are available as stereo pairs that are readily interpreted to produce a level II land-use map (Figure 12-5). Aerial photographs have the advantages of color or IR color and a higher spatial resolution than SPOT pan images. SPOT images cover a larger area (3600 km^2 versus 830 km^2) and are available for most of the world. The SPOT image (Figure 12-3) was acquired 7 years later than the aerial photograph (Figure 12-4); comparing the two images reveals some changes in land use during that period.

Figure 12-6 is a medium-altitude, black-and-white aerial photograph of central Los Angeles at a scale of 1:30,000 that was interpreted stereoscopically with overlapping photographs to produce the level III map (Figure 12-7).

The following definitions and descriptions of level I and level II categories are summarized from Anderson and others (1976), together with comments on some level III categories.

Table 12-2 Land-use and land-cover classification system

Level I	Level II	Level III
100 Urban or built-up	110 Residential	111 Single unit, low-density (less than 2 DUPA*) 112 Single unit, medium-density (2 to 6 DUPA) 113 Single unit, high-density (greater than 6 DUPA) 114 Mobile homes 115 Multiple dwelling, low-rise (2 stories or less) 116 Multiple dwelling, high-rise (3 stories or more) 117 Mixed residential
	120 Commercial and services	121 Retail sales and services 122 Wholesale sales and services (including trucking and warehousing) 123 Offices and professional services 124 Hotels and motels 125 Cultural and entertainment 126 Mixed commercial and services
	130 Industrial	131 Light industrial 132 Heavy industrial 133 Extractive 134 Industrial under construction
	140 Transportation	141 Airports, including runways, parking areas, hangars, and terminals 142 Railroads, including yards and terminals 143 Bus and truck terminals 144 Major roads and highways 145 Port facilities 146 Auto parking facilities (where not directly related to another land use)
	150 Communications and utilities	151 Energy facilities (electrical and gas) 152 Water supply plants (including pumping stations) 153 Sewage-treatment facilities 154 Solid-waste disposal sites
	160 Institutional	161 Educational facilities, including colleges, universities, high schools, and elementary schools 162 Religious facilities, excluding schools 163 Medical and health-care facilities 164 Correctional facilities 165 Military facilities 166 Governmental, administrative, and service facilities 167 Cemeteries
	170 Recreational	171 Golf courses 172 Parks and zoos 173 Marinas 174 Stadiums, fairgrounds, and race tracks
	180 Mixed	
	190 Open land and other	191 Undeveloped land within urban areas 192 Land being developed; intended use not known
200 Agriculture	210 Cropland and pasture	211 Row crops 212 Field crops 213 Pasture

Table 12-2 Land-use and land-cover classification system *(continued)*

Level I	Level II	Level III
	220 Orchards, groves, vineyards, nurseries, and ornamental horticultural areas	221 Citrus orchards 222 Noncitrus orchards 223 Nurseries 224 Ornamental horticultural 225 Vineyards
	230 Confined feeding operations	231 Cattle 232 Poultry 233 Hogs
	240 Other agriculture	241 Inactive agricultural land 242 Other
300 Rangeland	310 Grassland	
	320 Shrub and brushland	321 Sagebrush prairies 322 Coastal scrub 323 Chaparral 324 Second-growth brushland
	330 Mixed rangeland	
400 Forest land	410 Evergreen forest	411 Pine 412 Redwood 413 Other
	420 Deciduous forest	421 Oak 422 Other hardwood
	430 Mixed forest	431 Mixed forest
	440 Clearcut areas	
	450 Burned areas	
500 Water	510 Streams and canals	
	520 Lakes and ponds	
	530 Reservoirs	
	540 Bays and estuaries	
	550 Open marine waters	
600 Wetlands	610 Vegetated wetlands, forested	611 Evergreen 612 Deciduous 613 Mangrove
	620 Vegetated wetlands, nonforested	621 Herbaceous vegetation 622 Freshwater marsh 623 Saltwater marsh
	630 Nonvegetated wetlands	631 Tidal flats 632 Other nonvegetated wetlands
700 Barren land	710 Dry lake beds	
	720 Beaches	
	730 Sand and gravel other than beaches	
	740 Exposed rock	
800 Tundra	830 Bare-ground tundra	
900 Perennial snow or ice	910 Perennial snowfields	
	920 Glaciers	

*DUPA = dwelling units per acre.

Source: Modified from Anderson and others (1976); Florida Bureau of Comprehensive Planning (1976).

Urban or Built-Up (100)

This level I category comprises areas of intensive land use where much of the land is covered by structures and streets. The range of uses is shown by the level II and level III categories that are assigned to the "urban or built-up" category. As urban development expands, other level I categories ("agriculture," "forest," and "water") may be enclosed, and small patches will be included with urban areas at the level I classification. Where these other categories occur on the fringes of urban land, they will be mapped separately, except where they are surrounded and dominated by urban development. Where criteria for more than one category are met, the "urban or built-up" category takes precedence over others. On the Landsat image (Figure 12-1) there are some residential areas with sufficient tree cover to meet the "forest" criteria, but these are classed as "urban or built-up."

On Landsat TM images, the "urban or built-up" category may be recognized by the following criteria:

1. There is a dense network of streets, with only the major arteries recognizable. The streets may fall into a regular grid (Chicago), a radial pattern (New Orleans), or an irregular pattern (Boston).
2. The central business and industrial section has a blue signature caused by the absence of vegetation and by the concentration of roofs and pavement.
3. The central section is surrounded by residential areas with characteristic red signatures caused by landscape vegetation. Interspersed patches of vegetation are caused by parks, golf courses, and cemeteries.

The level I "urban or built-up" category is subdivided into a number of level II categories, which are described in the following sections.

Residential (110) On the SPOT image (Figure 12-3) and the high-altitude aerial photograph (Figure 12-4), there are extensive residential areas, but individual structures are difficult or impossible to resolve at this scale. The fine texture, the regular pattern of closely spaced streets, and the distribution around urban core areas are keys for recognizing this level II category. On normal color and IR color photographs, lawns and trees have distinctive signatures. At level II the residential areas may be separated from the various nonresidential buildings that are larger or less regular in size and shape.

At level III the "residential" category is subdivided into various classes of single-family and multiple-family occupancy. As shown in Table 12-2, single-family units are classed as low-, medium-, or high-density based on the number of *dwelling units per acre* (DUPA). DUPA may be estimated by using a 10-acre square (as shown at the bottom of Figure 12-7). Trace the square onto a transparent sheet and then position it over a residential area on the aerial photograph (Figure 12-6). Count the units within the square and divide by 10 to obtain DUPA. A 5-hectare template is also provided on Figure 12-7 for the metric system. For the three categories of residential density, the equivalent *dwelling units per hectare* (DUPH) as follows

Category 111-low-density has less than 5 DUPH (<2 DUPA).

Category 112-medium-density has 5 to 15 DUPH (2 to 6 DUPA).

Category 113-high-density has greater than 15 DUPH (>6 DUPA).

Areas of sparse residential land use, such as farmsteads and farm-labor housing, are assigned to a level I category such as "agriculture." Rural residential and recreational subdivisions, however, are included in the "residential" category because the land is primarily used for residences, even though it may have forest or rangeland types of cover. Housing facilities at military bases, resorts, colleges, and universities are assigned to the appropriate "institutional" (160) level III category.

Commercial and Services (120) Facilities in this level II category are used predominantly for the distribution and sale of products and services. Included are facilities that support the basic uses such as office buildings, warehouses, landscaped areas, driveways, and parking lots. Commercial areas may include some noncommercial areas that are too small to be mapped separately. Churches, schools, residences, and industry may be enclosed within commercial areas. Where these noncommercial areas exceed one-third of the total commercial area, the "mixed" category (180) is employed at level II. Facilities in the "commercial and services" category typically occur in three distinct settings: (1) concentrations in central urban cores, (2) strips along major streets and highways, and (3) shopping centers and malls adjacent to residential areas.

Facilities in the "commercial and services" category are distinguished from residential areas (category 110) by the criteria given earlier. Where "commercial and services" facilities consist of multistory buildings or shopping centers, they are distinguishable from the "industrial" category (130). In many areas, however, warehouses and wholesale stores of the "commercial and services" category are intermixed with industrial buildings (category 130) and cannot be distinguished on high-altitude aerial photographs. The area directly southeast of downtown Los Angeles (the east-central portion of Figures 12-3 through 12-5) includes both commercial and industrial facilities and is classified as "mixed" (180).

At Level III many of the "commercial and services" categories cannot be interpreted from images without the aid of ground information. The "mixed" category (126) is employed where more than one-third of the area is occupied by uses other than the predominant category.

Figure 12-1 Landsat TM band 3 (red) image of the Los Angeles region, used for level I land-use classification. Image was acquired July 3, 1985.

Industrial (130) This level II category designates manufacturing facilities, which are separated at level III into the "light industrial" (131) and "heavy industrial" (132) categories. Light industries are those that design, assemble, finish, process, and package products. Many light industries are concentrated in industrial parks, which may be located adjacent to airports or residential areas, or in open country.

Heavy industries use raw materials such as timber, iron ore, and crude oil. Pulp and lumber mills, steel mills, oil refineries and tank farms, chemical plants, and brickmaking plants are typical facilities. Food-processing plants (canneries, grain storage, and milling) are included in the "heavy industrial" category. Heavy industries are recognized on images by stockpiles of raw materials, waste-disposal areas, and transportation facil-

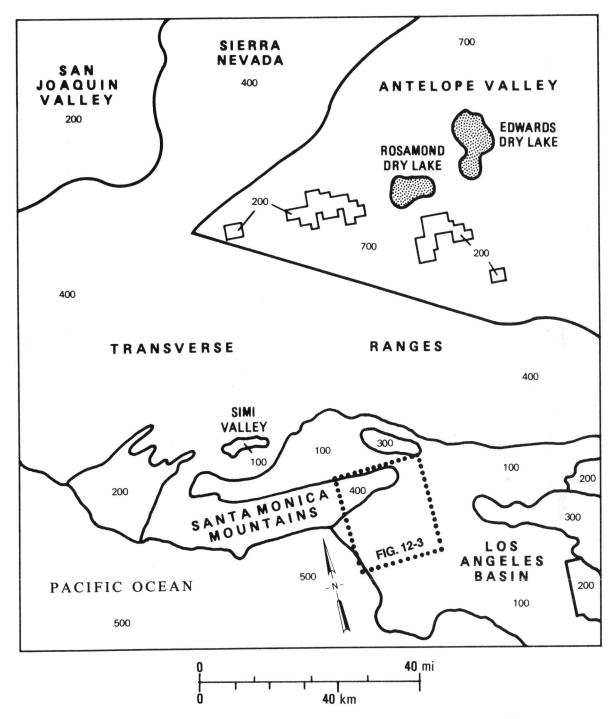

Figure 12-2 Level I land-use classification map interpreted from the Landsat image of the Los Angeles region (Figure 12-1). Categories are explained in Table 12-2.

ities for bulk shipments. The oil refinery adjacent to the coast in the southwest corner of the SPOT image (Figure 12-3) is recognizable by the numerous round storage tanks.

Extractive facilities such as mines, quarries, oil fields, sand pits, and gravel pits belong to the level III subdivision (category 133) of the "industrial" category. Some classification sys-

tems assign extractive facilities to the "barren land" (700) category, but this assignment ignores the industrial aspect of extractive activities.

Transportation (140) At level II, only the largest transportation facilities, such as airports, seaport facilities, and major

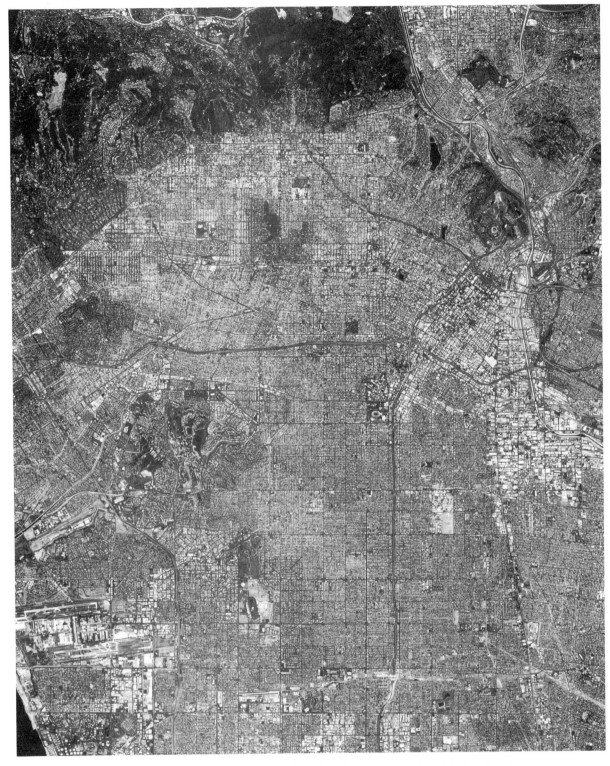

Figure 12-3 SPOT pan image (10-m resolution) of the western portion of the Los Angeles region, used for the level II land-use classification. Image was acquired July 20, 1986.

Figure 12-4 High-altitude aerial photograph of the western portion of the Los Angeles region, used for level II land-use classification. Photograph was acquired by NASA aircraft, September 2, 1979.

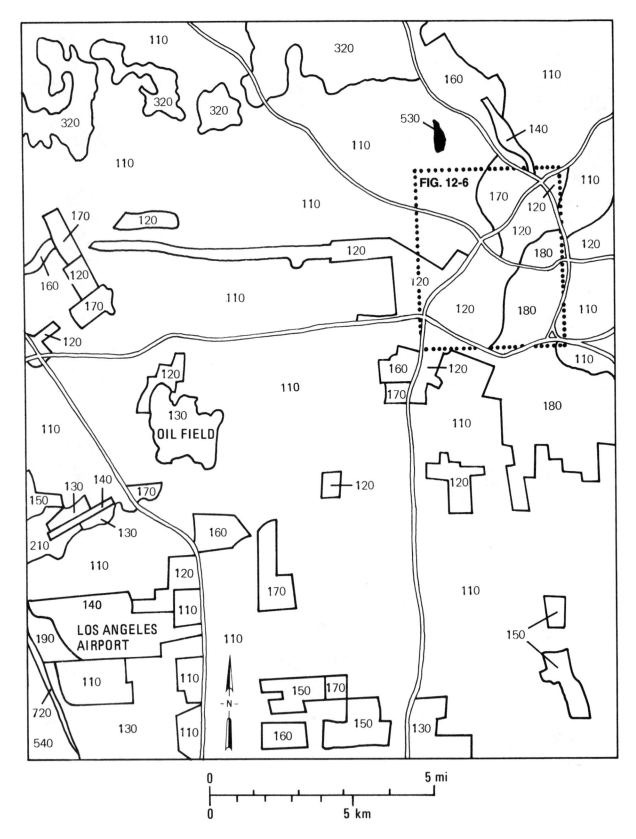

Figure 12-5 Level II land-use classification map interpreted from the SPOT image (Figure 12-3) and high-altitude photograph (Figure 12-4) for the western portion of the Los Angeles region. Categories are explained in Table 12-2.

highways, are recognizable, as shown in Figures 12-3 through 12-5. At this level, most of the transportation facilities are included with the associated level II categories, such as 120 or 130.

More detailed interpretation of transportation facilities is possible at level III, but only the larger areas are mapped. On the medium-altitude aerial photograph (Figure 12-6), it is possible to recognize railroad yards and terminals (category 142) that could not be distinguished on the high-altitude photograph.

Communications and Utilities (150) At level II this category is difficult to recognize. At level III, however, the categories shown in Table 12-2 are generally recognizable. The "energy facilities" category (151) includes electrical generating plants, natural-gas storage tanks, and related facilities. This category does not include coal mines or oil and gas fields, which belong to the "extractive" category, or oil refineries, which belong to the "heavy industrial" category (132). On Figure 12-6 (middle right), energy facilities can be recognized from the storage tanks for natural gas. Transmission lines for electricity and telephones as well as pipelines for gas, oil, and water are narrow and rarely constitute a dominant land use; therefore, these are not assigned a separate category.

Institutional (160) The level III categories listed in Table 12-2 describe the facilities assigned to the "institutional" subdivision. A few of the categories are recognizable by their characteristics on stereo aerial photographs. Many educational facilities may be recognized by the associated athletic facilities (ball fields and tracks). Cemeteries are recognizable by the expanses of lawns with driveways and the absence of other facilities. Supplemental ground information is generally required to identify the remaining level III institutional categories.

Recreational (170) Recreational land uses include the level III categories listed in Table 12-2. One problem in classification is that the "golf courses" and "parks and zoos" categories (171 and 172) may be confused with the open landscaped areas associated with some institutional facilities (category 160). On the Los Angeles images, Dodger Stadium and the surrounding parking lots are recognizable at both level II (northeast portion of Figures 12-3 and 12-4) and level III (north-central portion of Figure 12-6).

Mixed (180) This category is used for areas where the preceding level II categories (110 through 170) cannot be separated at the mapping scale. Where more than one-third of an area has an intermixture of two or more uses, the area is classified as "mixed." Where the intermixed uses occupy less than one-third of the area, the dominant land-use category is used.

Open Land and Other (190) Many areas in the "urban and built-up" category include tracts of vacant land not being used for any of the preceding activities. These tracts are not "rangeland" (category 300); therefore, they are classified as "unde-

veloped land within urban areas" (191) or as "land being developed; intended use not known" (192). An example of category 192 is shown in Figure 12-6 where the bright-toned area 1 km north of Dodger Stadium has been graded and terraced but the ultimate use is not apparent. Elsewhere the ultimate use of land being developed may be inferred from the images. For example, open land adjacent to residential areas that is being graded and laid out with a street grid can be assigned to the "residential" category (110).

Agriculture (200)

In some parts of the world, one can readily recognize land used for crops and orchards on satellite images by rectangular or circular patterns. The cultivation pattern is clearly shown in the San Joaquin Valley, in the northwest portion of the Landsat image. Rectangular dark patches within the Antelope Valley indicate areas where the desert is being reclaimed for irrigated farming. Fields with circular irrigation are also easily recognized. In much of Canada and the United States, agricultural fields are subdivisions of the basic land survey unit, which is a section 1 mile square containing 640 acres. In Russia, however, wheat fields are typically larger than the 40- and 160-acre fields of the United States and Canada. In other regions, such as India and Southeast Asia, individual fields are too small to be recognized on Landsat images and the patterns do not form a regular grid. Agricultural land (category 200) is distinguishable from urban land (category 100), which has indicators of population concentrations.

Cropland and Pasture (210) This level II category includes a wide variety of crops, which are subdivided into three level III categories. Row crops (category 211) are distinguished by their pattern from field crops (category 212), such as wheat and alfalfa, which uniformly cover the area. "Pasture" (category 213) refers to relatively small areas of grazing land commonly interspersed with croplands.

Ground-survey information is generally needed to identify specific crops. However, experienced interpreters who are familiar with a region and its crop cycles may derive accurate crop estimates from images alone. Crop types and yields have been estimated by digital processing of MSS images acquired at several dates during the growing season. The Large Area Crop Inventory Experiment (LACIE), sponsored by NASA and the U.S. Department of Agriculture, developed this technique to predict wheat production in major growing areas of the world during the late 1970s (Myers and others, 1983).

Orchards, Groves, Vineyards, Nurseries, and Ornamental Horticultural Areas (220) This aggregate category is employed at level II, and the individual components listed in Table 12-2 are used at level III. Knowledge of the region and ground information are generally needed to recognize most of the level III categories.

Figure 12-6 Medium-altitude aerial photograph of central Los Angeles, used for level III land-use classification. Photograph was acquired October 25, 1972.

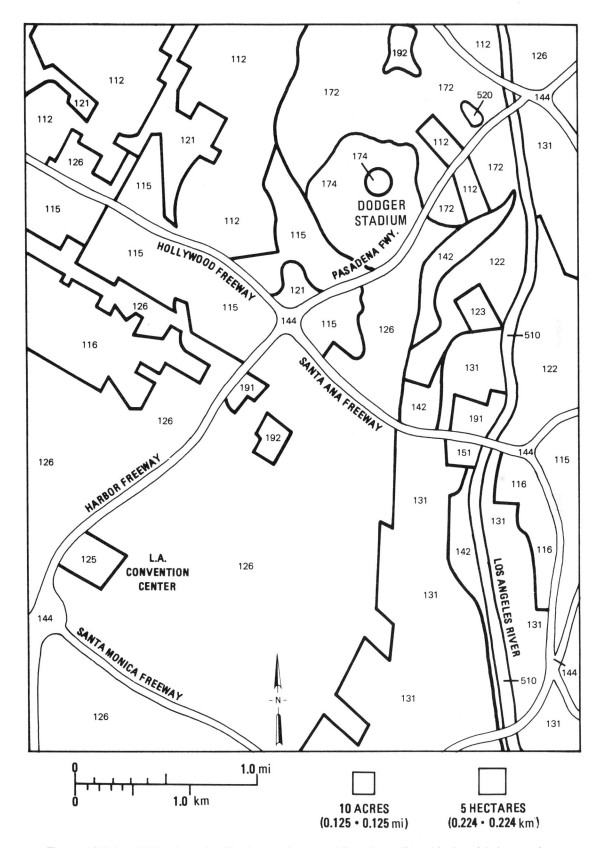

Figure 12-7 Level III land-use classification map interpreted from the medium-altitude aerial photograph for central Los Angeles (Figure 12-6). Categories are explained in Table 12-2.

Confined Feeding Operations (230) Stockyards, hog and cattle feedlots, confined dairy feeding operations, and large poultry farms constitute this category of land use. In spite of the relatively small areas occupied, waste products from concentrations of these animals cause environmental problems that justify a separate category. On high-altitude images, these facilities are recognized by the built-up appearance, access paths, waste-disposal areas, and lack of vegetation.

Other Agriculture (240) "Inactive agricultural land" (241) is a major level III subdivision of "other agriculture." Also at level III, the "other" category (242) is used for farmsteads, corrals, and other relatively small areas associated with the major agricultural activities.

Rangeland (300)

Rangeland is land covered by natural grasses, shrubs, and forbs, which include nonwoody plants such as weeds and flowers. Rangeland is capable of supporting native or domesticated grazing animals. Some rangelands have been modified by eradicating nonproductive plants, such as sagebrush and mesquite, and by planting grasses.

Grassland (310) This category has no level III subdivisions but includes a wide range of grass types that were summarized by Anderson and others (1976).

Shrub and Brushland (320) A wide range of plant communities make up this category. "Sagebrush prairies" (category 321) includes semiarid lands that also support shadscale, greasewood, and creosote bush. The "coastal scrub" category (322) includes extensive areas in southern Texas and Florida that do not qualify as wetlands. "Chaparral" (category 323) refers to a dense growth of evergreen shrubs that include manzanita, mountain mahogany, and scrub oaks. "Second-growth brushland" (category 324) occurs widely in the eastern United States and consists of former croplands or pastures (cleared from original forest land) that have now grown up in brush.

Mixed Rangeland (330) Where more than one-third of an area is an intermixture of grassland with shrub and brushland, the area is classified as "mixed rangeland."

Forest Land (400)

Forest lands have a *crown density* (also called the *crown closure percentage*) of 10 percent or more and support trees capable of producing timber or other wood products. The "forest" category also includes lands from which trees have been removed to a crown density of less than 10 percent but have not been developed for other uses. Lumbering, fire, and disease are some of the agents that reduce crown density. On the Landsat image (Figure 12-1), forest lands occur in the mountains of the Transverse Ranges and the southern end of the Sierra Nevada. In this and other regions, there is a transition from forest at higher elevations to brush of the "rangeland" category (300) at lower elevations. Without additional information, however, it is usually difficult to separate these two categories.

Evergreen forests (category 410) consist predominantly of trees that remain green all year. Deciduous forests (category 420) consist predominantly of trees that seasonally lose their leaves. Where more than one-third of an area is a mixture of deciduous trees with evergreen trees, the area is classified as "mixed forest" (430). "Clearcut areas" (440) and "burned areas" (450) are level II categories that are recognizable on high-altitude aerial photographs. In some areas the system may be expanded to recognize planted and cultivated forests.

Water (500)

Few comments are needed for this category. Lakes and ponds (category 520) are natural water bodies, whereas reservoirs (category 530) are artificially impounded.

Wetlands (600)

Most wetlands are located adjacent to water bodies and include marshes, swamps, tidal flats, and many river floodplains. Areas that are only seasonally flooded and do not support typical wetland vegetation are assigned to other categories. Cultivated wetlands such as rice fields and developed cranberry bogs are classified as agricultural land (category 200). Wetlands from which uncultivated products such as timber are harvested or that are grazed by cattle are retained in the "wetlands" category. Shallow water areas covered by floating vegetation (such as water lilies and water hyacinths) are classed as wetlands, but where aquatic vegetation is submerged the area is classed as water (category 500).

Vegetated Wetlands, Forested (610) Management and environmental planning requirements for wetlands are very different from those for dry land. For this reason, forested land occurring in swamps, marshes, and seasonally flooded bottomland is assigned to the "wetlands" category (600) rather than to "forest land" (400). The forested wetlands are divided into the level III categories of "evergreen" (611), "deciduous" (612), and "mangrove" (613).

Vegetated Wetlands, Nonforested (620) Nonforested wetlands are divided into the level III categories of "herbaceous vegetation" (621), "freshwater marsh" (622), and "saltwater marsh" (623). The "herbaceous vegetation" category includes nonwoody plants such as grasses, sedges, rushes, mosses, water lilies, and water hyacinths.

Nonvegetated Wetlands (630) Tidal flats are the major category of nonvegetated wetlands.

Barren Land (700)

Barren land has a limited ability to support life, and less than one-third of the area has vegetation cover. The surface is predominantly thin soil, sand, or rocks. Any vegetation present is more scrubby and widely spaced than that in the "shrub and brushland" category (320). Land that is barren because of human activity is assigned to the actual land use category such as "mine dumps" (133), "inactive agricultural land" (241), clear-cut forest land (440). On the Landsat image (Figure 12-1), the Antelope Valley is typical of barren land in the Mojave Desert. The margins of the Antelope Valley are covered by alluvial fans made up of gravel eroded from the adjacent mountains. Sand, thin soil, dry lake beds, and rock exposures cover the central portion of the valley. In this example, level II categories are readily mapped on an image at the scale of level I.

Dry Lake Beds (710) In many arid regions the floors of closed valleys are covered by salt and silt deposited in lakes that are normally dry (playas or salt flats). These deposits are readily identified on images by their location and distinctive bright tone.

Beaches (720) Beaches are the deposits of sand and gravel along shorelines.

Sand and Gravel Other Than Beaches (730) The sand of this category occurs as windblown sheets and dune fields. Mixtures of sand and gravel occur in floodplains and as barren alluvial fans surrounding mountains in arid areas.

Exposed Rock (740) This category includes area of bedrock exposure, volcanic deposits, and talus. When these rock types occur in tundra areas, they are assigned to the "bare-ground tundra" category (830).

Tundra (800)

"Tundra" refers to treeless regions beyond the limit of the boreal forest. Regions above the timberline in high mountain ranges are also classed as tundra. Tundra vegetation is low and dwarfed, and it commonly forms a complete mat. The presence of permafrost and the prevalence of subfreezing temperatures most of the year are characteristics of tundra. Where these conditions prevail, areas that would otherwise be assigned to "wetlands" (such as the Arctic Coastal Plain, Alaska) or to "barren land" (such as the Brooks Range, Alaska) are assigned to the "tundra" category (800). Late summer images are best for interpreting tundra categories.

Perennial Snow or Ice (900)

This category is used for areas where snow and ice accumulations persist throughout the year as snowfields or glaciers.

Perennial Snowfields (910) Perennial snowfields are accumulations of snow and firn (coarse, compacted granular snow) that did not entirely melt during previous summers. Snowfields lack flow features, which distinguishes them from glaciers.

Glaciers (920) This category of flowing ice consists of both ice caps and valley glaciers. Flowage is indicated by crevasses and by lateral and medial moraines that form dark streaks against the bright background of the glacial ice. The major problem in recognizing glaciers is the presence of snow that may obscure the moraines and crevasses.

Summary

The multilevel classification system was developed in the early 1970s, with Landsat MSS images and aerial photographs as data sources. Today additional satellite images are available with a wide range of spectral bands and spatial resolution. Current computer technology greatly exceeds that of the 1970s. Remarkably, the multilevel system is compatible with these new developments and provides a standard system for classifying land use and land cover.

DIGITAL CLASSIFICATION OF LAND USE, LAS VEGAS, NEVADA

The preceding examples of land-cover classification were manually interpreted using hard-copy images. Land cover can also be classified by computer processing of digital image data.

Plate 23A is a Landsat TM 2-3-4 image of much of Las Vegas, Nevada, and vicinity. The central part of the city is the dark area in the northwest part of this image. A runway of McCarran Airport is visible in the southwest corner. The east and southeast margins of the image are desert. Much of the image is a mixture of residential areas, commercial developments, and various types of landscape vegetation. Plate 23B is an unsupervised classification map prepared from the six TM bands of visible and reflected IR data. The classification produced 16 classes. The analyst combined seven of these with other classes to produce the nine categories listed here, with their display colors in Plate 23B:

Violet	Residential (category 110)
Orange	Commercial (120)
Black	Streets and parking lots (144 and 146)
Gray	Construction sites (192)
Blue	Open land (191)
Dark green	Irrigated vegetation (170 and 210)
Medium green	Mixed rangeland (330)
Light green	Shrub and brushland (320)
Yellow	Sand and gravel (730)

A. Seasat radar image (23.5-cm wavelength) acquired August 2, 1978.

B. Landsat image acquired August 10, 1978.

Figure 12-8 Seasat and Landsat images of Phoenix, Arizona, and vicinity.

Business and commercial activities (orange) are concentrated along major streets and include streets and parking lots (black). Residential areas (violet) occur as rectangles adjacent to commercial areas and in outlying areas. Three types of vegetation were classified. Irrigated vegetation (dark green) occurs in golf courses, parks, and irrigated agriculture; in the TM image (Plate 23A) this category has a bright red signature. Mixed rangeland (medium green) includes areas where houses and vegetation are intermingled in approximately equal proportions. Shrub and brushland (light green) occurs largely around the outskirts of

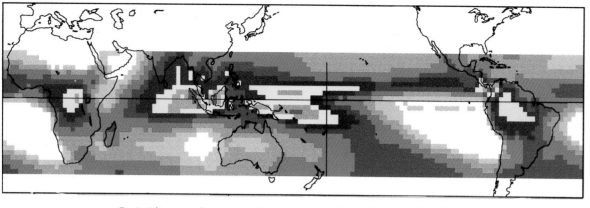

Satellite estimate of tropical rainfall (mm/day)

0.5   1   2   4   6   8   10   12

A. Mean annual rainfall (1986 to 1994) derived from cloud temperatures.

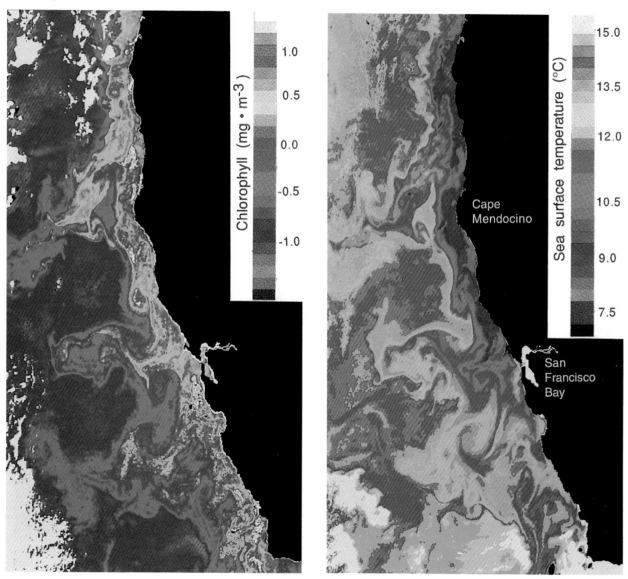

B. CZCS map of central California coast, July 7, 1981, showing chlorophyll concentration.

C. AVHRR thermal IR image of central California coast, July 8, 1981, showing sea temperatures.

Plate 17   Environmental remote sensing images. Image A, courtesy J. E. Janowiak, NOAA/NWS/NMC. Images B and C, from Abbott (1984, Figures 27-1 and 2).

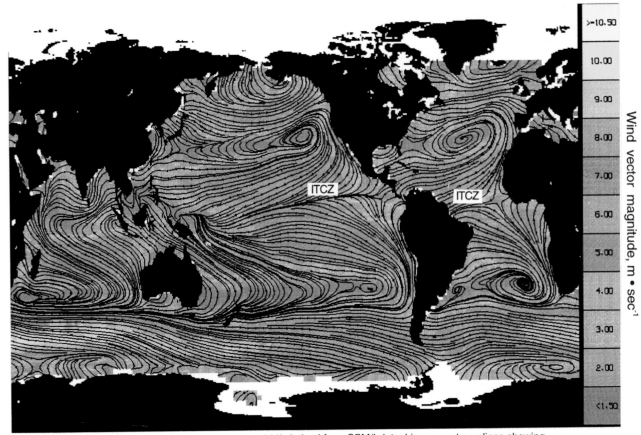

A. Average wind patterns (July 1987 through June 1988) derived from SSM/I data. Lines are streamlines showing wind direction. Colors show wind vector magnitude. Note location of intertropical convergence zone (ITCZ). From Atlas, Hoffman, and Bloom (1993, Figure 1). Courtesy R. Atlas, NASA Goddard Space Flight Center.

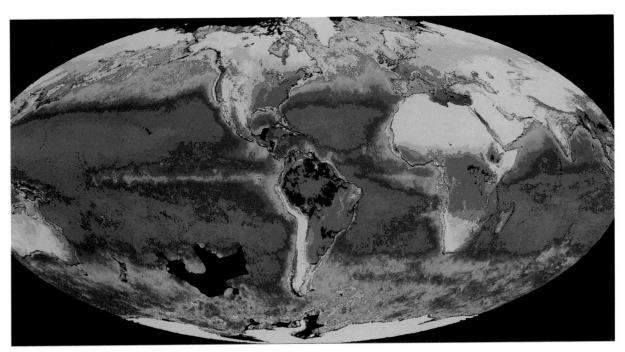

B. Global composite image. Productivity of oceans is shown by chlorophyll content (from CZCS). Vegetation cover of continents is shown by normalized difference vegetation index (from AVHRR).

Plate 18 Environmental remote sensing images.

A. Bands 1-2-3 = BGR.

B. Bands 4-5-7 = BGR.

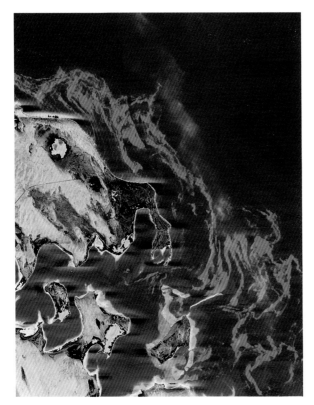

C. Bands 2-4-7 = BGR.

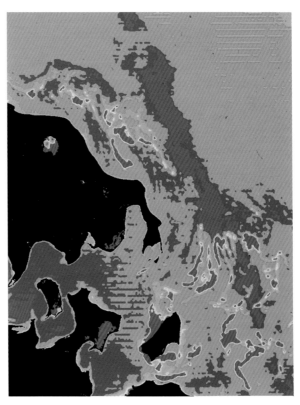

D. Thermal IR band 6 density-sliced. Red = warm; dark blue = cool.

Plate 19 Landsat TM subscenes of Arabian Gulf oil spill,
acquired February 16, 1991.

A. Landsat MSS image, Sudan.

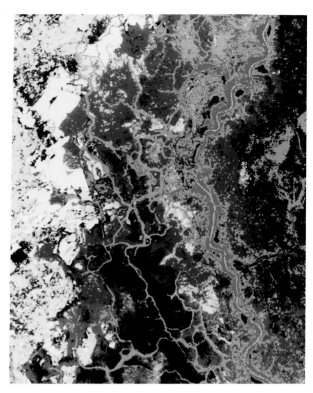

B. Supervised classification of MSS image, Sudan.

C. TM 2-4-7 image, Shaib Thamamah anomaly, Saudi Arabia.

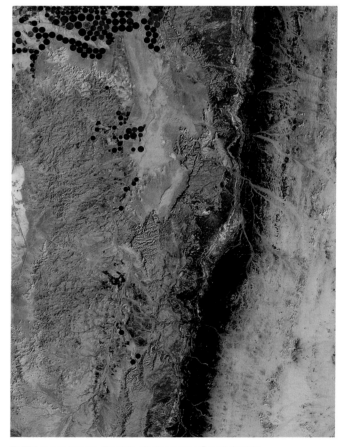

D. TM 2-4-7 image, Raghib oil discovery, Saudi Arabia.

Plate 20 Oil exploration images.

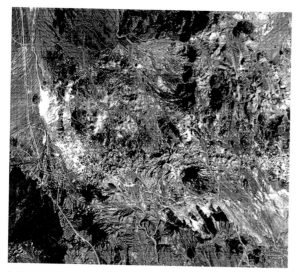

A. TM 1-2-3 normal color image.

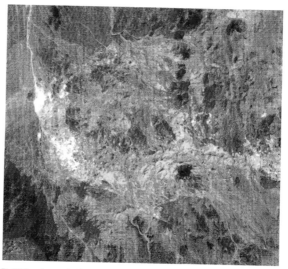

B. TM color ratio image. Ratio 5/7 = red, 3/1 = green, 3/5 = blue.

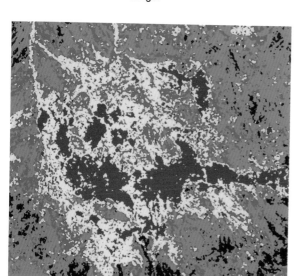

C. TM ratio 5/7 image with density slice. High ratio values shown in red.

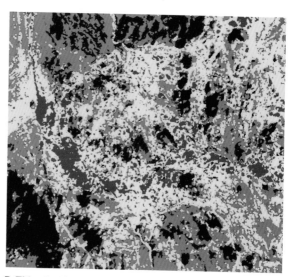

D. TM ratio 3/1 image with density slice. High ratio values shown in red.

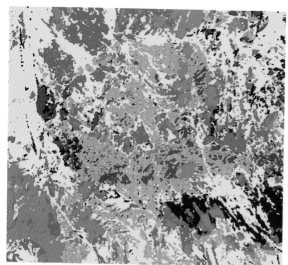

E. TM unsupervised classification map.

F. Color composite of AVIRIS endmember abundance images (Figure 11-18). Illite = blue, alunite = green, kaolinite = red.

Plate 21 Recognizing hydrothermally altered rocks at Goldfield Mining district, Nevada. Image F, courtesy F. A. Kruse, Analytical Imaging and Geophysics LLC, Boulder, Colorado.

Plate 22 Landsat TM 2-4-7 image merged with SPOT
pan image, Collahuasi mining district, Chile. Courtesy
J. D. Mancuso, Chevron Resources Company (Retired).

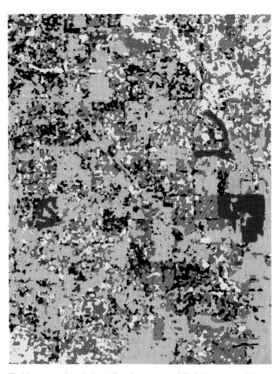

A. TM 2-3-4 image, Las Vegas, Nevada. Area is 10 km wide.

B. Unsupervised classification map of TM data, Las Vegas, Nevada. Colors are explained in Chapter 12.

C. Landsat MSS 1-2-4 image, Phoenix, Arizona, acquired in 1976. Area is 27 km wide.

D. Landsat MSS 1-2-4 image, Phoenix, Arizona, acquired in 1992 showing changes in land use.

Plate 23 Land use and land cover. Images C and D, courtesy R. H. Rogers, Environmental Research Institute of Michigan.

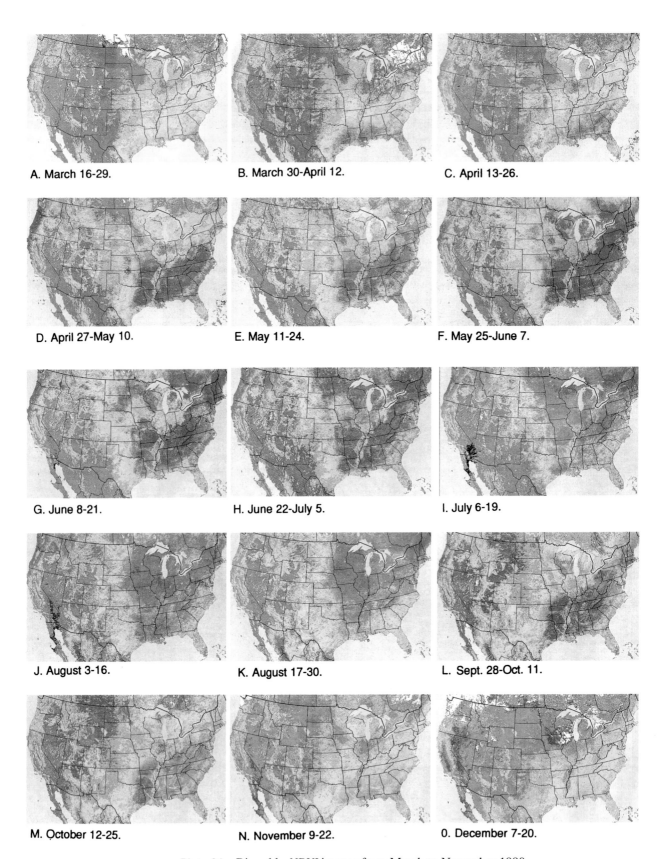

A. March 16-29.

B. March 30-April 12.

C. April 13-26.

D. April 27-May 10.

E. May 11-24.

F. May 25-June 7.

G. June 8-21.

H. June 22-July 5.

I. July 6-19.

J. August 3-16.

K. August 17-30.

L. Sept. 28-Oct. 11.

M. October 12-25.

N. November 9-22.

0. December 7-20.

Plate 24 Biweekly *NDVI* images from March to November 1990
of the conterminous United States highest and lowest *NDVI* values
are shown by dark green and light tan colors, respectively. From
Eidenshink (1992, Plate 1). Courtesy J. C. Eidenshink, U.S.
Geological Survey.

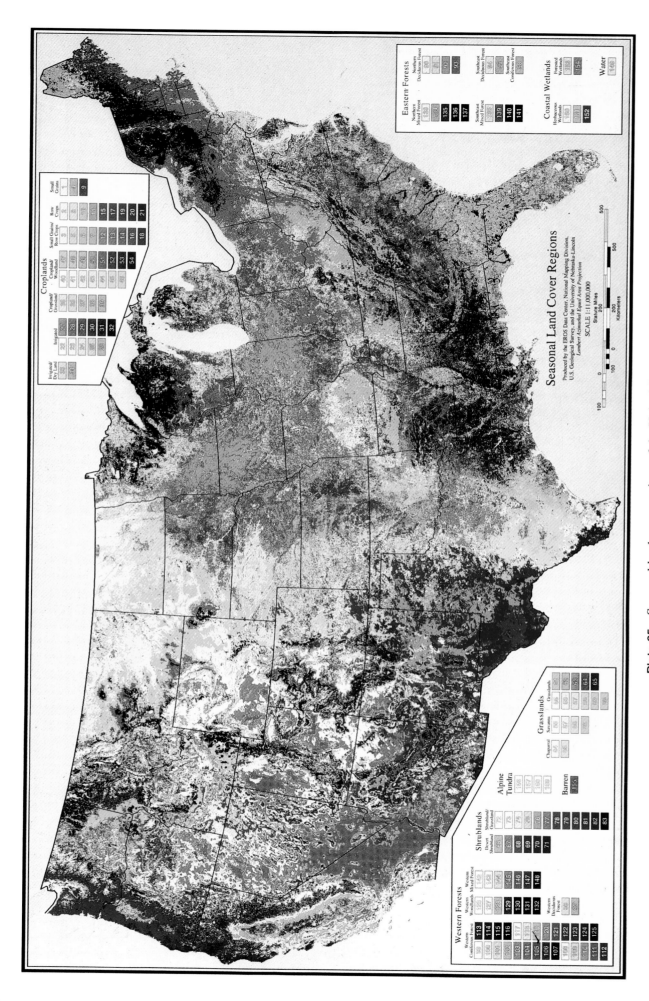

Plate 25 Seasonal land-cover regions of the United States from biweekly *NDVI* data for 1990. From U.S. Geological Survey (1993). Courtesy T. R. Loveland, U.S. Geological Survey.

A. Normal flow, May 27, 1989. TM 2-3-4 image.

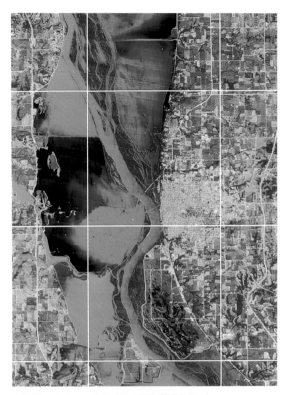

B. Flood stage, July 18, 1993. TM 2-3-4 image.

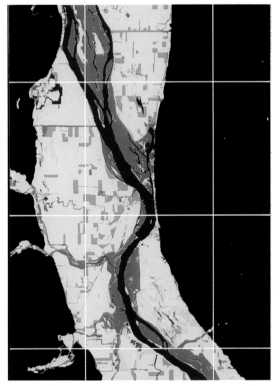

C. Classification map showing flooded and potentially flooded classes of land cover.

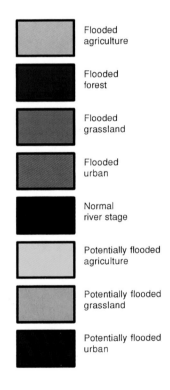

D. Explanation of classification map.

Flooded
agriculture

Flooded
forest

Flooded
grassland

Flooded
urban

Normal
river stage

Potentially flooded
agriculture

Potentially flooded
grassland

Potentially flooded
urban

Plate 26 GIS mapping of 1993 flood, Mississippi Valley.
Courtesy C. Erikson, ERDAS Corporation.

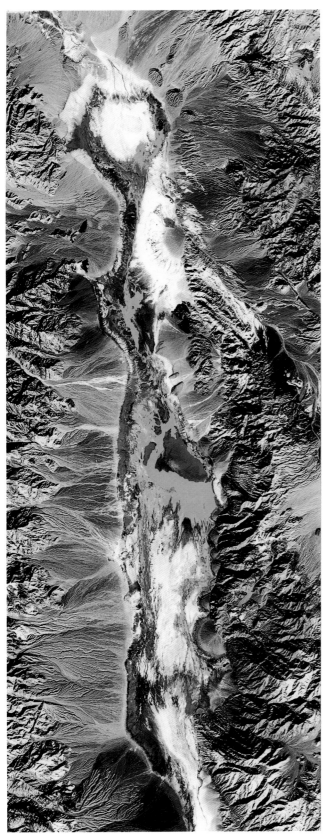

A. Landsat TM 2-4-7 image.

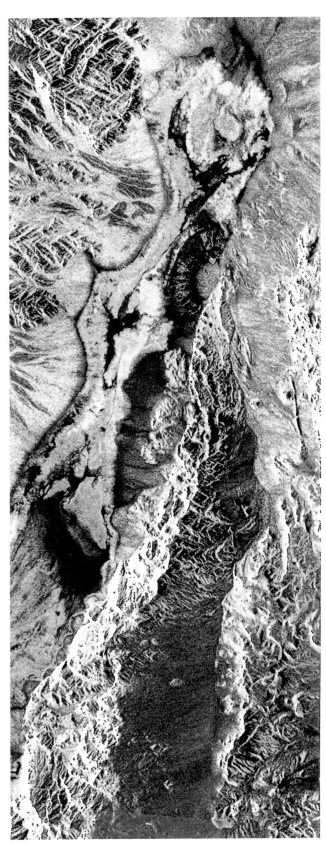

B. Color composite of SIR-C images (VV polarization).
Blue = X band, green = C band, red = L band.

Plate 27 Landsat and radar images, Death Valley.
Image B, courtesy A. Freeman and E. O'Leary, JPL.

A. TIMS image, western margin of Death Valley.

B. TM 2-4-7 image, western margin of Death Valley.

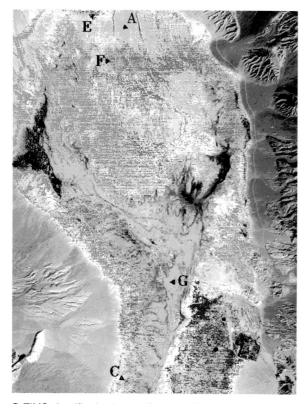

C. TIMS classification image, Cottonball Basin.

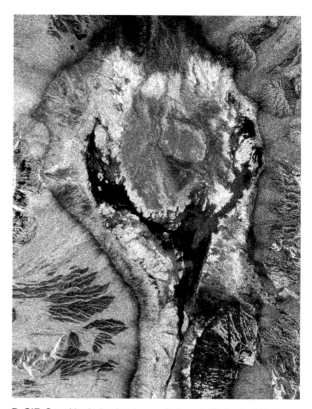

D. SIR-C multipolarization image (L-band), Cottonball Basin. Blue = VV, green = HV, red = HH.

Plate 28   TIMS images, Death Valley. Image A, from Kahle (1987, Figure 2). Courtesy A. B. Kahle, JPL. Image C, from Crowley and Hook (1996, Plate 1). Courtesy S. J. Hook, JPL. Image D, courtesy A. Freeman and E. O'Leary, JPL.

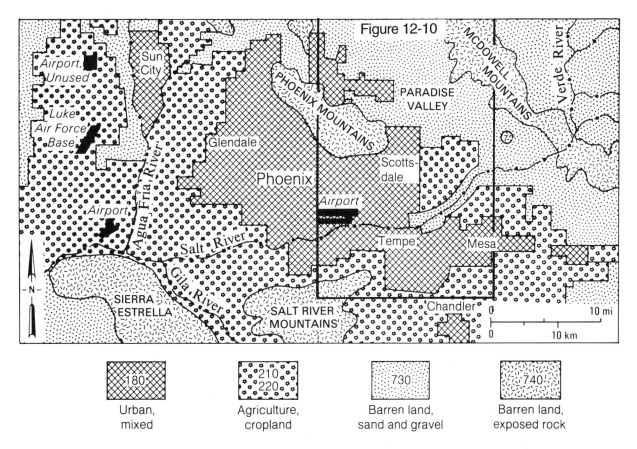

Figure 12-10 appears within the map image.

180	210 220	730	740
Urban, mixed	Agriculture, cropland	Barren land, sand and gravel	Barren land, exposed rock

Figure 12-9 Land use and land cover interpreted from the Seasat image of Phoenix, Arizona, and vicinity (Figure 12-8A).

Las Vegas. Sand and gravel (yellow) is unvegetated and has undergone little human disturbance. Open land (blue) has scattered vegetation and isolated buildings. Construction sites (gray) are a minor but distinctive category of open ground that is thoroughly disturbed by building activity. A good example occurs in the northeast corner of the images.

INTERPRETING LAND USE AND CHANGES IN PHOENIX, ARIZONA

The preceding examples of land-cover classification used images that recorded visible and reflected IR. Other wavelength regions are also useful for this purpose.

Land Use Interpreted from Radar Image

Figure 12-8 shows Seasat and Landsat images of the city of Phoenix, Arizona, and vicinity acquired in August 1978. Figure 12-9 shows level II categories of land cover that were interpreted from the Seasat image. Phoenix is located in a broad desert valley partly surrounded by mountain ranges. Much of

the desert was originally reclaimed for irrigated agriculture, but urban areas are expanding and replacing the croplands.

The "barren land" category is represented by sand and gravel (730) and exposed rock (740). In the Seasat image (Figure 12-8A), sand has a dark signature because the fine grain size forms smooth surfaces. Gravel occurs at the foot of the mountains and in the channels of the rivers and washes, all of which were dry when the images were acquired. Gravel has an intermediate to bright Seasat signature because of its coarse grain size and rough surface. Native trees and brush growing along the drainage channels also contribute to their bright radar signatures. The "exposed rock" category (740) is represented by the mountain ranges that are identified in Figure 12-9. The look direction for the Seasat image is toward the northeast; therefore, the southwest-facing slopes have bright highlights and the northeast-facing slopes have dark shadows. Radar layover causes the mountain crests to be displaced toward the west.

The cities and towns belong to the "mixed" category (180) and have intermediate to bright signatures on the Seasat image. One would expect the radar signature of the urban areas to be brighter than it is in Figure 12-8A because of the abundant

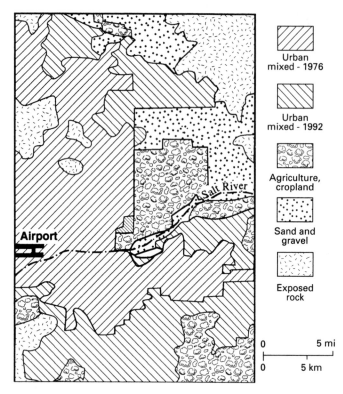

Figure 12-10 Changes in land use from 1976 to 1992 for the eastern portion of Phoenix, Arizona. Changes were interpreted from Landsat MSS images acquired in 1976 and 1992 (Plate 23C, D).

Urban
mixed - 1976

Urban
mixed - 1992

Agriculture,
cropland

Sand and
gravel

Exposed
rock

Airport

Salt River

0   5 mi

0   5 km

cotton, wheat, and alfalfa were the major crops in the eastern area. Vegetables, citrus fruits, and grapes along with some cotton and wheat were grown in the western area, which was more intensively irrigated. The relatively bright Seasat signatures of the western area may be caused by the different crops and soil moisture contents.

Changing Patterns of Land Use

Phoenix is one of the "sunbelt cities" in the southwestern United States that have experienced explosive urban growth in the past two decades. Repeated Landsat images are well suited to interpret the expansion of urban land use. Plate 23C,D shows MSS images acquired in 1976 and 1992 of the eastern portion of Phoenix and vicinity. The images coincide with a portion of the Seasat coverage as shown by the rectangle in Figure 12-9. Comparing the MSS images shows the extent and location of urban growth over a 13-year period. The map in Figure 12-10 shows level II land-use categories. The "mixed urban" category (180) is shown in contrasting patterns for 1976 and 1992. The "urban" category expanded by replacing adjacent agricultural land. The 1992 image clearly shows agricultural lands that are candidates for the next cycle of urban expansion. Repetitive images provide valuable information for regulating urban growth and planning the infrastructure of new schools, utilities, parks, and public services.

In this example, changes in land use were interpreted manually. Alternatively, each Phoenix image could have been digitally classified, such as in the example of Las Vegas. A change-detection algorithm applied to the registered 1976 and 1992 classifications would show graphically and quantitatively the changes in land use.

VEGETATION MAPPING WITH AVHRR IMAGES

Vegetation, both native and cultivated (agriculture), covers much of the earth and strongly influences the environment. Vegetation provides food, fiber, and building material. Until recently adequate data were lacking for mapping the composition, concentration, and dynamics of the world's vegetation. Now the advanced very high resolution radiometer (AVHRR) system on the NOAA environmental satellites provides worldwide coverage twice daily. The AVHRR was described in Chapter 4.

Normalized Difference Vegetation Index

Figure 12-11 shows spectral reflectance curves for soil and vegetation together with wavelength ranges of AVHRR bands 1 and 2, which were selected to record significant properties of vegetation. Band 1 (red, or R) records the absorption of red wavelengths by chlorophyll; lower values indicate higher chlorophyll content. Band 2 (reflected IR, or RIR) records the

corner reflectors in urban environments. The reduced brightness is explained by the orientation of the street patterns and buildings relative to the Seasat look direction. Street patterns in this Arizona image are oriented north-south and east-west; therefore, the northeast Seasat look direction is oblique to most buildings, which reduces the intensity of radar backscatter. This orientation effect was described in Chapter 6. The retirement community of Sun City (Figure 12-9) northwest of Phoenix is an exception to the orthogonal street pattern, because the streets are laid out in curved and circular patterns. The radar-bright patches within Sun City (Figure 12-8A) are caused by groups of houses oriented with some walls normal to the northeast radar look direction.

The dark linear features in the urban areas are major highways and irrigation canals. Airport runways (Figure 12-9) also have dark signatures. The narrow irregular dark features in Sun City are golf courses and parks with smooth lawns.

The irrigated agricultural areas surrounding Phoenix belong to the major categories of "cropland and pasture" (210) and "orchards, groves, vineyards, nurseries, and ornamental horticulture" (220). In the Seasat image, the fields east of Phoenix have a distinctly darker tone than the fields west of Phoenix. When the Seasat and Landsat images were acquired in 1978,

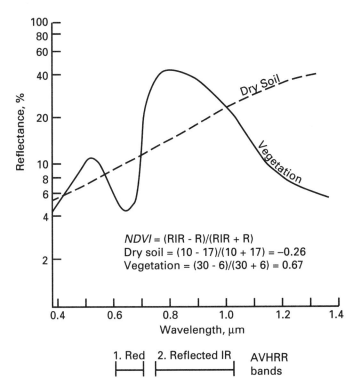

$$NDVI = \frac{(RIR - R)}{(RIR + R)}$$
Dry soil = (10 - 17)/(10 + 17) = -0.26
Vegetation = (30 - 6)/(30 + 6) = 0.67

1. Red   2. Reflected IR   AVHRR bands

Figure 12-11 Calculation of *NDVI* for vegetation and soil from AVHRR bands 1 and 2.

reflection of IR wavelengths by the cell structure of leaves; higher values indicate more vigorous growth. These bands may be combined in various mathematical formulas to produce vegetation indexes. Richardson and Everitt (1992) describe the eight more commonly used indexes. By far the most widely employed version is the *normalized difference vegetation index* (*NDVI*), which is defined as

$$NDVI = \frac{(RIR - R)}{(RIR + R)} \qquad (12\text{-}1)$$

Values for *NDVI* range from 1.0 to −1.0. Higher values indicate higher concentrations of green vegetation. Lower values indicate nonvegetated features, such as water, barren land, ice, snow, or clouds. For the vegetation spectrum in Figure 12-11, the *NDVI* is calculated as 0.67. For the dry soil spectrum, the *NDVI* is only −0.26. The *NDVI* is also useful because it largely compensates for differences in solar illumination.

The global image in Plate 18B shows concentrations of vegetation for the continents based on *NDVI* values. Dark-green signatures represent the highest values and orange the lowest. High concentrations of vegetation are shown in the equatorial regions of South America, Africa, and Southeast Asia. Vegetation patterns in the United States are clearly shown.

Vegetation Maps Using AVHRR

Until recently, AVHRR data at the full spatial resolution of 1.1 km were not readily available. Earlier studies used data that were resampled as Global Area Coverage (GAC) with 4-km pixels or as Global Vegetation Index (GVI) data with 16-km pixels. Tucker, Townshend, and Goff (1985) derived the *NDVI* from GAC data to map major vegetation types and seasonal changes for Africa over a 19-month period in 1982 and 1983. Townshend, Justice, and Kalb (1987) used GAC and GVI data of South America to evaluate different approaches for mapping land cover. Goward, Tucker, and Dye (1985) derived the *NDVI* from GVI data of North America at 3-week intervals from April through November 1982. Seasonal *NDVI* patterns were associated with major land-cover regions, and multidate images portray patterns of vegetation growth and senescence. Lloyd (1990) mapped worldwide vegetation cover by a supervised classification of multidate GVI data.

A few early analyses employed 1.1-km AVHRR data. Tucker, Gatlin, and Schneider (1984) used 1.1-km data of the Nile Delta acquired from May to October 1981. They noted changes in greenness that corresponded to known vegetation cycles and agricultural practices. Gervin and others (1985) compared 1.1-km data for the Washington region with Landsat MSS data. Rather than calculating the *NDVI*, they performed an unsupervised classification of level I land use using AVHRR bands 1 through 4 for a single image acquired in July 1981. The results were compared with the MSS classification. Overall accuracy was 72 percent for AVHRR and 77 percent for MSS.

Eidenshink (1992A) of the U.S. Geological Survey published an AVHRR mosaic of North America for the period August 11 to 20, 1990. Each 1.1-km pixel is shown with a color code for the highest *NDVI* value recorded during the 10-day period. This maximum value represents the peak of vegetation "greenness," which is a measure of photosynthetic activity. The mosaic is a graphic portrayal of vegetation vigor for that period on a continent-wide scale. Zhu and Evans (1994) of the U.S. Department of Agriculture Forest Service used 1.1-km *NDVI* data to classify the forests of the United States (including Alaska and Hawaii) into 25 categories. They also used this information to estimate the percentage of forest cover for the conterminous United States.

Biweekly *NDVI* Maps

Eidenshink (1992B) and associates at the EROS Data Center (EDC) compiled 19 biweekly *NDVI* images of the conterminous United States for the 1990 growing season (March 16 to December 20). Plate 24 shows 15 of these images that were selected to illustrate the seasonal variation of vegetation cover. These images, plus images for the period 1991 to 1995 and supporting information, are available on a CD-ROM from the EDC. The following section describes how Eidenshink (1992B) prepared the images.

Digital Processing The AVHRR data were resampled to 1.0-km pixels, which were processed at the EDC with the *Land Analysis System* (LAS) software described by Ailts and others (1990). Each biweekly *NDVI* image was produced by the following procedure.

1. **Scene selection** For each biweekly period all images without major cloud cover were selected, which typically amounted to 20 images. A single composite image was generated for each biweekly period, using the steps outlined below.
2. **Correcting for atmospheric scattering** The angular field of view for the AVHRR is 56° on either side of nadir. Toward either side of the image, the path length is much greater than at the nadir (directly beneath the satellite) and these longer paths are more severely affected by atmospheric scattering. In order to correct for atmospheric scattering, the relationship between solar illumination and satellite viewing geometry must be determined. These relationships are calculated from satellite orbital characteristics and used to correct for atmospheric scattering.
3. **Radiometric calibration** Performance of the AVHRR sensors is known to have degraded after launch because of exposure to the space environment. Data from bands 1 and 2 are calibrated using measurements of desert targets.
4. **Geometric registration** In order to produce a composite biweekly image the daily data sets (step 1) are registered to a common map projection; this ensures that each pixel is referenced to the correct ground location. A master reference image was prepared with a geographic error of less than 1.0 pixel. Automated correlation techniques are used to register all images to the reference image.
5. **Calculation of *NDVI*** For each daily mosaic, equation 12-1 was used to calculate the *NDVI* for each pixel.
6. **Image compositing** At this stage there are approximately 20 registered daily mosaics for each biweekly period. In other words, there are 20 *NDVI* values for each pixel in a biweekly period. For each pixel the maximum *NDVI* value is selected, which represents the peak greenness for the period.
7. **Products** For each biweekly period an image is produced that shows the maximum *NDVI* for each pixel. Plate 24 shows 15 of the images that are greatly reduced from the original scale (1:5,000,000). Maximum *NDVI* is shown in dark green; intermediate is yellow; minimum is brown and red. Water is light blue. Clouds are white. Statistical tables (not shown) are produced for each period and include a summary of the mean *NDVI* for each county in the United States.

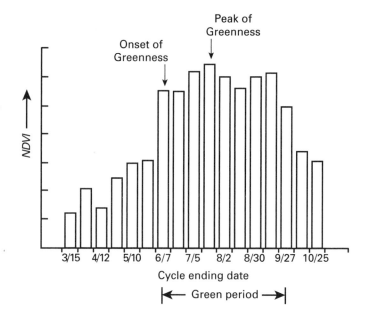

Figure 12-12 Derivation of greenness attributes from *NDVI* data of biweekly cycles. Total *NDVI* is the cumulative *NDVI* for the green period. From Loveland and others (1991, Figure 5).

The images in Plate 24 graphically show seasonal changes of vegetation patterns in the United States. In the spring of 1990 (Plate 24B) greening began in the Gulf and Pacific Coasts and expanded progressively inland. In mid-July and late August (Plate 24I,K) the northeastern United States was notably greener than the southeast portion, probably because of rainfall patterns. In the fall season, greenness diminished in a pattern that is essentially the reverse of the spring greening.

Wade and others (1994) used similar methods to prepare 12 biweekly *NDVI* maps of the USA for the growing seasons in both 1992 and 1993. In 1993 crops were affected by floods in the northeast USA and drought in the southeast. The year 1992 was normal. The two years were compared by computing difference images (Chapter 8). For each biweekly period the 1992 *NDVI* values were subtracted from the 1993 values on a pixel-by-pixel basis. The resulting difference maps graphically show the extent and severity of floods and droughts on crops.

Greenness Attributes from Biweekly NDVI Data Further processing of biweekly *NDVI* data derives additional vegetation information, called *greenness attributes*. Figure 12-12 is a histogram of the mean *NDVI* values for deciduous forests, plotted for each biweekly interval during the 1990 season. The following greenness attributes were derived from such plots and displayed as maps (U.S. Government Printing Office Document: 1993-556-415):

1. **Onset of greenness** The date when the *NDVI* first exceeds a threshold value, which occurs in early June for this exam-

ple. Onset of greenness corresponds to the emergence or greening of vegetation at the beginning of a growing season. Onset of greenness is effectively shown as a series of monthly maps. April is the time of major greening for much of the United States, followed by May for the northeast and north-central regions.

2. **Peak of greenness** The date when the maximum *NDVI* values occur, which is mid-July for Figure 12-12. Peak of greenness is the time of maximum photosynthetic activity, which is concentrated in May and June.

3. **Green period** The number of days that the *NDVI* exceeds a threshold level. The green period extends from early June to late September for the vegetation in Figure 12-12. The green period is basically the length of the growing season. There is a wide range in the green period for the conterminous United States. In the Gulf and Pacific Coasts the green period extends throughout the year, but is less than 3 months at higher elevations in the west and in the northern Great Plains.

4. **Total *NDVI*** The cumulative *NDVI* through the green period, which indicates total photosynthetic activity or net primary production.

Map of Seasonal Land-Cover Regions of the United States

Plate 25 is a map of the conterminous United States, called "Seasonal Land-Cover Regions," that was produced by scientists at the EDC and the University of Nebraska (Loveland and others, 1991). The land cover is classified according to the following hierarchy:

CATEGORIES—9 total
    Cover Types—23 total
        Seasonal classes—159 total
        (shown in colors in Plate 25)

The seasonal classes are the basic unit of the map. Each class represents a unique combination of vegetation/land cover types, seasonal properties, and relative primary production. The classes are grouped into cover types, which in turn are assigned to categories. A distinct hue is assigned to all the classes belonging to a cover type; increasing intensities of the hue represent increasing levels of annual primary production for individual classes. For example, in the "croplands" category, the "small grains" cover type includes the following classes in order of increasing productivity: (1) spring wheat (pale brown), (4) small grains (medium brown), and (9) winter wheat (dark brown).

The seasonal land-cover map is available at a scale of 1:7,500,000 as U.S. Government Printing Office Document 1993-556-415 from the U.S. Geological Survey in Denver, CO 80225 or Reston, VA 22092. Several additional maps are included with the document:

1. A level II classification version of Plate 25
2. A map showing the length of the green period for the conterminous United States
3. Monthly maps showing the onset of greenness
4. Monthly maps showing peak greenness

The following digital data are available in CD-ROM format:

1. 159 seasonal land-cover regions
2. Attributes that describe key spectral, vegetation, seasonal, and site properties of each seasonal land-cover region
3. Derived thematic land-cover sets, such as level II land cover
4. The biweekly AVHRR composite images and other source data used to develop the seasonal land-cover regions and attributes

The CD-ROMs are available from

EROS Data Center
Customer Services
Sioux Falls, SD 57198
Telephone: 605-594-6507

With a desktop computer and image-processing software, you can produce a range of output images from this geographically registered set of data. You can display relatively large-scale images of land-cover and greenness attributes for your state or county at 1-km resolution. With additional software you can process the data sets into different information combinations by employing the concept of geographic information systems.

GEOGRAPHIC INFORMATION SYSTEMS

A *geographic information system* (GIS) is an organized collection of computer hardware and software, with supporting data and personnel, that captures, stores, manipulates, analyzes, and displays all forms of geographically referenced information.

GIS Process

The GIS process consists of the following steps:

1. **Compile source data** The first step is to compile and prepare the various original geographic data sets, which may consist of multispectral images (Landsat, AVHRR), contour maps (topography), thematic maps (climate, soils, others), and/or tabular data (statistics). The tabular data and analog maps must be converted into digital format. Some data sets may consist of several adjacent maps that must be manually or digitally compiled into a mosaic that provides seamless coverage of the project area. At this stage the different data sets are in a raster format and cover the same area. The data sets may differ in scale, pixel size, and map projection.

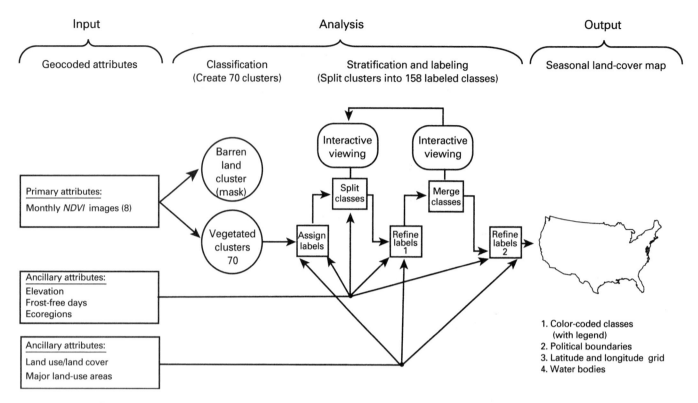

Input

Geocoded attributes

Primary attributes:
Monthly *NDVI* images (8)

Ancillary attributes:
Elevation
Frost-free days
Ecoregions

Ancillary attributes:
Land use/land cover
Major land-use areas

Analysis

Classification
(Create 70 clusters)

Stratification and labeling
(Split clusters into 158 labeled classes)

Barren land cluster (mask)

Vegetated clusters 70

Interactive viewing

Interactive viewing

Split classes

Merge classes

Assign labels

Refine labels 1

Refine labels 2

Output

Seasonal land-cover map

1. Color-coded classes (with legend)
2. Political boundaries
3. Latitude and longitude grid
4. Water bodies

Figure 12-13 GIS processing flow to produce seasonal land-cover classes. Modified from Brown and others (1993, Figure 1).

2. **Geocode source data** The next step is to *geocode* the data, which is the process of resampling each data set to a uniform pixel size that is registered to a geographic coordinate system. The compilation and geocoding of data normally require the major portion of time and effort in a GIS project. *Georeferencing* is an alternate term for geocoding.

3. **Derive attributes** After geocoding, some data sets, such as land-use/land-cover maps, are ready to use in the analysis stage. A data set that is suitable for analysis is called an *attribute*. Other data sets require additional processing to produce attributes. For example, a *digital elevation map* (DEM) is a raster array of pixels, each with an associated elevation value. The elevation information itself is an attribute, but it may be processed to provide additional useful attributes including steepness of slopes, orientation (aspect) of slopes, and solar illumination (the number of hours a pixel is exposed to the sun). Each spectral band of a multispectral image is an attribute. Combinations of bands may be processed to produce additional attributes such as band ratios and classification images. The collection of geocoded attribute data is often called *input data*.

4. **Analyze attribute data** The attribute data are digitally processed to produce the desired information, such as categories of land use. Several steps are normally required. In a

typical first step, selected sets of attribute data are classified using supervised and unsupervised algorithms. This classification process is also called *clustering* because it produces clusters of data with common characteristics. The initial clusters are rarely the desired final product, because the clusters may be too large and may contain diverse classes of data. The next step is to split the clusters into smaller homogeneous subdivisions; this process is also called *stratification*. Additional attributes may be employed in the splitting process, which may be repeated several times. The final step is to label each class with a descriptive name.

5. **Display results** The labeled classes are combined with ancillary information such as political boundaries or latitude and longitude coordinates. The final display of a GIS session is typically a map with the desired information shown in colors or patterns such as Plate 25. Tabulations of data are useful auxiliary information. The displayed results are also called *output data*.

Mapping Seasonal Land Cover with GIS

The following section describes how GIS was used to prepare an actual map, namely, the seasonal land-cover map (Plate 25). Figure 12-13 is a flowchart for the GIS process. The chart and

Table 12-3 Data sets and attributes employed in GIS processing of seasonal land-cover regions

Data set	Attributes	Source
Multitemporal *NDVI* data	Seasonal greenness	AVHRR images
Digital elevations	Height (20-ft resolution)	U.S. Defense Mapping Agency (1986)
Climate	Number of frost-free days	National Oceanic and Atmospheric Administration (1979)
Ecoregions	Name	Omernik (1987)
	Landform	Omernik and Gallant (1990)
	Potential natural vegetation	
	Land use	
	Soils	
Major land resource areas (MLRA)	Name	U.S. Department of Agriculture, Soil Conservation Service (1981)
	Land use	
	Elevation	
	Topography	
	Average annual precipitation	
	Average annual temperature	
	Average frost-free period	
Land use and land cover (LULC)	USGS classification levels	U.S. Geological Survey (1986)

Source: Brown and others (1993, Tables 1, 2).

the following description are summarized from the detailed account by Brown and others (1993). In Figure 12-13 the data flow from left to right in three major phases: (1) input, (2) analysis, and (3) output.

Input Phase The input phase includes all the steps in preparing the data for the analysis phase. Table 12-3 lists the data sets that were employed. The original data sets were in a wide range of digital and nondigital formats as reports, images, and maps at different scales and projections. In preparation for geocoding, the data sets were mosaicked and digitized. The biweekly AVHRR images of *NDVI* values (Plate 24) consist of 1-km pixels that are shown in a conic map projection. The other data sets were geocoded to match this pixel size and map projection. Walker and others (1996) described the process of preparing nondigital data for input to a GIS.

Table 12-3 lists the attributes that were derived from the geocoded data sets. The biweekly images of maximum *NDVI* (Plate 24) were resampled to derive eight monthly images for March through October 1990. In the input phase of Figure 12-13, the monthly *NDVI* images are shown as *primary attributes* because they are employed in the initial step of the analysis phase. Ancillary attributes are employed later in the analysis phase to modify results of the initial step. One set of ancillary attributes consists of elevation, number of frost-free days, and ecoregion. The second set consists of land use/land cover and major land-use areas.

Analysis Phase The analysis phase consists of a classification step and a subsequent stratification and labeling step.

1. **Classification** The maximum *NDVI* values are used to separate the pixels into two major clusters. One cluster contains pixels with consistently low *NDVI* values that represent nonvegetated, barren terrain (bare soil, water, clouds, snow, and ice). These pixels are set aside as a barren mask. The group of nonbarren pixels are processed with an unsupervised-classification algorithm to produce 70 clusters with seasonally distinct *NDVI* characteristics.

2. **Stratification and labeling** The two sets of ancillary attributes are used to assign labels to the 70 vegetated clusters. A review of statistical data for each cluster shows that 59 are multiclass clusters that include two or more cover types (such as agriculture plus deciduous forest). The next step is to stratify, or split, these multiclass clusters into distinct land-cover types, or classes. Figure 12-13 shows how three ancillary attributes (elevation, number of frost-free days, and ecoregion) are used in the stratification process.

Some examples illustrate the stratification process. Statistics show that original cluster 35 is a multiclass cluster consisting largely of winter wheat in the Great Plains and Pacific Northwest, but with a significant area of cool-season grasslands in California. Figure 12-14 is a *multitemporal NDVI spectrum* for cluster 35 that shows seasonal variation of the *NDVI*. The

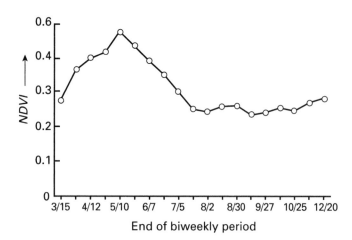

Figure 12-14 Multitemporal *NDVI* spectrum for the original cluster 35, which represents winter wheat and California grasslands. From Brown and others (1993, Figure 3).

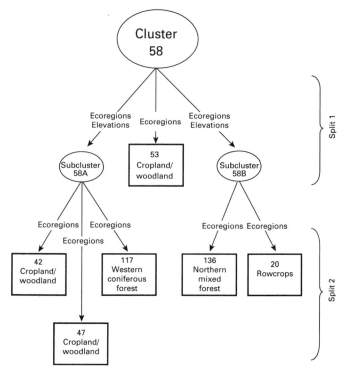

Figure 12-15 Stratification process used to split the original cluster 58 into six seasonal land-cover classes. Modified from Brown and others (1993, Figure 4).

annual greenness cycle of both winter wheat and California grasslands matches this spectrum. The winter wheat fields have a quick onset of greenness in April and May, senesce (ripen), and are harvested in late June. In California the maximum rainfall occurs in the winter and results in a similar phenologic cycle for grasses. This multiclass cluster is stratified in the following manner. The geographic distribution of pixels assigned to cluster 35 is viewed on an interactive display (Figure 12-13). The pixel map is merged with the ancillary-attribute map showing ecoregions. This combined display enables the analyst to split the cluster into two classes, based on geographic distribution. Pixels located within ecoregions along the Pacific Coast, with land cover of open woodland or chaparral, are stratified into the new class 62. The remaining pixels are assigned to new class 9. The next step is to label the new classes. Class 62 becomes "annual grasses, manzanita, oak, and white pine." The new class 9 is labeled "winter wheat."

Another original multiclass cluster includes both warm-season desert grasslands and alpine meadows. Desert grasslands that receive limited midsummer precipitation have a late onset of greenness, moderate peak greenness, and a short green period. These characteristics mimic the phenology of alpine meadows at high altitudes. The two classes are separated by incorporating the ancilliary attributes of elevation data and length of the frost-free period.

Twenty-seven original multiclass clusters contain only two classes and are adequately stratified with a single splitting operation. Thirty-two original clusters include three or more classes and require multiple splits. Figure 12-15 diagrams the stratification process for cluster 58, which includes natural, agricultural, urban, and mixed landscapes. The first split uses ecoregion and elevation attributes to produce three subclusters (53, 58A, 58B). Statistical data and interactive viewing show

that subcluster 53 is a discrete class labeled "cropland/woodland." Class 53 is located within the ecoregions of the southeastern plains and the mid-Atlantic and Gulf coastal plains with a land cover of oak, pine, soybeans, corn, cotton, and peanuts (Table 12-4). Interactive viewing of subclusters 58A and 58B shows that each contains multiple classes. Additional ecoregion evaluation is used to stratify subcluster 58A into classes 42, 47, and 117. Subcluster 58B is stratified into classes 20 and 136. Table 12-4 lists characteristics of the six classes that are stratified from the original cluster 58. Brown and others (1993) give details of the stratification process.

Once stratification and labeling are complete, smaller classes (<1000 pixels) are combined with similar larger classes to produce the 158 final seasonal land-cover classes shown in the map of seasonal land-cover regions (Plate 25).

Output Phase In the output phase, colors are assigned to the classes that are merged with political boundaries and map coordinates to produce the map in Plate 25. Because the data are geocoded, they are readily converted into alternate output displays. For example, the 158 seasonal land-cover classes were converted into 26 level II categories of land cover and printed as one of the maps in U.S. Government Printing Office Document 1993-556-415. The original concept of multilevel classification called for a spatial resolution of Landsat TM

Table 12-4 Seasonal land-cover classes stratified from cluster 58

Class no.	Cover type	Ecoregion	Primary vegetation types
20	Row crops	Northwest Great Lakes	Corn, soybeans, pasture
42	Cropland/woodland	Central and northeastern plains	Corn, soybeans, sorghum, irrigated agriculture, mixed woodlots
53	Cropland/woodland	Mid-Atlantic and Gulf coastal plains; southeast plains	Oak, pine, soybeans, corn, cotton, peanuts
47	Cropland/woodland	Appalachian Mountains; northeast uplands	Oak, hickory, pine, soybeans, corn, pasture
117	Western coniferous forest	Rocky Mountains	Ponderosa pine, lodgepole pine, western white pine, Douglas fir, aspen
136	Northern mixed forest	Northwest Great Lakes; northeast highlands	Oak, maple, ash, beech, birch, jack pine, red pine

Source: Modified from Brown and others (1993, Table 9).

(30 m) to recognize level II categories. However, a useful level II map was produced, despite the 1-km resolution of AVHRR data. Three factors contributed to this surprising result:

1. Multitemporal *NDVI* spectra from repeated images provide vital phenologic information that is lacking in land-cover classifications based on a single image, such as TM, despite its superior spatial (and spectral) resolution.
2. The seasonal land-cover map employed attributes in addition to remote sensing (Table 12-3) that greatly improve classification accuracy.
3. All attributes are used interactively, thanks to GIS technology. The analyst can concentrate on value-added interpretations, rather than managing data files.

Verification of Map Accuracy Verifying the accuracy of a classification map, such as Plate 25, is a complex issue. For instance, in cultivated regions the 1-km resolution of AVHRR exceeds the size of typical agricultural fields. Several different classes of crops may occur within a single AVHRR pixel. Therefore, any comparison between the AVHRR-derived map and crop statistics must be made on an aggregate, or large-area, basis rather than through a field-by-field analysis. Verification of accuracy is also inhibited by the limited availability of consistent independent information. Loveland and others (1991) did a preliminary evaluation of the map and concluded that the procedures used are, for the most part, acceptable. They also reported some unresolved issues. For example, the *NDVI* classification was clearly influenced by weather during 1990. California experienced drought, which undoubtedly affected the classification. The specific effects of climatic anomalies on seasonal classification maps are uncertain and require research.

AVHRR data were omitted from late November to March, which probably impaired classifications in the southeastern United States with its prolonged growing season. Barren lands, water bodies, and snow and ice are not separated because they lack chlorophyll-bearing vegetation. Some important classes, such as wetlands, typically occur in small patches that are not recognized by the coarse resolution of AVHRR.

Despite these and other problems, the ability to produce maps such as Plate 25 using satellite and ancillary data in a GIS environment is a major step forward in understanding our ecosystems. Scientists at the EDC and elsewhere are using this approach to prepare similar maps for the entire earth.

Sources of GIS Information

Like remote sensing, GIS is a new discipline; the term *GIS* was coined in the late 1960s. GIS technology and applications have expanded rapidly, in parallel with advances in remote sensing and computer technology and the increasing demand for environmental information. Table 12-5 lists representative GIS books published in the 1990s. Table 12-6 lists technical journals devoted to GIS.

Both government and commercial organizations are providing increasing amounts of data in digital format that are suitable for GIS use. Some major U.S. government sources are listed in Table 12-3. Many commercial vendors of GIS data and software systems advertise their products in the journals in Table 12-6.

ARCHAEOLOGY

In the context of this book, archaeology may be defined as the study of ancient land use. Remote sensing is becoming an accepted method for locating and mapping archaeologic sites. In

Table 12-5 Representative GIS books published in the 1990s

Author(s), date	Title	Publisher
J. C. Antenucci and others, 1991	GIS: a guide to the technology	Van Nostrand Reinhold, New York
W. J. Douglas, 1995	Environmental GIS	Lewis Publishers, Boca Raton, FL
Environmental Systems Research Institute, 1993	Understanding GIS: the Arc/Info method	John Wiley & Sons, New York
A. S. Fotheringham and P. Rogerson, eds., 1994	Spatial analysis and GIS	Taylor & Francis, Bristol, PA
R. Haines-Young and D. Green, eds., 1993	Landscape ecology and GIS	Taylor & Francis, Bristol, PA
A. I. Johnson, C. B. Pettersson, and J. L. Fulton, 1992	GIS and mapping: practices and standards	American Society for Testing Materials, Philadelphia, PA
G. Langrans, 1992	Time in GIS	Taylor & Francis, Bristol, PA
J. G. Lyon and J. McCarthy, eds., 1995	Wetland and environmental applications of GIS	Lewis Publishers, Boca Raton, FL
D. J. Maguire, M. F. Goodchild, and D. W. Rhind, eds., 1991	GIS: Vol. 1. Principles	John Wiley & Sons, New York
D. J. Maguire, M. F. Goodchild, and D. W. Rhind, eds., 1991	GIS: Vol. 2. Applications	John Wiley & Sons, New York
D. F. Marble and D. J. Peuquet, eds., 1990	Introductory readings in GIS	Taylor & Francis, Bristol, PA
P. M. Mather, ed., 1994	Geographical information handling	John Wiley & Sons, New York
D. Medyckyj-Scott and H. M. Hearnshaw, eds., 1993	Human factors in GIS	John Wiley & Sons, New York
M. Price, ed., 1994	Mountain environments and GIS	Taylor & Francis, Bristol, PA
W. J. Ripple, ed., 1994	The GIS applications book: examples in natural resources	American Society for Photogrammetry and Remote Sensing, Falls Church, VA
C. D. Tomlin, 1990	GIS and cartographic modeling	Prentice Hall, Englewood Cliffs, NJ
D. J. Unwin and H. M. Hearnshaw, eds., 1994	Visualization in GIS	John Wiley & Sons, New York
M. F. Worboys, ed., 1994	GIS: a computer perspective	Taylor & Francis, Bristol, PA
M. F. Worboys, ed., 1994	Innovations in GIS	Taylor & Francis, Bristol, PA

Table 12-6 GIS journals

Journal	Publisher
GIS World	GIS World, Inc. 155 E. Boardwalk Dr., Suite 250 Fort Collins, CO 80525
International Journal of GIS	Taylor & Francis Group 1900 Frost Road, Suite 101 Bristol, PA 19007

Europe many towns and roads of Roman and earlier times are now covered by agricultural fields. These sites commonly cause differences in character, moisture content, and vegetation cover of the overlying soils that are recognizable in IR color photographs.

The Lost City of Ubar

Ubar is the name of a legendary "lost" city in the Arabian Peninsula that most western historians regard as a myth. In the Islamic world, however, accounts of the city go back for thousands of years. According to those accounts Ubar was the major trading center for frankincense, a fragrant balm obtained from the sap of a desert tree. Frankincense was a highly prized commodity and was a gift from the Magi to the Christ child. It was used in perfume, medicine, incense, and especially in preparing bodies for cremation. Ubar virtually monopolized the frankincense trade, and the traders became fabulously wealthy, with a luxurious and decadent lifestyle. The Koran describes it as a "many-towered city . . . whose like has not been built in the entire land." In the fourth century A.D., however, the Emperor Constantine and the Roman Empire converted to Christianity; cremation went out of fashion and the lucrative frankincense market collapsed. According to the Koran, a windstorm sent by Allah buried the city under sand sometime between the first and fourth centuries A.D. Legend

Table 12-7 Issues of remote sensing journals devoted largely or wholly to AVHRR

Journal	Issue	Topic
International Journal of Remote Sensing	Vol. 7, no. 11, November 1986	Grasslands of Africa using AVHRR
Remote Sensing of Environment	Vol. 23, no. 2, November 1987	Remote sensing of arid rangeland
International Journal of Remote Sensing	Vol. 10, nos. 4 and 5, April/May 1989	Applications of AVHRR
Geocarto International	Vol. 7, no. 1, January 1992	Rangeland remote sensing
International Journal of Remote Sensing	Vol. 15, no. 17, November 1994	AVHRR global data sets for the land

therefore has it that Ubar was destroyed by God for a wicked lifestyle, but the real cause was probably economic.

Early in the 20th century, desert adventurers made futile attempts to locate Ubar. An Englishman named Bertram Thomas searched for the city in the Empty Quarter of the Arabian Peninsula on the basis of Bedouin folktales. Thomas drew a partial map of the caravan route he believed the original frankincense traders must have followed south from Mesopotamia. Thomas found traces of the caravan tracks, which he recorded on maps, but was unable to find the city. In the early 1950s Wendell Phillips of Phillips Petroleum Company attempted to search for Ubar in the dune fields but was defeated by the terrain. In the 1980s a documentary filmmaker named Nicholas Clapp studied Thomas's maps plus medieval Muslim histories and ancient maps, especially those of the Greek geographer Claudius Ptolemy, who lived in Egypt around A.D. 150. According to these sources, Ubar lay in a region along the border between Oman and Saudi Arabia.

Clapp learned of the sand-penetration capability of SIR-A radar images and approached Jet Propulsion Laboratory (JPL) scientists for help. During the SIR-B mission, images were acquired of the area, but the data were of marginal quality. The images of the gravel plains show traces of caravan tracks that must be ancient because portions are overlapped by dunes that required centuries to reach heights up to 200 m. The tracks occur in the area identified by Thomas. Because of the narrow swath width and low quality of the SIR-B images, the tracks could not be traced for any distance. An initial field expedition found some ancient tracks and concluded that, contrary to legend and Bertram Thomas, Ubar must be located south of the dune fields. At this point, it was decided to employ Landsat images.

Figure 12-16 is a Landsat TM subscene of the small village of Ash Shisar in west-central Oman that was digitally processed by R. Blom and R. Crippen of JPL. Figure 12-17 is an interpretation map. The terrain consists of alluvial deposits and rock outcrops. Three modern roads, shown by pronounced

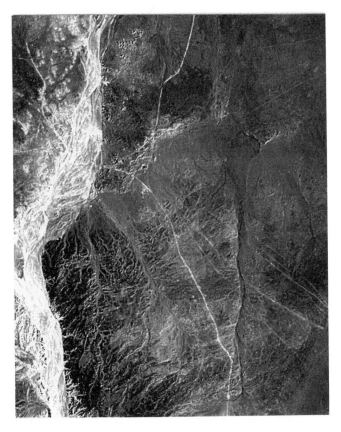

Figure 12-16 Landsat TM subscene in west-central Oman that covers the site of Ubar and the present village of Ash Shisar. Most of the area is a rocky gravel plain. Very fine bright lines are modern and ancient caravan trails. Courtesy R. J. Blom and R. E. Crippen, JPL.

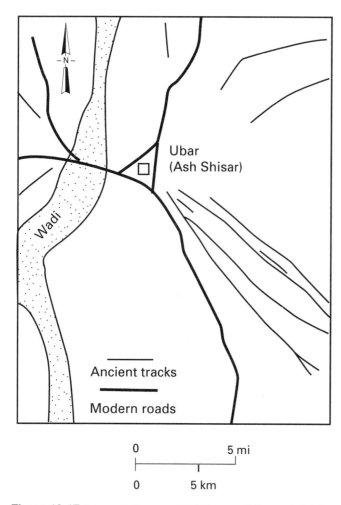

Figure 12-17 Interpretation map of TM image of Ubar and vicinity (Figure 12-16).

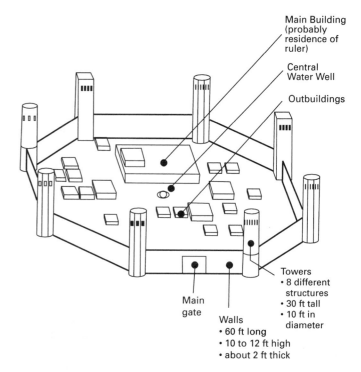

Figure 12-18 Archaeologic reconstruction of Ubar. Courtesy *Los Angeles Times*.

bright lines, intersect at Ash Shisar. Of much greater significance are the very fine, bright lines that converge on Ash Shisar. These fine lines represent both modern tracks and ancient trails. The number and extent of converging trails indicate a volume of ancient traffic that greatly exceeds the present activity. The initial field expedition had passed through Ash Shisar and noted that the town's ancient well is a large sinkhole with a vast amount of rubble collapsed in its center. Groves of frankincense trees grow in the Qara Mountains 160 km to the south.

These clues led to a second expedition to Ash Shisar that excavated the rubble pile. Archaeologists excavated the foundations of a fortress complex surrounded by a massive eightsided wall with 30-ft towers at each corner. A small city, dating back earlier than 2000 B.C., was located within the walls. Figure 12-18 is an archaeologic reconstruction of the fortress. Part of the complex was built over a limestone cavern that collapsed and destroyed Ubar, which perhaps gave rise to the legends. The ages of the youngest artifacts recovered indicate the

city collapsed between A.D 300 and 400. Most of this account was provided courtesy of R. J. Blom of JPL, who participated in the two expeditions. Aspaturian (1992) and Williams (1992) describe the history and discovery of Ubar.

The story of Ubar has several interesting facets. The actual remains of the fortress are not expressed on the TM image. Instead, the location was indicated by the pattern of ancient trails converging on the site that were revealed on the digitally enhanced TM image (Figure 12-16). The early maps and accounts of Ubar, which had largely been discounted as myths, proved to be surprisingly accurate. Remote sensing provided the key evidence for locating the ancient site.

Other Archaeologic Applications

In northern Arizona, cornfields cultivated by prehistoric Indians have been recognized in the pattern of alternating warm and cool strips of land in daytime thermal IR images. Fieldwork by Berlin and others (1977) showed that the warm strips correlate with bands of volcanic ash used as mulch by the Indian farmers. The low-density ash has a lower thermal inertia than adjacent soil and therefore warmer daytime radiant temperatures.

Archaeologists once thought that the Mayan civilization in Yucatan practiced a slash-and-burn system of agriculture, but this system could not have supported the estimated Mayan

	Percent reflectance		
Material	*Band 1*	*Band 2*	*NDVI*
Damp soil	5	11	———
Evergreen foliage	6	15	———

population. Aerial reconnaissance flights and ground investigations of the region were hampered by swampy terrain with dense vegetation cover and persistent cloud cover. Aircraft radar surveys (Adams, Brown, and Culbert, 1981) revealed an extensive network of canals by which the Maya drained the swamps for cultivation. The region is now overgrown with jungle, but on the radar images the canals are expressed as narrow gray lines surrounded by bright signatures of tropical vegetation. The canals are also covered by vegetation, but the canopy height is lower there than in the surrounding area, which accounts for the darker signature. Joyce (1992) describes additional archaeologic applications of remote sensing.

COMMENTS

Land use and land cover are readily interpreted using the multilevel classification system and remote sensing images (Landsat, SPOT, and aerial photographs). Radar images are also useful. These classifications cover areas ranging up to states as large as Alaska in size at scales of 1:1,000,000 and larger. Sequential images are used to recognize and project changes in land-use patterns. For national and global interpretations at smaller scales, AVHRR images are appropriate, especially for mapping vegetation (Table 12-7). Multitemporal *NDVI* spectra are used to classify vegetation based on seasonal variations. Geographic information systems (GIS) merge images with other attributes to produce specialized maps for many applications. Many archaeologic sites are recognizable on images.

QUESTIONS

1. For your area, assemble a set of images for levels I, II, and III (images from Landsat TM and MSS, a photograph from the National High Altitude Photography Program, and a larger-scale aerial photograph). Use Table 12-2, and prepare land-use maps for each of the three levels.
2. The kinds of images mentioned in question 1 are all acquired in the visible and reflected IR bands. Images acquired in other spectral bands may also be employed to analyze land use. Examine the thermal IR image of Ann Arbor, Michigan (Chapter 5) and the radar images of New Orleans and the Imperial Valley, California (Chapter 6). Evaluate the advantages and disadvantages of these images relative to visible-band images for land-use interpretation.
3. For your area, assume that the population will double in the next 5 years. Use the images and maps of question 1 as a base, and develop a plan for adding this population. You must provide for all the facilities required for this growth.
4. Calculate the normalized difference vegetation index for materials with the following reflectance characteristics on AVHRR:

REFERENCES

Adams, R. E. W., W. E. Brown, and T. P. Culbert, 1981, Radar mapping, archaeology, and ancient Maya land use: Science, v. 213, p. 1457–1463.

Ailts, B., D. Akkerman, B. Quirk, and D. Steinwand, 1990, LAS 5.0—an image processing system for research and production environments: Proceedings, American Society for Photogrammetry and Remote Sensing/American Congress of Surveying and Mapping, Annual Convention, v. 4, p. 1–12, Falls Church, VA.

Anderson, J. R., E. T. Hardy, J. T. Roach, and R. E. Witmer, 1976, A land use and land cover classification system for use with remote sensor data: U.S. Geological Survey Professional Paper 964.

Aspaturian, H., 1992, The road to Ubar: Caltech News, v. 26, p. 1–8, Pasadena, CA.

Berlin, G. L., J. R. Ambler, R. H. Hevly, and G. G. Schaber, 1977, Identification of a Sinagua agricultural field by aerial thermography, soil chemistry, pollen/plant analysis, and archaeology: American Antiquity, v. 42, p. 588–600.

Brown, J. F., T. R. Loveland, J. W. Merchant, B. C. Reed, and D. O. Ohlen, 1993, Using multisource data in global land cover characterization—concepts, requirements and methods: Photogrammetric Engineering and Remote Sensing, v. 59, p. 977–987.

Eidenshink, J. C., 1992A, The North American vegetation index map: U.S. Geological Survey, Reston, VA.

Eidenshink, J. C., 1992B, The 1990 conterminous U.S. AVHRR data set: Photogrammetric Engineering and Remote Sensing, v. 58, p. 809–813.

Florida Bureau of Comprehensive Planning, 1976, The Florida land use and cover classification system: Florida Bureau of Comprehensive Planning Report DSP-BCP-17-76, Tallahassee, FL.

Gervin, J. C., A. G. Kerber, R. G. Witt, Y. C. Lu, and R. Sekhon, 1985, Comparison of level I land cover accuracy for MSS and AVHRR data: International Journal of Remote Sensing, v. 6, p. 47–57.

Goward, S. N., C. J. Tucker, and D. G. Dye, 1985, North American vegetation patterns observed with the NOAA advanced very high resolution radiometer: Vegetation, v. 64, p. 3–14.

Joyce, C., 1992, Archaeology takes to the skies: New Scientist, v. 25, p. 42–46.

Lloyd, D., 1990, A phenological classification of terrestrial vegetation using shortwave vegetation index imagery: International Journal of Remote Sensing, v. 11, p. 2269–2279.

Loveland, T. R., J. W. Merchant, D. O. Ohlen, and J. F. Brown, 1991, Development of a land-cover characteristics data base for the conterminous U.S.: Photogrammetric Engineering and Remote Sensing, v. 57, p. 1453–1463.

Myers, V. I. and others, 1983, Remote sensing applications in agriculture *in* Colwell, R. N., ed., Manual of remote sensing, second edition: ch. 33, p. 2111–2228, American Society for Photogrammetry and Remote Sensing, Falls Church, VA.

National Oceanographic and Atmospheric Administration, 1979, Climatic atlas of the United States: U.S. Department of Commerce, National Oceanic and Atmospheric Administration, Environmental Data Services, Asheville, NC.

Omernik, J. M., 1987, Ecoregions of the conterminous United States: Annals of the American Association of Geographers, v. 77, p. 118–125.

Omernik, J. M. and A. L. Gallant, 1990, Defining regions for evaluating environmental resources: Proceedings, Global Natural Resources Monitoring and Assessment Symposium, p. 936–947, Bethesda, MD.

Richardson, A. J. and J. H. Everitt, 1992, Using spectral vegetation indices to estimate rangeland productivity: Geocarto International, v. 1, p. 63–69.

Townshend, J. R. G., C. O. Justice, and V. Kalb, 1987, Characterization and classification of South American land cover types: International Journal of Remote Sensing, v. 8, p. 1189–1207.

Tucker, C. J. , J. A. Gatlin, and S. R. Schneider, 1984, Monitoring vegetation in the Nile Delta with NOAA-6 and NOAA-7 AVHRR imagery: Photogrammetric Engineering and Remote Sensing, v. 50, p. 53–61.

Tucker, C. J., J. R. G. Townshend, and T. E. Goff, 1985, African land-cover classification using satellite data: Science, v. 227, p. 369–375.

U.S. Defense Mapping Agency, 1986, Defense Mapping Agency specifications for digital terrain elevation data (DTED), second edition: Washington, DC.

U.S. Department of Agriculture, Soil Conservation Service, 1981, Land resource regions and major land resource areas of the United States: Agriculture Handbook 296, Washington, DC.

U.S. Geological Survey, 1986, Land use and land cover digital data from 1:250,000- and 1:100,000-scale maps: U.S. Geological Survey Data Users Guide 4, Reston, VA.

Wade, G., R. Mueller, P. Cook, and P. Doralswarmy, 1994, AVHRR map products for crop conditions assessment—a geographic information systems approach: Photogrammetric Engineering and Remote Sensing, v. 60, p. 1145–1150.

Walker, J. D. and others, 1996, Development of geographic information systems—oriented databases for integrated geological and geophysical applications: GSA Today, v. 6, no. 3, p. 1–7.

Williams, R. J., 1992, In search of a legend—the lost city of Ubar: Point of Beginning, v. 17, p. 10–18.

Zhu, Z. and D. L. Evans, 1994, U.S. forest types and predicted percent forest cover from AVHRR data: Photogrammetric Engineering and Remote Sensing, v. 60, p. 525–531.

ADDITIONAL READING

Bonham-Carter, G. F., 1994, Geographic information systems for geoscientists: Pergamon Press, Tarrytown, NY.

Coppin, P. R. and M. E. Bauer, 1996, Digital change detection in forest ecosystems with remote sensing imagery: Remote Sensing Reviews, v. 13, p. 207–234.

Curran, P. and others, 1990, Remote sensing of soils and vegetation: Taylor and Francis, Bristol, PA.

Ebert, J. I. and others, 1983, Archaeology, anthropology, and cultural resources management in Colwell, R. N., ed., Manual of remote sensing, second edition: ch. 26, p. 1233–1304, American Society for Photogrammetry and Remote Sensing, Falls Church, VA.

Eden, M. J. and J. T. Parry, 1986, Remote sensing and tropical land management: John Wiley & Sons, New York, NY.

Jensen, J. R. and others, 1983, Urban/suburban land use analysis in Colwell, R. N., ed., Manual of remote sensing, second edition: ch. 30, p. 1571–1666, American Society for Photogrammetry and Remote Sensing, Falls Church, VA.

Kessler, B. L., 1994, Glossary of GIS terms in W. J. Ripple, ed., The GIS applications book: p. 26–34, American Society for Photogrammetry and Remote Sensing, Falls Church, VA.

Lark, R. M., 1995, Components of accuracy of maps with special reference to discriminant analysis on remote sense data: International Journal of Remote Sensing, v.16, p. 1461–1480.

Loelkes, G. L., G. E. Howard, E. L. Schwartz, P. D. Lambert, and S. W. Miller, 1983, Land use/land cover and environmental photointerpretation keys: U.S. Geological Survey Professional Paper 1600.

Loveland, T. R., and D. O. Ohlen, 1991, A strategy for large-area land characterization—the conterminous U.S. example: Proceedings, U.S. Geological Survey Global Research Symposium Forum, Reston, VA.

Townshend, J. R. G., C. O. Justice, W. Li, C. Gurney, and J. McManus, 1991, Global land cover classification by remote sensing—present capabilities and future possibilities: Remote Sensing of Environment, v. 35, p. 243–255.

Welch, R., M. Remillard, and J. Alberts, 1992, Integration of GPS, remote sensing, and GIS for coastal resource management: Photogrammetric Engineering and Remote Sensing, v. 58, p. 1571–1578.

Z. Zhu and L. Yang, 1996, Characteristics of the 1 km AVHRR data set for North America: International Journal of Remote Sensing, v. 17, p. 1915–1924.

13

NATURAL HAZARDS

Earthquakes, landslides, volcanic eruptions, fires, and floods are natural hazards that kill thousands of people and destroy billions of dollars of habitat and property each year. These losses will increase as world population increases and more people reside in areas that are subject to these hazards. Dams can control flood hazards, and proper engineering design can reduce landslide risks. Aside from steps such as these, there is little that people can do to prevent the occurrence of natural hazards. However, the following actions will minimize their effects:

1. Analyze the risk that natural hazards will occur in a given area. Examples are to identify faults or volcanoes that have the potential for earthquakes or eruptions. In addition to recognizing hazards, risk analysis should delineate areas on the basis of their relative susceptibility to damage.
2. Provide advance warning for specific hazardous events, which is not practical for many hazards. Satellite images provide timely warnings of floods and severe weather. Volcanic eruptions in Hawaii and elsewhere have been predicted on the basis of ground movements.
3. Assess the damage caused by a hazardous event. An early evaluation of damage caused by floods and earthquakes is essential for carrying out rescue, relief, and rehabilitation efforts.

Remote sensing is becoming increasingly valuable for analyzing, warning about, and assessing damage related to natural hazards.

EARTHQUAKES

Earthquakes are caused by the abrupt release of strain that has built up in the earth's crust. Most zones of maximum earth-quake intensity and frequency occur at the boundaries between the moving plates that form the crust of the earth. Major earthquakes also occur within the interior of crustal plates such as those in China, Russia, and the southeast United States. Much research has been done to predict earthquakes, using non–remote sensing technologies, but results to date are inconclusive. *Seismic risk analysis,* however, is an established discipline that estimates the geographic distribution, frequency, and intensity of seismic activity without attempting to predict specific earthquakes. This analysis is essential for locating and designing dams, power plants, and other projects in seismically active areas.

One method of seismic risk analysis is based on the study of *historic earthquakes,* which are those recorded by humans. These records cover some 2000 years in Japan and 3000 years in China but are briefer in other regions. In southern California, for example, the earliest historic earthquake was recorded in 1769. Beginning in the 1930s earthquakes have been recorded by instruments called *seismographs.* Both the historic and instrumental records are too brief to make valid predictions of earthquakes.

The second method of seismic risk analysis is based on the recognition of *active faults,* which are defined as breaks along which movement has occurred in late Quaternary, or Holocene, time (the past 11,000 years). Remote sensing analyses and field studies of active faults provide a geologic record that extends our instrumental and historic records. Surface faulting during large shallow earthquakes is more universal than has been recognized; analysis of this geomorphic evidence and radiometric age dating of earlier events are two techniques that have not been fully utilized (Allen, 1975). Remote sensing images are now facilitating the recognition and analysis of active faults, as shown by examples from California, China, and the seafloor.

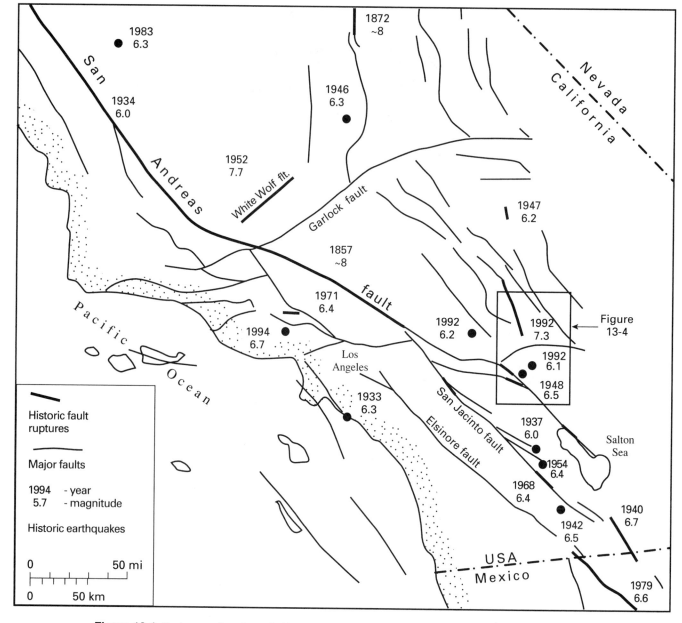

Figure 13-1 Fault map of southern California showing dates and magnitudes of historic earthquakes (magnitude >6.0). Darker lines indicate historic fault ruptures. Dots indicate earthquakes with no surface ruptures. Modified from Hutton and others (1991, Figure 2).

Southern California

Southern California is an ideal region to demonstrate remote sensing for seismic risk analysis:

1. The region is seismically active.
2. The Southern California Seismic Network (SCSN) records details of earthquakes, with an on-line computer catalog extending back to 1932.

3. Many active faults are well exposed in the mountains and desert.
4. A variety of images are available and have been interpreted by investigators.

Figure 13-1 is a map of the region showing the major faults, many of which are active. The heavy lines identify fault segments that have ruptured in historic time. The dates and magnitudes of major historic earthquakes (having magnitudes of 6.0

and greater on the Richter scale) are shown. Many of these earthquakes are associated with the active fault breaks. Dots indicate earthquakes that did not rupture the surface, although several occurred along the traces of active faults.

Characteristics of Active Faults Evidence for Holocene movement includes (1) historic earthquakes, (2) rock units younger than 11,000 years that are faulted, and (3) certain topographic features caused by faulting. Figure 13-2 is a diagram of an active fault zone that may be hundreds of kilometers in length and several kilometers in width. A *fault trace* is the surface expression of an individual fault. Figure 13-2 also shows typical topographic features formed by active strike-slip faults, while Figure 13-3 is a satellite photograph of typical features along the active Garlock strike-slip fault in the Mojave Desert of California. These topographic features are formed by horizontal and vertical displacements along faults. *Sag ponds* are lakes that occupy structural depressions within the fault zone. In arid environments these are dry lakes. *Shutter ridges* are topographic ridges that have been offset laterally to shut off drainage channels. Other narrow fault blocks are called *benches* and *linear ridges*. *Scarps* are the surface expression of fault planes; they may cut a topographic ridge to form a *faceted ridge*. *Springs* form where faults block the movement of groundwater, causing it to emerge at the surface. In arid terrain, *vegetation anomalies* are strips of vegetation that are concentrated along faults because of shallower groundwater. *Linear valleys* result from increased erosion of fractured rocks along a fault. *Offset drainage channels* are especially significant because they also indicate the sense and amount of lateral displacement along a strike-slip fault. The left-lateral, strike-slip displacement of the Garlock fault is indicated by the offset drainage channels in Figure 13-3. The presence of these topographic features shows that a fault is active; had the features formed before the Holocene epoch, most of them would have been obliterated by erosion and deposition.

Landsat and SPOT images are well suited for recognizing the continuity and regional relationships of faults as well as many local details. Stereo viewing of SPOT images and aerial photographs provides detailed information on topographic features formed by faulting. Highlights and shadows on low-sun-angle aerial photographs can emphasize topographic scarps associated with active faults, as shown in Chapter 2. Thermal IR images of arid and semiarid areas may record the presence of active faults with little or no surface expression, such as the San Andreas and Superstition Hills faults, which were interpreted in Chapter 5. Radar images have highlights and shadows that enhance the expression of faults, even in forested terrain, as shown in Chapters 6 and 7.

Relationship between Faults and Earthquakes Allen (1975) of Caltech noted the following relationships between faulting and earthquakes, using California as an example:

1. Virtually all large earthquakes (with magnitudes greater than 6.0) have resulted from ruptures along faults that had been recognized by field geologists prior to the events. A notable exception is the 1994 Northridge earthquake, which occurred along a fault that is concealed beneath a cover of sedimentary rocks.
2. All of these faults have a history of earlier displacements in Quaternary and possibly Holocene times.
3. All the earthquakes have been relatively shallow, not exceeding about 20 km in depth. Most earthquakes larger than magnitude 6.0 have been accompanied by surface faulting, as have many of the smaller events.
4. The larger earthquakes have generally occurred on the longer faults, although there has been sufficiently wide variation to indicate caution in blindly applying any single formula for this relationship.
5. Generally only a small segment of the entire length of a fault zone has broken during any single earthquake, although there are some conspicuous and significant exceptions.

These relationships are clearly seen on images that have been merged with earthquake data and maps of active faults.

Landsat Regional Mosaic Plate 13 is a digital mosaic of Landsat TM images of southern California prepared by R. E. Crippen of Jet Propulsion Laboratory (JPL). The base image is a color composite of TM band 7 shown in red, band 4 in green, and the average of bands 1 and 2 shown in blue. Crippen used GIS methods to merge two attributes with the mosaic:

1. Traces of active faults were digitized from the map by Jennings (1975) and registered to the mosaic, where they are shown as yellow lines.
2. A digital database of earthquakes recorded from 1970 to 1995 by the SCSN was registered to the mosaic. In the earthquake scale, magnitudes of 5 or greater are shown by white circles of increasing diameter, 2 to 4 in red circles, and 1 in red dots.

Plate 13 shows the relationship between active faults and earthquakes that was summarized earlier. Most of the major earthquakes (large white circles) are associated with active faults (shown in yellow), which form topographic lineaments on the image. A few large earthquakes occurred in the greater Los Angeles Basin, where the faults are concealed beneath relatively young sediments. Many of the smaller earthquakes (shown in red) are aftershocks of major events and are clustered around the major epicenters. Many other small events are concentrated in linear belts along faults and represent small releases of energy.

Landers, California, Earthquake

Early on the morning of June 28, 1992, millions of southern Californians were awakened by the largest earthquake in the

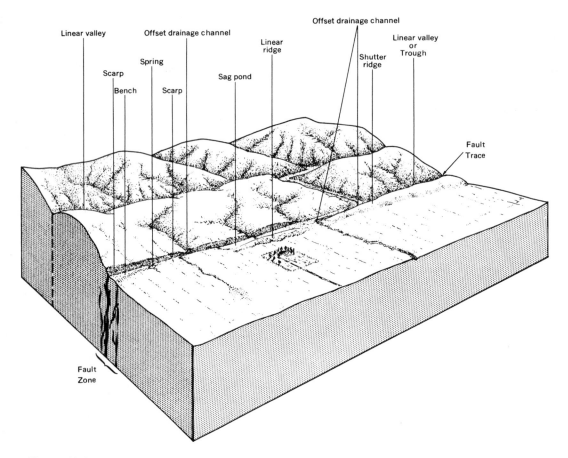

Figure 13-2 Typical topographic features along an active strike-slip fault. From Vedder and Wallace (1970).

western United States in the past 40 years. The magnitude 7.3 quake began at the town of Landers in the Mojave Desert (Figure 13-4) and caused ruptures to the north and northwest along active faults. Fortunately, the strongest shaking occurred in sparsely inhabited regions, but one person was killed and 400 were injured. Property damage exceeded $100 million. Three hours after the Landers earthquake, a second damaging quake (magnitude 6.5) occurred on a separate fault near Big Bear Lake, 35 km west of Landers. These events were preceded on April 23, 1992, by the Joshua Tree earthquake and succeeded by many aftershocks. The SCSN provided a detailed record of the seismic activity (Hauksson and others, 1993). The earthquake was studied by many earth scientists, including Sieh and others (1993), which is the source of the following summary.

Surface Faulting

Figure 13-4 shows the active faults and the earthquakes (magnitude greater than 1.8) recorded from January 1 to August 18, 1992, in the Landers region. The correspondence between earthquakes and faults agrees with Allen's assessments, cited earlier. The Landers earthquake resulted from right-lateral slip on faults within a broad zone that is 70 km long. The total length of the overlapping fault strands is 85 km. All but the Landers fault had been mapped before the quake. Landforms that are characteristic of active faults (Figure 13-2) were present along the faults prior to the earthquake. These landforms are less common and more eroded than seismic landforms along the San Andreas fault, where ruptures occur every one to two centuries. This comparison suggests that the last major ruptures in the Landers region occurred at least several thousand years ago. This long interval, during which stress accumulated, may account for the high stress drop of the Landers earthquake. All of the Landers fault and most of the Homestead Valley fault ruptured, but only portions of the other faults were offset in the Landers earthquake. Up to several meters of right-lateral offset were typical; the maximum slip of 6 m equals that of the largest surficial strike-slip dislocation of the 20th century in the Western Hemisphere. Vertical displacements were also common and in several places exceeded 1 m.

Figure 13-3 Satellite photograph of the topographic features along the left-lateral, strike-slip Garlock fault, in the Mojave Desert of southern California. The following features indicate that the fault is active: D = depression, SR = shutter ridge, OC = offset channel, LR = linear ridge, LV = linear valley, FR = faceted ridge. From Merifield and Lamar (1975, Figure 2). Courtesy P. M. Merifield, UCLA.

Aside from these large breaks, most of the surface offsets are smaller and occur as multiple cracks in zones up to several hundred meters wide that cross the desert floor (Mori and others, 1992). SPOT and radar images were digitally processed to help analyze surface effects of the earthquake.

SPOT Images Figure 13-5 shows subscenes of SPOT panchromatic images that are located north of the area shown in Figure 13-4. The images cover a portion of the Emerson fault, which ruptured during the Landers earthquake. The image in Figure 13-5A was acquired eleven months before the earthquake and shows no evidence of the Emerson fault. A geologic sketch of the area (Figure 13-6) shows the alluvial deposits that concealed the fault trace. The image in Figure 13-5B was acquired one month after the Landers earthquake and was digi-

tally processed by R. E. Crippen to enhance the trace of the rupture as a dark line that trends northwest across the image. Figure 13-5C is an enlargement of the central portion of the post-earthquake image. The fault ruptures were very subtle in the field, where Crippen found that individual cracks ranged from millimeters to only a few centimeters in width with little, if any, vertical displacement. The cracks are visible on the image, however, despite the 10-m resolution of SPOT. Their visibility is attributed to the digital image enhancement and to the abundance of cracks and the near-vertical view of SPOT which recorded the shadowed interiors of the open cracks.

Crippen and Blom also processed the SPOT images to analyze the dynamics of the Landers earthquake. The images were digitally matched on one side of the fault at subpixel levels by shifting the postquake image relative to the prequake image,

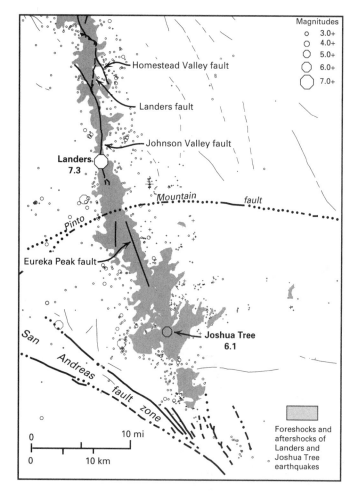

Figure 13-4 Seismicity (January 1 to August 18, 1992) for the Landers earthquake of June 28, 1992. Solid lines are exposed faults; dotted lines are faults concealed by young deposits; dashed lines are inferred faults. Modified from Yeats, Sieh, and Allen (1996, Figure 8-51). Courtesy K. Sieh, California Institute of Technology.

A. Image acquired July 27, 1991 (before earthquake).

B. Image acquired July 25, 1992 (after earthquake).

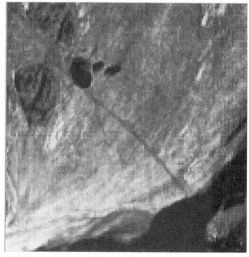

C. Enlarged central portion of image B.

Figure 13-5 SPOT pan images of the Emerson fault, which ruptured during the Landers earthquake of June 28, 1992. From Crippen (1992). Courtesy R. E. Crippen, JPL.

using correlation analysis methods. By rapidly alternating the enlarged images on a computer display screen, subpixel displacements along the fault appeared as motion, which confirmed the fault location and right-lateral sense of displacement. By using statistical measurements at several points in the images, Crippen and Blom (1992) were also able to quantify the displacements, thereby measuring offsets along the fault.

The ready availability of SPOT images makes this a practical technique for analyzing earthquakes with surface ruptures. Crippen (1992) gives details of the processing technique.

Radar Interferograms Chapter 6 described how interferograms showing topography are generated from radar images that are simultaneously acquired by two spatially separated antennas. Massonnet and others (1993) processed ERS-1 images of the Landers region to produce an interferogram of the dis-

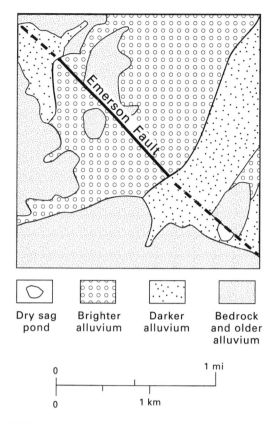

Dry sag pond | Brighter alluvium | Darker alluvium | Bedrock and older alluvium

0   1 mi

0   1 km

Figure 13-6 Geologic sketch of SPOT image of Emerson fault acquired after the Landers earthquake (Figure 13-5B).

placement field caused by the earthquake. *Displacement field* refers to all the vertical and horizontal shifts in the region of an earthquake, not just the fault ruptures.

A pair of ERS-1 images acquired April 24 and August 7, 1992, span the earthquake date and provide adequate orbital separation for preparing an interferogram. Such an interferogram shows both the pre-earthquake topography plus the displacement field caused by the earthquake. Digital terrain data obtained before the earthquake were used to remove the topographic information from the interference pattern. The result is a *residual interferogram,* which shows only the displacement field caused by the earthquake. Figure 13-7 is the residual interferogram, which is a contour map of the change in range distance, or the component of displacement that points toward the satellite. Each gray-scale fringe corresponds to 28 mm of displacement. For comparison, Massonnet and others (1993) used field data and a dislocation model to calculate a synthetic interferogram (Figure 13-8) that models the predicted displacement field. The synthetic stereogram closely matches the actual interferogram; the two versions agree to within two fringes (56 mm). Peltzer, Hudnut, and Feigl (1994) used the interferogram to analyze details of displacement gradients in the vicinity of the fault ruptures. The worldwide availability of images from ERS-1 and JERS-1 makes the interferometric method practical for other areas where digital topographic data are available. In the absence of topographic data, two prequake images could be used to generate topographic data.

Radar Image of Calico Active Fault, California

The Calico fault is one of the active faults that strike northwest across the western Mojave Desert and have right-lateral, strike-slip displacement. Figure 13-9A is a Landsat image of the western portion of the Troy Valley, which is crossed by the Calico fault. The trace of the fault is obscure on the Landsat image and is also obscure on an enlarged TM color image (not shown). Figure 13-9B is a Seasat radar image of the same area. The Calico fault is a distinct northwest-trending radar lineament formed by the boundary between brighter signatures on the west and darker signatures on the east. Radar roughness criteria (Chapter 6) for Seasat predict a surface roughness of greater than 6 cm for the bright terrain and less than 1 cm for the dark terrain. In the field the bright terrain consists of hummocky sand dunes that have been stabilized by desert shrubs. The dunes terminate abruptly eastward against relatively smooth desert terrain. The map in Figure 13-10 shows these relationships.

Figure 13-11 is a west-to-east section across the fault that explains the lineament in Figure 13-9B. The lineament is not caused by surface displacement along the fault; rather, it is due to the effect of the fault on the water table, which is the boundary between dry soil near the surface and deeper soil that is permanently saturated by groundwater. In the Troy Valley, unobstructed groundwater would flow eastward in the direction of the surface slope. The Calico fault, however, forms a barrier to groundwater movement, and the water table is shallow along the west side of the fault. The shallow water table supports an unusually dense growth of desert shrubs and trees (creosote bush and mesquite) along the west side of the fault. The prevailing westerly wind moves sand along the desert surface. The vegetation interrupts the wind flow, causing the sand to accumulate in dunes along the vegetated west side of the fault trace. Continued growth of vegetation stabilizes the dunes.

The complex relationship that formed the radar lineament is not unique to the Calico fault. An identical relationship is shown in radar images and in the field at the Mesquite Lake fault near the town of Twenty Nine Palms, California.

China

Much of China is seismically active, but it is inadequately understood by Western earth scientists because of its vast size and limited access. Images from satellites have helped improve our knowledge. Tapponnier and Molnar (1977, 1979) used mosaics of Landsat MSS images to interpret a number of major faults that appear to be active. For example, the active Altyn Tagh fault trends northeast for almost 2000 km along the south

Figure 13-7 Radar interferogram of the Landers, California, region prepared from ERS-1 images acquired April 24, 1992 (before quake) and August 7, 1992 (after quake). White lines that show fault breaks were added from field maps. Each cycle of gray shading represents a radar range difference of 28 mm between the two image dates. There are at least 20 cycles (560-mm total range difference) in the interferogram patterns on either side of the fault. From Massonnet and others (1993, Figure 3A). Courtesy G. Peltzer, JPL

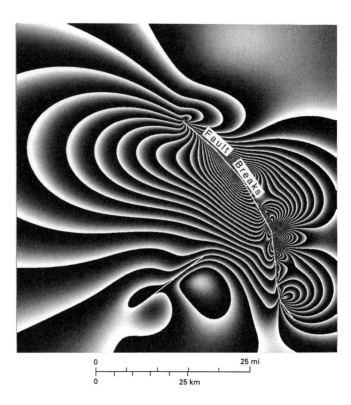

Figure 13-8 Synthetic interferogram modeled from fault displacements caused by the Landers earthquake. The displacement data were measured in the field. Each cycle of gray shading represents the same interval as the radar interferogram (28 mm) shown in Figure 13-7. The two interferograms agree to within two gray cycles (56 mm). From Massonnet and others (1993, Figure 3B). Courtesy G. Peltzer, JPL.

margin of the Tarim Basin. Major left-lateral displacement accommodates much of the northeastward movement of the India plate relative to the Asia plate. Little else was known about this important fault until SPOT images became available. Peltzer, Tapponnier, and Armijo (1989) selected seven SPOT panchromatic images based on the earlier MSS interpretations. The SPOT images of the Altyn Tagh fault show spectacular long-term, left-lateral offsets of drainage channels, alluvial fans, and glacial deposits that imply slip rates of 2 to 3 cm per year.

Avouac and Peltzer (1993) interpreted SPOT panchromatic images of the south margin of the Tarim Basin, which is north of the Altyn Tagh fault. Figure 13-12 is an image of the Hotan-Qira fault system, which cuts young alluvial fan surfaces that slope northward into the Tarim Basin. The sun is shining from the southeast at a relatively low elevation. The northeast-trending faults are indicated by linear highlights and shadows. The shadows are caused by northwest-facing fault scarps, and the highlights are caused by southeast-facing scarps. Enlargements

of the image (not shown) depict many smaller faults that are not visible at the scale of Figure 13-12. Figure 13-13 is an interpretation map and cross section for the image.

Sequences of river terraces and alluvial fan surfaces are seen in the image. Offsets of these features are used to determine the amount and rate of movement along the faults. Field investigations (Avouac and Peltzer, 1993) show that the highest scarps reach 20 m with recent offsets about 2 m high. Total vertical offset of the Hotan-Qira fault system is 70 m. The minimum rate of subsidence is 3.5 ± 2 mm $\cdot$ yr^{-1}.

Active Faults on the Seafloor

Active faults on the seafloor are imaged by side-scan sonar systems that are described in Chapter 9. Figure 13-14 is a mosaic of side-scanning sonar images of a portion of the Blanco fault zone in the Pacific Ocean 200 km off the coast of Oregon. The image was acquired by the U.S. Geological Survey (1986) as part of the survey of the offshore Exclusive Economic Zone (EEZ). Water depths are approximately 3000 m. Bright signatures are strong sonar returns, and dark signatures are weak returns.

The Blanco fracture zone is a linear trench trending west-northwest that is bounded by steep fault scarps shown in the map and cross section of Figure 13-15. The south-facing scarp along the north margin is marked by a narrow linear shadow. Bathymetric contours (not shown) indicate that the scarp reaches a height of 300 m. The south margin of the trench is formed by a series of north-facing scarps with strong highlights. The seafloor adjacent to the fracture zone is crossed by narrow ridges trending north-northeast that represent small fractures. Seamounts, of possible volcanic origin, have circular to irregular outlines. The sinuous Cascadia sea channel flows southward into the Blanco Trench, follows the north margin for 70 km, then follows the south margin for 50 km. The dark signature of the channel is caused by the fine-grained sediments, which do not scatter energy back to the sonar detector.

Summary

The preceding examples were selected from many investigations around the world to demonstrate the utility of various remote sensing systems for recognizing active faults with earthquake potential on both land and seafloor. Local offsets along faults and regional displacement fields are measured from images, which are also used to evaluate the damage caused by earthquakes.

LANDSLIDES

Landslides occur on the land and on the seafloor in areas underlain by unstable materials. Each year landslides cause extensive property damage. Individual landslide events are difficult or impossible to predict by remote sensing or other methods. The most effective way to prevent landslide damage is to identify unstable areas and avoid building in their vicinity. Where construction must be done in unstable areas, knowledge of the potential for landslides can be used to stabilize the foundations.

Stereo pairs of aerial photographs (black-and-white, normal color, and IR color) have long been used to recognize slides and slide-prone terrain. Rib and Liang (1978) published an extensive collection of stereo pairs together with interpretation criteria. High soil moisture lubricates unstable material and is a major factor in landslides. Thermal IR images have been used to recognize damp ground associated with landslides in California (Blanchard, Greeley, and Goettleman, 1974, Figure 1). Evaporative cooling of the damp ground produces a cool signature on aircraft thermal IR images. Satellites, radar, and side-scanning sonar have expanded our capability to recognize unstable terrain where slides occur, as shown by the following examples.

Blackhawk Landslide, Southern California

The prehistoric Blackhawk landslide originated on the north flank of Blackhawk Ridge in the San Bernardino Mountains and moved northward for 9 km into the Mojave Desert. The slide has a maximum width of 3.2 km and includes a volume of 2.7×10^9 m^3 of crushed rock (Shreve, 1968). In form and structure the Blackhawk slide is similar to the smaller and well-known historic slides at Elm in Switzerland, at Frank in Alberta, Canada, and on the Sherman Glacier in Alaska and to the great prehistoric slide at Saidmarreh in Iran. These similarities make the Blackhawk slide a good example for remote sensing analysis.

Figure 13-16 shows two remote sensing images and a map of the Blackhawk slide. The slide resulted from erosion that over-steepened the north slope of Blackhawk Ridge and caused the highly fractured bedrock to collapse along an arcuate scarp (Figure 13-16C). The falling rock debris trapped a cushion of air during its descent that lubricated the mass, enabling it to flow several kilometers onto the desert floor (Shreve, 1968). In the satellite photograph (Figure 13-16A) the headward scarp of the original debris fall is shadowed, which enhances the characteristic crescent shape. The distal lobe of the slide has a hummocky appearance that is typical for landslide deposits. The toe and lateral margins of the slide form a pressure ridge 15 to 30 m high that stands slightly above the surface of the slide. A northwest-flowing ephemeral stream has deposited younger alluvium over the central portion of the slide. In the photograph, the slide debris is darker than the stream and desert alluvium. The radar image (Figure 13-16B) has a coarser resolution than the photograph but portrays major features of the slide. The coarse debris and hummocky surface are rough and produce a bright signature relative to the darker signature of the finer-grained alluvium. The pressure ridges have narrow, bright signatures.

Plate 14A is a perspective view of the Blackhawk slide that was digitally composited from Landsat TM and SPOT pan images, plus digital terrain data. This view graphically shows (1) the source area on the north flank of the San Bernardino Mountains, (2) the steep slope down which the debris initially fell, and (3) the abrupt transition to the gentle slope of the Mojave Desert.

Submarine Landslides, Mississippi Delta

The delta of the Mississippi River consists of unconsolidated, water-saturated, fine-grained sediment. The submerged, gentle depositional slopes around the margin of the delta are unstable and may collapse to form submarine landslides and related features. Because the delta is an oil-producing region, it is important to map slide-prone areas before installing production platforms and seafloor pipelines. The turbid water prevents remote sensing at visible wavelengths. The Coastal Studies Institute of Louisiana State University and the U.S. Geological Survey

A. Landsat image acquired 1978.

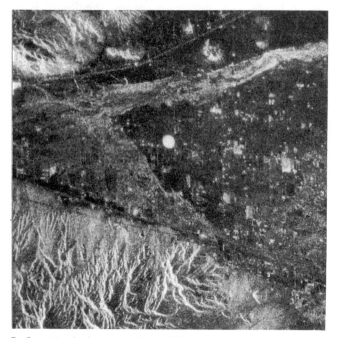

B. Seasat radar image acquired 1978.

Figure 13-9 Images of the Calico fault in the Troy Valley, Mojave Desert, California.

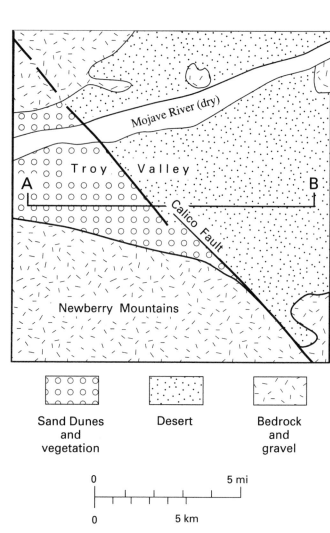

Sand Dunes and vegetation   Desert   Bedrock and gravel

0 ——————— 5 mi

0 ——————— 5 km

Figure 13-10 Interpretation map for the Seasat image of the Calico fault, Troy Valley, California (Figure 13-9B).

acquired parallel strips of side-scanning sonar images that were digitally processed and compiled into the mosaic of Figure 13-17A. In this mosaic bright signatures record weak sonar returns from shadow zones or smooth surfaces. Dark signatures record strong returns from scarps facing the sonar pulse and from rough surfaces.

The interpretation map (Figure 13-17B) shows that the slides originate at lobate slump areas where the sediment collapses to form irregular blocks. The slump material moves downslope through narrow, steep-sided channels (chutes) that merge at junctions. At the junctions, the slump material may leave the chute and form spillover deposits. Extensive systems of slumps and chutes have been mapped on sonar images of the submerged portions of the Mississippi Delta.

Piper and others (1985) acquired and interpreted sonar images near the epicenter of the 1929 Grand Banks earthquake on the seafloor south of Newfoundland. The earthquake triggered a major submarine landslide and turbidity currents. The sonar images clearly show the following features: landslide scarps, slumps, debris flows, a lineated seafloor, channels, and gullies. Sonar images can identify unstable areas that should be avoided for offshore engineering projects.

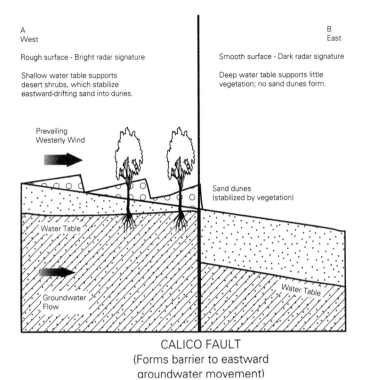

A
West

Rough surface - Bright radar signature

Shallow water table supports
desert shrubs, which stabilize
eastward-drifting sand into dunes.

B
East

Smooth surface - Dark radar signature

Deep water table supports little
vegetation; no sand dunes form.

Prevailing
Westerly Wind

Sand dunes
(stabilized by vegetation)

Water Table

Groundwater
Flow

Water Table

CALICO FAULT
(Forms barrier to eastward
groundwater movement)

Figure 13-11 Diagrammatic section across the Calico fault to explain the tonal lineament on the Seasat image (Figure 13-9B).

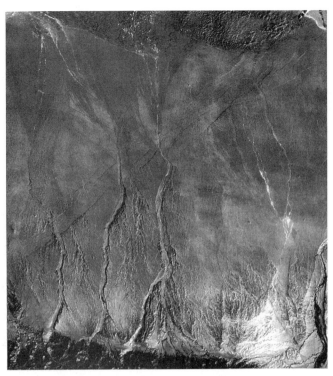

Figure 13-12 SPOT pan image of the Hotan-Qira zone of active faults in southern Xinjiang, China. The sun was shining toward the northwest (upper left corner). From Avouac and Peltzer (1993, Figure 3A). Courtesy G. Peltzer, JPL.

LAND SUBSIDENCE

Land subsidence, which is downward vertical movement, is a different type of hazard from landslides, in which lateral movement predominates. Subsidence results from removal of underlying support and can have serious effects on man-made structures.

Surface subsidence of up to several meters in depth can be caused by extraction of subsurface fluids. Where subsidence occurs in populated and industrialized coastal areas, the resulting flooding can be a major problem. A classic example is the subsidence associated with the Long Beach oil field in southern California, which was stopped through injection of water to replace the extracted oil.

In the past 30 years, up to 2.5 m of land subsidence has occurred in metropolitan Houston, Texas, and is related to the following causes (Verbeek and Clanton, 1981):

1. Active geologic processes of faulting and compaction of young sedimentary deposits
2. Extraction of oil and gas
3. Widespread pumping of groundwater from shallow depths, which is thought to be a major cause of subsidence

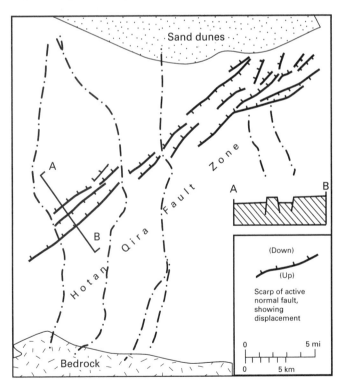

Figure 13-13 Interpretation map and cross section of the SPOT image of the Hotan-Qira fault system (Figure 13-12). Modified from Avouac and Peltzer (1993, Figure 3A).

Figure 13-14 Mosaic of side-scanning sonar images of the Blanco fault zone in the Pacific Ocean off northern California. From U.S. Geological Survey (1986).

The rate of subsidence in Houston is increasing and is concentrated along active faults, which have caused extensive damage to buildings, roads, and especially underground utilities. Clanton and Verbeek (1981) acquired oblique aerial photographs with normal color and IR color film and interpreted them using the following criteria to identify traces of active faults:

1. Fault scarps
2. Sag ponds along fault traces
3. Differences in drainage patterns on opposite sides of faults
4. Linear tonal anomalies caused by higher soil moisture on the downthrown sides of faults
5. Vegetation anomalies

These criteria have been used to prepare maps showing the locations of active faults.

In areas of limestone terrain, groundwater can dissolve the bedrock to produce underground caverns that may collapse and cause surface subsidence. In Florida these collapse areas, called *sinkholes*, are up to several hundred meters wide and 10 m deep and may form in a few hours. Areas of incipient sinkholes have anomalously high surface moisture. A few aircraft thermal IR images show cool areas that may be indicators of future sinkholes.

VOLCANOES

Based on their activity, volcanoes may be classified as follows:

Active Erupted at least once in historic time
Dormant No historic eruptions, but probably capable of erupting
Extinct Incapable of further eruptions

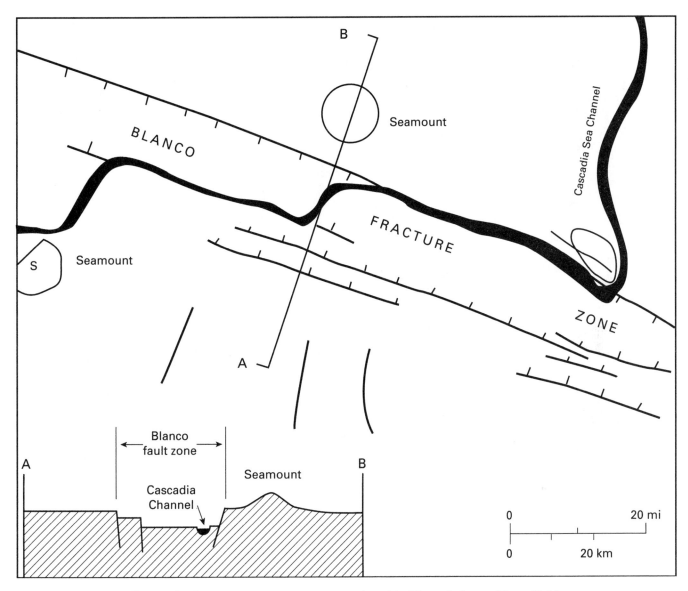

Figure 13-15 Interpretation map and cross section of the Blanco fault zone (Figure 13-14).

Table 13-1 lists the following categories of volcanic hazards:

1. **Pyroclastic eruption** Explosive eruptions produce clouds of ash and coarser ejecta. Of particular danger are glowing cloud avalanches, such as the eruption that destroyed Martinique (Table 13-1). Ash plumes are a hazard to jet aircraft.
2. **Slope failure** The over-steepened flanks of volcanoes can result in massive rapid landslides. The slope failures may trigger eruptions by removing overburden that confined the magma.
3. **Mudflows** (also called *lahars*) Thick accumulations of ash on steep flanks of volcanoes mix with rain or snowmelt to

produce high-velocity mudflows that destroy everything in their path.
4. **Toxic gas** Carbon monoxide and other lethal gases are by-products of eruptions.

Table 13-1 lists representative occurrences of these hazards. Despite the danger, many regions of active volcanoes are densely populated, including Japan, Java, and Italy. Predicting eruptions months or years in advance has not been particularly successful. Predictions a few days or hours in advance are notably more exact. Figure 13-18 shows the four major warning criteria, which are summarized below:

A. Satellite photograph.

B. Aircraft X-band radar image.

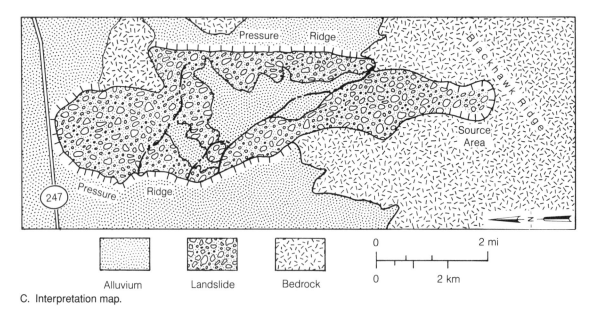

C. Interpretation map.

Figure 13-16 Blackhawk landslide, San Bernardino Mountains, California.

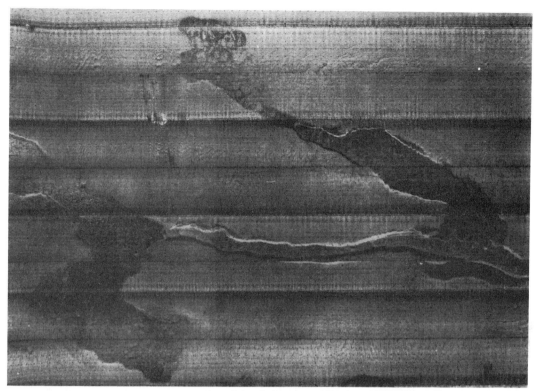

A. Mosaic of side-scanning sonar images. From Prior, Coleman, and Garrison (1979, Figure 2). Courtesy D. B. Prior, Louisiana State University.

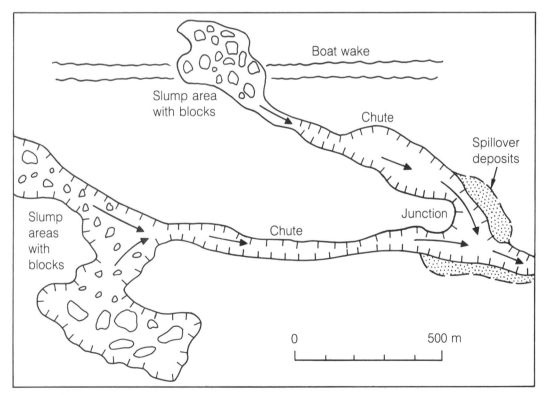

B. Interpretation map.

Figure 3-17 Submarine landslides, offshore Mississippi Delta, Gulf of Mexico.

Table 13-1 Volcanic hazards

Pyroclastic eruption	*Slope failure*	*Mudflows (lahars)*	*Toxic gas*

El Chichon, Mexico, 1982
  1000 dead
	Mount St. Helens, USA, 1980
	  57 dead*
		Ruiz, Colombia, 1985
		  22,000 dead†
			Lake Nyos, Cameroon, 1986
			  1700 dead

Mt. Pelee, Martinique, 1902
  29,025 dead
	Ili Werung, Indonesia, 1979
	  500 dead
		Irazu, Costa Rica, 1964
		  1 dead*
			Lake Monoun, Cameroon, 1984
			  37 dead

Vesuvius, Italy, A.D. 79
  2000 dead
	Unzen, Japan, 1792
	  14,524 dead
		Kelut, Indonesia, 1919
		  5000 dead
			Sinila crater, Indonesia, 1979
			  142 dead

*Hazard warning acted upon locally.

†Hazard warning issued but not acted upon.

Source: Rothery (1989).

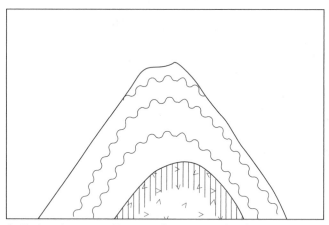

A. Earthquakes caused by expanding magma chamber.

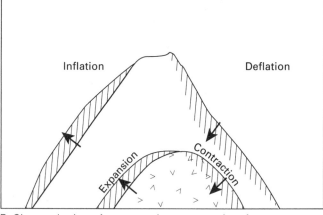

B. Changes in shape from expansion or contraction of chamber.

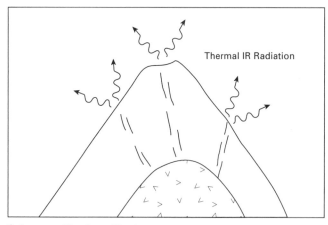

C. Increased heating of fractures.

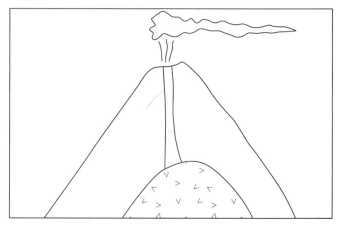

D. Increased eruption of ash and gas.

Figure 13-18 Warning criteria for monitoring active volcanoes. From Foxworthy and Hill (1982).

A. Earthquakes caused by expanding magma chamber are detected by seismometers.

B. Changes in the shape of a volcano caused by an expanding or contracting magma chamber are recorded by tiltmeters and by repeated GPS (global positioning system) surveys. Radar interferometry is now being used.

C. Increased emission of heat and gas at fractures can be detected on thermal IR images from aircraft or satellites.

D. Increased eruption of ash is monitored by visual observation and images.

The last three criteria may be detected by remote sensing systems, as described in the following examples.

Mount St. Helens, Washington

Mount St. Helens, in the southwestern part of the state of Washington, erupted several times in the 1800s and has long been recognized as an active volcano. The major explosive eruption of May 18, 1980, was thoroughly monitored and analyzed. The first warning was a strong earthquake at shallow depth beneath the volcano on March 20, 1980. On March 27, explosive hydrothermal activity began at the summit, accompanied by the formation of a small crater, ground fracturing, and the beginning of a topographic bulge high on the north flank. Strong seismic activity and relatively mild steam-blast eruptions continued intermittently into mid-May. During that time the new crater gradually enlarged and the bulge became larger. On the morning of May 18, an earthquake caused great avalanches of rock debris at the over-steepened bulge on the north flank. This unloading led to a northward-directed lateral blast, partly driven by steam explosions, that devastated an area of nearly 600 km^2. A vertical column of ash extended more than 25 km high, causing ash falls more than 1500 km to the east. Pyroclastic flows occurred on the north flank. Melting snow and ice contributed to catastrophic mudflows. Smaller eruptions have occurred intermittently since the main blast. U.S. Geological Survey reports by Lipman and Mullineaux (1981) and by Foxworthy and Hill (1982) document the events.

Judged by its volume of ejecta (0.6 km^3), Mount St. Helens was a minor event compared with the 1883 eruption of Krakatoa, Indonesia (26 km^3) or the 1815 eruption of Tambora, Indonesia (46 km^3). Because of the advance warnings, authorities had restricted access to Mount St. Helens, and only about 60 people were killed in the eruption. Property damage is estimated at $2 to $3 billion. The fact that the relatively moderate eruption of Mount St. Helens caused so much damage is reason to continue improving methods of volcano monitoring and prediction.

Aerial Photographs Numerous remote sensing images were acquired before and after the main eruption. A pre-eruption aerial photograph (Figure 13-19A) shows the symmetric form of the mountain, which lacked a summit crater. The May 18 blast created a large crater and blew out the north rim, as shown in the post-eruption photograph (Figure 13-19B). The map (Figure 13-20) shows the major destructional features and volcanic deposits. Timber north of the volcano was flattened by the blast. The deposits blocked the outlet of Spirit Lake; as a result the lake is larger in the post-eruption photograph and has a gray signature, which is caused by floating logs. The east, south, and west flanks of the volcano are modified by mudflows. The upper slopes were scoured by the flowing mud, and the lower slopes were covered by mudflow deposits. In the post-eruption photograph (Figure 13-19B), steam clouds conceal a lava dome that formed on the floor of the crater.

Thermal IR Images Aircraft thermal IR images (8 to 14 μm) were acquired before and after the main eruption. The initial explosive event occurred on March 27, 1980. Figure 13-21A is an IR image of the resulting crater acquired on May 16, 2 days before the main blast. The image has not been geometrically corrected for scanner distortion. The image was processed to emphasize the hottest temperatures (>12°C), which have very bright signatures. Figure 13-21B is an interpretation map. Dark patches are the hot spots that coincide with deep pits, many of which were the source of steam plumes. Kieffer, Frank, and Friedman (1981) inferred from the image data that a temperature of 400°C occurred at a depth of only 40 m and was caused by intense hydrothermal circulation. The cluster of hot spots on the north flank of the volcano mark the bulge that was subsequently obliterated by the May 18 blast.

Figure 13-21C is a post-eruption image that has not been corrected for scanner distortion; hence the semicircular crater has an oval outline in the image. Thermal patterns in the image correlate with features shown in the geologic map (Figure 13-20). Flanks of the volcano are cool (dark signature). Pyroclastic flow deposits with intermediate temperatures (gray signatures) fill the crater and the depression on the north flank that was created by the blast. The highest radiant temperatures (brightest signatures) were identified by density-slicing the data and are shown in black in the interpretation map (Figure 13-21D). The lava dome formed in August is the hot oval on the crater floor. A semicircular pattern of hot spots occurs along the inner wall of the crater. The irregular hot spots in the southeast floor of the crater are fumaroles of steam and gas. A linear hot feature extending south from the rim is an igneous dike 15 m wide. Other hot linear features occur in the new pyroclastic deposits and represent fractures probably caused by compaction. A northwest-trending alignment of hot spots is probably related to a major structural feature cutting across the core of the volcano (Friedman and others, 1981).

Landsat and SPOT Images of Volcanoes

Landsat and SPOT images have long been used to study morphology, structure, and distribution of volcanoes. De Silva and

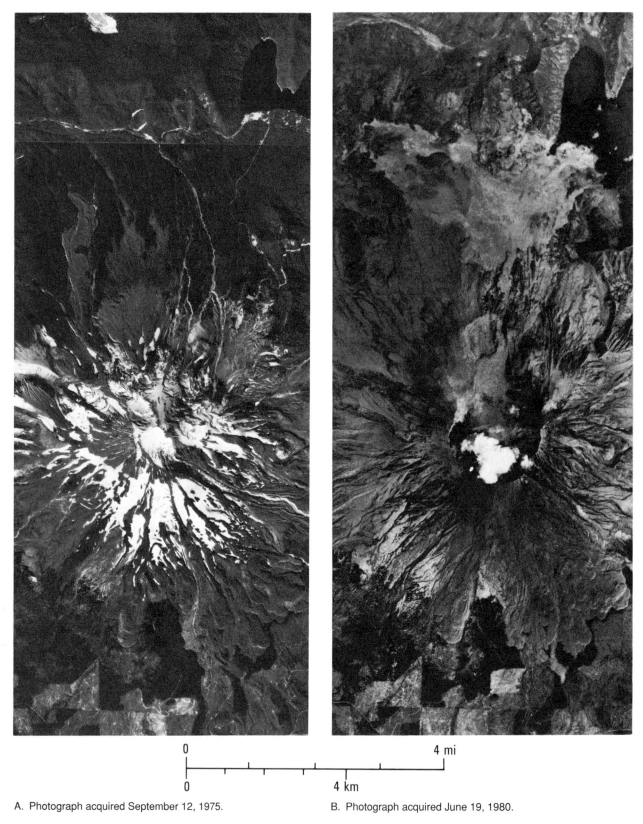

A. Photograph acquired September 12, 1975.   B. Photograph acquired June 19, 1980.

Figure 13-19 Aerial photographs of Mount St. Helens, Washington, acquired before and after the May 18, 1980, eruption.

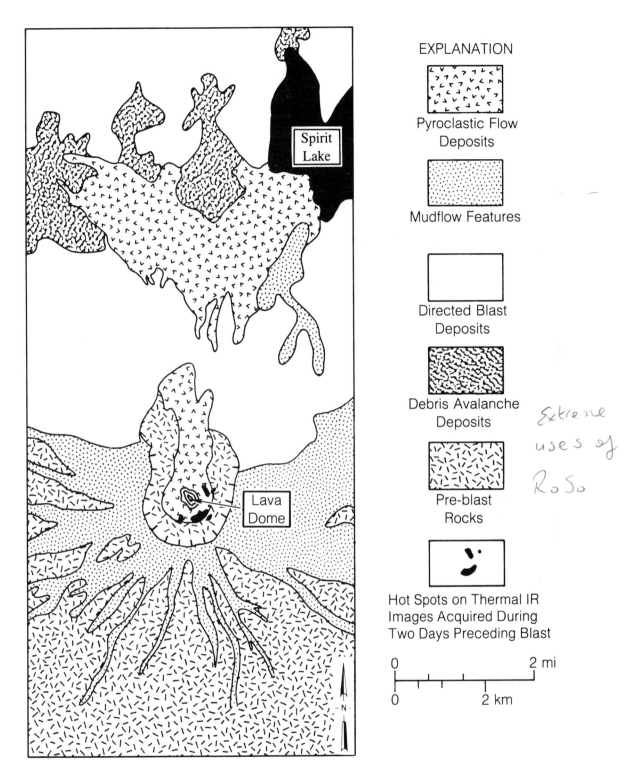

EXPLANATION

Pyroclastic Flow
Deposits

Mudflow Features

Directed Blast
Deposits

Debris Avalanche
Deposits

Pre-blast
Rocks

Hot Spots on Thermal IR
Images Acquired During
Two Days Preceding Blast

Spirit Lake

Lava Dome

0   2 mi

0   2 km

N

Extreme uses of RoSo

Figure 13-20 Volcanic features and deposits of the 1980 eruption of Mount St. Helens (Figure 13-19B). From Lipman and Mullineaux (1981, Plate 1).

A. May 16, 1980, image.

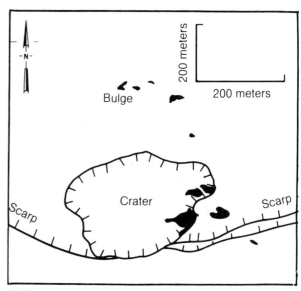

B. May 16, 1980, map.

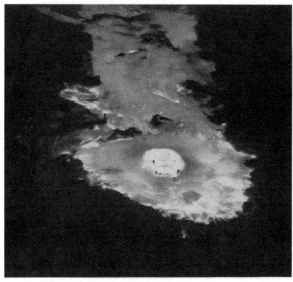

C. August 20, 1980, image.

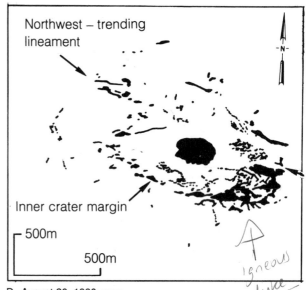

D. August 20, 1980, map.

Figure 13-21 Nighttime thermal IR images (8 to 14 μm) of the summit of Mount St. Helens acquired before and after the eruption. Maps show highest temperatures in black. From Kieffer, Frank, and Friedman (1981, Figures 158 and 174); Courtesy H. H. Kieffer and J. D. Friedman, U.S. Geological Survey.

Francis (1991) compiled an atlas of volcanoes in the central Andes that includes 90 color images.

Figure 13-22 shows SPOT images of the Nevado Sabancaya volcano in southern Peru that were interpreted by Chorowicz and others (1992). The volcano was active in the mid-18th century but is thought to have been quiet until November 1986, when fumarolic emissions and weak seismic activity began. The image in Figure 13-22A was acquired before this resumption of volcanism. The undisturbed summit is at an elevation of almost 6000 m and has a permanent cap of ice and snow. The

image in Figure 13-22B was acquired 3 years after volcanism resumed. The summit crater was enlarged by melting of the ice cap from increased heating, not from an explosion. On May 28, 1990, an eruption began with increased fumaroles, seismic activity, and an ash plume over 10 km high. Figure 13-22C is a panchromatic image of July 8, 1990, that shows the east-drifting ash plume. Much of the ice cap is mantled by dark ash. Small mudflows have occurred. Larger-scale versions of the image show additional melting and deepening of the summit ice cap and fallen ice blocks.

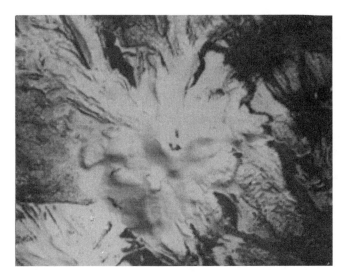

A. SPOT XS image, July 21, 1986.

B. SPOT XS image July 1, 1989.

C. SPOT pan image July 8, 1990.

Figure 13-22 SPOT images of Nevado Sabancaya, Peru. Eruptions occurred November 1986 and May 28, 1990. Images cover 8 km in east–west direction. From Chorowicz and others (1992, Figure 3). Courtesy J. Chorowicz, URA-CNRS, Paris.

Aircraft Thermal IR Images—Rabaul, New Britain, PNG

Figure 13-23 is a map of the town of Rabaul and vicinity, at the northeast end of the island of New Britain in Papua New Guinea. This region has a long history of volcanic eruptions. Simpson Harbor occupies a large collapse caldera formed by a series of ignimbrite eruptions over the past 18,000 years. Ignimbrite eruptions are incandescent clouds of ash and ejecta that move rapidly down the flanks of a volcano with destructive effects. Dating of the ignimbrite deposits shows the interval between eruptions ranges from 2000 to 3600 years. The last major eruption occurred about 1400 years ago (Johnson and others, 1995). A number of volcanic centers are associated with the caldera (Figure 13-23). Both Tavurvur and Vulcan erupted in 1878 and 1937. Poorly documented eruptions occurred in 1767 and 1791 (Johnson and others, 1995). Because of this history the Australian Bureau of Mineral Resources conducted an aircraft thermal IR survey of the area in 1973, which was described by Perry and Crick (1976). Figure 13-24 shows images of the two volcanoes that have significant warm signatures. The horizontal lines are the flight path. The wavy lines are profiles of the radiant temperature along the flight path that were measured with a radiometer aimed downward from the survey aircraft. The temperature scale is shown at the left margin of the images. Rabalankaia volcano (Figure 13-24A) has concentric warm rings that coincide with bare ground on the floor and walls of the summit crater. Tavurvur volcano (Figure 13-24B) has an irregular pattern of warm spots that partially correspond to bare ground. The radiometer profile crossed the warm areas and recorded radiant temperatures up to 27°C.

On September 19, 1994, explosive eruptions occurred at Tavurvur and Vulcan (Figure 13-23). Ash falls up to a meter thick caused massive damage to Rabaul, which was evacuated. The 1973 image (Figure 13-24B) clearly shows thermal features at Tavurvur, but none were present at Vulcan. Rabalankaia has conspicuous thermal signatures, but no record of historic eruptions, which should not downgrade its potential for future eruptions. This survey points out both the promise and uncertainty of predicting eruptions. The IR survey identified one of the two future eruptive sites, but two decades passed before the eruption occurred.

Satellite Thermal IR Images—Lascar Volcano, Chile

Landsat TM images are widely used to monitor volcanoes. Representative studies include Erta 'Ale, Ethiopia (Rothery, 1989);

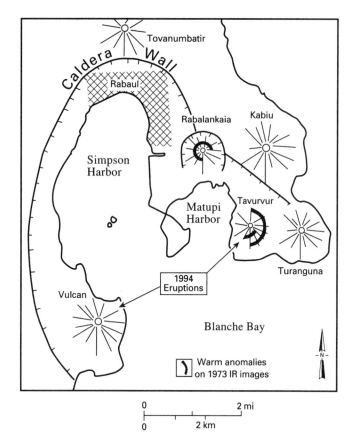

Figure 13-23 Map of Rabaul and vicinity, New Britain, PNG.

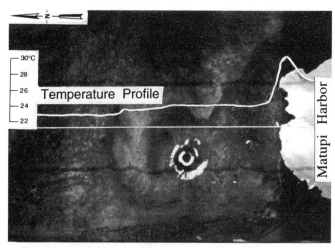

A. Rabalankaia volcano.

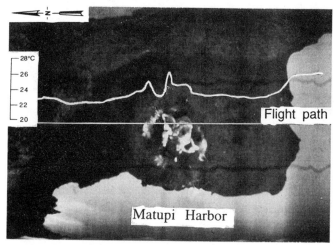

B. Tavurvur volcano.

Figure 13-24 Nighttime thermal IR images (8 to 14 μm) of volcanoes near Rabaul, New Britain, PNG. Bright signatures are warm radiant temperatures. The images are 5 km wide from left to right. From Perry and Crick (1976, Figure 5). Courtesy W. J. Perry, Bureau of Mineral Resources, Australia.

Barren Island volcano, Andaman Sea (Reddy, Bhattacharya, and Srivastav, 1993); Stromboli and Vulcano, Italy (Gaonac'h and others, 1994); Mount Erebus, Antarctica (Rothery and Francis, 1990). The repeated coverage by Landsat was used to analyze cycles of eruption at Lascar volcano.

Lascar is the most active volcano in the central Andes of northern Chile. The region is sparsely populated and the record of volcanic activity is meager. Local residents recall periodic events dating back to 1955. Glaze and others (1989) relied largely on satellite images to document an eruption in 1986. Oppenheimer and others (1993) analyzed volcanism at Lascar using Landsat TM bands 5 and 7 in the reflected IR region. Chapter 5 pointed out that these bands record the high radiant temperatures typical of fresh lava. Fifteen images acquired between December 1984 and April 1992 recorded the evolution of thermal features and spanned two eruptions. Figure 13-25 shows four band 7 images (three daytime, one nighttime) of the summit that were recorded in 1989. In all the images, a large central cluster of very bright pixels is a lava dome. Figure 13-26 is a map based on the images that shows the dome and other features of the summit. Field observations in early 1990 showed the dome to be a circular body of blocky lava approxi-

mately 200 m in diameter. At night the dome was peppered with many glowing sites in arcuate chains and clusters located mostly near the margins. Thermal IR radiometers recorded radiant temperatures ranging from 500° to almost 800°C for the incandescent sites. The arrows on the images (Figure 13-25) indicate three persistent small hot spots beyond the margins of the dome that are probably high-temperature fumaroles. The map (Figure 13-26) shows the relationship of the hot spots to the dome and the crater walls.

Figure 13-27 plots the radiance of band 7 for the central dome for the images acquired from 1984 to 1992. Variations in radiance correspond to periods of dome growth that were punctuated by explosive eruptions (1986, 1990) which produced major ash columns. High radiance values in late 1989

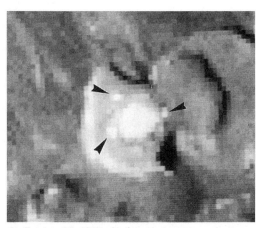

A. October 27, 1989, daytime.

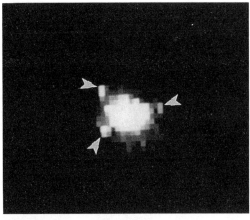

B. November 17, 1989, nighttime.

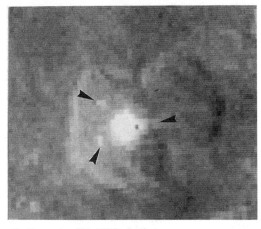

C. November 28, 1989, daytime.

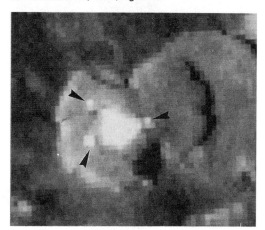

D. December 14, 1989, daytime.

Figure 13- 25 Landsat TM band 7 images of the summit of the Lascar volcano, Chile. From Oppenheimer and others (1993, Figure 13). Courtesy C. Oppenheimer (The Open University) and L. Glaze (NASA Goddard Space Center).

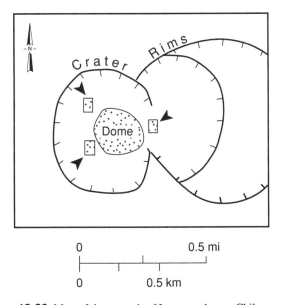

Figure 13-26 Map of the summit of Lascar volcano, Chile.

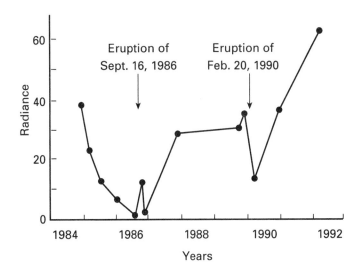

Figure 13-27 Spectral radiance of TM band 7 for the summit of the Lascar volcano from 1984 to 1992. From Oppenheimer and others (1993, Figure 18).

A. Pre-eruption image, acquired March 1991.

B. Post-eruption image, acquired November 1991.

Figure 13-28 Aircraft X-band radar images of Mount Pinatubo, Philippines, acquired before and after the June 15, 1991, eruption. Courtesy Intera Information Technologies Corporation.

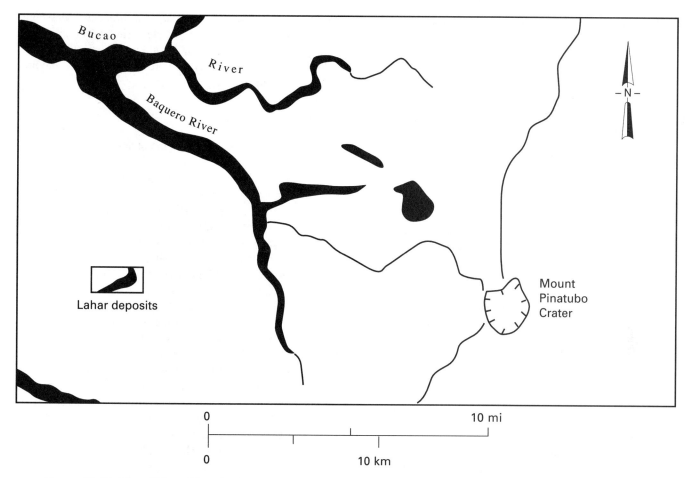

Figure 13-29 Map of Mount Pinatubo showing mudflows (black) interpreted from post-eruption radar image (Figure 13-28B).

include measurements from the images of Figure 13-25. Similar high values in late 1987 indicate the dome was present then, although it was not confirmed by field observations until early 1989. High values in early 1985 indicate an earlier appearance of the dome. Aerial photographs taken in January 1987, however, show no evidence of a dome, which suggests that it had been destroyed by the 1986 eruption. These results at Lascar demonstrate the capability and value of the TM for repetitive monitoring of active volcanoes in remote areas.

Radar Images, Mount Pinatubo, Philippines

In tropical regions, volcano monitoring at visible wavelengths is impractical because of persistent clouds. Radar images penetrate not only clouds, but also volcanic steam and ash. Figure 13-28 compares aircraft X-band images acquired before and after the recent eruption of Mount Pinatubo on the island of Luzon, 100 km northwest of Manila. Figure 13-28A was acquired in March 1991, prior to the latest eruptive cycle. The volcano came to life on April 2, 1991. Small tremors and minor explosions lasted until June 15, when a major explosion oc-

curred. The summit was lowered 145 m, and huge ash plumes were expelled. Ash falls up to 200 m thick accumulated in valleys 12 to 18 km from the volcano. On July 18 rain from heavy typhoons mixed with the ash to form hot mudflows (lahars) several meters thick that flowed down the volcano flanks and filled valleys tens of kilometers away. Figure 13-28B is a post-eruption image that clearly shows changes caused by the volcanic activity. Figure 13-29 is a map that identifies major features. The summit is now a wide, deep caldera. Mudflow deposits in the valleys have dark radar signatures because of their high moisture content and smooth surface. Clark Field, a U.S. Air Force base, is located 25 km east of Mount Pinatubo, beyond the area shown in the images. The ash falls damaged the base beyond the point of repair. The bright circle is a circular array of metal communications antennas.

Radar Interferograms of Mount Etna

Figure 13-18B shows that expansion and contraction of the magma chamber cause inflation and deflation of the volcano. Monitoring these changes in shape can be used to predict eruptions. Making timely and precise measurements of these minor

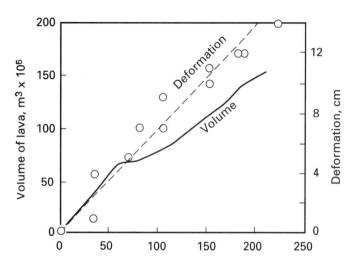

Figure 13-30 Chart showing 1992 volcanic activity at Mount Etna, Italy. Deformation was measured with satellite radar interferometry. Note the correlation between deformation (dashed line) and the volume of lava flows (solid line). From Massonnet and others (1995, Figure 5).

topographic differences is, however, very difficult. Interferograms from satellite radar data can greatly simplify this task. The compilation of radar interferograms was described in Chapter 6. In Figure 13-7, an interferogram was combined with digital topographic data to create a residual interferogram that shows the terrain displacement caused by the Landers earthquake. Massonnet, Briole, and Arnaud (1995) used this method to monitor deformation of Mount Etna, Italy, during a recent eruption.

Mount Etna is one of the world's most active and best studied volcanoes. The last eruption began December 14, 1991, and ended October 23, 1993, for a total of 473 days. Lava flowed at a constant rate during this period and erupted a volume of approximately 3×10^8 m^3, which caused deflation of the volcano. A stereo pair of SPOT panchromatic images were used to compile a digital elevation model of the volcano prior to the eruption. A series of interferograms of the west flank of Mount Etna were compiled from 13 ERS-1 satellite images acquired on northbound orbits; 16 images of the east flank were acquired on southbound orbits. The interferograms were combined with the digital elevation data to produce a series of 12 reliable measurements of deformation that cover the second half of the eruption. In Figure 13-30 these interferogram data are plotted as circles showing subsidence (in centimeters) of the volcano summit, beginning with the first interferogram on May 17, 1992. The error of individual measurements is 1 cm or less. Interferograms acquired after the eruption stopped show no deformation. The solid line shows the volume of lava

produced, as inferred from field measurements. Massonnet, Briole, and Arnaud (1995) illustrate their interferograms and describe the limitations and applications of this technique.

Radar interferograms can help predict eruptions by monitoring deformation prior to an eruption.

Volcanic Ash Plumes

Volcanic eruptions can inject huge volumes of ash into the atmosphere. After some eruptions, such as Krakatoa and more recently Mount Pinatubo, the ash remains in the atmosphere for years and causes worldwide cooling. The ash plumes are also hazards for aviation. The tiny particles of volcanic glass are ingested by jet engines, where they melt and are deposited on turbine vanes, which stalls the engine. Acidic aerosol particles also etch windshields and other surfaces. At aircraft altitudes the plumes are difficult to see and are not detected by aircraft radars. One incident occurred at midday on December 15, 1989, when a KLM Boeing 747 enroute from Amsterdam to Anchorage at an altitude of 7.5 km flew through an ash plume from the Redoubt volcano, Alaska. All four engines shut down for 12 min, and the jet descended steeply for 4 km before the crew managed to restart the engines after seven or eight tries. The aircraft came within about 1.5 km of the mountaintops before landing safely at Anchorage. Damage to the jet reportedly exceeded $50 million (Kienle and others, 1990). Several dozen such incidents have occurred over the past two decades in regions of active volcanism. The ash plume from the 1982 eruption of the Galunggung volcano on the Indonesian island of Java caused engine failures and emergency landings of two Boeing 747 commercial flights.

The plumes are detectable on satellite images by their distinctive shape and orientation relative to volcanic sources. AVHRR images are particularly useful because of their daily worldwide coverage with daytime and nighttime images in the visible and thermal IR bands. Figure 13-31A is an AVHRR thermal IR image of the ash plume from the Redoubt volcano that was acquired one day after the KLM aircraft encountered the same plume. Air traffic controllers could use images such as this to direct aircraft away from these hazards.

Figure 13-31B is an AVHRR band 4 thermal IR image of the ash plume from the September 1994 eruption at Rabaul, described earlier in this chapter. The Rabaul plume exceeds 500 km in length in this image, whereas the Redoubt plume is only 50 km in length. In both images the thickest portions of the plumes have the coldest temperatures (brightest signatures), and the thinner margins have apparently warmer temperatures. In reality, the thinner portions of the plumes have the same cold temperature as the thicker portions; thermal energy radiated from the earth's surface is transmitted through the thinner plumes to produce an apparent warmer temperature.

The images of volcanic plumes (Figure 13-31) include meteorologic clouds that have the same range of temperatures as the plumes, which can make identification difficult. Prata

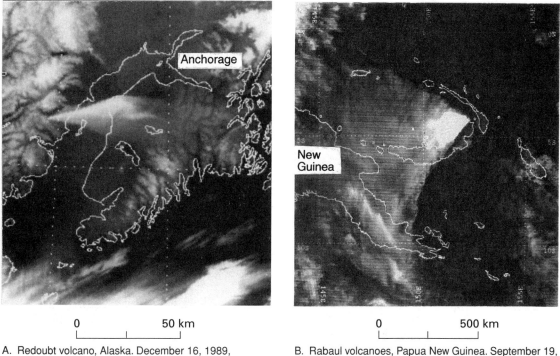

0   50 km
└──┴──┘

A. Redoubt volcano, Alaska. December 16, 1989, at 1223 GMT.

0   500 km
└──┴──┘

B. Rabaul volcanoes, Papua New Guinea. September 19, 1994, at 0859 GMT. From Rose and others (1995, Figure 2).

Figure 13-31 Thermal IR images of volcanic ash plumes recorded by AVHRR band 4 (10.3 to 11.3 μm). Bright signatures represent cool temperatures. Times given as Greenwich mean time (GMT). Courtesy W. I. Rose and D. J. Schneider, Michigan Technological University.

(1989) noted that volcanic plumes contain droplets of sulfuric acid (H_2SO_4) formed by the hydration of sulfur dioxide gas (SO_2). Droplets of sulfuric acid absorb energy at the shorter wavelengths of AVHRR band 4 (10.3 to 11.3 μm) more strongly than at the longer wavelengths of band 5 (11.5 to 12.5 μm); therefore, band 4 minus band 5 is a negative value for volcanic plumes. Droplets of water do not absorb the shorter wavelengths; therefore, band 4 minus band 5 is zero or a positive value for clouds. These relationships are used to differentiate volcanic plumes from clouds on digitally processed images of AVHRR band 4 minus band 5. Wen and Rose (1994), using AVHRR bands 4 and 5, developed a method for estimating the total mass and size range of particles in plumes.

Many other plumes have been monitored on satellite images. The Tobalchik volcano on the Kamchatka Peninsula of eastern Russia erupted on July 6, 1975. Jayaweera, Seifert, and Wendler (1976) measured the length and orientation of the plume on 14 NOAA images acquired from July 9 through August 17, 1975. The most spectacular image was acquired July 18, when the length of the plume was at least 960 km. From the distance between the plume and its shadow, which was measured from the visible image, and from the known sun elevation, the height of the plume on July 28 was estimated at 6.5 km.

FLOODS

Periodic floods in the Mississippi River Valley have been monitored by successive generations of remote sensing technology. The 1973 floods were analyzed with Landsat MSS images by the U.S. Geological Survey (Deutsch and Ruggles, 1974). A decade later the floods of 1983 were analyzed by comparing Landsat TM images acquired before and during the high water (Sabins, 1987, Figures 11-16, 11-17). The 1993 floods were analyzed with radar images and GIS processing of digital terrain data and repeated TM images.

Radar Images, 1993 Mississippi River Floods

Figure 13-32 shows ERS-1 satellite radar images acquired before and during the 1993 floods of the Missouri and Mississippi Rivers upstream from St. Louis, Missouri. Dark signatures in both images are smooth surfaces such as calm water, highways, and airport runways. In the preflood image of May (Figure 13-32A), some stretches of river have gray signatures due to ripples caused by wind or currents. The brightest signatures in the May image are urban areas that are suitably oriented to produce strong backscatter, as described for New Orleans in Chapter 6. These bright signatures are lacking in the

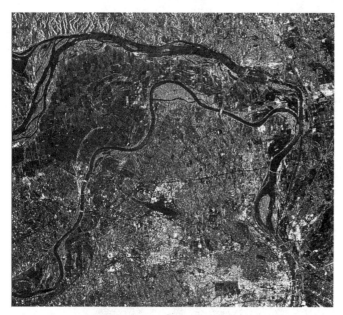

A. Preflood image, acquired May 30, 1993.

B. Flood-stage image, acquired July 14, 1993.

Figure 13-32 ERS-1 satellite radar images acquired before and during floods of 1993 along the Mississippi and Missouri Rivers near St. Louis, Missouri. Courtesy EOSAT Corporation.

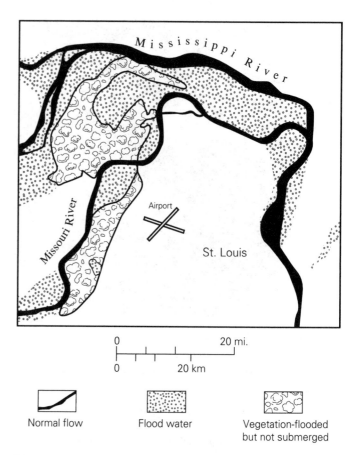

Figure 13-33 Map of flood conditions along Mississippi and Missouri Rivers interpreted from ERS-1 images (Figure 13-32).

July image because the viewing geometry is slightly different. In the flood-stage image of July (Figure 13-32B), the brightest signatures are vegetated areas adjacent to the rivers in the western portion of the image that are relatively dark in the May image. This vegetation was flooded, but not covered by water, in the July image. Much of the incident radar energy was scattered from the above-water vegetation. Some energy pene-

trated the vegetation and was multiply scattered from the water surface and plant stalks to produce the bright signatures. Figure 13-33 is an interpretation map of the two images. Water in the normal stage is shown in black; floodwaters are shown as a stippled pattern; vegetation that is flooded but not covered by water is shown as the vegetation pattern.

Brakenridge and others (1994) interpreted an ERS-1 image acquired July 16, 1993, of the Mississippi River in southern Iowa, 200 km upstream from St. Louis. The margins of the flood were interpreted from the image and transferred to a detailed topographic map. From this display the elevation of the water was measured on the east and west banks of the floodplain, which are 10 km apart. The floodwaters were consistently 1.2 to 2.4 m higher on the west. This elevation difference is attributed to strong inflow from the Des Moines River, which enters the west side of the valley. These radar measurements of flood height are not as accurate as those from flood gauges, but an image provides an instantaneous record of conditions for a broad stretch of the floodplain.

These examples illustrate some applications of radar images for monitoring floods. The all-weather capability enables im-

GIS Images of the 1993 Mississippi Flood Several groups conducted GIS analyses of the 1993 floods for different applications. Scientists at the EROS Data Center (EDC) determined the maximum flood extent from Landsat images. The flood outlines were merged with other attributes to evaluate impacts of the flood on wildlife habitat, fish spawning areas, and nesting sites of bald eagles. Potential flooding of known toxic waste sites was also evaluated. The EOSAT Corporation registered a TM image with an ERS-1 radar image of the St. Louis region. Both images were acquired on the same day during the peak of the flood. The two images were merged using the IHS method described in Chapter 8. The merged image shows spectral characteristics of vegetation and urban areas (from the TM) plus the outline of floodwaters (from radar).

Figure 13-34 is a shaded relief model of the Mississippi Valley at Quincy, Illinois, that was produced from digital terrain data. The broad, flat floodplain is cut by channels that contain the river during normal stages. Eroded upland areas bound the floodplain. Plate 26A,B shows subscenes of TM 2-3-4 images acquired during normal flow and during flood stage. In these IR color images, vigorous vegetation has red signatures. Fallow fields and the city of Quincy on the east bank have light blue signatures. In the normal-flow image (Plate 26A) the relatively clear water in the river has a dark-blue signature. In the flood-stage image (Plate 26B) the floodwaters are predominantly light blue because of suspended silt. Vigorous vegetation is abundant in the upland areas during the flood season because of heavy rainfall.

The ERDAS Corporation registered the TM images and terrain data and used them as input for a GIS analysis of the flood. Plate 26C,D is the output map and legend. The normal-flow image was classified to produce a map showing categories of land cover in the floodplain. The classification map was compared with the flood-stage image to define four categories of flooded land cover: agriculture, forest, grassland, and urban. The terrain data were used to define "potentially flooded" categories that would be inundated if future water levels exceeded the level shown on the flood-stage image of July 18, 1993 (Plate 26B). This example illustrates how GIS analysis of timely remote sensing images can be used to analyze flood hazards. This analysis can readily be expanded to the entire Mississippi River system and to other flood-prone areas.

FOREST AND RANGE FIRES

Remote sensing and GIS are becoming significant tools for managing forest and range resources, which includes combating fires.

U. S. Forest Service Experience

The U.S. Forest Service has employed remote sensing and GIS in three stages: before, during, and after a fire.

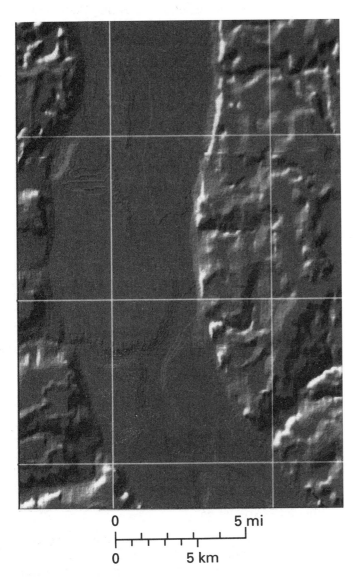

0   5 mi

0   5 km

Figure 13-34 Shaded relief model of the Mississippi River at Quincy, Illinois, computed from digital terrain data. Model is illuminated from the west. This data set is one attribute used in the GIS analysis of the 1993 flooding in the Mississippi River Valley. Courtesy C. Erikson, ERDAS Corporation.

ages to be acquired despite cloud cover that prevents imaging in the visible and reflected IR regions recorded by aerial photographs and Landsat. Radar images record differences in roughness that indicate different flood conditions.

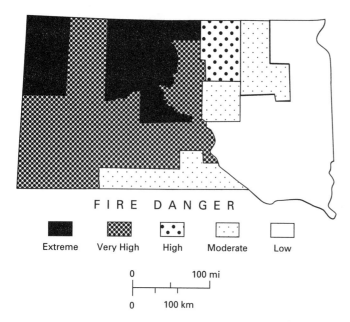

FIRE DANGER

| Extreme | Very High | High | Moderate | Low |

0   100 mi

0   100 km

Figure 13-35 Map of the fire danger for the grasslands in South Dakota for June 6, 1988. The map was prepared from *NDVI* data and weather information. From Eidenshenk and others (1989, Figure 3B).

Before a Fire During a fire season, it is important to assess the changing potential for fires in order to plan suppression measures. In the Great Plains of the United States the National Weather Service issues daily grassland fire danger warnings during periods of fire hazard. A fire danger index is calculated by integrating daily weather conditions (humidity, wind speed, cloud cover, and temperature) with an estimate of the *percentage of green composition* (*PGC*), which represents the fuel condition. State forestry agencies estimate *PGC* by clipping vegetation from predetermined sample sites and measuring the percentage of green components. A shortcoming of the warning index is that the *PGC* information is measured at fewer than three sites within a state and at intervals of 2 weeks or longer. A greater number of sample sites and reduced time between measurements could improve the accuracy of fire danger warnings. Eidenshenk and others (1989, 1990) demonstrated that *NDVI* data from AVHRR could be converted into *PGC* information for grasslands of the northern Great Plains. Fire danger ratings based on *PGC* derived from *NDVI* compared favorably with ratings based on traditional field measurements of *PGC*. During the 1988 fire season weekly *NDVI* maps were generated for the northern Great Plains, using methods described in Chapter 12. These maps were converted into *PGC* values and combined with daily weather data to produce maps of the fire danger index. Figure 13-35 is a fire danger index map using *NDVI* data for grasslands in South Dakota for June 6, 1988. These maps are significant improvements over the traditional maps.

During a Fire Once a fire breaks out, its progress must be monitored in order to deploy personnel and equipment for suppression. Forest fires are routinely monitored by aircraft thermal IR scanners that record the fire front, despite dense smoke cover (Chapter 5). GPSs (Chapter 8) are used in the aircraft and by fire crews to coordinate their efforts.

During the 1994 fire season in the Pacific Northwest the U.S. Forest Service employed AVHRR images and GIS in the suppression of several major wildfires. Greer (1994) and Prevedel (1995) described this effort. Daily AVHRR images of the western United States were received directly from the satellites by an antenna in San Diego, California, where the data were processed in preliminary fashion and transmitted over a network to the Forest Service office in Ogden, Utah. AVHRR band 3 (3.55 to 3.93 μm, thermal IR) data were used to monitor fires because this band penetrated the smoke plumes and recorded the high temperatures of the fires. Despite the coarse spatial resolution (1.1 km), fires less than 10 hectares (ha) in area were detected. From each regional image, subareas were extracted that covered individual fire complexes. Data for each subarea were geocoded and registered to geographic attributes (latitude-and-longitude grid, political boundaries, topography, lakes and drainage, federal land boundaries). The combined data sets were available over a local network to local fire-fighting centers. The goal was to provide the information within 1 hour after the satellite pass. The images were used for daily planning and coordination of fire fighting. In addition, the images were used for public information and to compile a historic record of each fire complex.

After a Fire After a fire is extinguished, the area must be surveyed to assess damage, plan rehabilitation efforts, and monitor recovery of the forest. Remote sensing and GIS play vital roles in all three aspects. The U.S. Forest Service is using TM images acquired before and after a wildfire to assess damage to vegetation and monitor its recovery (Greer, 1994).

Other Fire Experience

During the 1990 fire season 1.28 million ha of boreal forest were burned in central Alaska. Kasischke and others (1993) used seasonal AVHRR images to map the extent of the burned terrain. *NDVI* images were prepared for the periods of June 15 to 30 and August 1 to 15. The *NDVI* values were normalized to one another to compensate for seasonal changes. The images were registered, and a third image showing difference in *NDVI* for the two dates was prepared. Pixels with *NDVI* difference values greater than 23 were classed as burned terrain and assigned a distinctive color on the base image. The number and areas of fire scars determined from AVHRR were compared with statistics from the Alaska Fire Service. The *NDVI*-derived map detected 89.5 percent of fires greater than 2000 ha with no false alarms. The AVHRR total burned area, however, was only 61 percent of the area mapped in the field. Some fires occurred after the August 15 cutoff date for AVHRR; use of later images could improve the accuracy of the satellite data.

One of the largest and most destructive fires in recent history, called the Great China Fire, occurred in May 1987 in northeast China. Over 200 lives and 50,000 homes were lost. Extensive areas in southeast Siberia were also burned. Cahoon and others (1994) used satellite images to document the occurrence, growth, and extent of the fires. The fires were first recognized in Siberia on a May 2 thermal IR image acquired by the DMSP satellite (Chapter 4). Color composite images of AVHRR bands 1, 2, and 4 were used to distinguish among smoke, clouds, water, and burned and unburned vegetation. One of these images, acquired on May 6, showed extensive smoke plumes from the Siberia fires, plus the first evidence of four new fires in China. One of these fires was controlled early, but the remaining three expanded and eventually coalesced into a single burned expanse. Images showed that between May 7 and 8, strong winds spread the three fires 60 to 70 km eastward, burning 350,000 ha. Despite vigorous suppression efforts the fires burned for 3 more weeks and consumed in excess of 1.3 million ha of prime forest. Images acquired after the fires showed an additional 13 million ha of burned forest in Siberia.

Kaufman, Tucker, and Fung (1990) developed a method to assess trace gases and particulates emitted from fires in the tropics, using AVHRR data. For the 3-month dry season of 1987, an area in Brazil (6.5° to 15.5°S; 55° to 67°W) had 240,000 fires that emitted 1×10^{14} g of particulates (ash), 7×10^{12} g of CH_4, 2×10^{14} g of CO, and 1×10^{15} g of CO_2.

COMMENTS

Remote sensing images are used extensively to monitor the effects of natural hazards including earthquakes, land movements, volcanoes, floods, and fires. Predicting hazardous events is much more difficult, but progress is being made. Images show the locations of active faults that are likely sites of earthquakes. Images that are merged with earthquake data and fault traces clearly display areas that are prone to seismic activity. Radar interferograms show the deformation caused by earthquakes and volcanic eruptions. Interferograms also have the potential to identify terrain deformation that may be a precursor to earthquakes or eruptions. Floods are monitored with Landsat and radar images. Merging these images with digital topographic data provides quantitative information on the volume and extent of flooding. Maps of *NDVI* from AVHRR data can identify areas of dry vegetation that are prone to range and forest fires.

QUESTIONS

1. How would you interpret the following images for evidence of active faults? stereo pairs of aerial photographs, nighttime thermal IR images, and SIR images

2. How would you interpret the same set of images for evidence of landslides?
3. Plan a remote sensing system using available aircraft and satellite technology to monitor volcanoes such as those of the Cascade Range in Washington and Oregon.
4. Describe an all-weather satellite system for monitoring the extent of flooded terrain.

REFERENCES

Allen, C. R., 1975, Geological criteria for evaluating seismicity: Geological Society of America Bulletin, v. 86, p. 1041–1057.

Avouac, J. P. and G. Peltzer, 1993, Active tectonics in southern Xinjiang, southern China—analysis of terrace riser and normal fault scarp degradation along the Hotan-Qira fault system: Journal of Geophysical Research, v. B12, p. 21,773–21,807.

Blanchard, M. B., R. Greeley, and R. Goettleman, 1974, Use of visible, near-infrared, and thermal infrared remote sensing to study soil moisture: Proceedings of the Ninth International Symposium of Remote Sensing of Environment, p. 693–700, Environmental Research Institute of Michigan, Ann Arbor, MI.

Brakenridge, G. R., J. C. Knox, E. D. Paylor, and F. J. Magilligan, 1994, Radar remote sensing aids study of the great flood of 1993: EOS Transactions of American Geophysical Union, v. 75, p. 521–527.

Cahoon, D. R., B. J. Stocks, J. S. Levine, W. R. Cofer, and J. M. Pierson, 1994, Satellite analysis of the severe 1987 forest fires in northern China and southeastern Siberia: Journal of Geophysical Research, v. 99, p. 18,627–18,638.

Chorowicz, J. and others, 1992, SPOT satellite monitoring of the eruption of Nevado Sabancaya volcano (southern Peru): Remote Sensing of the Environment, v. 42, p. 43–49.

Clanton, U. S. and E. R. Verbeek, 1981, Photographic portrait of active faults in the Houston metropolitan area, Texas in Etter, E. M., ed., Houston area environmental geology—surface faulting, ground subsidence, and hazard liability: p. 70–113, Houston Geological Society, Houston, TX.

Crippen, R. E., 1992, Measurements of subresolution terrain displacements using SPOT panchromatic imagery: Episodes, v. 15, p. 56–61.

Crippen, R. E. and R. G. Blom, 1992, The first visual observations of fault movements from space—the 1992 Landers earthquake: EOS, Transactions of the American Geophysical Union, v. 73, p. 364.

De Silva, S. L. and P. W. Francis, 1991, Volcanoes of the central Andes: Springer-Verlag, Berlin.

Deutsch, M. and F. H. Ruggles, 1974, Optical data processing and projected applications of the ERTS-1 imagery covering the 1973 Mississippi River floods: Water Resources Bulletin, v. 10, p. 1023–1039.

Didier, M., P. Briole, and A. Arnaud, 1995, Deflation of Mt. Etna monitored by spaceborne radar interferometry: Nature, v. 375, p. 567–570.

Eidenshenk, J. C., R. E. Burgan, and R. H. Haas, 1990, Monitoring fire fuels condition using time series of advanced very high resolution radiometer data: Proceedings of Second International Symposium on Advanced Technology in Natural Resource Management, Washington, DC.

Eidenshenk, J. C., R. H. Haas, D. M. Zokaites, and D. O. Ohlen, 1989, Integration of remote sensing and GIS technology to monitor fire danger in the northern Great Plains: Proceedings of the National GIS Conference, p. 944–956, Ottawa, Canada.

Foxworthy, B. L. and M. Hill, 1982, Volcanic eruptions of 1980 at Mount St. Helens, the first 100 days: U. S. Geological Survey Professional Paper 1249.

Friedman, J. D., D. Frank, H. H. Kieffer, and D. L. Sawatzky, 1981, Thermal infrared surveys of the May 18 crater, subsequent lava domes, and associated deposits in Lipman, P. W. and D. L. Mullineaux, eds., The 1980 eruption of Mount St. Helens, Washington: U.S. Geological Survey Professional Paper 1250, p. 279–293.

Gaonac'h, H., J. Vandemeulebrouck, J. Stix, and M. Halbwachs, 1994, Thermal infrared satellite measurements of volcanic activity at Stromboli and Vulcano: Journal of Geophysical Research, v. 99, p. 9477–9485.

Glaze, L. S., P. W. Francis, S. Self, and D. W. Rothery, 1989, The 16 September 1986 eruption of Lascar volcano, north Chile—satellite investigations: Bulletin of Volcanology, v. 51, p. 149–160.

Greer, J. D., 1994, GIS and remote sensing for wildland fire suppression and burned area restoration: Photogrammetric Engineering and Remote Sensing, v. 60, p. 1059–1064.

Hauksson, E., L. M. Jones, K. Hutton, and D. Eberhart-Phillips, 1993, The 1992 Landers earthquake sequence—seismological observations: Journal of Geophysical Research, v. 98, p. 19,835–19,858.

Hutton, L. K., L. M. Jones, E. Hauksson, and D. D. Given, 1991, Seismotectonics of southern California in Slemmons, D. B. and others, eds., Neotectonics of North America: Geological Society of America Decade Map, v.1, p. 133–152.

Jayaweera, K. O. L. F., R. Seifert, and G. Wendler, 1976, Satellite observations of the eruption of Tolbachik volcano: EOS, Transactions of the American Geophysical Union, v. 57, p. 196–200.

Jennings, C. W., comp., 1975, Fault map of California with locations of volcanoes, thermal springs, and thermal wells: California Division of Mines and Geology, California Geologic Data Map Series Map 1, Sacramento, CA.

Johnson, R. W. and others, 1995, Taking petrologic pathways toward understanding Rabaul's restless caldera: EOS, Transactions of the American Geophysical Union, v. 76, p. 171–188.

Kasischke, E. S. and others, 1993, Monitoring of wildfires in boreal forests using large area AVHRR NDVI composite image data: Remote Sensing of the Environment, v. 45, p. 61–71.

Kaufman, Y. J., C. J. Tucker, and I. Fung, 1990, Remote sensing of biomass burning in the tropics: Journal of Geophysical Research, v. 95, p. 9927–9939.

Kieffer, H. H., D. Frank, and J. D. Friedman, 1981, Thermal infrared surveys at Mount St. Helens prior to the eruption of May 18 in Lipman, P. W. and D. L. Mullineaux, eds., The 1980 eruption of Mount St. Helens, Washington: U.S. Geological Survey Professional Paper 1250, p. 257–277.

Kienle, J., K. G. Dean, H. Garbiel, and W. I. Rose, 1990, Satellite surveillance of volcanic ash plumes, application to aircraft safety: EOS Transactions, American Geophysical Union, v. 71, p. 266.

Lipman, P. W. and D. L. Mullineaux, eds., 1981, The 1980 eruption of Mount St. Helens, Washington: U.S. Geological Survey Professional Paper 1250.

Massonnet, D., M. Rossi, C. Carmona, F. Adragna, G. Peltzer, K. Feigl, and T. Rabaute, 1993, The displacement field of the Landers earthquake mapped by radar interferometry: Nature, v. 364, p. 138–142.

Massonnet, D., P. Briole, and A. Arnaud, 1995, Deflation of Mount Etna monitored by spaceborne radar interferometry: Nature, v. 375, p. 567–570.

Merifield, P. M. and D. L. Lamar, 1975, Active and inactive faults in southern California viewed from Skylab: NASA Earth Resources Survey Symposium, NASA TM X-58168, v. 1, p. 779–797.

Mori, J., K. Hudnut, L. Jones, E. Hauksson, and K. Hutton, 1992, Rapid response to Landers quake: EOS, v. 73, p. 417–418.

Newhall, C. G. and R. S. Punongbayan, 1996, Fire and mud—the eruption and lahars of Mount Pinatubo, the Philippines: University of Washington Press, Seattle, WA.

Oppenheimer, C., P. W. Francis, D. A. Rothery, and R. W. T. Carlton, 1993, Infrared image analysis of volcanic thermal features—Lascar volcano, Chile, 1984–1992: Journal of Geophysical Research, v. 98, p. 4269–4286.

Peltzer, G., K. W. Hudnut, and K. L. Feigl, 1994, Analysis of coseismic surface displacement gradients using radar interferometry—new insights into the Landers earthquake: Journal of Geophysical research, v. 99, p. 21,971–21,981.

Peltzer, G., P. Tapponnier, and R. Armijo, 1989, Magnitude of late Quaternary left-lateral displacements along the north edge of Tibet: Science, v. 246, p. 1285–1289.

Perry, W. J. and I. H. Crick, 1976, Aerial thermal infrared survey, Rabaul area, New Britain, Papua New Guinea, 1973 in Johnson, R. W., ed., Volcanism in Australasia: Elsevier Scientific Publishing Co., New York, NY.

Piper, D. J. W. and others, 1985, Sediment slides and turbidity currents on the Laurentian Fan—sidescan sonar investigation near the epicenter of the 1929 Grand Banks earthquake: Geology, v. 13, p. 538–541.

Prata, A. J., 1989, Observations of volcanic ash clouds in the 10–12 μm window using AVHRR/2 data: International Journal of Remote Sensing, v. 10, p. 751–761.

Prevedel, D. A., 1995, Project Sparkey—a strategic wildfire monitoring package using AVHRR satellites and GIS: Photogrammetric Engineering and Remote Sensing, v. 61, p. 271–278.

Prior, D. B., J. M. Coleman, and L. E. Garrison, 1979, Digitally acquired undistorted side-scan sonar images of submarine landslides, Mississippi River Delta: Geology, v. 7, p. 423–425.

Reddy, C. S. S., A. Bhattacharya, and S. K. Srivastav, 1993, Nighttime TM short wavelength infrared data analysis of Barren Island volcano, South Andaman, India: International Journal of Remote Sensing, v. 14, p. 783–787.

Rib, H. T. and T. Liang, 1978, Recognition and identification in Schuster, R. L. and R. J. Krizek, eds., Landslides—analysis and control: National Academy of Sciences Special Report 176, ch. 3, p. 34–69, Washington, DC.

Rose, W. I. and others, 1995, Ice in the 1994 Rabaul eruption cloud—implications for volcano hazard and atmospheric effects: Nature, v. 375, p. 477–479.

Rothery, D. A., 1989, Volcano monitoring by satellite: Geology Today: v. 5, p. 128–132.

Rothery, D. A. and P. W. Francis, 1900, Short wavelength infrared images for volcano monitoring: International Journal of Remote Sensing, v. 11, p. 1665–1667.

Sabins, F. F., 1987, Remote sensing—principles and interpretation, second edition: W. H. Freeman & Co., New York, NY.

Shreve, R. L., 1968, The Blackhawk landslide: Geological Society of America Special Paper No. 108, Boulder, CO.

Sieh, K. and others, 1993, Near-field investigations of the Landers earthquake sequence, April to July, 1992: Science, v. 260, p. 171–176.

Tapponnier, P. and P. Molnar, 1977, Active faulting and tectonics in China: Journal of Geophysical Research, v. 82, p. 2905–2930.

Tapponnier, P. and P. Molnar, 1979, Active faulting and Cenozoic tectonics of the Tien Shan, Mongolia, and Baykal regions: Journal of Geophysical Research, v. 84, p. 3425–3459.

U.S. Geological Survey, 1986, Atlas of the Exclusive Economic Zone, western United States: U.S. Geological Survey, Investigations Report I-1792, Washington, DC.

Vedder, J. G. and R. E. Wallace, 1970, Map showing recently active breaks along the San Andreas and related faults between Cholame Valley and Tejon Pass, California: U.S. Geological Survey, Miscellaneous Geologic Investigations, Map I-574.

Verbeek, E. R. and U. S. Clanton, 1981, Historically active faults in the Houston metropolitan area, Texas in Etter, E. M., ed., Houston area environmental geology—surface faulting, ground subsidence, and hazard liability: p. 28–68, Houston Geology Society, Houston, TX.

Wen, S. and W. I. Rose, 1994, Retrieval of sizes and total masses of particles in volcanic clouds using AVHRR bands 4 and 5: Journal of Geophysical Research, v. 99, p. 5421–5431.

Yeats, R. S., K. Sieh, and C. A. Allen, 1996, Geology of earthquakes: Oxford University Press, New York, NY.

Costa, J. E. and G. F. Wieczorek, 1987, Debris flows/avalanches—process, recognition, and mitigation: Geological Society of America, Reviews in Engineering Geology, v. 7.

Hays, W. W., ed., 1981, Facing geologic and hydrologic hazards, earth science considerations: U.S. Geological Survey Professional Paper 1240-B.

Mintzer, O., 1983, Engineering applications in Colwell, R. N., ed., Manual of remote sensing, second edition: ch. 32, p. 1955–2109, American Society for Photogrammetry and Remote Sensing, Falls Church, VA.

Newhall, C. G. and D. Dzurisin, 1988, Historical unrest at large calderas of the world: U.S. Geological Survey, Bulletin 1598.

Pereira, A. C. and A. W. Setzer, 1996, Comparison of fire detection in sanannas using AVHRR'S channel 3 and TM images: International Journal of Remote Sensing, v. 17, p. 1925–1937.

Reiter, L., 1990, Earthquake hazard analysis: Columbia University Press, New York, NY.

Rosenfeld, C. A., 1980, Observations on the Mount St. Helens eruption: American Scientist, v. 68, p. 494–509.

Slosson, J. E., A. G. Keene, and J. A. Johnson, eds., 1992, Landslides/landslide mitigation: Geological Society of America, Reviews in Engineering Geology, v. 9.

Tazieff, H. and J. C. Sabroux, eds., 1983, Forecasting volcanic events: Developments in Volcanology: v. 1, p. 1–635.

Wright, T. L. and T. C. Pierson, 1992, Living with volcanoes: U.S. Geological Survey, Circular 1073.

ADDITIONAL READING

Barrett, E. C., K. A. Brown, and A. Micallef, 1991, Remote sensing for hazard monitoring and disaster assessment—marine and coastal applications in the Mediterranean region: Gordon and Breach Science Publishers, Philadelphia, PA.

CHAPTER

COMPARING IMAGE TYPES:
SUMMARY

This chapter compares images from the full range of wavelength regions that cover the same site—Death Valley, California. Images are compared at two scales: regional images covering the entire valley, and local images covering Cottonball Basin at the north end of the valley.

DEATH VALLEY

Death Valley in the Mojave Desert of California is a useful site for comparing images for the following reasons:

1. From the early days of remote sensing Death Valley has been used as a test site for virtually all remote sensing systems. Table 14-1 lists the characteristics of the images that are compared in this chapter.
2. A wide range of materials are well exposed. Table 14-2 lists the major materials and some of their characteristics.
3. The geology is well known.

Earlier in this book, radar images (Chapter 6) of Death Valley were interpreted.

Terrain and Materials

Death Valley is an elongate, closed basin bounded on the west by the Panamint Range and on the east by the Funeral and Black Mountains (Figure 14-1). The floor of the valley includes the lowest elevation in the United States, 86 m below sea level, at Badwater. The high point of the Panamint Range is only 30 km west of Badwater and has an elevation of 3368 m. Hunt and Mabey (1966), Drewes (1963), Troxel and Wright (1987), and others have described the geology of the area. The

materials at Death Valley belong to the major categories of bedrock, alluvial fan deposits, and basin deposits (Table 14-2).

Bedrock in the mountain ranges consists of a wide range of igneous, sedimentary, and metamorphic rocks that have been folded and faulted. The Landsat image (Plate 27A) shows the rugged topography and variegated colors that characterize this terrain.

Alluvial fan deposits occur at the base of the mountains from which they were derived. The fans slope gently toward the basin and are cut by stream channels that only flow during the infrequent rains. As shown in the TM image and map, the fans on the west margin of the valley are broad and extensive, while those on the east margin are small. This asymmetry is caused by continuing subsidence of the eastern valley floor along the zone of active border faults at the foot of the Black Mountains. The fans at the foot of the Panamint Range form broad, coalescing depositional surfaces; at the foot of the Black Mountains, older fans are down-dropped along the border faults and are buried by younger gravels. Alluvial fans consist of two different materials: coarse gravel and desert pavement. Coarse gravel ranges in size from boulders to sand and is characterized by cobbles of bedrock. These young deposits occur along the dry streams that merge into active fans. Desert pavement consists of older fan deposits that have been weathered to a relatively smooth surface crudely resembling a tile surface. Because of the prolonged weathering, the pavement materials are coated with desert varnish, which imparts a dark signature.

Basin deposits on the floor of Death Valley are predominantly salts that formed during evaporation of a saline lake that formerly occupied the basin. Halite (sodium chloride or rock salt), plus carbonate and sulfate deposits are the major minerals. Weathering and recrystallization produce a range of

Table 14-1 Characteristics of remote sensing images of Death Valley

System	Date	Wavelength	Spatial resolution	Property recorded
Visible and reflected IR regions				
Large format camera	Oct. 1984	0.5 to 0.7 μm	~5 m	Albedo and topography
Landsat TM bands 1–5, 7	Nov. 1982	0.45 to 2.35 μm	30 m	Reflected solar energy
SPOT panchromatic	Aug. 1991	0.51 to 0.73 μm	10 m	Albedo and topography
Thermal IR region				
Landsat TM band 6	Nov. 1982	10.5 to 12.5 μm	120 m	Radiant temperature
Aircraft images	Mar. 1977	7.9 to 13.5 μm	10 m	Radiant temperature
TIMS	1982, 1994	8.0 to 12.0 μm	18 m	Spectral emissivity
Radar				
Aircraft (X band)	July, 1976	3.0 cm	10 m	Roughness
Seasat (L band)	Sept. 1978	23.5 cm	25 m	Roughness
SIR-B (L band)	Oct. 1984	23.5 cm	40 m	Roughness
SIR-C (X, C, L bands)	1994	3, 6, 23 cm	25 m	Roughness

Table 14-2 Materials at Death Valley

Material	Roughness, cm	Color	Comments
Bedrock			
Bedrock	Very high	Varied	Panamint, Funeral, and Black Mountains
Alluvial fan deposits			
Coarse gravel	12	Medium gray	Active channels
Desert pavement	1.0	Black	Older surfaces on west side of valley
Basin deposits			
Rough halite	29	Brown	Weathered and recrystallized
Intermediate halite	6	White to brown	Periodic flooding reduces roughness
Carbonate and sulfate deposits	2.0	Light gray	Puffy surface with intermixed sand
Sand and fine gravel	1.0	Gray	Irregular narrow belt
Floodplain deposits	0.2	White	Brine-saturated silt and clay
Vegetation			
Vegetation	High	Green	Furnace Creek Ranch, scattered drainages

surface roughness on the saline deposits. The valley floor is bordered by an irregular belt of sand and fine gravel that is too narrow to show in Figure 14-1. The belt is shown, however, in the regional images (Figure 14-2). Runoff from rainfall forms floodplain deposits of silt and clay that are saturated with brine. The flat surfaces of the floodplains have a thin coating of salt evaporated from the brine.

The major structural features are faults and turtlebacks, which are shown in the geologic map (Figure 14-1). The major faults are the Keane Wonder fault, the Furnace Creek fault, and the border fault along the west margin of the Black Mountains.

At the surface these are high-angle faults that bring bedrock of the upthrown blocks against unconsolidated gravel deposits of the downthrown blocks. Recent movement along the border faults causes the following features: (1) slightly eroded fault scarps formed on the basement rocks; (2) wineglass-shaped canyons along the west face of the Black Mountains; and (3) fault scarps cutting the recent gravel deposits. *Turtlebacks* are anticlinal arches of metamorphic rocks once covered by volcanic breccia that has largely been removed by erosion. Three turtlebacks occur along the west margin of the Black Mountains and are identified in Figure 14-1. Wright, Otton,

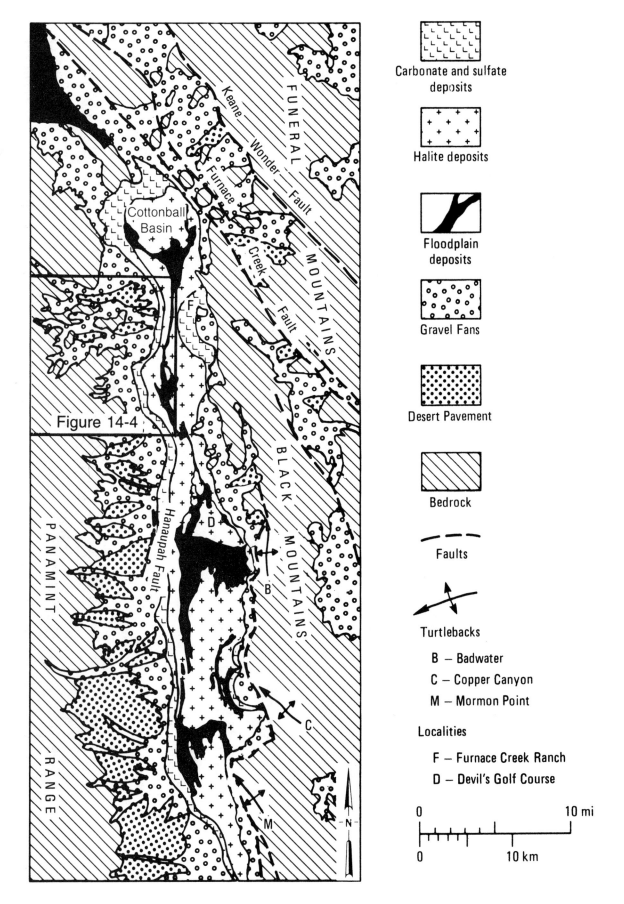

Figure 14-1 Regional map of Death Valley, California.

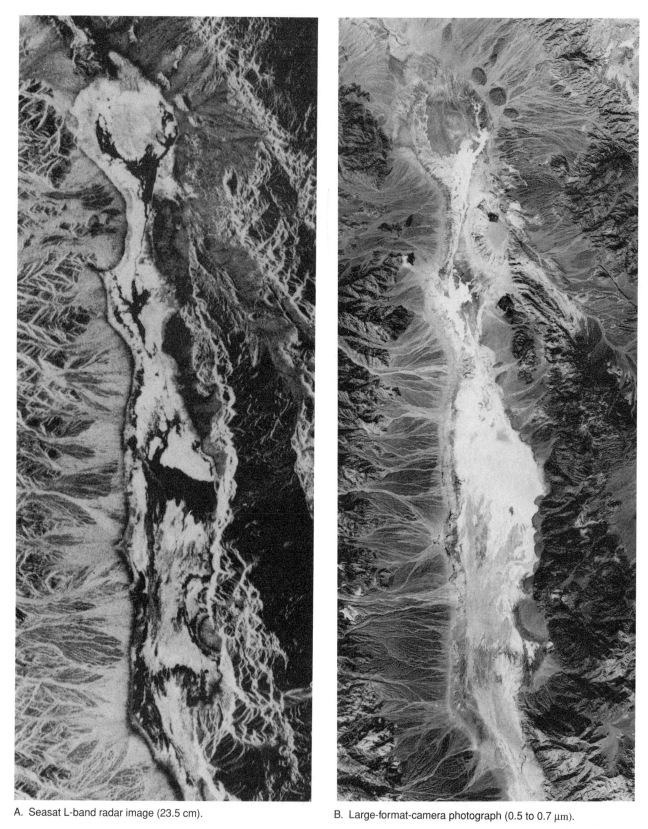

A. Seasat L-band radar image (23.5 cm).

B. Large-format-camera photograph (0.5 to 0.7 μm).

Figure 14-2 Regional satellite images of Death Valley.

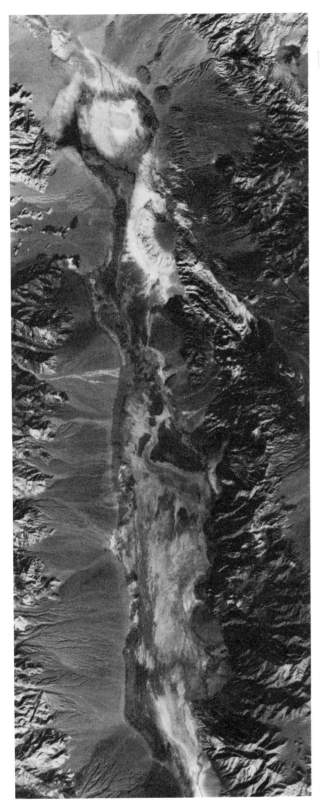

C. Landsat TM band 5 image (1.55 to 1.75 μm).

D. Landsat TM band 6 image (10.40 to 12.50 μm).

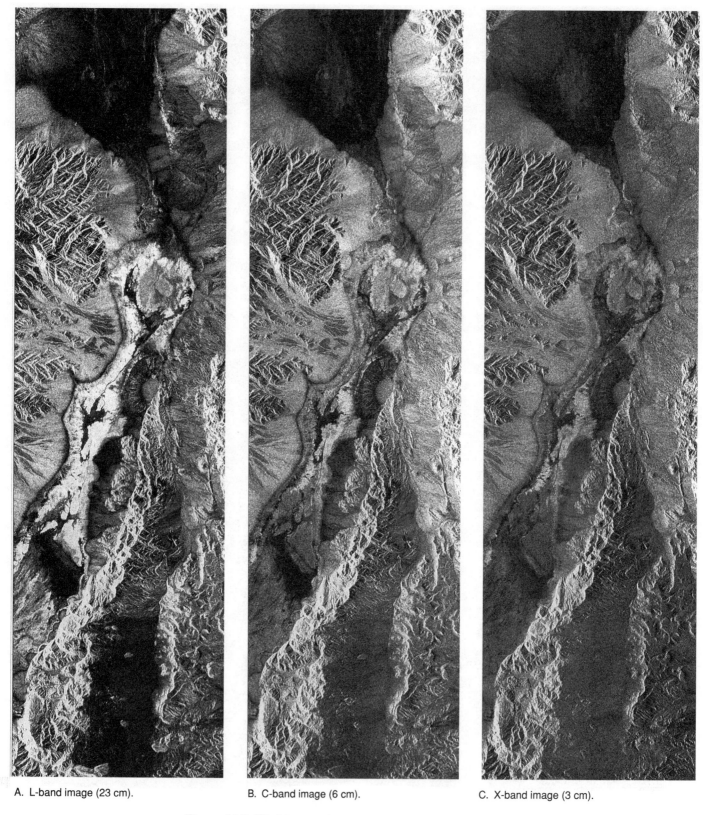

A. L-band image (23 cm).   B. C-band image (6 cm).   C. X-band image (3 cm).

Figure 14-3 SIR-C images of Death Valley with 40° depression angle. All images are VV parallel-polarized. The images are 25 km wide. Courtesy A. Freeman and E. O'Leary, JPL.

TABLE 14-3 Signatures of materials on regional images

Material	TM 2-4-7	Thermal IR, daytime	Seasat radar	SIR-C (X, C, L)
Bedrock				
Bedrock	Variegated	Highlights and shadows	Highlights and shadows	Highlights and shadows
Alluvial fan deposits				
Coarse gravel	Gray	Cool	Rough	Light orange
Desert pavement	Orange, red	Warm	Smooth	Medium gray
Basin deposits				
Rough halite	White, blue	Medium	Rough	Dark orange
Intermediate halite	White, blue	Medium	Intermediate	Gray
Carbonate and sulfate deposits	Magenta	Warm	Intermediate	Medium orange
Sand and fine gravel	Dark	Warm	Smooth	Medium to dark gray
Floodplain deposits	Blue	Cool	Smooth	Dark gray to black
Vegetation				
Vegetation	Green	Cool	Rough	Orange

and Troxel (1974) reviewed various theories for the origin of these features.

Comparison of Regional Images

Figure 14-2 shows regional images that were acquired by the large format camera (LFC), Landsat TM bands 5 and 6, and Seasat. These images represent the visible, reflected IR, thermal IR, and radar spectral regions. In addition, Figure 14-3 shows three SIR-C multispectral radar images. The seven images in this comparison were acquired from 1982 (TM) to 1994 (SIR-C). The terrain has not changed during this time, aside from shifts in the floodplains (Crowley and Hook, 1996, Plate 3). The scale and coverage of the geologic map (Figure 14-1) is comparable to that of these regional images. Table 14-3 lists the materials and compares their signatures on the different images.

Large-Format-Camera Photograph Figure 14-2B is a minus-blue panchromatic photograph from the LFC acquired from a Space Shuttle mission in 1984. This photograph has the finest spatial resolution (~5 m) of all the regional images, as shown by the detailed expression of drainage patterns along the west margin of the valley. LFC photographs were also acquired as stereo pairs (not shown here), which is a useful capability.

Landsat TM Visible and Reflected IR Images Figure 14-2C is a Landsat TM band 5 (reflected IR) image acquired in 1982. Plate 27A is a Landsat TM image of bands 2, 4, and 7 in blue, green, and red that was digitally enhanced using the contrast stretch, IHS, and nondirectional edge-enhancement techniques described in Chapter 8. This color combination is effective for imaging arid terrains around the world. Different materials within the bedrock and alluvial fans are shown by variations in color and texture in the color image. The few occurrences of vegetation are shown by green signatures, such as the irrigated lawns and trees at Furnace Creek Ranch (Figure 14-1) and small patches of native shrubs elsewhere in the image. Comparing Plate 27A with the black-and-white image of TM band 5 (Figure 14-2C) illustrates the superiority of color images. Although the LFC photograph (Figure 14-2B) has a higher spatial resolution, the broad spectral range of the TM color image is more effective for discriminating materials in bedrock, alluvial fans, and deposits of the valley floor.

Landsat TM Thermal IR Image Figure 14-2D is a 1982 TM daytime thermal IR image with a spatial resolution of 120 m. Bedrock on the east and west margins of the image is shown by highlights (warm signatures) and shadows (cool) caused by solar heating and shading of the rugged topography. Highlights and shadows are minimal in the low-relief terrain of the alluvial fans and valley floor. Along the west margin of the valley, the desert pavement has a distinctly warmer signature than the coarse gravel. The desert pavement is coated with dark desert varnish that absorbs visible energy. The resulting increase in kinetic temperature causes warm signatures in the thermal IR image. Within the valley floor the floodplain deposits have conspicuous cool signatures caused by evaporative cooling of the moist silt and mud. The carbonate and sulfate deposits around the margins of the valley floor have distinct warm signatures caused by their puffy texture and low density,

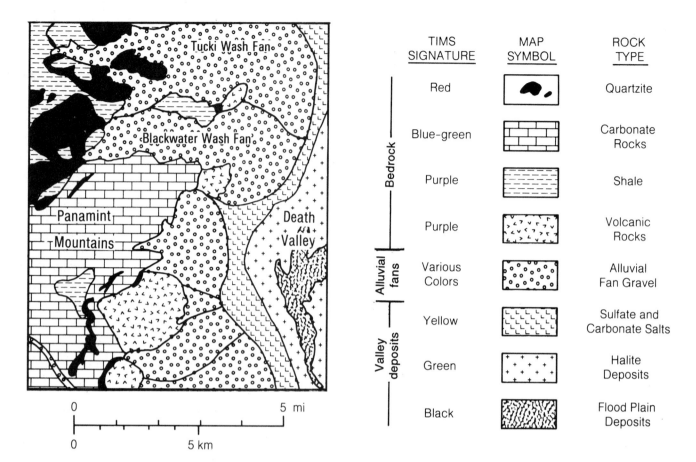

TIMS SIGNATURE	MAP SYMBOL	ROCK TYPE
Bedrock		
Red		Quartzite
Blue-green		Carbonate Rocks
Purple		Shale
Purple		Volcanic Rocks
Alluvial fans		
Various Colors		Alluvial Fan Gravel
Valley deposits		
Yellow		Sulfate and Carbonate Salts
Green		Halite Deposits
Black		Flood Plain Deposits

Figure 14-4 Interpretation map of TIMS image of northwest portion of Death Valley. After Gillespie, Kahle, and Palluconi (1984, Figure 3).

which results in a low thermal inertia for these deposits. Chapter 5 noted that materials with low thermal inertias are relatively warm in the daytime and cool at night.

Thermal IR Multispectral Image Plate 28A is a color composite of three bands of imagery acquired by the thermal IR multispectral scanner (TIMS), which was described in Chapter 5. The aircraft data were processed at Jet Propulsion Laboratory (JPL) using a technique that suppresses temperature information while emphasizing spectral emissivity features (Kahle, 1987). The resulting images for TIMS bands 1 (8.4 µm), 3 (9.2 µm), and 5 (10.7 µm) were composited in blue, green, and red, respectively, to produce the color image. The image was acquired from an aircraft in 1982 with ground resolution cells of 18 by 18 m. The image covers a portion of the northwest margin of Death Valley shown in Figure 14-1. Figure 14-4 is an interpretation map that shows the rock types with their color signatures on the TIMS image. For comparison, Plate 28B is a TM 2-4-7 image of the area covered by the TIMS image. The location of Figure 14-4 and the color images are shown in Figure 14-1.

Bedrock is exposed in the western portion of the TIMS image and consists of volcanic rocks and slightly metamorphosed quartzite, carbonate rocks, and shale. In the TIMS image (Plate 28A), quartzite outcrops and the adjacent detritus have a conspicuous red color. Quartzite consists of silica, which has a strong spectral response at the wavelengths recorded by band 5. This band is shown in red, which accounts for the red signature of quartzite outcrops in Plate 28A. Carbonate rocks are blue-green. Shale and volcanic rocks are purple. These bedrock units are more readily distinguished in the TIMS image than in the corresponding Landsat image (Plate 28B).

The alluvial fans along the west side of the valley consist of gravel eroded from bedrock. The lithology of the gravel in each fan is determined by the type of bedrock source. The volcanic rocks (purple signature), carbonate rocks (green signature), and quartzite (red signature) form alluvial fans with matching signatures.

Deposits on the valley floor are shown in the eastern portion of the TIMS image. A belt with a distinctive yellow signature represents deposits of sulfate and carbonate salts. The green

Table 14-4 Roughness criteria for SIR-C bands; depression angle (γ) is 40°

Roughness category	X band $\lambda = 3$ cm	C band $\lambda = 6$ cm	L band $\lambda = 23$ cm
Smooth	$h < 0.21$	$h < 0.42$	$h < 1.60$
Intermediate	$h = 0.21$ to 1.19	$h = 0.42$ to 2.38	$h = 1.60$ to 9.11
Rough	$h > 1.19$	$h > 2.38$	$h > 9.11$

signature represents halite (rock salt) deposits. The very dark signature represents moist floodplain deposits. Evaporative cooling at all TIMS wavelengths results in the dark signature. The multispectral information content of the TIMS image provides greater discrimination of thermal spectral properties than TM band 6 (Figure 14-2D), which records a single broad band from 10.4 to 12.5 μm.

Seasat Radar Image Figure 14-2A is a Seasat L-band (23.5-cm wavelength) image acquired with a look direction toward the east. The steep depression angle (70°) causes topographic features in the bedrock to show severe layover toward the west. The low-relief terrain of the alluvial fans and valley floor is undistorted. The geometry and wavelength of Seasat are well suited for distinguishing materials of the fans and valley floors based on differences in surface roughness, as described and illustrated in Chapter 6. Table 14-2 lists the roughness of the materials, and Table 14-3 lists their radar signatures.

The Seasat image (Figure 14-2A) and the thermal IR image (Figure 14-2D) record different reactions between materials and electromagnetic energy of different wavelengths. Despite these differences, some materials are distinguishable on both images. Floodplain deposits have similar patterns and dark signatures on both images. The dark radar signature is caused by the smooth surface. The dark IR signature is caused by evaporative cooling. Older alluvial fan deposits have similar outlines on both images, but the signatures are reversed. The dark signature on Seasat is due to the smooth surface. The bright IR signature is due to radiation of solar energy absorbed by the dark coating of desert varnish.

SIR-C Radar Images Figure 14-3 shows SIR-C images acquired at wavelengths of 3 cm (X band), 6 cm (C band), and 23 cm (L band). All images were acquired at the same depression angle (40°). Table 14-4 lists roughness criteria for the three SIR-C bands. Using these criteria and the surface roughness of materials (Table 14-1), one can identify materials on the different SIR-C images.

Plate 27B is a color composite image of the SIR-C bands. The X-band image is shown in blue, C-band image in green, and L-band image in red.

Structural Features on Regional Images Structural features (faults and turtlebacks) are well expressed in the Landsat

2-4-7 image of Plate 27A. The Copper Canyon turtleback in the southeast portion of the image is shown by its topography and the contrast between the dark red of the turtleback surface and the bright orange of the overlying breccia. Faults are shown by the abrupt truncation of topographic trends and the juxtaposition of contrasting materials, such as bedrock against alluvial deposits. At this midlatitude the sun elevation was sufficiently low to produce shadows and highlights that emphasize the major faults. The SIR-C images enhance the relatively smooth, rounded surfaces of the turtlebacks that contrast with the rugged topography of the Black Range. The fault scarps along the west margin of the Black Range have strong highlights because of the east look direction. The LFC photograph provides maximum spatial detail for structural mapping, but the lack of color hampers identification of materials.

Local Images

Figure 14-5 compares a series of larger-scale local images of Cottonball Basin, located at the north end of Death Valley (Figure 14-1). The SPOT panchromatic image with a ground resolution cell of 10 m provides maximum resolution. TM bands 4 and 7 compare the shorter- and longer-wavelength bands in the reflected IR region. Figure 14-5D is a nighttime thermal IR image acquired from an aircraft. The Tucki alluvial fan in the southwest portion of the image includes areas of desert pavement and coarse gravel (Figure 14-5G), but these materials are indistinguishable on the nighttime image. The solar heating that caused the warm signature of desert pavement on the daytime thermal image (Figure 14-2D) was dissipated before the nighttime image was acquired.

TIMS Image Plate 28C is a TIMS aircraft image of Cottonball Basin acquired April 6, 1994. The following interpretation of the image is summarized from Crowley and Hook (1996). Each of the six bands was digitally processed to extract spectral emissivity information (Chapter 5). Instead of producing a color composite image of three selected bands, an alternate approach to extracting information was employed. An unsupervised classification was applied to all six bands of emissivity data. The area was restricted to the floor of Cottonball Basin and excluded the adjacent alluvial fans and bedrock. The classification resulted in the classes shown in color on Plate 28C. The background, shown in gray tones, is a radiant temperature

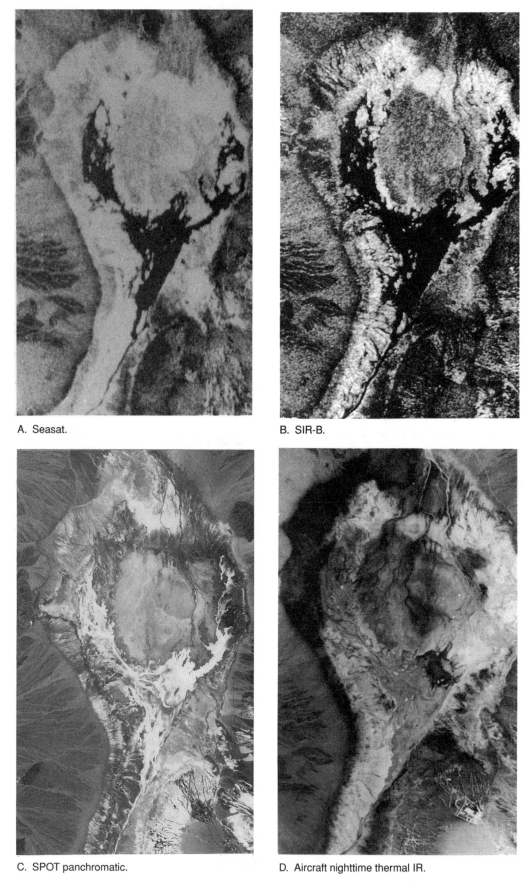

A. Seasat.

B. SIR-B.

C. SPOT panchromatic.

D. Aircraft nighttime thermal IR.

Figure 14-5 Local images and map of Cottonball Basin, Death Valley.

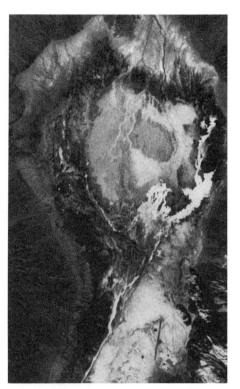

E. Landsat TM band 4.

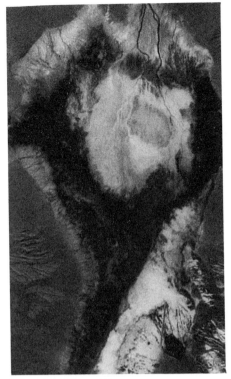

F. Landsat TM band 7.

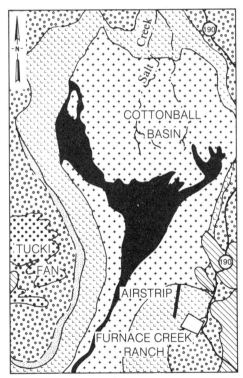

G. Map and legend.

SYMBOL	MATERIAL
	Carbonate and sulfate deposits
	Halite deposits
	Floodplain deposits
	Sand and fine gravel
	Alluvial fan gravel
	Desert pavement
	Bedrock

0   4 mi

0   4 km

image derived from the TIMS bands. The next step was to identify the material represented by each class, or color. Spectra were plotted for each class using values for the six TIMS bands. Field spectra and samples were obtained from localities in Cottonball Basin that represent each of the unsupervised classes. Comparing these data with the TIMS spectra resulted in the following identifications:

Color on Plate 28C	Material
Red	Gypsum
Orange	Silty halite and carbonate deposits
Yellow	Thenardite (sodium sulfate)
Green	Quartz-rich gravel and floodplain deposits
Light green	Mixed silicate and evaporite minerals
Cyan	Massive halite and silty halite
Dark blue	Clay (illite and muscovite) and alluvial deposits

This analysis of TIMS data illustrates the use of multispectral thermal IR data for distinguishing and classifying materials.

SIR-C Multipolarization Image Plate 27B is a SIR-C multispectral regional image that shows three wavelengths of parallel-polarized imagery composited in color. Plate 28D is an enlarged multipolarization image of Cottonball Basin that shows three polarizations of L-band imagery in color; VV polarization is shown in blue, HH in red, and HV in green. Chapter 6 pointed out that incident radar energy is depolarized by multiple reflections that are primarily caused by vegetation. At Cottonball Basin the only vegetation occurs at Furnace Creek Ranch and vicinity, which is shown in the southeast corner of the map in Figure 14-5G. In the multipolarization image, vegetation has a green signature because the cross-polarized HV component is shown in green. The absence of green signatures elsewhere in the image indicates that

1. there is no other vegetation in the scene;
2. none of the other materials depolarizes incident energy to the degree of vegetation.

Dark signatures on the multipolarization image correspond to smooth materials (floodplain, desert pavement) with essentially no backscattering or dark signatures at all polarizations. White signatures around the margin of Cottonball Basin are materials that strongly backscatter all three polarizations at similar intensities. Very rough halite, with its jagged surface and cavities, is a likely cause. The center and western margin of the basin have a purplish red signature, which indicates strong HH and weaker VV returns. The materials are halite with intermediate roughness and carbonate and sulfate deposits. A belt with light-blue signatures occurs between the floodplain deposits (dark) and carbonate and sulfate deposits (purplish red) on the west margin. The light-blue signature corresponds to halite. Fieldwork is needed to determine how these deposits differ from halite with bright and purplish red signatures.

SUMMARY

Death Valley has been imaged by remote sensing systems that extend from the visible through the microwave spectral regions. We can objectively compare and evaluate the information provided by each image. Details of evaluations may differ for other environments, but signatures on all images are determined by the wavelength of energy and the interaction with materials at that wavelength. Photographs (panchromatic, normal color, or IR color) from aircraft or satellites provide maximum spatial resolution and convenient stereo coverage. The principal limitation is that photographs do not record the longer wavelengths of the reflected region where many important interactions occur. SPOT and IRS images have the finest spatial resolution (10 and 20 m) of current satellite systems, with worldwide coverage of digital data and stereo capability. These images are restricted to the visible and photographic band of the reflected IR spectral region. Landsat TM images cover the visible and reflected regions with adequate spatial resolution (30 m) for many applications. Color images can be produced from the optimum combination of bands for different environments. For semiarid and arid terrain, such as Death Valley, TM bands 2, 4, and 7 in blue, green, and red are optimum. For tropical terrain with hazy atmosphere, bands 4, 5, and 7 are optimum.

Two types of thermal IR systems are available. The most common type records a single broad band of imagery within the range from 8 to 14 μm. The signatures are determined by thermal inertia, evaporative cooling, and emissivity of materials. The second type (TIMS) records multiple narrow bands that are processed to identify specific materials. These multispectral IR images are especially sensitive to variations in silica content.

Radar signatures are determined by two terrain characteristics: (1) topography that is expressed as highlights and shadows, and (2) surface roughness that is expressed as variations in brightness. Radar signatures are also strongly influenced by the look direction and depression angle of the system. Experimental multispectral radar systems (SIR-C) provide images at different wavelengths and polarizations. Future evaluations will determine the usefulness and applications of these images.

COMMENTS

This chapter is an appropriate conclusion for this book. We began with general principles of remote sensing and then analyzed systems and images for each of the electromagnetic regions. The digital processing of images was followed by applications of remote sensing in diverse fields. This final chapter compared and evaluated images representing the range of electromagnetic regions that are employed in remote sensing.

QUESTIONS

1. For your local area, list the major terrain categories (types of vegetation, rocks, land use, land cover, and so forth). For each category, predict the signature in the following images: Landsat TM IR color image, Seasat radar image, and daytime and nighttime thermal IR images.
2. Explain the reason for each predicted signature.
3. Acquire and interpret various types of remote sensing images of your area. For each terrain category, compare the actual image signatures with your predictions.
4. Explain any differences between your predictions and the actual signatures. Field checking should help resolve these discrepancies.

REFERENCES

Crowley, J. K. and S. J. Hook, 1996, thermal infrared multispectral scanner (TIMS) study of playa evaporite minerals in Death Valley, California: Journal of Geophysical Research, v. 101, p. 643–660.

Drewes, H., 1963, Geology of the Funeral Peak Quadrangle, California, on the east flank of Death Valley: U.S. Geological Survey Professional Paper 413.

Gillespie, A. R., A. B. Kahle, and F. D. Palluconi, 1984, Mapping alluvial fans in Death Valley, California, using multichannel thermal infrared images: Geophysical Research Letters, v. 11, p. 1153–1156.

Hunt, C. B. and D. R. Mabey, 1966, Stratigraphy and structure of Death Valley, California: U.S. Geological Survey Professional Paper 494-A.

Kahle, A. B., 1987, Surface emittance, temperature, and thermal inertia derived from thermal infrared multispectral scanner (TIMS) data for Death Valley, California: Geophysics, v. 52, p. 858–874.

Troxel, B. W. and L. A. Wright, 1987, Tertiary extensional features, Death Valley region, eastern California: Centennial Field Guide—Cordilleran Section, Geological Society of America, p. 121–132.

Wright, L. A., J. K. Otton, and B. W. Troxel, 1974, Turtleback surfaces of Death Valley viewed as a phenomenon of extensional tectonics: Geology, v. 2, p. 53–54.

ADDITIONAL READING

Crowley, J. K., 1993, Mapping playa evaporite minerals with AVIRIS data—a first report from Death Valley, California: Remote Sensing of Environment, v. 44, p. 337–356.

Kahle, A. B., J. P. Schieldge, M. J. Abrams, R. E. Alley, and C. J. LeVine, 1981, Geologic applications of thermal inertia imaging using HCMM data: Jet Propulsion Laboratory Publication 81-55, Pasadena, CA.

Kruse, F. A., A. B. Lefkoff, and J. B. Dietz, 1993, Expert system-based mineral mapping in northern Death Valley, California/Nevada, using the airborne visible/infrared imaging spectrometer (AVIRIS): Remote Sensing of Environment, v. 44, p. 309–336.

Sabins, F. F., 1984, Geologic mapping of Death Valley from thematic mapper, thermal infrared, and radar images: International Symposium on Remote Sensing of the Environment, Proceedings of the third thematic conference, Remote Sensing for Exploration Geology, p. 139–152, Environmental Research Institute of Michigan, Ann Arbor, MI.

APPENDIX

BASIC GEOLOGY FOR REMOTE SENSING

On images of land areas, the terrain is a direct expression of the geology of the area. Many interpreters are concerned with nongeologic subjects (such as land use, environment, or forestry), but an understanding of the geology will contribute to the overall understanding of the image. This brief review emphasizes the major rock types and geologic structures that are expressed on images. Additional information is given in general geology texts, such as Press and Siever (1993).

ROCK TYPES

Rocks belong to three major categories—sedimentary, igneous, and metamorphic—which are described in the following sections.

Sedimentary Rocks

Material that has been transported and deposited by water or wind forms sedimentary rocks characterized by layers, called *beds* or *strata,* formed during deposition. The surfaces separating strata are called *bedding planes*. Outcrops of sedimentary strata typically have a banded appearance on images. *Sedimentary rocks* are divided into the broad categories of clastic and chemical rocks.

Clastic Rocks Erosion of older rocks produces fragments and particles that are transported and deposited to form *clastic rocks*. After deposition, the fragments are compressed and cemented to form rocks. The consolidated rocks are classified on the basis of particle size before consolidation, as follows:

Consolidated rocks	*Unconsolidated particles*
Conglomerate	Boulders and gavel
Sandstone	Sand
Siltstone	Mud
Shale	Clay

Clastic rocks differ in their resistance to erosion; sandstone and conglomerate typically form ridges, but shale and siltstone form valleys.

Chemical Rocks Minerals dissolved in water may be removed from solution to form *chemical rocks,* either by chemical precipitation or by uptake into organisms whose shells and skeletons form sediments after death. Algae and shellfish remove calcium carbonate from seawater as the mineral calcite, which accumulates to form the rock called *limestone*. Half of the calcium atoms in calcite may be replaced by magnesium to form the mineral and rock called *dolomite*. Evaporation of seawater produces deposits of gypsum, anhydrite, and salt. Because of its low density, salt may migrate upward into the overlying strata to form cylindrical plugs called *salt domes*. Open spaces between the grains of sedimentary rocks are called *pores* and contain fresh or salt water or, less commonly, oil and gas.

Igneous Rocks

Igneous rocks are rocks that have cooled from molten material, called *magma*. Igneous rocks are assigned to the classes of intrusive and extrusive rocks.

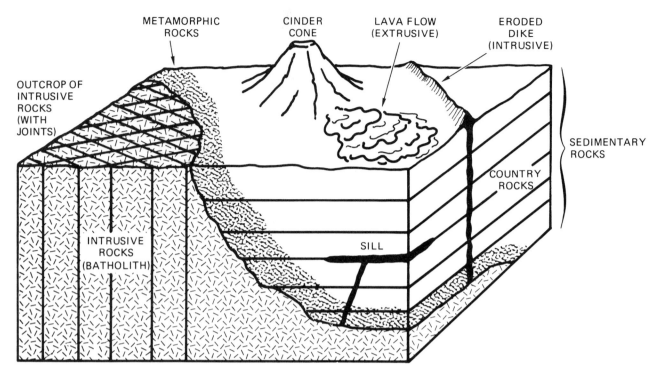

Figure A-1 Intrusive and extrusive rocks.

Intrusive Rocks Magma that invades country rocks (areas of older rock) cools to form *intrusive rocks*. Based on the relationship to the country rock, intrusive rocks are classed as batholiths, dikes, or sills (Figure A-1). *Batholiths* are large, irregularly shaped masses of intrusive rock that cut across the structure of the country rock. Erosion of the overlying country rocks exposes outcrops of the batholith rock. The outcrops are commonly cut by intersecting fractures, called *joints,* that give a distinctive appearance to these rocks on images. Typical batholithic rocks crop out in the Peninsular Ranges of southern California, which are illustrated on Landsat images in Chapter 3.

As shown in Figure A-1, country rocks are commonly layered or stratified. Tabular bodies of igneous rock intruded parallel with the layers are *sills,* such as the Palisades sill that crops out along the west bank of the Hudson River. Tabular bodies of intrusive rock that cut across the layers of country rock are called *dikes.* Erosion of the overlying country rock may expose dikes in the form of ridges, such as the Chinese Wall at Yellowstone National Park. Some dikes are less resistant to erosion than the surrounding country rock and weather to form depressions. On images, dikes are recognized by their linear shape and cross-cutting relationships to the country rock.

As the intruded magma slowly cools, silicate minerals form relatively coarse crystals that are visible to the unaided eye.

There are two major types of intrusive rocks, classed according to silica content.

Granite	High silica content and typically light gray to pink.
Gabbro	Relatively low silica content and high content of iron- and magnesium-bearing minerals. Gabbro is dark and has a higher density than granite.

Extrusive Rocks Magma may reach the surface as a liquid called *lava,* which cools to form lava flows. The lava may also be explosively ejected into the air as cinders and ash that can accumulate as volcanoes, or *cinder cones* (Figure A-1). Because lava cools rapidly at the surface, the resulting *extrusive rocks* are very fine grained. Three major categories of extrusive rocks are as follows:

Rhyolite	The extrusive equivalent of granites. Rhyolite has a high silica content and is typically pink.
Andesite	A rock having intermediate silica content. Andesite volcanoes form the "Ring of Fire" around the Pacific Ocean and make up the Andes of South America and the Cascade Range of northwestern United States. Mount St. Helens and Pinatubo (Chapter 13) are andesite volcanoes.

Basalt   The extrusive equivalent of gabbro. Basalt has a low silica content and high content of iron-and-magnesium-bearing minerals. Basalt is dark and has a relatively high density. Flows of basalt with rough surfaces are called *aa;* those with smooth, ropy surfaces are called *pahoehoe.*

Metamorphic Rocks

Heat and pressure may transform igneous and sedimentary rocks into *metamorphic rocks* (Figure A-1). The heat and pressure associated with intrusive rocks may convert original country rock into the following kinds of metamorphic rocks:

Original rock	*Metamorphic equivalent*
Sandstone	Quartzite
Shale and siltstone	Schist
Limestone	Marble

Intensive regional compression and deep burial can also produce metamorphic rocks.

STRUCTURAL GEOLOGY

Stresses within the earth's crust may deform the rocks to produce geologic structures that may be mapped on various types of images. Most sedimentary rocks were deposited as horizontal strata, but later uplift and tilting caused the strata to become inclined.

Strike and Dip

Geologists use the terms *strike* and *dip* to describe the orientation and degree of inclination. The line formed by the intersec-

tion between an inclined surface and a horizontal plane is the *strike* of the inclined surface (Figure A-2). The orientation of the strike is described by its geographic azimuth, in this case N45°E. *Dip* is measured in the vertical plane oriented normal to the strike and measures the inclination below horizontal of the inclined surface in degrees. In Figure A-2, the dip is 30° toward the southeast. The strike-and-dip symbol is used to record the *attitudes* of structures on geologic maps. Exposed bedding planes of dipping strata are called *dipslopes.* Erosion of dipping strata produces ledges called *antidip scarps* (Figure A-2) that face the opposite direction from dipslopes.

The ability to recognize strike and dip is fundamental for interpreting geologic structure from remote sensing images. Except for highly deformed areas, beds generally dip less than 45°, and the following criteria apply. Dipslopes are relatively broad and are traversed by relatively long streams that flow in the direction of dip. Antidip scarps are narrow and have a few short drainage channels that flow opposite to the dip direction. The orientation of shadows and highlights is an important key for interpreting images acquired with inclined illumination, such as low-sun-angle aerial photographs (Chapter 2) and radar images (Chapter 6). If the scene in Figure A-2 were photographed in the morning with sun shining from the southeast, the dipslope would form a bright expanse and the shadowed antidip scarp would form a narrow dark band; in the afternoon the highlights and shadows would be reversed.

Folds

Compressive stresses form folds called anticlines and synclines. As shown in Figure A-3, in *anticlines* the beds dip away from the axis; in *synclines* the beds dip toward the axis. Note the orientation of the strike-and-dip symbols and the attitude of the dipslopes. The *plunge* ("nose") of a fold is marked by arcuate outcrop patterns. Erosion exposes older beds in the center of anticlines and younger beds in the center of synclines. On geologic maps the axes of folds are shown by long lines with

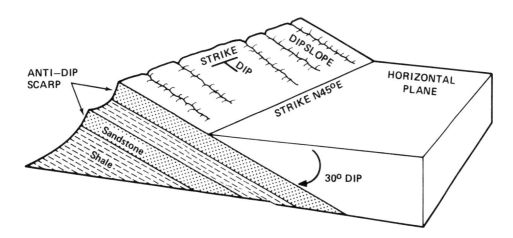

Figure A-2 Strike and dip of inclined beds.

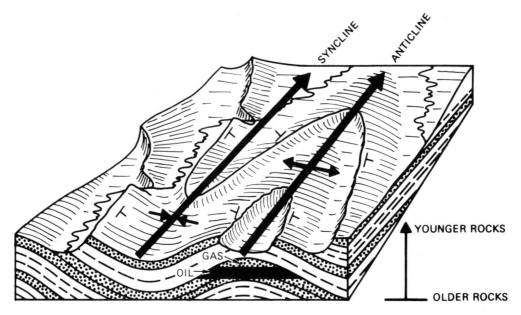

Figure A-3 Anticline, syncline, and gas and oil fields.

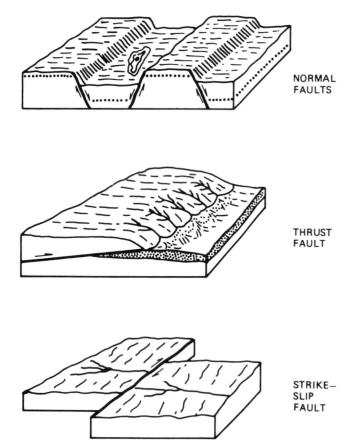

NORMAL
FAULTS

THRUST
FAULT

STRIKE–
SLIP
FAULT

Figure A-4 Types of faults.

short crossing arrows that point away from the crest of anticlines and toward the center of synclines (Figure A-3). Chapter 2 shows aerial photographs of the Alkali anticline and syncline in Wyoming.

Organic matter contained in shales (*source rocks*) generates oil and gas, which migrate into the pore spaces of sandstones and limestones (*reservoir rocks*). Rocks in the subsurface are also saturated with water. Because hydrocarbons are less dense than water, oil and gas float on the water, becoming concentrated in the high points of structures, such as crests of anticlines. Such concentrations may form oil fields (Figure A-3). Chapter 3 shows a Landsat TM image of anticlinal oil fields in the Thermopolis area of Wyoming. Chapter 10 shows radar images of anticlinal oil fields in Papua New Guinea. Oil fields also occur along faults and in ancient reefs, sandbars, and river channels.

Joints and Faults

Joints, mentioned earlier in connection with batholiths, are intersecting sets of fractures along which no movement has occurred (Figure A-1). Erosion along joints produces linear depressions that are readily interpreted on images. *Faults* are fractures along which appreciable movement has occurred. Strike-and-dip terminology is also used to describe the attitude of fault surfaces. Faults are assigned to the following categories according to the kind of relative movement: normal faults, thrust faults, and strike-slip faults.

Normal Faults The relative movement of rocks on either side of a fault is indicated by arrows, as shown in Figure A-4. *Normal faults* dip less than 90°, and the rocks on the upper side of the fault have moved downward relative to those on the lower side. Topographic escarpments caused by faults are called *fault scarps*. In the Basin and Range Province of the United States, normal faults separate the uplifted mountain ranges from the intervening down-dropped basins, as shown in the Landsat mosaic of Nevada in Chapter 3. The uplifted blocks are called *horsts,* and the down-dropped basins are called *graben.*

Thrust Faults *Thrust faults* are horizontal or gently dipping faults in which the rocks overlying the fault have moved up and over the rocks below the fault (Figure A-4). Thrust faults are common in the Papuan Fold and Thrust Belt (Chapter 10) and in the Central Range of Irian Jaya (Chapter 7).

Strike-Slip Faults Faults along which the rocks have moved laterally are *strike-slip faults* (Figure A-4). Because of their steep dip, strike-slip faults have linear surface traces, as seen by the traces of the San Andreas, Garlock, and other faults in the Landsat mosaic of southern California (Plate 13). The offset of features on opposite sides of a strike-slip fault is used to determine the relative displacement. In Figure A-4 an observer standing on the stream and looking across the fault at the offset channel will note that the lateral displacement of the stream is toward the right, making this a right-lateral, strike-slip fault. In left-lateral faults the features are displaced to the left.

REFERENCES

Press, F. and R. Siever, 1994, *Understanding earth,* W. H. Freeman and Co., N.Y.

GLOSSARY

The following glossary includes terms, abbreviations, and acronyms commonly employed in remote sensing. The glossary definitions refer to the applications for which the terms are used in this text and omit applications outside the field of remote sensing. Definitions of geologic and geographic terms may be found in any standard text on those subjects or in the *Glossary of Geology*, edited by R. L. Bates and J. A. Jackson and published by the American Geological Institute.

absorption band Wavelength interval within which electromagnetic radiation is absorbed by the atmosphere or by other substances.

absorption features Downward excursions in spectral reflectance curves.

absorptivity Capacity of a material to absorb incident radiant energy.

active remote sensing Remote sensing methods that provide their own source of electromagnetic radiation to illuminate the terrain. Radar is one example.

additive primary colors Blue, green, and red. Filters of these colors transmit the primary color of the filter and absorb the other two colors.

adiabatic cooling Refers to decrease in temperature with increasing altitude.

advanced very high resolution radiometer (AVHRR) Multispectral scanner on NOAA polar-orbiting environmental satellites that acquires five spectral bands of data (0.55 to 12.50 μm) with a ground resolution of 1.1 km. Data are widely used for regional and global mapping of vegetation.

aerial magnetic survey Geophysical survey that records variations in the earth's magnetic field.

air base Ground distance between optical centers of successive overlapping aerial photographs.

airborne laser fluorosensor (ALF) Active UV scanner system for detecting oil slicks.

airborne visible/infrared imaging spectrometer (AVIRIS) Hyperspectral scanner that acquires 224 images in the spectral region from 0.4 to 2.4 μm.

albedo (A) Ratio of the amount of electromagnetic energy reflected by a surface to the amount of energy incident upon the surface.

Almaz Russian unmanned satellite that acquired S-band radar images.

along-track scanner Scanner with a linear array of detectors oriented normal to flight path. The *IFOV* of each detector sweeps a path parallel with the flight direction.

alteration Changes in color and mineralogy of rocks surrounding a mineral deposit that are caused by the solutions that formed the deposit. Suites of alteration minerals commonly occur in zones.

amplitude For waves, the vertical distance from crest to trough.

analog display A form of data display in which values are shown in graphic form, such as curves. Differs from digital displays, in which values are shown as arrays of numbers.

analog mosaic Mosaic compiled manually by assembling individual images.

angular beam width In radar, the angle subtended in the horizontal plane by the radar beam.

angular field of view Angle subtended by lines from a remote sensing system to the outer margins of the strip of terrain that is imaged by the system.

angular resolving power Minimum separation between two resolvable targets, expressed as angular separation.

anomaly An area on an image that differs from the surrounding normal area. For example, a concentration of vegetation within a desert scene constitutes an anomaly.

antenna Device that transmits and receives microwave energy in a radar system.

aperture Opening in a remote sensing system that admits electromagnetic radiation to the film or detector.

Apollo U.S. lunar exploration program of satellites with crews of three astronauts.

apparent thermal inertia (ATI) An approximation of thermal inertia calculated as 1 minus albedo divided by the difference between daytime and nighttime radiant temperatures.

ASA index Index of the American Standards Association designating film speed, or sensitivity to light. Higher values indicate higher sensitivity. The ASA index is being replaced by the ISO index.

atmosphere Layer of gases that surrounds some planets.

atmospheric correction Image-processing procedure that compensates for effects of selectively scattered light in multispectral images.

atmospheric scattering Multiple interactions between light rays and the gases and particles in the atmosphere.

atmospheric window Wavelength interval within which the atmosphere readily transmits electromagnetic radiation.

attitude Angular orientation of a remote sensing system with respect to a geographic reference system.

automatic gain control (AGC) Electronic device that reduces saturation of image data.

AVHRR Advanced very high resolution radiometer.

AVIRIS Airborne visible/infrared imaging spectrometer.

azimuth Geographic orientation of a line given as an angle measured in degrees clockwise from north.

azimuth direction In radar images, the direction in which the aircraft or spacecraft is heading. Also called *flight direction.*

azimuth resolution In radar images, the spatial resolution in the azimuth direction.

background Area on an image or the terrain that surrounds an area of interest, or target.

backscatter In radar, the portion of the microwave energy scattered by the terrain surface directly back toward the antenna.

backscatter coefficient A quantitative measure of the intensity of energy returned to a radar antenna from the terrain.

band A subdivision within an electromagnetic region. For example, the visible region is subdivided into the blue, green, and red bands.

banding A defect in scanner images in which alternating groups of scan lines are brighter or darker overall than adjacent groups.

bandpass The wavelength interval recorded by a detector.

bandwidth The wavelength interval recorded by a detector. Also called *spectral resolution.*

base-height ratio Air base divided by aircraft height. This ratio determines vertical exaggeration on stereo models.

batch processing Method of data processing in which data and programs are entered into a computer that carries out the entire processing operation with no further instructions.

bathymetry Configuration of the seafloor.

beam A focused pulse of energy.

binary Numerical system using the base 2.

bit Contraction of *binary digit*, which in digital computing represents an exponent of the base 2.

blackbody An ideal substance that absorbs all the radiant energy incident on it and emits radiant energy at the maximum possible rate per unit area at each wavelength for any given temperature. No actual substance is a true blackbody, although some substances, such as lampblack, approach its properties.

brightness Magnitude of the response produced in the eye by light.

brute-force radar See **real-aperture radar.**

byte A group of eight bits of digital data.

calibration Process of comparing an instrument's measurements with a standard.

calorie Amount of heat required to raise the temperature of 1 g of water by 1°C.

camera Framing system that records images on photographic film.

camouflage detection photographs Another term for *IR color photograph.*

cardinal point effect In radar, very bright signatures caused by optimally oriented corner reflectors, such as buildings.

cathode ray tube (CRT) A vacuum tube with a phosphorescent screen on which images are displayed by an electron beam.

CCD Charge-coupled detector. Tiny light-sensitive detectors.

CCT Computer-compatible tape.

CD-ROM Compact disk read-only memory. Storage medium for digital data.

centerpoint The optical center of a photograph.

change-detection images A difference image prepared by digitally comparing images acquired at different times. The gray tones or colors of each pixel record the amount of difference between the corresponding pixels of the original images.

charge-coupled detector (CCD) A device in which electrons are stored at the surface of a semiconductor.

chlorosis Yellowing of plant leaves resulting from an imbalance in the iron metabolism caused by excess concentrations of copper, zinc, manganese, or other elements in the plant.

circular scanner Scanner in which a faceted mirror rotates about a vertical axis to sweep the detector *IFOV* in a series of circular scan lines on the terrain.

classification Process of assigning individual pixels of an image to categories, generally on the basis of spectral reflectance characteristics.

coastal zone color scanner (CZCS) Multispectral scanner on NOAA polar-orbiting satellite designed to measure chlorophyll concentrations in the oceans.

color composite image Color image prepared by combining three individual images in blue, green, and red.

color ratio image Color composite image prepared by combining three ratio images.

combination image A color image composited from two or more different images, such as Landsat and SPOT.

complementary colors Two primary colors (one additive and the other subtractive) that produce white light when added together. Red and cyan are complementary colors.

computer-compatible tape (CCT) General term for magnetic data tapes that can be read by most computers.

conduction Transfer of electromagnetic energy through a solid material by molecular interaction.

contact print A reproduction from a photographic negative in direct contact with photosensitive paper.

contrast enhancement Image-processing procedure that improves the contrast ratio of images. The original narrow range of digital values is expanded to utilize the full range of available digital values. Also called *contrast stretch.*

contrast ratio On an image, the ratio of reflectance between the brightest and darkest parts of the image.

convection Transfer of heat through the physical movement of heated matter.

corner reflector Cavity formed by two or three smooth planar surfaces intersecting at right angles. Electromagnetic waves (especially radar) entering a corner reflector are reflected directly back toward the source.

COSMIC Computer Software Management and Information Center, University of Georgia. This facility distributes computer programs developed by U.S. government-funded projects.

cross-polarized A radar return in which the electric field vibrates in a direction normal to the direction of the transmitted pulse. Cross-polarized images may be HV (horizontal transmit, vertical return) or VH (vertical transmit, horizontal return).

cross-track scanner Scanner in which a faceted mirror rotates about a horizontal axis to sweep the detector *IFOV* in a series of parallel scan lines oriented normal to the flight direction.

CRT Cathode ray tube.

cycle One complete oscillation of a wave.

CZCS Coastal zone color scanner.

Defense Meteorologic Satellite Program (DMSP) Polar-orbiting meteorologic satellites operated by U. S. Defense Department to acquire global images in the visible and reflected IR regions.

ΔT Difference between maximum and minimum radiant temperatures during a diurnal cycle.

densitometer Optical device for measuring the density of photographic transparencies.

density, of images Measure of the opacity, or darkness, of a negative or positive transparency.

density, of materials (ρ) Ratio of mass to volume of a material, typically expressed as grams per cubic centimeter.

density slicing Process of converting the continuous gray tones of an image into a series of density intervals, or slices, each corresponding to a specific digital range. The density slices are then displayed either as gray tones or as colors.

depolarized Refers to a change in polarization of a transmitted radar pulse as a result of various interactions with the terrain surface.

depression angle (γ) In radar, the angle between the imaginary horizontal plane passing through the antenna and the line connecting the antenna and the target.

detectability Measure of the smallest object that can be discerned on an image.

detector Component of a remote sensing system that converts electromagnetic radiation into a recorded signal.

developing Chemical processing of an exposed photographic emulsion to produce an image.

dielectric constant Electrical property of matter that influences radar returns. Also called *complex dielectric constant.*

difference image Image prepared by subtracting the digital values of pixels in one image from those in a second image to produce a third set of pixels. This third set is used to form the difference image.

diffuse reflector Surface that scatters incident radiation nearly equally in all directions.

digital display A form of data display in which values are shown as arrays of numbers.

digital image processing Computer manipulation of digital images.

digital mosaic Mosaic generated by a computer from digital records of images.

digital number (DN) Value assigned to a pixel in a digital image.

digital perspective images Three-dimensional views produced by merging digital topographic data with image data.

digitization Process of converting an analog display into a digital display.

digitizer Device for scanning an image and converting it into numerical format.

directional filter Mathematical filter for image processing that enhances linear features oriented in a designated direction.

distortion On an image, changes in shape and position of objects with respect to their true shape and position.

diurnal Daily.

Doppler principle Describes the change in observed frequency of electromagnetic or other waves caused by movement of the source of waves relative to the observer.

dwell time Time required for a detector *IFOV* to sweep across a ground resolution cell.

Earth Observation Satellite (EOS) International program led by NASA to launch polar-orbiting environmental satellites in late 1990s.

EDC EROS Data Center.

edge enhancement Image-processing technique that emphasizes the appearance of edges and lines.

Ektachrome A Kodak color positive film.

electromagnetic energy Energy that travels at the speed of light in a harmonic wave pattern.

electromagnetic radiation Energy propagated in the form of an advancing interaction between electric and magnetic fields.

electromagnetic spectrum Continuous sequence of electromagnetic energy arranged according to wavelength or frequency.

emission Process by which a body radiates electromagnetic energy. Emission is determined by kinetic temperature and emissivity.

emissivity (ε) Ratio of radiant flux from a body to that from a blackbody at the same kinetic temperature.

emulsion Suspension of photosensitive silver halide grains in gelatin that constitutes the image-forming layer on photographic film.

energy flux Radiant flux. The amount of energy reflected or radiated from a surface.

enhancement Process of altering the appearance of an image so that the interpreter can extract more information.

EOS Earth Observation Satellite.

EOSAT A company that distributes Landsat and other images.

EROS Earth Resource Observation System.

EROS Data Center (EDC) Facility of the U.S. Geological Survey at Sioux Falls, South Dakota, that archives, processes, and distributes images.

ERS European remote sensing satellite that acquires C-band radar images.

ERTS Earth Resource Technology Satellite, now called Landsat.

evaporative cooling Temperature drop caused by evaporation of water from a moist surface.

***f*-number** Representation of the speed of a lens determined by the focal length divided by diameter of the lens. Smaller numbers indicate faster lenses.

***f*-stop** Focal length of a lens divided by the diameter of the lens's adjustable diaphragm. Smaller numbers indicate larger openings, which admit more light to the film.

far range The portion of a radar image farthest from the aircraft or spacecraft flight path.

film Light-sensitive photographic emulsion and its base.

film speed Measure of the sensitivity of photographic film to light. Larger numbers indicate higher sensitivity.

filter, digital Mathematical procedure for modifying values of numerical data.

filter, optical A material that, by absorption or reflection, selectively modifies the radiation transmitted through an optical system.

filter kernel A digital filter consisting of an odd number of lines and pixels.

flight path Line on the ground directly beneath a remote sensing aircraft or spacecraft. Also called *flight line*.

fluorescence Emission of light from a substance stimulated by exposure to radiation from an external source.

focal length In cameras, the distance from the optical center of the lens to the plane at which the image of a distant object is brought into focus.

focal plane In a remote sensing system, the plane at which the image is sharply defined.

foreshortening In radar images, the geometric displacement of the top of objects toward the near range relative to their base. Also called *layover*.

format Size and scale of an image.

forward overlap The percent of duplication by successive photographs along a flight line.

framing system Remote sensing system that instantaneously acquires an image of an area, or frame, of terrain.

frequency (ν) Number of wave oscillations per unit time or the number of wavelengths that pass a point per unit time.

Fuyo Nonstandard term for JERS.

GCP Ground-control point.

geographic information system (GIS) Integrated computer hardware and software that captures, stores, analyzes, and displays geographically referenced information.

geometric correction Image-processing procedure that corrects spatial distortions in an image.

geostationary Refers to satellites traveling at the angular velocity at which the earth rotates; as a result, they remain above the same point on earth at all times.

Geostationary Operational Environmental Satellite (GOES) An NOAA satellite program that acquires visible and thermal IR images for meteorologic purposes.

geothermal Refers to heat from sources within the earth.

global positioning system (GPS) A portable device that determines latitude and longitude from navigation satellites.

Goddard Space Flight Center (GSFC) The NASA facility at Greenbelt, Maryland, that is also a Landsat ground receiving station.

GMT Greenwich mean time. A universal 24-hour system for designating time.

GOES   Geostationary Operational Environmental Satellite.

GOES precipitation index (GPI)   An algorithm that estimates rainfall from cloud cover on satellite images.

gossan   Outcrop of iron oxide formed by the weathering of metallic sulfide ore minerals.

granularity   Graininess of developed photographic film that is determined by the texture of the silver grains.

gravity survey   Geophysical technique that measures variations in density of rocks in the subsurface.

gray scale   A sequence of gray tones ranging from black to white.

ground control point (GCP)   A geographic feature of known location that is recognizable on images and can be used to determine geometric corrections.

ground range   On radar images, the distance from the ground track to an object.

ground-range image   Radar image in which the scale in the range direction is constant.

ground receiving station   Facility that records image data transmitted by a satellite, such as Landsat.

ground resolution   The ability to resolve terrain features on images.

ground resolution cell   Area on the terrain that is covered by the *IFOV* of a detector.

ground swath   Width of the strip of terrain that is imaged by a scanner system.

GSFC   Goddard Space Flight Center.

harmonic   Refers to waves in which the component frequencies are whole-number multiples of the fundamental frequency.

heat capacity (*c*)   Ratio of heat absorbed or released by a material to the corresponding temperature rise or fall. Expressed in calories per gram per degree centigrade. Also called *thermal capacity*.

Heat Capacity Mapping Mission (HCMM)   NASA satellite orbited in 1978 to record daytime and nighttime visible and thermal IR images of large areas.

highlights   Areas of bright tone on an image caused by strong reflections from topographic features that face the energy direction.

histogram   Statistical chart showing the distribution of data points as a function of some attribute, such as brightness.

hue   In the IHS system, represents the dominant wavelength of a color.

hyperspectral scanner   A special type of multispectral scanner that records many tens of bands of imagery at very narrow bandwidths.

IFOV   Instantaneous field of view.

IHS   Intensity, hue, and saturation system of colors.

image   Pictorial representation of a scene recorded by a remote sensing system. Although *image* is a general term, it is commonly restricted to representations acquired by nonphotographic methods.

image swath   See **ground swath**.

imaging spectrometer   Synonym for **hyperspectral scanner.**

incidence angle   In radar, the angle formed between a line normal to the target and another connecting the antenna and the target.

incident energy   Electromagnetic radiation impinging on a surface.

index of refraction (*n*)   Ratio of the wavelength or velocity of electromagnetic radiation in a vacuum to that in a substance.

India Remote Sensing (IRS) Satellite   Earth resources satellite launched by India to record multispectral visible and reflected IR images.

instantaneous field of view (*IFOV*)   Solid angle through which a detector is sensitive to radiation. In a scanning system, the solid angle subtended by the detector when the scanning motion is stopped.

intensity   In the IHS system, brightness ranging from black to white.

interactive processing   Method of image processing in which the operator views preliminary results and can alter the instructions to the computer to achieve desired results.

interferograms   In radar, images that record interference patterns created by superposing images acquired by two antennas that are separated by a short distance.

interferometry   The field of physics that deals with the interaction between superposed wave trains.

International Satellite Cloud Climatology Project (ISCCP)   Project that maps distribution and density of cloud cover, from which rainfall can be estimated.

interpretation   The process in which a person extracts information from an image.

interpretation key   Characteristic or combination of characteristics that enables an interpreter to identify an object on an image.

IR   Infrared region of the electromagnetic spectrum that includes wavelengths from 0.7 µm to 1 mm.

IR color photograph   Color photograph in which the red-imaging layer is sensitive to photographic IR wavelengths, the green-imaging layer is sensitive to red light, and the blue-imaging layer is sensitive to green light. Also known as *camouflage detection photographs* and *false-color photographs*.

ISO index Index of the International Standards Organization, designating film speed in photography. Higher values indicate higher sensitivity.

isotherm Contour line connecting points of equal temperature. Isotherm maps are used to portray surface-temperature patterns of water bodies.

JERS Unmanned Japanese Earth Resources Satellite that acquires radar and multispectral images.

Johnson Space Flight Center A NASA facility in Houston, Texas.

JPL Jet Propulsion Laboratory, a NASA facility at Pasadena, California, operated under contract by the California Institute of Technology.

kernel Two-dimensional array of digital numbers used in digital filtering.

kinetic energy The ability of a moving body to do work by virtue of its motion. The molecular motion of matter is a form of kinetic energy.

kinetic temperature Internal temperature of an object determined by random molecular motion. Kinetic temperature is measured with a contact thermometer.

Kodachrome A Kodak color positive film.

L band Radar wavelength region from 15 to 30 cm.

LACIE Large Area Crop Inventory Experiment.

Landsat A series of unmanned NASA earth resource satellites that acquire multispectral images in the visible and IR bands.

Laplacian filter A digital filter used in nondirectional edge enhancement.

large format camera (LFC) A camera with film size of 23 by 46 cm that was carried on the Space Shuttle in October 1984.

latent image Invisible image produced by the photochemical effect of light on silver halide grains in the emulsion of film. The latent image is not visible until after photographic development.

layover In radar images, the geometric displacement of the top of objects toward the near range relative to their base. Also called *foreshortening*.

lens One or more pieces of glass or other transparent material that form an image by refraction of light.

LFC Large format camera.

light Electromagnetic radiation ranging from 0.4 to 0.7 μm in wavelength that is detectable by the human eye.

light meter Device for measuring the intensity of visible radiation and determining the appropriate exposure of photographic film in a camera.

line-pair Pair of light and dark bars of equal widths. The number of such line-pairs aligned side by side that can be distinguished per unit distance expresses the resolving power of an imaging system.

lineament Linear topographic or tonal feature on the terrain and on images, maps, and photographs that may represent a zone of structural weakness.

linear Adjective that describes the straight line-like nature of features on the terrain or on images and photographs.

Linear imaging self-scanning (LISS) system Along-track multispectral scanner carried by India Remote Sensing Satellite.

lineation The one-dimensional alignment of internal components of a rock that cannot be depicted as an individual feature on a map.

look direction Direction in which pulses of microwave energy are transmitted by a radar system. The look direction is normal to the azimuth direction. Also called *range direction*.

low-sun-angle photograph Aerial photograph acquired in the morning, evening, or winter when the sun is at a low elevation above the horizon.

luminance Quantitative measure of the intensity of light from a source.

Magellan NASA unmanned satellite that acquired complete radar coverage of Venus.

magnetic survey Geophysical technique for measuring variations in magnetism of rocks in the subsurface.

map projection A systematic representation of the curved surface of the earth on a plane.

Meteosat Geostationary meteorologic satellite program operated by European Space Agency to acquire visible and reflected IR images of Europe and Africa.

microwave Region of the electromagnetic spectrum in the wavelength range from 0.1 to 30 cm.

minimum ground separation Minimum distance on the ground between two targets at which they can be resolved on an image.

minus-blue photographs Black-and-white photographs, acquired using a filter that removes blue wavelengths to produce higher spatial resolution.

modular optoelectric multispectral scanner (MOMS) An along-track scanner carried on the Space Shuttle that records two bands of data.

modulate To vary the frequency, phase, or amplitude of electromagnetic waves.

modulation transfer function (MTF) A method of describing spatial resolution.

mosaic Composite image or photograph made by piecing together individual images or photographs covering adjacent areas.

MSS Multispectral scanner system of Landsat that acquires images at four wavelength bands in the visible and reflected IR regions.

multiband camera Optical system that simultaneously acquires photographs of the same scene at different wavelengths.

multipolarization image Color image composited from radar images of three different polarizations.

multispectral classification Identification of terrain categories by digital processing of data acquired by multispectral scanners.

multispectral system Framing or scanning system that simultaneously acquires images of the same scene at different wavelengths.

multitemporal spectrum Plot showing variation of a characteristic as a function of time. For example, a multitemporal *NDVI* spectrum shows variation of *NDVI* during a growing season.

nadir Point on the ground directly in line with the remote sensing system and the center of the earth.

NASA National Aeronautic and Space Administration.

navigation satellite A satellite that transmits data used by global positioning systems to determine latitude and longitude.

NDVI Normalized difference vegetation index.

near range Refers to the portion of a radar image closest to the aircraft or satellite flight path.

negative photograph Photograph on film or paper in which the relationship between bright and dark tones is the reverse of that of the features on the terrain.

NHAP National High Altitude Photography Program of the U.S. Geological Survey.

NOAA National Oceanic and Atmospheric Administration.

noise Random or repetitive events that obscure or interfere with the desired information.

nondirectional filter Mathematical filter that enhances all orientations of linear features equally.

nonsystematic distortion Geometric irregularities on images that are not constant and cannot be predicted from the characteristics of the imaging system.

normal color film Film in which the colors are essentially true representations of the colors of the terrain.

normalized difference vegetation index (NDVI) A measure of vegetation vigor computed from multispectral data.

NSSDC National Space Science Data Center.

oblique photograph Photograph acquired with the camera intentionally directed at some angle between horizontal and vertical orientations.

operational linescan system (OLS) Scanner on DMSP that acquires visible and reflected IR images.

OPS Optical sensor on JERS satellite that records multispectral visible and reflected IR images.

orbit Path of a satellite around a body such as the earth, under the influence of gravity.

orthophotographs Aerial photographs that have been scanned into digital format and computer-processed to remove radial distortion.

overlap Extent to which adjacent images or photographs cover the same terrain along a flight line.

pan images Black-and-white images acquired by SPOT with 10-m resolution.

panchromatic film Black-and-white film that is sensitive to all visible wavelengths.

parallax Displacement of the position of a target in an image caused by a shift in the observation position.

parallel-polarized Describes a radar pulse in which the polarization of the return is the same as that of the transmission. Parallel-polarized images may be HH (horizontal transmit, horizontal return) or VV (vertical transmit, vertical return).

passive remote sensing Remote sensing of energy naturally reflected or radiated from the terrain.

path-and-row index System for referencing and locating Landsat and other images acquired in sun-synchronous orbits.

pattern Regular repetition of tonal variations on an image or photograph.

periodic line dropout Defect on scanner images in which no data are recorded at repetitive intervals, causing a pattern of black lines on the image.

photodetector Device for measuring energy in the visible region.

photogeology Interpretation of geologic features from images.

photograph Image that records the interaction between light and a photosensitive emulsion.

photographic IR Short-wavelength portion (0.7 to 0.9 μm) of the IR band that is detectable by film.

photographic UV Long-wavelength portion of the UV band (0.3 to 0.4 μm) that is transmitted through the atmosphere and is detectable by film.

photomosaic Composite of many adjacent photographs.

photon Minimum discrete quantity of radiant energy.

picture element In a digitized image, the area on the ground represented by each digital number. Commonly contracted to *pixel*.

pitch Rotation of an aircraft about the horizontal axis normal to its longitudinal axis that causes a nose-up or nose-down attitude.

pixel Contraction of *picture element*.

polarimetric images A sequence of radar images that record a range of polarizations from parallel-polarized to cross-polarized.

polarization The direction in which the electrical field vector of electromagnetic radiation vibrates.

positive photograph Photographic image in which the tones are directly proportional to the terrain brightness.

previsual symptom Vegetation stress that is recognizable on IR film before it is detected by the eye or normal color photographs. Stressed vegetation initially loses its ability to reflect photographic IR energy, which reduces the red signature on IR color photographs.

primary colors A set of three colors that in various combinations will produce the full range of colors in the visible spectrum. There are two sets of primary colors, additive and subtractive.

principal-component (PC) image Digitally processed image produced by a transformation that recognizes maximum variance in multispectral images.

principal point Optical center of an aerial photograph.

printout Analog display of computer data typically on paper.

pulse Short burst of electromagnetic radiation transmitted by a radar antenna.

pulse length Duration of a burst of energy transmitted by a radar antenna, measured in microseconds.

pushbroom scanner An alternate term for an **along-track scanner.**

radar Acronym for *radio detection and ranging*. Radar is an active form of remote sensing that operates in the microwave and radio wavelength regions.

radar shadow Dark signature on a radar image representing no signal return. A shadow extends in the far-range direction from an object that intercepts the radar beam.

Radarsat Canadian unmanned earth-orbiting satellite that acquires C-band radar images.

radian Angle subtended by an arc of a circle equal in length to the radius of the circle; 1 rad = 57.3°.

radiant energy peak (λ_{max}) Wavelength at which the maximum electromagnetic energy is radiated at a particular temperature.

radiant flux Rate of flow of electromagnetic radiation measured in watts per square centimeter.

radiant temperature Concentration of the radiant flux from a material.

radiation Propagation of energy in the form of electromagnetic waves.

radiometer Device for quantitatively measuring radiant energy, especially thermal radiation.

random line dropout In scanner images, the loss of data from individual scan lines in a nonsystematic fashion.

range direction See **look direction.**

range resolution In radar images, the spatial resolution in the range direction, which is determined by the pulse length of the transmitted microwave energy.

raster array Arrangement of digital data in lines and pixels.

ratio image An image prepared by processing digital multispectral data as follows: for each pixel, the value for one band is divided by that of another. The resulting digital values are displayed as an image.

Rayleigh criterion In radar, the relationship among surface roughness, depression angle, and wavelength that determines whether a surface will respond in a rough or smooth fashion to the incident radar pulse.

real-aperture radar Radar system in which azimuth resolution is determined by the transmitted beam width, which in turn is determined by the physical length of the antenna and by the wavelength.

real time Refers to images or data made available for inspection simultaneously with their acquisition.

recognizability Ability to identify an object on an image.

rectilinear Refers to images with no geometric distortion. The scales in the orthogonal directions are identical.

reflectance Ratio of the radiant energy reflected by a body to the energy incident on it.

reflectance peaks Upward excursions in spectral reflectance curves.

reflectance spectrometer Instrument that records percent reflectance as a function of wavelength.

reflected energy peak Wavelength (0.5 μm) at which maximum amount of energy is reflected from the earth's surface.

reflected IR Electromagnetic region from 0.7 to 3.0 μm that consists primarily of reflected solar radiation.

reflectivity Ability of a surface to reflect incident energy.

refraction Bending of electromagnetic rays as they pass from one medium into a medium with a different index of refraction.

registration Process of geometrically adjusting two images so that equivalent geographic points coincide.

relief Vertical irregularities of a surface.

relief displacement Geometric distortion on vertical aerial photographs. The tops of objects appear in the photograph to be radially displaced from their bases outward from the photograph's centerpoint.

remote sensing The science of acquiring, processing, and interpreting images that record the interaction between electromagnetic energy and matter.

repeat cycle For earth satellites in sun-synchronous orbits, the number of days between repeated orbits.

resolution target Series of regularly spaced alternating light and dark bars used to evaluate the resolution of images.

resolving power The ability to distinguish closely spaced targets.

Reststrahlen band The absorption of thermal IR energy as a function of silica content.

return In radar, a pulse of microwave energy reflected by the terrain and received at the radar antenna. The strength of a return is referred to as *return intensity*.

return-beam vidicon (RBV) A system in which images are formed on the photosensitive surface of a vacuum tube; the image is scanned with an electron beam and transmitted or recorded.

roll Rotation of an aircraft that causes a wing-up or wing-down attitude.

roll compensation system Component of an airborne scanner system that measures and records the roll of the aircraft. This information is used to correct the imagery for distortion due to roll.

rough criterion In radar, the relationship among surface roughness, depression angle, and wavelength that determines whether a surface will scatter the incident radar pulse in a rough or intermediate fashion.

roughness In radar, the average vertical relief of small-scale irregularities of the terrain surface. Also called *surface roughness*.

satellite An object in orbit around a celestial body.

saturation In the IHS system, represents the purity of color. Saturation is also the condition in which energy flux exceeds the sensitivity range of a detector.

scale Ratio of distance on an image to the equivalent distance on the ground.

scan line Narrow strip on the ground that is swept by the *IFOV* of a detector in a scanning system.

scan skew Distortion of scanner images caused by forward motion of the aircraft or satellite during the time required to complete a scan.

scanner See **scanning system.**

scanner distortion Geometric distortion that is characteristic of cross-track scanner images.

scanning system An imaging system in which the *IFOV* of one or more detectors is swept across the terrain.

scattering Multiple reflections of electromagnetic waves by particles or surfaces.

scattering coefficient curves Display of scatterometer data in which relative backscatter is shown as a function of incidence angle.

scatterometer Nonimaging radar device that quantitatively records backscatter of terrain as a function of incidence angle.

scene Area on the ground that is covered by an image or photograph.

Seasat NASA unmanned earth-orbiting satellite that acquired L-band radar images in 1978.

seismic surveys Geophysical technique for mapping subsurface geology. Mechanical devices or explosions transmit sonic energy into the subsurface, where it is reflected from rock layers. Reflected energy is recorded and displayed as maps and cross sections.

sensitivity Degree to which a detector responds to incident electromagnetic energy.

sensor Device that detects electromagnetic radiation and converts it into a signal that can be recorded and displayed as either numerical data or an image.

shadows Dark signatures caused by topographic features that block incident energy. Shadows are prominent on radar images and low-sun-angle photographs.

Shuttle imaging radar (SIR) Series of three radar projects (SIR-A, B, C) deployed on the Space Shuttle.

side-looking airborne radar (SLAR) An airborne side-scanning system for acquiring radar images.

side-scanning sonar Active system for acquiring images of the seafloor using pulsed sound waves.

side-scanning system A system that acquires images of a strip of terrain parallel with the flight or orbit path but offset to one side.

sidelap Extent to which images acquired on adjacent flight lines or orbits cover the same terrain.

signal Response of a detector to incident energy.

signature Set of characteristics by which a material or an object may be identified on an image or photograph.

silver halide Silver salts that interact with visible energy and convert to metallic silver when developed.

skylight Component of light that is strongly scattered by the atmosphere and consists predominantly of shorter wavelengths.

slant range In radar, an imaginary line running between the antenna and the target.

slant-range distance Distance measured along the slant range.

slant-range distortion Geometric distortion of a slant-range image.

slant-range image In radar, an image in which objects are located at positions corresponding to their slant-range distances from the aircraft flight path. On slant-range images, the scale in the range direction is compressed in the near-range region.

smooth criterion In radar, the relationship among surface roughness, depression angle, and wavelength that determines whether a surface will scatter the incident radar pulse in a smooth or intermediate fashion.

software Programs that control computer operations.

sonar Acronym for *sound navigation ranging*. Sonar is an active form of remote sensing that employs sonic energy to image the seafloor.

space oblique Mercator (SOM) projection The map projection in which Landsat images are acquired.

Space Shuttle U.S. manned satellite program in the 1980s and 1990s, officially called the Space Transportation System (STS).

Space Station International program led by NASA to construct permanent manned earth-orbiting satellites.

spatial resolution The ability to distinguish between closely spaced objects on an image. Commonly expressed as the most closely spaced line-pairs per unit distance that can be distinguished.

special sensor microwave/imager (SSM/I) Scanner on DMSP that acquires passive images in the microwave region.

spectral reflectance Reflectance of electromagnetic energy at specified wavelength intervals.

spectral reflectance curves Plots that show percent reflectance as a function of wavelength.

spectral resolution Range of wavelengths recorded by a detector. Also called *bandwidth*.

spectral sensitivity Response, or sensitivity, of a film or detector to radiation in different spectral regions.

spectrometer Device for measuring intensity of radiation radiated or reflected by a material as a function of wavelength.

spectrum Continuous sequence of electromagnetic energy arranged according to wavelength or frequency.

specular Refers to a surface that is smooth with respect to the wavelength of incident energy.

SPOT Système Probatoire d'Observation de la Terre. Unmanned French earth resource satellite program.

stationary scanners Scanners that operate from a fixed position, rather than a moving aircraft or satellite.

Stefan-Boltzmann constant $5.68 \times 10^{-12} \, \text{W} \cdot \text{cm}^{-2} \cdot {}^{\circ}\text{K}^{-4}$.

Stefan-Boltzmann law States that radiant flux of a blackbody is equal to the temperature to the fourth power times the Stefan-Boltzmann constant.

stereo base Distance between a pair of correlative points on a stereo pair that are oriented for stereo viewing.

stereo model Three-dimensional visual impression produced by viewing a pair of overlapping images through a stereoscope.

stereo pair Two overlapping images or photographs that may be viewed stereoscopically.

stereoscope Binocular optical device for viewing overlapping images or diagrams. The left eye sees only the left image, and the right eye sees only the right image.

subscene A portion of an image that is used for detailed analysis.

subtractive primary colors Yellow, magenta, and cyan. When used as filters for white light, these colors remove blue, green, and red light, respectively.

sun-synchronous Satellite polar orbit pattern that covers the earth during a cycle that lasts from days to weeks. On subsequent cycles the orbit paths are repeated at the same local sun time.

sunglint Bright reflectance of sunlight caused by ripples on water.

supervised classification Digital-information extraction technique in which the operator provides training-site information that the computer uses to assign pixels to categories.

surface phenomenon Interaction between electromagnetic radiation and the surface of a material.

surface roughness See **roughness**.

synthetic-aperture radar (SAR) Radar system in which fine azimuth resolution is achieved by storing and processing data on the Doppler shift of multiple return pulses in such a way as to give the effect of a much longer antenna.

synthetic stereo images Stereo images constructed through digital processing of a single image. Topographic data are used to calculate parallax.

system Combination of components that constitute an imaging device.

systematic distortion Geometric irregularities on images that are caused by known and predictable characteristics.

target Object on the terrain of specific interest in a remote sensing investigation.

telemeter To transmit data by radio or microwave links.

terrain Surface of the earth.

texture Frequency of change and arrangement of tones on an image.

thematic mapper (TM) A cross-track scanner deployed on Landsat that records seven bands of data from the visible through the thermal IR regions.

thermal capacity (*c*) See **heat capacity.**

thermal conductivity (*K*) Measure of the rate at which heat will pass through a material, expressed in calories per centimeter per second per degree centigrade.

thermal crossover On a plot of radiant temperature versus time, the point at which temperature curves for two different materials intersect.

thermal diffusivity (*k*) Governs the rate at which temperature changes within a substance, expressed in centimeters squared per second.

thermal inertia (*P*) Measure of the response of a material to temperature changes, expressed in calories per square centimeter per square root of second.

thermal IR IR region from 3 to 14 μm that is employed in remote sensing. This spectral region spans the radiant power peak of the earth.

thermal IR image Image acquired by a scanner that records radiation within the thermal IR region.

thermal IR multispectral scanner (TIMS) Airborne scanner that acquires multispectral images within the 8- to 14-μm band of the thermal IR region.

thermal model Mathematical expression that relates thermal and other physical properties of a material to its temperature. Models may be used to predict temperature for given properties and conditions.

thermography Medical applications of thermal IR images. Images of the body, called *thermograms*, have been used to detect tumors and monitor blood circulation.

TIMS Thermal IR multispectral scanner.

TM Thematic mapper.

tone Each distinguishable shade of gray from white to black on an image.

topographic inversion An optical illusion that may occur on images with extensive shadows. Ridges appear to be valleys, and valleys appear to be ridges. The illusion is corrected by orienting the image so that shadows trend from the top margin of the image to the bottom.

topographic reversal A geomorphic phenomenon in which topographic lows coincide with structural highs and vice versa. Valleys are eroded on crests of anticlines to cause topographic lows, and synclines form topographic highs.

total ozone mapping spectrometer (TOMS) Satellite system that measures ozone concentration by comparing narrow spectral bands in the UV region.

Tracking and Data Relay Satellite (TDRS) Geostationary satellite used to communicate between ground receiving stations and satellites such as Landsat.

trade-off As a result of changing one factor in a remote sensing system, there are compensating changes elsewhere in the system; such a compensating change is known as a trade-off.

training site Area of terrain with known properties or characteristics that is used in supervised classification.

transmissivity Property of a material that determines the amount of energy that can pass through the material.

transparency Image on a transparent photographic material.

transpiration Expulsion of water vapor and oxygen by vegetation.

travel time In radar, the time interval between the generation of a pulse of microwave energy and its return from the terrain.

unsupervised classification Digital information extraction technique in which the computer assigns pixels to categories with no instructions from the operator.

UV Ultraviolet region of the electromagnetic spectrum, ranging in wavelengths from 0.01 to 0.4 μm.

vegetation anomaly Deviation from the normal distribution or properties of vegetation. Vegetation anomalies may be caused by faults, trace elements in soil, or other factors.

vertical exaggeration In a stereo model, the extent to which the vertical scale appears larger than the horizontal scale.

vidicon Framing system that acquires images on a photosensitive electronically charged surface.

visible energy Energy at wavelengths from 0.4 to 0.7 μm that is detectable by the human eye. Also called *light.*

volume phenomenon Interaction between electromagnetic energy and the internal properties of matter.

volume scattering In radar, interaction between electromagnetic radiation and the interior of a material.

watt (W) Unit of electrical power equal to rate of work done by one ampere under a potential of one volt.

wavelength (λ) Distance between successive wave crests or other equivalent points in a harmonic wave.

Wien's displacement law Describes the shift of the radiant power peak to shorter wavelengths as temperature increases.

X band Radar wavelength region from 2.4 to 3.8 cm.

XS images Multispectral images acquired by SPOT with 20-m resolution.

yaw Rotation of an aircraft about its vertical axis so that the longitudinal axis deviates left or right from the flight line.

R E F E R E N C E S   C I T E D
I N   C O L O R   P L A T E S

Abbott, M. R., 1984, Northern California/ Cape Mendocino, *in* W. A. Hovis, ed., Nimbus-7 CZCS coastal zone color scanner imagery for selected coastal regions: NASA Goddard Space Flight Center, Greenbelt, MD.

Atlas, R., R. N. Hoffman, and S. C. Bloom, 1993, Surface wind velocity over the oceans, *in* R. J. Gurney, J. L. Foster, and C. L. Parkinson, eds., Atlas of satellite observations related to global change: Cambridge University Press, p. 129–139, Cambridge, Great Britain.

Crowley, J. K. and S. J. Hook, 1996, Thermal infrared scanner (TIMS) study of playa evaporite minerals in Death Valley, California: Journal of Geophysical Research, v. 101, n. B1, p. 643–660.

Eidenshink, J. C., 1992, The 1990 conterminous U. S. AVHRR data set: Photogrammetric Engineering and Remote Sensing, v. 58, p. 809–813.

Harris, P. M. and W. S. Kowalik, eds., 1994, Satellite images of carbonate depositional settings: American Association of Petroleum Geologists, Tulsa, OK.

Hook, S. J., C. D. Elvidge, M. Rast, and H. Watanabe, 1991, An evaluation of short-wavelength-infrared (SWIR) data from the AVIRIS and GEOSCAN instruments for mineralogic mapping at Cuprite, Nevada: Geophysics, v. 56, p. 1432–1440.

Kahle, A. B., 1987, Surface emittance, temperature, and thermal inertia derived from thermal infrared multispectral scanner (TIMS) for Death Valley, California: Geophysics, v. 52, p. 858–874.

Rossow, W. B., 1993, Clouds, *in* R. J. Gurney, J. L. Foster, and C. L. Parkinson, eds., Atlas of satellite observations related to global change: Cambridge University Press, p. 141–163, Cambridge, Great Britain.

Schoeberl, M. A., 1993, Stratospheric ozone depletion, *in* R. J. Gurney, J. L. Foster, and C. L. Parkinson, eds., Atlas of satellite observations related to global change: Cambridge University Press, p. 59–65, Cambridge, Great Britain.

U.S. Geological Survey, 1993, Seasonal land cover map of the conterminous U.S.: Reston, VA.

INDEX

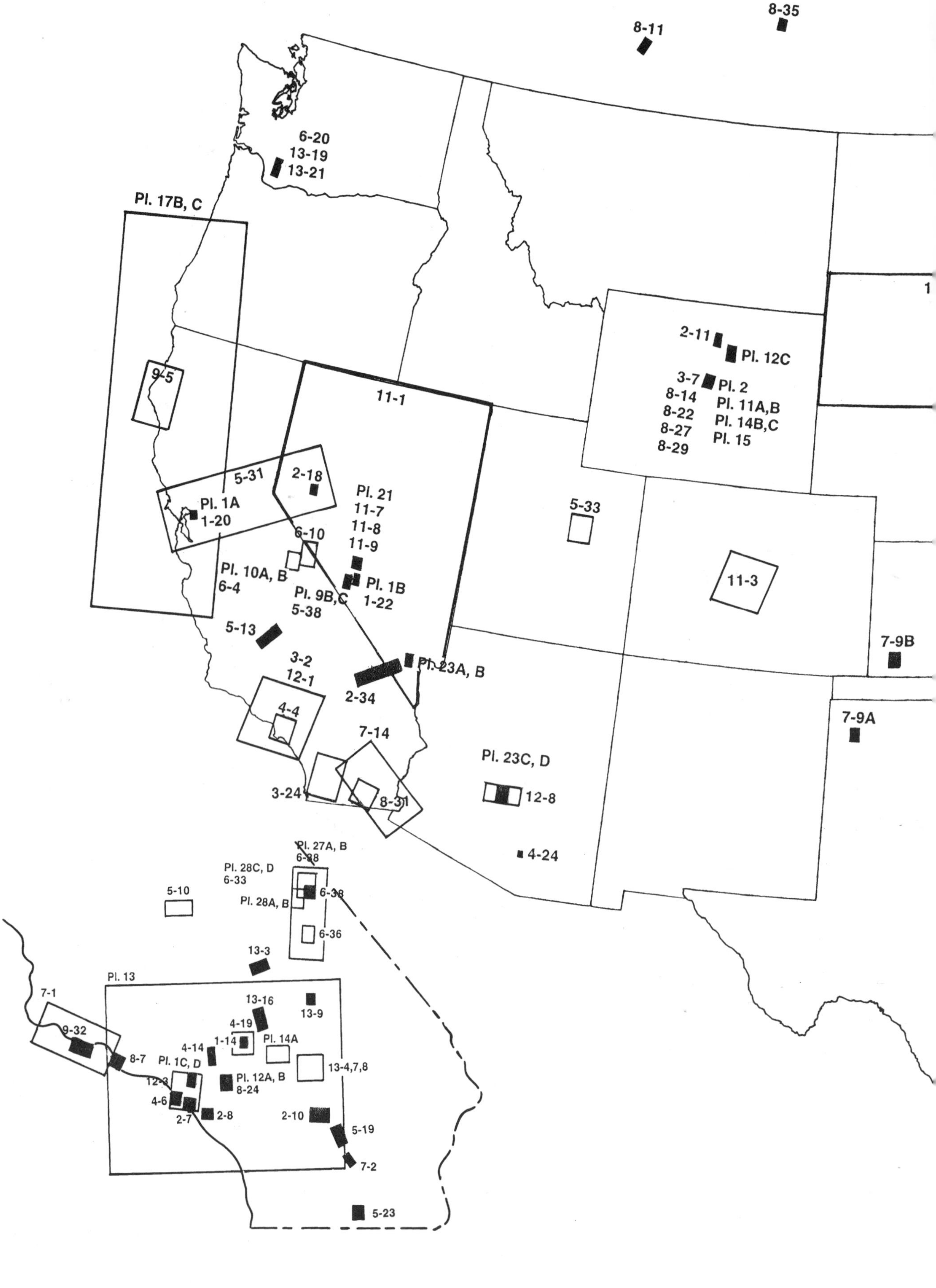

8-35

8-11

6-20
13-19
13-21

Pl. 17B, C

9-5

11-1

2-11 ■ Pl. 12C

3-7 ■ Pl. 2
8-14   Pl. 11A,B
8-22   Pl. 14B,C
8-27   Pl. 15
8-29

1

5-31

2-18

Pl. 21
11-7
11-8
11-9

5-33

Pl. 1A
1-20

6-10

Pl. 10A, B
6-4

Pl. 9B,C
5-38

Pl. 1B
1-22

11-3

5-13

3-2
12-1

Pl. 23A, B

2-34

7-9B

4-4

7-14

7-9A

3-24

8-31

Pl. 23C, D

12-8

4-24

Pl. 27A, B
6-28

Pl. 28C, D
6-33

5-10

Pl. 28A, B

6-38

6-36

13-3

Pl. 13

13-16

13-9

7-1

4-19

9-32

8-7

4-14 1-14

Pl. 14A

13-4,7,8

Pl. 1C, D

12-3

Pl. 12A, B
8-24

4-6

2-7 2-8

2-10

5-19

7-2

5-23